THE VEGETABLE PATHOSYSTEM

Ecology, Disease Mechanism, and Management

THE VEGETABLE PATHOSYSTEM

Ecology, Disease Mechanism, and Management

Edited by

Mohammad Ansar, PhD

Abhijeet Ghatak, PhD

Apple Academic Press Inc.
4164 Lakeshore Road
Burlington ON L7L 1A4
Canada

Apple Academic Press Inc.
1265 Goldenrod Circle NE
Palm Bay, Florida 32905
USA

First issued in paperback 2021

© 2020 by Apple Academic Press, Inc.

Exclusive worldwide distribution by CRC Press, a member of Taylor & Francis Group

No claim to original U.S. Government works

ISBN 13: 978-1-77463-454-7 (pbk)
ISBN 13: 978-1-77188-776-2 (hbk)

Library and Archives Canada Cataloguing in Publication

Title: The vegetable pathosystem : ecology, disease mechanism and management / edited by Mohammad Ansar, PhD, Abhijeet Ghatak, PhD.

Names: Ansar, Mohammad, editor. | Ghatak, Abhijeet, editor.

Series: Innovations in horticultural science.

Description: Series statement: Innovations in horticultural science | Includes bibliographical references and index.

Identifiers: Canadiana (print) 20190113642 | Canadiana (ebook) 20190113677 | ISBN 9781771887762 (hardcover) | ISBN 9780429022999 (ebook)

Subjects: LCSH: Plant diseases.

Classification: LCC SB731 .V44 2019 | DDC 571.9/2—dc23

Library of Congress Cataloging-in-Publication Data

Names: Ansar, Mohammad, editor. | Ghatak, Abhijeet, editor.

Title: The vegetable pathosystem : ecology, disease mechanism and management / editors: Mohammad Ansar, Abhijeet Ghatak.

Description: 1st edition. | Palm Bay, Florida : Apple Academic Press, 2020. | Includes bibliographical references and index. | Summary: "Key Features: Provides an overview of phytopathogens that affect vegetables in various environmental conditions Presents a comprehensive guide to identifying and addressing numerous diseases for individuals in the fields of horticulture Discusses how to manage vegetables affected by specific pathogens Offers eco-friendly approaches to prevent postharvest diseases"-- Provided by publisher.

Identifiers: LCCN 2019020806 (print) | LCCN 2019980697 (ebook) | ISBN 9781771887762 (hardcover) | ISBN 9780429022999 (ebook other)

Subjects: LCSH: Vegetables--Diseases and pests. | Plant diseases.

Classification: LCC SB608.V4 V47 2019 (print) | LCC SB608.V4 (ebook) | DDC 635/.0496--dc23

LC record available at https://lccn.loc.gov/2019020806

LC ebook record available at https://lccn.loc.gov/2019980697

Apple Academic Press also publishes its books in a variety of electronic formats. Some content that appears in print may not be available in electronic format. For information about Apple Academic Press products, visit our website at **www.appleacademicpress.com** and the CRC Press website at **www.crcpress.com**

About the Book Series:
Innovations in Horticultural Science

Editor-in-Chief:
Mohammed Wasim Siddiqui, PhD
Assistant Professor-cum- Scientist
Bihar Agricultural University | www.bausabour.ac.in
Department of Food Science and Post-Harvest Technology
Sabour | Bhagalpur | Bihar | P. O. Box 813210 | INDIA
Contacts: (91) 9835502897
Email: wasim_serene@yahoo.com | wasim@appleacademicpress.com

The horticulture sector is considered as the most dynamic and sustainable segment of agriculture all over the world. It covers pre- and postharvest management of a wide spectrum of crops, including fruits and nuts, vegetables (including potatoes), flowering and aromatic plants, tuber crops, mushrooms, spices, plantation crops, edible bamboos etc. Shifting food pattern in wake of increasing income and health awareness of the populace has transformed horticulture into a vibrant commercial venture for the farming community all over the world.

It is a well-established fact that horticulture is one of the best options for improving the productivity of land, ensuring nutritional security for mankind and for sustaining the livelihood of the farming community worldwide. The world's populace is projected to be 9 billion by the year 2030, and the largest increase will be confined to the developing countries, where chronic food shortages and malnutrition already persist. This projected increase of population will certainly reduce the per capita availability of natural resources and may hinder the equilibrium and sustainability of agricultural systems due to overexploitation of natural resources, which will ultimately lead to more poverty, starvation, malnutrition, and higher food prices. The judicious utilization of natural resources is thus needed and must be addressed immediately.

Climate change is emerging as a major threat to the agriculture throughout the world as well. Surface temperatures of the earth have risen significantly over the past century, and the impact is most significant on agriculture. The rise in temperature enhances the rate of respiration, reduces cropping periods, advances ripening, and hastens crop maturity, which adversely affects crop productivity. Several climatic extremes such as droughts, floods, tropical cyclones, heavy precipitation events, hot extremes, and heat waves cause a negative impact on agriculture and are mainly caused and triggered by climate change.

In order to optimize the use of resources, hi-tech interventions like precision farming, which comprises temporal and spatial management of resources in horticulture, is essentially required. Infusion of technology for an efficient utilization of resources is intended for deriving higher crop productivity per unit of inputs. This would be possible only through deployment of modern hi-tech applications and precision farming methods. For improvement in crop production and returns to farmers, these technologies have to be widely spread and adopted. Considering the above-mentioned challenges of horticulturist and their expected role in ensuring food and nutritional security to mankind, a compilation of hi-tech cultivation techniques and postharvest management of horticultural crops is needed.

This book series, Innovations in Horticultural Science, is designed to address the need for advance knowledge for horticulture researchers and students. Moreover, the major advancements and developments in this subject area to be covered in this series would be beneficial to mankind.

Topics of interest include:

1. Importance of horticultural crops for livelihood
2. Dynamics in sustainable horticulture production
3. Precision horticulture for sustainability
4. Protected horticulture for sustainability
5. Classification of fruit, vegetables, flowers, and other horticultural crops
6. Nursery and orchard management
7. Propagation of horticultural crops
8. Rootstocks in fruit and vegetable production
9. Growth and development of horticultural crops
10. Horticultural plant physiology
11. Role of plant growth regulator in horticultural production
12. Nutrient and irrigation management
13. Fertigation in fruit and vegetables crops
14. High-density planting of fruit crops
15. Training and pruning of plants
16. Pollination management in horticultural crops
17. Organic crop production

18. Pest management dynamics for sustainable horticulture
19. Physiological disorders and their management
20. Biotic and abiotic stress management of fruit crops
21. Postharvest management of horticultural crops
22. Marketing strategies for horticultural crops
23. Climate change and sustainable horticulture
24. Molecular markers in horticultural science
25. Conventional and modern breeding approaches for quality improvement
26. Mushroom, bamboo, spices, medicinal, and plantation crop production

---

BOOKS IN THE INNOVATIONS IN HORTICULTURAL SCIENCE SERIES

- **Spices: Agrotechniques for Quality Produce**
  Amit Baran Sharangi, PhD, S. Datta, PhD, and Prahlad Deb, PhD
- **Sustainable Horticulture, Volume 1: Diversity, Production, and Crop Improvement**
  Editors: Debashis Mandal, PhD, Amritesh C. Shukla, PhD, and Mohammed Wasim Siddiqui, PhD
- **Sustainable Horticulture, Volume 2: Food, Health, and Nutrition**
  Editors: Debashis Mandal, PhD, Amritesh C. Shukla, PhD, and Mohammed Wasim Siddiqui, PhD
- **Underexploited Spice Crops: Present Status, Agrotechnology, and Future Research Directions**
  Amit Baran Sharangi, PhD, Pemba H. Bhutia, Akkabathula Chandini Raj, and Majjiga Sreenivas
- **Advances in Pest Management in Commercial Flowers**
  Editors: Suprakash Pal, PhD, and Akshay Kumar Chakravarthy, PhD

About the Editors

Mohammad Ansar, PhD, is an Assistant Professor at the Department of Plant Pathology of Bihar Agricultural University (BAU), India. He established the Plant Virology Laboratory at BAU and acts as a coordinator of virology research. He has gained expertise as vegetable pathologist and has been assisting in the research and helping farmers by providing management advice for various vegetable crops. Dr. Ansar is associated with a program in which he actively performs the research work of the All-India Coordinated Researches on vegetable crops. His major research activities involve diagnosis, characterization, and epidemiology of plant viruses. Recently, he is investigating the viral complexity in papaya and attempting to characterize the associated viral populations. He has earned a project funded by the Department of Science and Technology, Government of India, where his work is to characterize the whitefly- and thrips-transmitted viruses with biological and molecular tools. Dr. Ansar teaches the undergraduate and postgraduate courses and is also actively engaged in BSc and MSc student guiding. He has been the author or co-author of many articles in research journals, book chapters, and conference papers. He has recently published an edited book *The Phytopathogen: Evolution and Adaptation*. He is also associated with numerous peer-reviewed journals as a member of editorial boards and happily reviews various research articles. He obtained his postgraduate degree in Plant Pathology from CS Azad University of Agriculture and Technology, Kanpur, India, and received his PhD degree in Plant Pathology from GB Pant University of Agriculture and Technology, Pantnagar, India.

Abhijeet Ghatak, PhD, is an Assistant Professor of Plant Pathology at Bihar Agricultural University (BAU), India, and is the author or co-author of many peer-reviewed journal articles, book chapters, and conference papers. He has an edited book published from the *Apple Academic Press, USA* to his credit. He has been serving as an editorial board member of an international peer-reviewed journal *Journal of*

Post-Harvest Technology, and performs as reviewer for *Indian Phytopa-thology*. At BAU, he is holding many projects including the international collaboration with CIMMYT, Mexico. Dr Ghatak is engaged in researches on nanotechnology and plant health management with implications of phytopathological and epidemiological tools. He is a member of the Core Research Group and Research Monitoring Committee at BAU, and has been providing appropriate direction and assistance to sensitize priority and evaluation of research in the university. BAU has recognized Dr Ghatak as the coordinator of host-parasite interaction research. He supervises students of undergraduate and postgraduate standards, and has been involved in class teaching for over 5 years. He earned the postgraduate degree in Mycology and Plant Pathology from Banaras Hindu University, Varanasi, India, and acquired the doctorate degree in Plant Pathology from GB Pant University of Agriculture and Technology, Pantnagar, India. Dr Ghatak has achieved experience on working at the International Rice Research Institute, Philip-pines, where he established a relationship between neck blast and leaf blast epidemics in rice.

Contents

Contributors.. *xiii*

Abbreviations .... *xvii*

Preface ....*xxi*

Introduction.. *xxiii*

Acknowledgments...*xxxi*

1. **Phytophtora Blight in Potato: A Challenge Ahead in the Climate Change Scenario**.. 1

   Mehi Lal, Saurabh Yadav, V. K. Dua, and Santosh Kumar

2. **Pathosystem in Cowpea: An Overview** 23

   Santosh Kumar, Erayya, Md. Nadeem Akhtar, Mahesh Kumar, and Mehi Lal

3. **A Population Genetics Perspective on Ecology and Management of** *Phytophthora* **Spp. Affecting Potato, Tomato, and Pointed Gourd in India** 39

   Sanjoy Guha Roy

4. *Phytoplasma* **in Vegetable Pathosystem: Ecology, Infection Biology, and Management**.. 49

   Nagendran Krishnan, Shweta Kumari, Koshlendra Kumar Pandey, Awadhesh Bahadur Rai, Bijendra Singh, and Karthikeyan Gandhi

5. *Ralstonia solanacearum*: **Pathogen Biology, Host–Pathogen Interaction, and Management of Tomato Wilt Disease** 85

   Bhupendra S. Kharayat and Yogendra Singh

6. **Nematodes: Pest of Important Solanaceous Vegetable Crops and Their Management**.. 125

   Ajay Kumar Maru, Tulika Singh, and Poonam Tapre

7. **Effect of Fungal Pathogen on Physiological Function of Vegetables**..... 163

   Nishant Prakash, *Erayya*, *Srinivasraghvan A.*, and Supriya Gupta

8. **Foilar Fungal Pathogens of Cucurbits** 203

   Ashok Kumar Meena, Shankar Lal Godara, Prabhu Narayan Meena, and Anand Kumar Meena

9. Yellow Vein Mosaic of Okra: A Challenge in the
Indian Subcontinent.. 229
Amit Kumar, R. B. Verma, Randhir Kumar, and Chandan Roy

10. Disease Dynamics and Management of Vegetable Pathosystems 255
Abhijeet Ghatak

11. Mechanism of Microbial Infection in Vegetables Diseases..................... 271
Vinod Upadhyay and Elangbam Premabati Devi

12. Virus Diseases: An Inimitable Pathosystem of Vegetable Crop............. 313
Mohammad Ansar, Aniruddha Kumar Agnihotri, and Srinivasaraghavan A.

13. Begomovirus Diversity and Management in Leguminous
Vegetables and Other Hosts 343
Muhammad Naeem Sattar and Zafar Iqbal

14. Biology and Molecular Epidemiology of Begomovirus Infection on
Cucurbit Crops... 385
Bhavin S. Bhatt, Fenisha D. Chahwala, Sangeeta Rathore, Bijendra Singh, and
Achuit K. Singh

15. Diversity of Potyviruses and Their Extent in Vegetable Pathosystem .. 409
Bhavin S. Bhatt, Sangeeta Rathore, Brijesh K. Yadav, Fenisha D. Chahwala,
Bijendra Singh, and Achuit K. Singh

16. Soilborne Microbes: A Culprit of Juvenile Plants in Nurseries............. 427
Puja Pandey and Bindesh Prajapati

17. Biological Control of Postharvest Diseases in Vegetables..................... 457
Pardeep Kumar and R. K. Singh

18. Postharvest Handling and Diseases and Disorders in Bulb Vegetables 483
Sangeeta Shree and Amrita Kumari

19. Management of Soilborne Diseases of Vegetable Crops Through Spent
Mushroom Substrate 511
Durga Prasad and Srinivasaraghavan A.

Color insert of illustrations ....A – H

Index... 527

Contributors

Aniruddha Kumar Agnihotri
Department of Plant Pathology, Bihar Agricultural University, Sabour 813210, Bihar, India

Md. Nadeem Akhtar
Plant Pathology Section, KVK, Agwanpur, Saharsa, Bihar Agricultural University,
Sabour 854105, Bihar, India

Mohammad Ansar
Department of Plant Pathology, Bihar Agricultural University, Sabour 813210, Bihar, India.
E-mail: ansar.pantversity@gmail.com

Bhavin S. Bhatt
Department of Environmental Science, Shree Ramkrishna Institute of Computer Education and
Applied Sciences, Surat 395001, Gujarat, India

Fenisha D. Chahwala
School of Life Sciences, Central University of Gujarat, Gandhinagar 382030, Gujarat, India

Elangbam Premabati Devi
Department of Plant Pathology, Wheat Research Station, Sardarkrushinagar Dantiwada
Agricultural University (SDAU), Vijapur 382870, Gujarat, India

V. K. Dua
Division of Crop Production, ICAR – Central Potato Research Institute, Shimla, India

Erayya
Department of Plant Pathology, Bihar Agricultural University, Sabour 813210, Bihar, India

Karthikeyan Gandhi
Department of Plant Pathology, Tamil Nadu Agricultural University, Coimbatore 641003,
Tamil Nadu, India

Abhijeet Ghatak
Department of Plant Pathology, Bihar Agricultural University, Sabour 813210, India.
E-mail: ghatak11@gmail.com

Shankar Lal Godara
Department of Plant Pathology, College of Agriculture, Swami Keshwanand Rajasthan
Agricultural University, Bikaner, India

Supriya Gupta
Department of Plant Pathology, G.B. Pant University of Agriculture and Technology, Pantnagar, India

Zafar Iqbal
Akhuwat-Faisalabad Institute of Research, Science and Technology, University Park,
Faisalabad, Pakistan
Department of Plant Pathology, University of Florida, Florida, United States of America

Bhupendra S. Kharayat
Division of Plant Pathology, ICAR-Indian Agricultural Research Institute, Pusa, New Delhi 110012, India

Nagendran Krishnan
ICAR – Indian Institute of Vegetable Research, Varanasi 221305, Uttar Pradesh, India.
E-mail: krishnagendra@gmail.com

Amit Kumar
Department of Horticulture (Vegetable and Floriculture), Bihar Agricultural University,
Sabour 813210, Bihar, India

Mahesh Kumar
Department of Molecular Biology and Genetic Engineering, Bihar Agricultural University,
Sabour 813210, Bihar, India

Pardeep Kumar
KVK, Siddharthnagar, Narendra Dev University of Agriculture and Technology, Faizabad 224229,
Uttar Pradesh, India. E-mail: drpardeepviro@gmail.com

Randhir Kumar
Department of Horticulture (Vegetable and Floriculture), Bihar Agricultural University,
Sabour 813210, Bihar, India

Santosh Kumar
Department of Plant Pathology, Bihar Agricultural University, Sabour 813210, Bihar, India.
E-mail: santosh35433@gmail.com

Amrita Kumari
Department of Horticulture (Vegetable and Floriculture), Bihar Agricultural University, Sabour,
Bhagalpur 813210, Bihar, India

Shweta Kumari
ICAR – Indian Institute of Vegetable Research, Varanasi 221305, Uttar Pradesh, India

Mehi Lal
Plant Protection Section, Central Potato Research Institute Campus, Modipuram, Meerut 250110,
Uttar Pradesh, India
ICAR – Central Potato Research Institute Regional Station, Modipuram, Uttar Pradesh, India.
E-mail: mehilalonline@gmail.com

Ajay Kumar Maru
Department of Nematology, Anand Agricultural University, Anand 388110, Gujarat, India.
E-mail: maruajay@gmail.com

Anand Kumar Meena
Department of Plant Pathology, College of Agriculture, Swami Keshwanand Rajasthan
Agricultural University, Bikaner, India

Ashok Kumar Meena
Department of Plant Pathology, College of Agriculture, Swami Keshwanand Rajasthan
Agricultural University, Bikaner, India

Prabhu Narayan Meena
Department of Plant Pathology, College of Agriculture, Swami Keshwanand Rajasthan
Agricultural University, Bikaner, India

Koshlendra Kumar Pandey
Division of Crop Protection, ICAR – Indian Institute of Vegetable Research, Varanasi 221305,
Uttar Pradesh, India

Puja Pandey
Department of Plant Pathology, Anand Agricultural University, Anand 388110, Gujarat.
E-mail: pujapandey41124@gmail.com

Bindesh Prajapati
Agricultural Research Station, S. D. Agricultural University, Aseda, Banaskantha, Gujarat, India

Nishant Prakash
Department of Plant Pathology, Krishi Vigyan Kendra, Lodipur Farm, Arwal, India

Durga Prasad
Department of Plant Pathology, Bihar Agricultural University, Sabour 813210, Bihar, India.
E-mail: dp.shubh@gmail.com

Awadhesh Bahadur Rai
Division of Crop Improvement, ICAR – Indian Institute of Vegetable Research,
Varanasi 221305, Uttar Pradesh, India

Sangeeta Rathore
School of Life Sciences, Central University of Gujarat, Gandhinagar 382030, Gujarat, India

Chandan Roy
Department of Plant Breeding and Genetics, Bihar Agricultural University, Sabour, Bhagalpur, India

Sanjoy Guha Roy
Department of Botany, West Bengal State University, Barasat, Kolkata 700126, India.
E-mail: s_guharoy@yahoo.com

Muhammad Naeem Sattar
Department of Environment and Natural Resources, College of Agriculture and Food Science,
King Faisal University, Alhafuf, Kingdom of Saudi Arabia. E-mail: naeem.sattar1177@gmail.com

Sangeeta Shree
Department of Horticulture (Vegetable and Floriculture), Bihar Agricultural University, Sabour,
Bhagalpur 813210, Bihar, India. E-mail: sangeetashreee@gmail.com

Achuit K. Singh
ICAR—Indian Institute of Vegetable Research, Varanasi 231304, Uttar Pradesh, India.
E-mail: achuits@gmail.com

Bijendra Singh
Division of Crop Improvement, ICAR – Indian Institute of Vegetable Research,
Varanasi 221304, Uttar Pradesh, India

R. K. Singh
Department of Plant Pathology, R. V. S. Krishi Vishwa Vidyalaya, College of Agriculture,
Indore 452001, Madhya Pradesh, India

Tulika Singh
Department of Nematology, Anand Agricultural University, Anand 388110, Gujarat, India

Yogendra Singh
Department of Plant Pathology,G.B. Pant University of Agriculture and Technology, Pantnagar 263145,
Udham Singh Nagar, Uttarakhand, India

Srinivasaraghavan A.
Department of Plant Pathology, Bihar Agricultural University, Sabour 813210, Bihar, India

Poonam Tapre
Department of Nematology, Anand Agricultural University, Anand 388110, Gujarat, India

Vinod Upadhyay
Department of Plant Pathology, Regional Agricultural Research Station, Assam Agricultural University, Gosaaigaon 783360, Assam, India. E-mail: vinodupadhyay148@gmail.com

R. B. Verma
Department of Horticulture (Vegetable and Floriculture), Bihar Agricultural University, Sabour 813210, Bihar, India

Brijesh K. Yadav
School of Life Sciences, Central University of Gujarat, Gandhinagar 382030, Gujarat, India

Saurabh Yadav
Division of Plant Protection, ICAR – Central Potato Research Institute Regional Station, Modipuram, Uttar Pradesh, India

Abbreviations

AM	arbuscular mycorrhiza
AMP	antimicrobial peptides
ASM	acibenzolar-S-methyl
AY	aster yellows
BCA	biocontrol control agent
BYMD	bean yellow mosaic disease
CA	controlled atmospheres
CABM	cowpea aphid-borne mosaic
CaMV	cauliflower mosaic virus
CBSD	cassava brown streak virus disease
CDPK	calmodulin like protein kinases
CEVd	citrus exocortis viriod
CI	chilling injury
CLCuD	cotton leaf curl disease
CMD	cassava mosaic disease
CMV	cucumber mosaic virus
CP	coat protein
CPG	casamino acid-peptone-glucose-agar
CPGM	cowpea golden mosaic
CR	common region
CWDEs	cell wall-degrading enzymes
EFE	ethylene-forming enzyme
ELVd	eggplant latent viriod
endoPG	endopolygalacturonase
EPS	extracellular polysaccharide
GA	gentisic acid
GDH	glutamate dehydrogenase
HPR	host–plant resistance
HR	hypersensitive reaction
HST	host-specific toxins
HSVd	hop stunt viriod
IAA	indoleacetic acid
IBs	inclusion bodies
ICY	Iranian cabbage yellows

IDM	integrated disease management
IP	incubation period
ISR	induced systemic resistance
LA	lesion area
MA	modified atmospheres
MAP	modified atmosphere packaging
MAPK	mitogen activated protein kinase
MLOs	mycoplasma-like organisms
MP	movement protein
MYMD	mung bean yellow mosaic disease
NGS	next generation sequencing
NO	nitric oxide
NSP	nuclear shuttle protein
ORFs	open reading frames
OY-M	onion yellows mild strain
PCR	polymerase chain reaction
PGPR	plant growth-promoting rhizobacteria
PK	protein kinases
PPN	plant–parasitic nematodes
PRR	pattern recognition receptors
PRSV	papaya ringspot virus
PTI	PAMP-triggered immunity
PVY	potato virus Y
QTL	quantitative trait loci
RCA	rolling circle amplification
RCPs	representative concentration pathways
RH	relative humidity
RISC	RNA-induced silencing complexes
RLKs	receptor-like kinases
RNPs	ribonucleo proteins
ROS	reactive oxygen species
SA	salicylic acid
SAMs	sphinganine-analogue mycotoxins
SAR	systemic acquired resistance
SBM	Southern bean mosaic
SC	surface coat
SEM	scanning electron microscopy
siRNA	small interfering RNAs

SMC	spent mushroom compost
SMS	spent mushroom substrate
SMV	soybean mosaic virus
SPV	sweet potato virus
SqMV	squash mosaic virus
SqMV	Squash mosaic virus
ss-DNA	single-stranded DNA
SSRs	single sequence repeats
TAL	transcription activator-like
TCA	tri carboxylic acid
TEM	transmission electron microscopy
ToLCuD	tomato leaf curl disease
TrAP	transcription activator protein
TRSV	tobacco ring spot virus
TYLCV	Tomato yellow leaf curl virus
TZC	tetrazolium chloride
VPD	vapor pressure difference
VPg	virus-coated protein
WMV	watermelon mosaic virus
YVM	yellow vein mosaic
YVMD	yellow vein mosaic disease
YVMV	yellow vein mosaic virus
ZYMV	zucchini yellow mosaic virus

Preface

Vegetable crops share great significance in Indian agriculture by virtue of fulfilling the nutritional necessity. The group "vegetables" has a wide range of species and, therefore, has great adaptability to diverse ecology around the globe. Fluctuating weather, particularly warm and humid conditions, are conducive for the appearance of large numbers of diseases. Incessant cropping and agronomic interventions are also contributing to the emerging scenario. Diseases can have a devastating impact on the productivity of vegetable crops. They can completely kill or significantly reduce crops quality, which may incur great losses in the marketing channel. There are several diseases, in the majority that are caused by fungi, bacteria, or viruses.

This book looks into the significant role of various pathogens making a great impact on vegetable crops. The book consists of chapters that encompass major thematic areas like environmental influence on disease development, mechanisms of disease development, and diversity among the pathogens. Moreover, it also covers storage losses in bulb crop under storage. Under the climate change scenario, *Phytophthora* blight of vegetables is creating a great challenge and causes crop loss, even after several resistant varieties are grown. The population genetics perspective, adaptive to diverse ecology, as a control strategy developed in different crops is dealt with this compilation. Suppression of soilborne microbes by the different antagonistic fungal pathogen is also highlighted here.

The infection process is one of the important attributes of pathogens, which incites the abnormality in plants. Different chapters focus on the infection process and biology of different vegetable pathogens, like *Ralstonia*, *Phytoplasma*, and Begomoviruses. The effect of pathogens on the physiological function of vegetables is also covered under the subject. The plant pathogen community and its population dynamics comprise an important stricture in understanding the leaf- and root-infecting diseases. The appearance of exotic or newly introduced isolates of a pathogen may involve vegetables having specific resistance; suddenly or slowly breaking down their resistance is a major concern that is associated with a particular vegetable pathosystem. Diversity among the plant pathogen is a great concern of the vegetable pathosystem. The appearance of new pathotypes or strains leads to the recurrent appearance of diseases, which may be challenging for sustainable management over the seasons.

In this book, the vector-borne viruses associated with diverse vegetable cropping systems are taken into consideration. Various species of Begomoviruses-infecting leguminous vegetables have been given a portrait in the text. Diversity of Potyviruses and their extent in vegetable pathosystem is also included in the content. However, exploration of plant virus outset may not be enough for explicated evolution in nature. Other events such as crop domestication, long-distance dispersal, and the utilization of natural habitats seem to influence plant virus spread and have evolved with the vegetable pathosystem. Additionally, space is also occupied by postharvest diseases and disorders in bulb crops. Ecofriendly approaches will also draw the attention of readers in order to control postharvest diseases. Waste management (spent mushroom) is also covered in the book for the management of soilborne microbes associated with the pathosystem.

This book is organized by considering the specific issues in vegetable crops, such as immensely focusing on the pathogenic microorganisms and complete pathosystems.

The book will be of great importance in the academic and research sectors. Different courses in plant pathology are offered at various levels of undergraduate and postgraduate lessons in various disciplines. The subject of this book provides a better understanding of plant pathogens associated with vegetables, which will assist in scholars' comprehension. The information presented on every pathosystem is incorporated as definite lessons in postgraduate classes. The book is illustrated in several aspects with labeled figures, tables, and citation of examples. Our compilation will be helpful to generate eagerness among the research scholars in order to formulate their research programs. The commencement of this volume specifically discusses the pathosystem and focuses on several aspects of intriguing areas related to the scientific community. The information presented in the book will directly or indirectly link pathologists, strategists, and program planners with an understanding of pathogen biology, epidemiology, and handling the vegetable pathosystem in order to formulate sustainable management options.

We would be looking to receive the readers' remarks, suggestions, and research input on different aspects of this book. The given information will be helpful in compiling revised editions of the book.

— **Mohammad Ansar**
Abhijeet Ghatak

Introduction

Vegetable production has increased on a per capita basis, which has grown 60% over the last two decades. This trend is noticeable in developing countries. Vegetable crops occupy 1.1% of the world's total agricultural area; Europe and the Central Asian region share 12% of the total global area, with 14% of production. China is the largest producer of fresh vegetables, with a production value of about 25.25 billion US dollars in 2013. Moreover, other leading vegetable-producing countries, such as India, Vietnam, Nigeria, and the Philippines, are playing a significant role in global production. As a result of intensive coverage and reasonable production, biotic stresses affect quality products globally. Diseases are significant constraints on vegetable productivity. Emerging diseases pretence an evolved threat to profitable production across the globe. Plant pathogens and their carriers/vectors have evolved and moved across national and international boundaries; occasionally they may be natural and sometimes altered by adopted crop practices.

THE VEGETABLE PATHOSYTEM: A SUBSYSTEM OF ECOSYSTEM

Vegetables are attacked by plenty of parasitic microbes right from seedling to the harvesting stage. Every stage of the crop is targeted by a specific pathogen in making a parasitic relationship. The vegetable pathosystem is a subsystem of an ecosystem by explaining the phenomenon of parasitic relation. The parasite links a part of its existence by inhabiting one host species and acquiring the nutrients. The parasite may thus be the lower and higher group of fungi, bacteria, phytoplasma, viruses, viroid, and nematodes. However, vegetable crops have resistance property to a parasitic infection. On the other hand, a parasite has the property to make a parasitic relationship. Therefore, parasitism is the interface of these two biological traits. In another aspect, the pathosystem perception is the parasitism that is considered at the higher levels and ecological aspects of the system in terms of populations. Both populations exhibit genetic diversity and genetic flexibility in the case of pathogenic relationships in wild vegetables. On the contrary, in a cultivated vegetable pathosystem, the host population usually exhibits genetic equivalence and inflexibility, such as pure lines, hybrid varieties, though the pathogen population is supposed to be an equivalent uniformity. This difference showed that a wild pathosystem can respond to selection pressures but a crop pathosystem is not achieved.

PARASITIC INTERACTION IN VEGETABLE PATHOSYSTEM

SOILBORNE MICROBES AND THEIR ROLE IN VEGETABLE PATHOSYSTEM

Soil is an ingenious reservoir for several plant pathogenic microbes. Interactions between the root and soilborne pathogens exhibit a great temporal and spatial variation. It is considered as the pathogen showing dominance that leads in the pathosystem. Generally, the fungi are considered prime soilborne microbes (pathogenic/antagonistic). Bacteria, nematodes, and to some extent viruses also pose significant impact in relation to productivity of vegetables. Soilborne microbes can be a major limiting factor in the cultivation of vegetable crops. Broadly, four groups of plant pathogens affect the crop, but only two are key players in the soil, that is, fungi and nematodes. However, a few groups of bacteria are found to be soilborne, but no longer is survival established in the soil, possibly because of its nonspore-bearing ability. Bacteria require an opening to enter inside the plant and cause infection, such as *Ralstonia solanacearum*, a cause of bacterial wilt of tomato and *Agrobacterium tumefaciens*, for crown gall. Dispersal of soilborne bacteria through irrigation water is always a major concern to establish in a new area. They enter into a plant system through injury caused by nematodes, root-feeding insects. Fungi and oomycetes are the most prominent soilborne microbes. Fungi are eukaryotic, filamentous, multicellular, heterotrophic organisms that form a network of hyphae-derived nutrients from the surrounding substrate. Oomycetes group produces swimming spores (zoospores) and contain cellulose in their cell walls. The mechanisms of parasitism are necrotrophic by producing enzymes and toxins in hyphae. Extreme environmental conditions in the soil are not favorable for most of the fungal growth, like high or low temperatures or severe drought conditions. The survival of pathogens, the soil is as defiant propagules like chlamydospores, sclerotia, thick-walled conidia, or hyphae. Moreover, it survives on plant roots and crop residues. Under favorable conditions, when a seed or root comes within reach of the dormant propagule, the fungus is stimulated to germinate and grow up toward the plants. The fungal germtube or zoospore can attach to the surface of the root, penetrate and infect the epidermal cells of the roots, or attack the budding shoots and radicles of the seedlings. The fast-growing pathogen *Pythium* spp. can colonize seeds and embryos before they emerge. The majority of soilborne fungi affect young, juvenile roots in contrast to secondary roots. After the root damage, the fungus remains in the cortex and reproduces spores within the root tissue. The hyphae continue to spread up the root, internally or externally or can move in close proximity. A specialized group of pathogen population present in the soil causes wilt complexities (e.g., *Fusarium oxysporum*, *Verticillium dahliae*), which penetrates through the endodermis into the vascular tissue and move up the xylem to above-ground parts.

Soil provides shelter for nematodes complex which is worm-like eukaryotic invertebrate and most likely among the numerous fauna on the planet. They are free-living and parasitize the vegetable crops particularly solanaceous family. Some feed on the surface of a root known as migratory ectoparasitic, some pierce and shift in inner parts of the root (migratory endoparasites). A group of nematode established at a feeding site of the root remains there for reproduction (sedentary endoparasites). The root-knot nematodes are the vital pests which parasitize vegetable roots belonging to solanaceous, cucurbitaceous, leguminous, and several other root and bulb crops. The appearance of symptoms by root-knot nematode on the above-ground plant part showed poor growth, a reduced quality, yields and resistance to other stresses. A greater population of root-knot nematode can destroy the plants leading to total crop failure.

FOLIAR FUNGAL PATHOSYSTEM OF VEGETABLE CROPS

Fungi are the leading plant pathogens responsible for severe diseases of vegetables. They affect the plants by destroying the cells and/or inducing stress. Pathogen propagules survive on infected seed, soil, crop residues, volunteer crops, and weeds. They are dispersed by air current, water splash, contaminated soil, animals, machinery tools, seedlings, and planting materials. The fungal pathogens enter plants through stomata, wounds, harvesting, hail, insects, and mechanical damage. Most of the fungi exhibited foliar diseases like downy and powdery mildews, and white blister, leaf spots are some of the highly prevalent diseases. A group of the fungal pathogen is associated with a wide variety of vegetables which includes anthracnose, Botryitis rots, rusts, Rhizoctonia, Sclerotinia, and Sclerotium rots.

Climatic conditions directly affect the foliar fungal pathogen, which leads to severe epidemics. The environmental changes are likely to affect pathogen growth and survival rates and adapt host susceptibility, which makes an impact on the vegetable pathosystems. The impact of climatic variations will vary by different pathosystem and geographical expanse. These variations may affect not only the favorable set for infection but also the specificity and mechanism of infection in pathosystem. In a vegetable pathosystem, variation in temperature, CO_2 and ozone level, drought and precipitation influence the biology of pathogens and their infectivity in plants and survival. The microclimate around plants changes due to abiotic conditions that will also, affect the vulnerability of plants to infection.

BACTERIAL PATHOSYSTEM IN VEGETABLE CROP

The plant pathogenic bacteria cause many serious diseases in vegetable crops all over the world. They are one of the major microorganisms in the vegetable pathosystem that cause leaf spots, blight, and vascular wilts. Commonly, they are single-celled microbes that may be epiphytes or endophytes. Several of the bacteria are residents and some are short-lived. Plant pathogenic bacteria survive in nature, usually in debris on the soil surface, internally or externally in seed, soil, and perennial weed hosts. Simply, plant pathogenic bacteria spread from one host to another, but do not always result in infection. For successful infection of bacterial pathogen, a certain temperature is required, for example, *Pseudomonas syringae* pv. *phaseolicola* causes infection below 22°C and *Xanthomonas campestris* pv. *phaseoli* more than 22°C on dry bean (*Phaseolus vulgaris*). Under growth conditions, both bacteria infect simultaneously in which the day and night temperature gap facilitates disease development in susceptible hosts. Bacterial wilt caused by *Ralstonia solanacearum* is a major problem across the world in heavy soils and in low-lying areas that keep moist soil for a long period of time. After the removal of plants, infected root parts remain in the soil, which provides an opportunity to bacteria for infection of a new solanaceous host. The bacterium stays alive in the soil for long periods without host roots. Moreover, it can survive for days to years in water, moist soils, or profound soil stratum (> 75 cm) that form a reservoir of inoculum. High temperatures 30–35°C and soil moisture favor wilt development. However, the populations of bacteria decrease rapidly at < 4°C. In the case of race 3, biovar 2 can still endure, which is a physiologically latent and viable state.

Phytoplasmas infect both monocot and dicot crops, including vegetables all over the world. They are dedicated wall-less bacteria that differ from Gram-positive, are obligate parasites and perpetuate in phloem tissues and insect hemolymph. It was previously known that plants exhibited typical yellowing symptoms in viral disease on the basis of symptomatology, their mode of infection, and vector transmissibility. Later, it was demonstrated that it was caused by pleomorphic prokaryote, which was a breakthrough in the discovery of a novel group of pathogens. The phytoplasmal infection in vegetable crops was reported worldwide, like tomato big bud, and its distribution was noted throughout the world. A varying range of symptoms was reported in plants by phytoplasma, namely, phyllody, virescence, witches' broom, yellowing, and stunting. Insect vectors select the host range of phytoplasma and are dispersed mostly by phloem-feeding insects of the order Hemiptera that belongs to different families, for example, Cicadellidae (leafhoppers), and Fulgoridae (planthoppers) from infected to healthy plants. Insect vectors quietly act as the alternate host in this pathosystem. They vary with the species of phytoplasma interaction, like *Hishimonus sellatus*, and alone can be able to transmit six different types of phytoplasma. In a few instances, a phytoplasma may also be transmitted by more than one species of insect vectors. All through overwintering of phytoplasma in perennial plants or insect–vector, their interface with insect hosts changed, namely, fitness may enhance or decline.

PLANT–VIRUS VECTOR: A COMPLEX PATHOSYSTEM IN VEGETABLE

Among various pathosystems, viruses are a major cause of reduced productivity. Viruses cause considerable loss of yield and quality produce by inciting mosaics, curling, stunting, and wilting in vegetable crops. The virus genus belongs to different families interacting by various means, like an insect vector, which played a significant role in dissemination and dispersal. Generally, the complex relationship between the virus, vegetable crop, and the insect vector makes difficulty in the formulation of a sustainable management approach. The molecular identity of viruses and consideration on their epidemiology is the aspect to assess the severity and economic impact of the diseases and efficient management strategy. The effect of interactions between virus, host, vector, and ecological changes decides the epidemic temperament. In vector-transmitted viruses, whitefly-transmitted geminiviruses, aphid-transmitted potyviruses, and thrips-transmitted orthotospoviruses encompass a wide range of solanaceous, cucurbitaceous, and leguminous vegetables. Begomo viruses are the leading and most important genus within the family geminiviridae with single-stranded DNA either a monopartite (a single DNA) or a bipartite (with two DNA components:

DNA-A and DNA-B) genome organization, infecting dicotyledonous plants. The monopartite virus is associated with betasatellite components. Detailed work has been initiated on these viruses, for example, sequence analysis, phylogeny, infectivity, functions of viral proteins, virus–host interactions, virus-derived transgenic resistance. A number of economically important crops are infected by begomoviruses, which covered more than 300 species. Therefore, begomoviruses need great attention in the vegetable pathosystem in order to understand the biology of virus and whiteflies transmission.

Potyviruses in vegetables are widely distributed, which causes economic damage. Members of the family *Potyviridae* comprise monopartite and bipartite plant viruses. Potyvirus is the largest genus of the plant virus group with 160 species, which contains some economically important viruses, such as *Potato virus Y, Bean yellow mosaic virus, Papaya ringspot virus* w strain in cucurbits. Based on amino acid sequences of coat proteins, the family *Potyiviridae* is grouped into eight genera, namely Potyvirus, Poacevirus, Ipomovirus, Macluravirus, Rymovirus, Tritimovirus, Brambyvirus, and Bymovirus. Viruses pass vertically from inherited parent to progeny. Both asexual and sexual reproduction plays a role like vegetative propagation. In sexual reproduction, the transmission of viruses is found as a consequence of seed infection. The family Potyviridae is transmitted by various organisms. Various genera of Potyvirus and Macluravirus are vectored by aphids in a nonpersistent and noncirculative mode. The aphid population survives year round in different vegetables, which may lead to transmission of the virus from one species to another by short probing.

NEMATODE–PLANT ROOT INTERACTION IN VEGETABLES

Vegetable crops are more prone to plant-parasitic nematodes in which several species incur greater damage. They induce a disease complex by interrelating with other parasites and breaking resistance in plants. Existence of nematodes is found in all habitats but is often unseen because the majority of them are microscopic in size. Soil-inhabiting nematodes can also be grouped on the basis of their feeding habits. Several important feeding groups of nematodes commonly occur in most soils. Additionally, nematodes have to face challenges such as insatiable predators, variation in soil temperature, moisture level, and the loss of its host plant. For the survival of the nematode population, these obstructions ought to be overcome. The nematode population avoids these biotic and abiotic barriers by utilizing a combination of behavioral and physiological endurance approaches. Nematodes make parasitic relation on above- and below-ground parts of plants in various ecologies. They feed on all parts (roots, stems, leaves, flowers, and seeds) of plants by various ways, but all use a stylet. Based on its size and shape, it is used to classify nematodes and also can be used to infer their mode of feeding. In the vegetable pathosystem, the root-knot nematode (*Meloidogyne* spp.), sugar beet cyst nematode (*Heterodera schactii*), onion stem and bulb nematode (*Ditylenchus dipsaci*), pea cyst nematode (*Heterodera goettingiana*), potato cyst nematode (*Globodera rostochiensis* and *G. pallida*) have a great role.

Interaction of nematodes with soilborne fungi and bacteria causes greater damage in vascular diseases of vegetables. The incidence of *Rhizoctonia solani* increases when tomato plants are inoculated with *M. incognita*. Similarly, the incidence of *Fusarium oxysporum* also increases. Bacterial wilt caused by *R. solanacearum* primarily infects host plants through their roots by feeding damage by insects and nematodes. The bacterial wilt incidence is 100% found prior inoculation with *Meloidogyne incognita*.

MICROBIAL INTERACTION IN POSTHARVEST

Postharvest loss is one of the main concerns in vegetables across the world. Postharvest diseases are caused by various factors, including the type of commodity, susceptibility to diseases, and environmental conditions (temperature, relative humidity, etc.). Moreover, maturity, ripening stage, and postharvest hygiene are also deciding factors under the pathosystem. Postharvest losses may take place at any time while handling, from harvest to consumption. Under certain conditions, some fungal genera produce mycotoxins, such as *Penicillium, Alternaria,* and *Fusarium.* Bulb crops are affected by a number of pests and diseases like thrips, mites nematodes, and various fungi-causing bulb rot. Several postharvest pathogens spoil the quality of bulbs and reduce the storage life. Abiotic and biotic factors are responsible for postharvest losses in the bulb, particularly onion and garlic. Generally, decaying is caused by several species of the fungi, such as *Alternaria, Botrytis, Botryosphaeria, Colletotrichum, Diplodia, Monilinia, Penicillium, Phomopsis, Rhizopus,* and *Sclerotinia.* Some bacteria also induce decaying in bulbs, for example, *Erwinia* and *Pseudomonas.*

Acknowledgments

We are thankful to Dr. Mohammed Wasim Siddiqui (Chief Editor, *Journal of Post Harvest Technology*) for helping in the preparation of the matters. We also appreciate the encouragement received from the Chairman, Department of Plant Pathology, Bihar Agricultural University. This book was generously developed with Bihar Agricultural University Communication Number 380/2018.

CHAPTER 1

Phytophtora Blight in Potato: A Challenge Ahead in the Climate Change Scenario

MEHI LAL[1*], SAURABH YADAV[1], V. K. DUA[2], and SANTOSH KUMAR[3]

[1]Division of Plant Protection, ICAR – Central Potato Research Institute Regional Station, Modipuram, Uttar Pradesh, India

[2]Division of Crop Production, ICAR – Central Potato Research Institute, Shimla, India

[3]Division of Plant Pathology, Bihar Agricultural University, Sabour, Bihar, India

*Corresponding author. E-mail: mehilalonline@gmail.com

ABSTRACT

Potato is an important crop in the group of vegetable crops. It is being affected by various microorganisms apart from abiotic factors. Fungi, bacteria, viruses, nematodes, and phytoplasma are the main biotic factors, which causes tremendous loss to the potato crop. The pathogen, *Phytophthora infestans* (Mont.) de Bary, is a major fungus-like organism causing late blight which leads to 80% loss under improper management. The disease is severely challenged by various climatic factors, viz., temperature, relative humidity, rainfall, and CO_2 which are behaving in erratic manner in the past few years. The pathogen has adapted itself to higher temperature (up to 28°C) resulting in greater chances of spreading to larger area or persistence extended. Moreover, the level of resistance and susceptibility of the potato cultivars will also be changed in climate change scenario. Therefore, in the current perspective, control strategies should be modified or improved to sustainable management. Application of suitable fungicides coupled with suitable planting date and resistant cultivars taken into consideration. Moreover, advisories of forecasting models could also assist in the management of the late blight. Effects of temperature, CO_2, and rainfall on the pathogen are focused. Impact of these factors which are influencing the management strategies is discussed in this chapter.

1.1   INTRODUCTION

Potato (*Solanum tuberosum*) is a prominent vegetable crop and is grown in about 130 countries. It is the third most important food crop after rice and wheat (CIP Annual Report, 2015). It can be grown in temperate, tropical, and subtropical regions of the world. Potato is the world's most important vegetable crop, with nearly 400 million tons produced worldwide every year, lending to stability in food supply and socioeconomic impact (Halterman et al., 2016). It is estimated that the increase in crop yield are 10–20% for C3 crops and 0–10% for C4 crops (Ainsworth and Long, 2005). Over the years, potato production has increased substantially in the developing countries. Today, China is the main producer of potato producing about 20% of global production (Staubli et al., 2008). It has been forecasted that the yield of potato will decrease by 18–32% due to changes in the climate. Past few years have witnessed a steady increase in national and international concern over the sustainability of the global environment. Climate change has emerged as the most prominent of the global environment issues, due to its harmful impact on the socioeconomic conditions of the human being. Climate change implies that the average conditions (mean and/or variability) are changing over time and may never return to those previously experienced (Coakley, 1988). Global climate has changed ever since the industrial revolution. The atmospheric concentrations of the greenhouse gases, carbon dioxide (CO_2), methane (CH_4), and nitrous oxide (N_2O), have increased since 1750 due to human activity. In 2011, the concentrations of these greenhouse gases were 391 ppm, 1803 ppb, and 324 ppb, exceeding the preindustrial levels by about 40, 150, and 20%, respectively (IPCC, 2013). Global surface temperature change by the end of the 21st century is likely to exceed 1.5°C relative to 1850–1900 for all Representative Concentration Pathways (RCPs) scenarios except RCP2.6. It is likely to exceed 2°C for RCP6.0 and RCP8.5 and more likely than not to exceed 2°C for RCP4.5. Warming will continue beyond 2100 under all RCP scenarios except RCP2.6. Warming will continue to exhibit interannual to decadal variability and will not be regionally uniform (IPCC, 2013). It is expected to have serious consequences for mankind and the environment. The critical threshold is to be around 2°C increase of temperature (IPCC, 2007). CO_2 and temperature interactions are recognized as a key factor in determining plant damage from pests in future decade; likewise CO_2 and precipitation interaction will also be important. Globally, atmospheric CO_2 has been increased, and in northern latitudes mean temperature at many locations has increased to about 1.0–1.4°C with accompanying changes in pest and pathogen incidence and farming practices (Gregory, 2009).

Pace of climate change and the unpredictability of its characteristics are of great concern with respect to the pathogens and insect pests that reduce crop yield. The classic disease triangle recognizes the role of climate in the plant diseases, as no virulent pathogen can incite disease on a highly susceptible host if prevailed climatic conditions are not favorable. For epidemic conditions, fourth factor is time. How many times favorable conditions existed, it is also affecting the diseases' development. All four factors (host, pathogen, environment, and time) are represented in disease tetrahedron or disease pyramid. Sometimes, a new component, human, is used in place of time in disease pyramid, but it should be considered a distinct new fifth component that influences the development of plant diseases directly and indirectly (Agrios, 2006). The climate change and global warming with increases in temperature, moisture, and CO_2 levels can impact all three legs of the plant disease triangle in various ways. Climate change could "alter stages and rate of development of the pathogen, modify host resistance, and result in changes in physiology of host– pathogen interaction" (Garrett et al., 2006). Climate influences all stages of host and pathogen life cycles as well as development of disease. Climate change and global warming will allow survival of plant and pathogens outside their existing geographical range. "The lack of action on climate change not only risks putting prosperity out of reach of millions of people in the developing world, it also threatens to roll back decades of sustainable development" (http://www.sciencemag. org/news/2012/11, Kim,19 November, 2012; a foreword to the report, Turn Down the Heat: Why a 4°C Warmer World Must be Avoided). The climatic factors including changes in temperature, rainfall and other atmospheric composition along with predominantly elevated CO_2 levels would accelerate the reproduction time of many plant pathogens and pests, thereby increasing their infection pressure on crop plants (Boonekamp, 2012).

The impact of climate change on the occurrence and activity of pathogens, pests, and diseases of agricultural crops have recently been considered more seriously. This is documented by the fact that renowned scientific journals in plant pathology (e.g., *Annual Review of Phytopathology*) published the papers focused on the implications of climate change for plant disease occurrence and management (Garrett et al., 2006). The range of plant pathogens and insect pests are mainly constrained by temperature, and the frequency and severity of weather events affecting the timing, intensity, and nature of outbreaks of most organisms (Yang and Scherm, 1997). It is well established that short- and long-term changes in climate will impact on which strains of pathogens prove most aggressive via their ability to grow and sporulate under particular environmental conditions. The change in environmental factors, particularly temperature and leaf wetness would influence the rate of progress at any stage of plant disease development (Agrios, 1997; Coakley et al., 1999). The late blight of potato, being a climate-dependent disease, might be more affected in climate change scenario.

As far as Indian scenario is concerned, losses in potato production due to late blight ranged between 5% and 90% depending upon various climatic conditions, with an average of 15% across the country (Collins, 2000). Tuber yield decline was significantly higher in unmanaged crop, which could go as high as 90% of total productivity in hilly regions. In the years, when disease appeared early in crop season due to heavy rains, late blight continued to build up throughout the crop duration resulting in crop losses up to 75% (Singh and Bhat, 2005). However, Lal et al. (2015) recently reported that yield losses were ranged 10–20% due to late blight during 2013–2014 in major potato growing states of the India. Potato late blight had become more severe in the recent years due to influx of a new population containing the A2 mating type especially in the United States (Goodwin et al., 1995). The worldwide late blight disease is re-emerging; therefore, this disease is constantly observed by the late blight researchers (Fry et al., 2015). The late blight disease is considered as an emerging disease; it is not only having importance in worldwide crop production but also pose severe threat on a local level, especially on small farms in developing countries (Subbarao et al., 2015). *Phytophthora infestans* is the most widely studied oomycete; about 1230 papers has beenpublished in the last 10 years (2005–2014) and is one of the top 10 oomycete pathogens studied using molecular techniques (Kamoun et al., 2015)

Late blight of potato caused by *P. infestans* still remains the leading threat to potato cultivation globally. This notorious pathogen is known to break host resistances across *Solanum* spp. and has the ability to develop resistance to chemical fungicides in a very short span of time. The pathogen possesses this quality owing to very large genome size of about 240 Mbps (Hass et al., 2009) which is largely composed of transposons. Consequently, it is also expected that this pathogen will quickly adapt itself to changed climatic conditions and will continue to play havoc in potato-growing regions across the globe. There are some reports that the *P. infestans* is introduced in different places by seeds. Pamella et al. (2004) reported that the introduction (40%), weather (41%), and farming techniques (19%) are the important factors for fungi in spreading emerging infectious disease. Fungal diseases such as those caused by *Ustilago* and *Sclerotia* species are not expected to be affected by climate change because of their monocyclic nature. However, polycyclic diseases such as those caused by *Colletotrichum, Peronospora, Phytophthora,* and *Puccinia* species, each additional disease cycle multiplies inoculums many fold, so an increased duration of growing season in Ontario (Canada) would be expected to result in an increased number of disease cycles and inoculums. Primary inoculums will be increased for disease established in tomato late blight whereas potential duration for epidemic of potato late blight will be increased in Ontario (Boland et al., 2004). Zargarzadeh et al. (2008) observed that there were significant correlations between climatic factors, particularly temperature and rainfall, with the incidence of the late blight. The late blight appearance, favorable period, and delay in appearance of late blight in western Uttar Pradesh were predicted using base line for the year 2000. It is clearly indicated that earliest late blight appearance during the potato crop season was predicted during 13 October to 1 November in baseline year 2000 and is expected to be delayed by 0–8 days in 2020 and 10–21 days in 2055 (Fig. 1.1) (Dua et al., 2015).

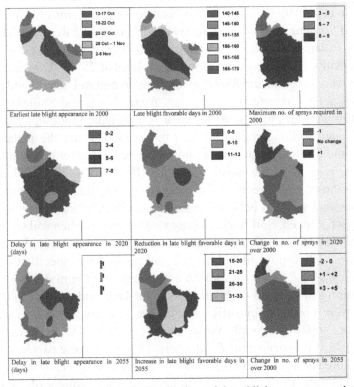

FIGURE 1.1 (See color insert.) Distribution of late blight appearance, late blight favorable days, and delay in late blight appearance in western Uttar Pradesh in years 2000, 2020, and 2055, respectively.

1.2   INFLUENCE OF CLIMATE CHANGE ON POTATO LATE BLIGHT

1.2.1   TEMPERATURE

Temperature is an important factor for affecting the growth and development of any pathogens. A change in environmental temperature is definitely affecting the pathogens. Moreover, its duration will have greater impact on pathogens. Temperature change might lead to emergence of new races of the pathogens hitherto not active but might cause sudden outbreak of the disease. *P. infestans* is an environmentally linked pathogen. For the indirect germination of sporangia (5–10 zoospores/sporangium), <20°C temperature (optimum 12–13°C) is needed, whereas >20°C (optimum 24°C) favors for direct germination (one sporangium gives rise to a germ tube). Optimum temperature for mycelial growth is 18°C. Even at optimum temperature, the percentage of direct germination of sporangium is usually much lower than the percentage of indirect germination of sporangium (Crosier, 1934; Goodwin et al., 1994). A subset of 47 isolates of *P. infestans* from India was analyzed for their thermal adaptation at 25°C and 28°C. Result revealed that 53.33% isolates of *P. infestans* were resistant (fast growing) and 46.67% were sensitive (slow growing) from UP isolates at 25°C. Whereas, 14.29% isolates were resistant and 85.71% were sensitive from Uttarakhand at 25°C. Hundred percent isolates were resistant from Shimla and Shillong regions, while 100% isolates were sensitive at 28°C from Uttar Pradesh, Uttarakhand, Shimla, and Shillong regions. It can be concluded that the isolates from Shimla and Shillong regions were more adapted to higher temperature than the other regions of the isolates in India (IISR, 2014–2015). The change in temperature will directly influence penetration, infection, reproduction, dispersal, and survival between seasons and other critical stages in the life cycle of a pathogen. Soil-borne diseases may increase when soil temperature increases. If climate change causes a gradual shift of cropping regions, pathogens will follow their host, as pathogens have specificity to their host. In case of *P. infestans*, potato and tomato are the major hosts. Some sporulation and necrosis on bittersweet nightshade and petunia on detached leaves were observed by different lineages, namely, US-22, US-23, and US-24 of *P. infestans* (Seidl Johnson and Gevens, 2014). These two may act as a new host for survival of *P. infestans*. Therefore, these two hosts should be monitored in future regularly for their change in the level of susceptibility. The susceptible host has an important role in epidemiology of the disease. Epidemiological components, namely, incubation period (IP), latent period, lesion area (LA) sporulation, have significant role in climate change as each component is linked to each other. IP of *P. infestans* was short at higher temperature, and the shortest IP was observed at 28°C (Becktell et al., 2005). Generally, IP declined exponentially and LA increased exponentially with increase in temperature (Mizubuti and Fry, 1998).

Becktell et al. (2005) studied sporulation of US-17, a new tomato adapted lineage, at 28°C and found that sporulation was almost nil, whereas the sporulation of US-1 isolates at 27°C was found similar to that recorded at its optimum temperature. It might have happened due to the location of tomato-producing areas at lower altitudes, which may have resulted in the higher fitness of US-1 on tomato over the longer time of co-existence of US-1 with tomato plants and the subsequent adaptation of US-1 isolates to higher temperatures. Mariette et al. (2016) reported that the local adaptation of *P. infestans* against temperature could result from selection for increased survival between epidemics, when *P. infestans* isolates are exposed to more extreme climatic conditions than during epidemics. It is also observed that different thermal responses among two clonal lineages sympatric in Western Europe, with lower performances of lineage 13_A2 compared to 6_A1, especially at low temperatures. The experiments conducted at CIP, Peru, to work out the risk of late blight especially in the number of pesticide sprays at global level in climate change scenario, it revealed that with rise in global temperature of 2°C, there will be lower risk of late blight in warmer areas (<22°C) and higher risk in cooler areas (>13°C). Earlier onset of warm temperatures could result in an early appearance of late blight disease in temperate regions with the potential for more severe epidemics and increased number of fungicide applications needed for its control. Another experiments carried out in Finland predicted that for each 1°C warming, late blight would occur 4–7 days earlier, and the susceptibility period extended by 10–20 days (Kaukoranta, 1996). This would result in 1–4 additional fungicide applications, increasing both cost of cultivation and environmental risk. Analysis of historical data from 1948 to 1999 revealed that late blight risk over a standardized growing season from 1 May to 30 September increased in the Upper Great Lakes region of the United States. Predominant genotypes of *P. infestans* (e.g., US-8) in the United States appear more tolerant to temperatures close to 0°C and their survival in warming conditions may explain their supremacy (Baker, 2004).

A positive correlation was observed between IP and LA development of late blight disease of potato (CPRI, 2011–2012). It was observed that up to 15°C IP was more and LA was less, while from 18 to 25°C IP was less and LA was more and an incremental increase in temperature thereafter resulted in increased IP with decrease in LA (CPRI, 2012–2013). These results clearly indicate that the temperature has an important role in LA development. Considering an increase in temperature at West Bengal in India during the potato-growing season, the late blight, disease severity is expected to be reduced by 5–7% from 1981–2010 levels to 2031–2040 (APN Project Report, 2011). In other parts of India also, the late blight scenario may change drastically with climate change. Earlier, late blight was not a serious problem in the states of Punjab, Haryana, and parts of Uttar Pradesh, primarily due to suboptimal temperature regimes during December–January. However, disease outbreaks will become more intense with increase in ambient temperature coupled with high relative humidity (RH). Such scenarios had been witnessed during warmer years, that is, 1997–1998 and 2006–2007, when average crop losses in this region exceeded 40%. During the recent past years also severity of late blight increased due to the occurrence of rainfall at advanced stage of crop growth. States like Madhya Pradesh, Gujarat, and Central Uttar Pradesh which are comparatively less affected by late blight may also witness frequent outbreaks of the disease under the climate change scenario (Singh and Bhat, 2010). Increase in both temperature and RH has added new dimensions to late blight disease cycle across the world. Under such a situation, *P. infestans* attacks potato stems more often than foliage. In fact, in recent years, it has been observed that "stem blight" is more common than the foliar blight. This phase of the disease is more serious than the foliar stage as it affects stem and subsequently tubers also. The survivability of inoculums is more in stem blight than foliar blight. During 2014–2015 and 2015–2016, it was observed that field symptoms of late blight was observed even at higher temperature (>27°C). It seems that pathogen had adopted higher temperature (CPRI, 2015–2016).

In a recent study conducted using JHULSACAST model, it was found that if the host physiology does not change even under the favorable RH in Punjab, late blight is expected to be delayed by 0–6 days in 2020 and 12–14 days in 2055 scenarios, over the baseline of 2000. In western Uttar Pradesh, earliest late blight appearance during the potato crop season is expected to be delayed by 0–8 days in 2020 and 10–21 days in 2055. In Punjab, there were an average of 105 late blight favorable days out of 182 suitable potato growth days in the year 2000; the number is likely to be increased to 135 and 140 days in 2020 and 2055, respectively, in Punjab. In western Uttar Pradesh, potato-growing season was warmer which would decrease late blight favorable days by 7 and 27 in 2020 and 2055, respectively. The number of sprays required to control late blight in potato seed crop would be 7.3 and 8 in future scenario (2020 and 2055) in comparison to 6.5 in baseline (2000) in

Punjab. In contrast, there would be no change in the number of sprays in the year 2020 (over baseline year 2000) in western Uttar Pradesh; however, due to further increase in the temperature in year 2055, it could be reduced by two numbers of sprays (Dua et al., 2015).

1.2.2 RAINFALL/HIGH HUMIDITY

RH is an important component to initiate the infection caused by *P. infestans*. Generally, about 85% RH is required to initiate infection. An increase in rainfall and high humidity will result in elevated threat of potato diseases, such as late blight (*P. infestans*), especially when combined with longer growing seasons. In Upper Great Lakes region of the United States, increases in total precipitation and increase in the number of days with precipitation over the years with warmer climate is supposed to be the reason for the increased risk of potato late blight infection, subsequent yield and economic losses (Baker et al., 2005). The United Kingdom, the unusually wet season (high humidity) during 2007, coincided with the prevalence of epidemiologically fit new pathotypes of *P. infestans*, resulted in an unprecedented number of outbreaks of late blight (www.eucablight.org). Indirect or secondary consequences were observed in Canada for outbreaks of potato late blight due to extreme environment condition. During 1994–1996, epidemic conditions were reported due to various genotypes of *P. infestans* from distant regions which were associated with the unusual tropical storm tracks moving up the eastern seaboard of the United States (Peters et al., 1999).

In India, in Lahaul valley of Himachal Pradesh, which was earlier free from late blight due to lack of precipitation, has now experienced attack of late blight due to occurrence of rainfall. However, hotter and drier summers which are likely in United Kingdom may reduce the importance of late blight, although earlier disease onset may prevent this advantage. An empirical climate disease model has suggested that under the climate change scenario of 1°C increase with 30% reduction in precipitation in Germany will decrease potato late blight to a mere 16% of its current level (Singh et al., 2013). In India, late blight attacks are likely to increase in North-Western plains with the increase in temperature. The disease is likely to be reduced in Eastern plains including West Bengal. Hilly regions may get less prolonged attack of late blight with the increase in temperature regime and decrease in precipitation. In general, high RH and occurrence of rainfall after the appearance of the late blight disease, the disease spreads in very short time and subsequently it causes economic losses of the potato yield (Fig. 1.2).

1.2.3   ELEVATED CO$_2$

CO_2 is an important component in global warming. Elevated CO_2 levels can impact both the host and pathogen in various ways. It was reported that the higher growth rates of leaves and stems observed for plants grown under high CO_2 concentrations may result in denser canopies with higher humidity that favor pathogens (http://www.climateandforming.org). The elevation of CO_2 (400–700 ppm) and/or ozone (ambient or twofold ambient) resulted in a change in susceptibility of potato plants infected with *P. infestans*. The main effect was that a rise in CO_2 caused a significantly enhanced resistance of the susceptible potato cultivar "'Indira" toward *P. infestans*, whereas ozone had no significant effect. The effect of N-fertilization in combination with CO_2 treatment on the resistance of potato against *P. infestans* showed CO_2 induced an increase in resistance correlated with an increased C/N ratio in potato leaves. The lower C/N ratio, due to higher N-concentrations, decreased in resistance against *P. infestans* (Osswald et al., 2006). In other studies, the potato cultivar Indira grown under normal conditions showed a high degree of susceptibility to *P. infestans*, whereas it developed resistance after exposure to 700 ppm CO_2 (Plessl et al., 2007). It was observed that an increased tolerance of tomato plants to *Phytophthora* root rot when grown at elevated CO_2 (Ywa et al., 1995). Mitchell et al. (2003) reported that elevated CO_2 increased the pathogen load of C_3 grass, perhaps due to increased leaf longevity and photosynthetic rate. Carter et al. (1996) reported that after running the data of 1961–1990 in a simulation model showed that the changes in climate and CO_2 by 2050 are expected to increase the occurrence of late blight in all the regions of Finland (Carter et al., 1996). Due to the current climatic conditions, the risk of late blight will increase in Scotland during the first half of the potato-growing season and decrease during the second half. Therefore, advancing the start of the potato-growing season by 1 month proved to be an effective strategy from both an agronomic and late blight management perspective (Skelsey, 2016).

1.2.4   OZONE (O$_3$)

Ozone is a minor gas in comparison to CO_2, which affects global warming. It affects both host tissues and pathogens. The potato cultivar Indira showed a slightly more susceptiblity to *P. infestans*, at double ambient of O_3 than ambient O_3 (Plessl et al., 2007) In Russia, potatoes cv. Detskoselskii was grown with and without UV-B-radiation at doses corresponding to 25%

destruction of the ozone layer. It was observed that the incidence of *P. infestans* was significantly reduced by UV-B (Zyabliskaya and Parshikov, 1998). More focused research is needed particularly on O_3 against late blight of potato.

1.3   IMPACT OF CLIMATE CHANGE ON LATE BLIGHT MANAGEMENT

The fungicides, host resistance, and biocontrol strategies are adopted for late blight management. These strategies will also be affected due to variation in climatic condition. Efficacy of fungicides will be mainly affected by elevated temperature and erratic rain fall, while host resistance and bioagents will also be affected by increasing temperature, CO_2, and O_3. Moreover, if the pathogen has adapted a wider range of temperature, then old forecasting model need to be relooked. All aspects need to be considered for better management of late blight under climate change scenarios.

1.3.1   TEMPERATURE

Temperature is the most important abiotic factor which affects the late blight management strategies. Temperature and precipitation can alter fungicides residue dynamics in the foliage, and degradation of product can be modified. Penetration, translocation, and mode of action of systemic fungicides are affected due to the CO_2-enriched atmosphere and different temperature and precipitation conditions (Ghini et al., 2008). Systemic fungicides could have been affected negatively by physiological changes like slow uptake rates, smaller stomata opening or thicker epicuticular waxes in crop plants grown under higher temperature. These fungicides could affect positively by increasing plant metabolites rates that could increase fungicides uptake (Petzoldt and Seaman, 2006).). In an experimental field, daytime temperatures was always above 24°C, even reaching above 30°C on some occasions. When cymoxanil fungicide was applied against late blight of potato, the efficacy of fungicide was reduced. It is believed that high temperatures limited the efficacy of cymoxanil in the field experiments (Mayton et al., 2001). Kapsa (2008) reported a negative correlation between the efficacy of contact fungicides (fluazinam and mancozeb) and rainfall. However, propamocarb-HCl + chlorothalonil and cymoxanil + mancozeb correlated significantly negatively with air temperature. Cymoxanil is known to degrade rapidly, and high temperatures may have accelerated the degradation (Klopping and Delp, 1980). In one of our experiments to study the efficacy of fungicides at elevated temperatures, potted plants were sprayed with different fungicides at different concentrations and exposed to 25°C and 30°C. Leaves were detached at regular interval and challenge inoculated with *P. infestans*. At 25°C, *metalaxyl 8% + mancozeb 64% WP* fungicide showed degradation after 14 days of spraying though the rate of degradation was very low compared to control, while at 30°C degradation was more. At 25°C, *cymoxanil 8% + mancozeb 64% WP* showed no degradation even after 14 days, while at 30°C, degradation was found at 14 days after spraying. No degradation was recorded in dimethomorph 50% WP fungicide, even after 14 days of spraying at 25°C and 30°C (Table 1.1) (Lal and Yadav, 2014; Yadav et al., 2016).

The effect of temperature on host resistance was studied at 20°C and 25°C. Two late blight differentials (R1 and R1.2.3) and two cultivars (Kufri Girdhari and K. Jyoti) were grown at the above-mentioned temperatures and then detached leaves were challenge inoculated with *P. infestans*. Results revealed that at higher temperature, more lesion size was produced (IISR, 2012). The studies revealed that in spite of a delay in the initial development and a reduction in host penetration, the established colonies develop faster under increased CO_2 concentration (Ghini et al., 2008). The most likely impact of climate change will be felt in three areas: in losses from plant diseases, in the efficacy of disease management strategies and in the geographical distribution of plant diseases (Chakraborty et al., 2000). Increase of 3°C temperature from ambient temperature had increased potato yield loss caused by late blight in the 3-year long controlled-environment study, reducing potential benefits from yield increases due to warmer temperatures (Kaukoranta, 1996). The cultivar Alpha is moderately susceptible to late blight, exhibited resistance in northern (New York, USA) than in southern latitudes (Taluca, Mexico). It could be influenced in the aerial environment particularly, temperature and the presence of local strain of *P. infestans* (Parker et al., 2009). Expressions of resistance (horizontal and vertical resistance) were also affected by temperature; however, their relative resistance ranking among tested cultivars was same in four different experiments with different temperatures and photoperiods (Rubio-Covarrubias, 2005). The warmer temperatures and a reduction in the occurrence of sustained periods of humid weather could result in easier control of potato late blight disease in the northern Netherlands (Schaap et al., 2011).

1.3.2 RAINFALL

In general, rainfall may affect the efficacy of fungicides. Frequent rainfall may affect the efficacy of contact fungicides due to washing off from leaf surface (Kocmankova et al., 2009). Effect of rain on the efficacy of fungicide upon its deposition on leaves, the interaction of precipitation frequency, intensity, and fungicides dynamics are complex, and for certain fungicides, precipitation following application may result in enhanced disease control because of a redistribution of the active ingredient on the foliage (Schepers, 1996). The clear-cut effect of simulated rain was observed on the efficacy of Mancozeb-based and Metalaxyl-based fungicides against tomato late blight (Rani et al., 2013). Late blight disease is difficult to manage when abundance moisture is prevailed in field; in an irrigated field with inoculum originating from infected seed tubers, incidence of late blight increased from 0.2% to 70% over a 4-week period after canopy closure even with nine applications of efficacious fungicides (Johnson et al., 2003). Chen and McCarl (2001) conducted a regression analysis between pesticide usage and climate variations in several US locations, with climate data provided by National Oceanic Atmospheric Administration. The average cost of the pesticide usage for corn, cotton, soybean, and potatoes also increases as temperature increases. The climate change models projected at worldwide that an increase in the frequency of intense rainfall events which could result in increased fungicide washoff and reduced control (Fowler and Hennessy, 1995). Alternations in plant morphology or physiology, resulting from growth in a CO_2-enriched atmosphere or from different temperature and precipitation conditions, can affect the penetration, translocation, and mode of action of systemic fungicides.

1.3.2.1 FORECASTING SYSTEMS

Prediction of forecasting system is an important component for management of late blight. However, most of the forecasting systems developed, a few years ago, are in need of updation as the pathogen has adopted a wider range of temperature. Forecasting system based on non-linear responses to the temperature and leaf wetness offers more potential to represent the impact of climate change and variability on disease epidemiology (Bourgeois et al., 2004). Forecasting of late blight is an important aspect of the late blight management, and forecasting of rainfall is also helpful for managing the late blight in Columbia Basin (Johnson et al., 2015).

In India, JHULSACAST and INDO-BLIGHT CAST models have been developed for prediction for late blight appearances. Lal et al. (2014) developed a model for yield loss assessment using cv. Kufri Bahar for assessing losses caused by *P. infestans*.

1.4 MITIGATION STRATEGIES

- Use of new molecules with higher efficacy at elevated temperature for disease management. New molecule (rain fastness) with higher efficacy should be used in high rainfall regions.
- New forecasting models for prediction of appearance of diseases at spatial scale, that is, region, state, or country. For example, INDO-BLIGHTCAST for forecasting first appearance of late blight in India (http://cpri.ernet.in).
- Adjustment of date of sowing to avoid favorable environmental conditions for *P. infestans*.
- Selection of bioagents having wide range of adaptability (elevated temperature, CO_2, rainfall, humidity, etc.) in climate change scenario.
- Use of high-resistant cultivar to late blight and cultivar disease resistant should not be affected at higher temperature.
- Management of diseases through integration of all the existing strategies.

1.5 FUTURE IMPACT

In climate change scenarios, every organism needs to adjust to the changed environmental conditions in order to survive and perpetuate. These conditions would have a direct bearing on pathogen characteristics such as frequency of generations, proportion of sexual reproduction, and rate of adaptation. Definitely *P. infestans* would change its behavior and accordingly plant protection measures need to be adopted. In case of host characteristics, that is, host anatomy, host physiology, and life span would also be affected. The new genotype might have more virulence to adapt to the changed climatic condition. The varietal resistance against late blight of potato might behave in different ways in case of climate change scenarios. Symptoms of late blight, especially stem blight, are more observed now days. There is a possibility that more number of sprays could be required in high rainfall areas.

1.6 CONCLUSION

Systematic research on climate change particularly involving plant diseases is still in nascent stage. Most of the efforts had been made: how plant diseases are affected under climate change scenarios using single atmospheric constituent or meteorological variable on the host, pathogen, or the interaction of the two under controlled conditions. However, interactions are more complex in the actual condition, where multiple climatological and biological factors are varying simultaneously in a changing environment. Moreover, some fungicides have shown sensitivity against late blight of potato at higher temperature; similarly, more fungicides should be tested at higher temperature for their efficacy. Host resistance is also affecting at higher temperature; therefore, large number of genotypes should be screened at higher temperature subsequently for their resistance against late blight of potato.

FIGURE 1.2 Symptomatic variations in late blight disease of potato. (A) Whitsh sporulation in case of high moisture. (B) Effect of higher temperature on radial growth of *P. infestans*. (C) Effect of higher rainfall on the incidence of late blight of potato. (D) Late blight affected tubers. (E) Late blight affected field.

TABLE 1.1 Effect of Temperature on the Efficacy of Fungicides Against *P. infestans*.

Treatments	Conc. (%)	Lesion size (cm²) and sporulation (×10⁴) at different temperatures after different days of fungicides spraying																	
		1*DAS						7*DAS						14*DAS					
		20°C		25°C		30°C		20°C		25°C		30°C		25°C			30°C		
		LS	SP*	LS	SP	LS	SP	LS	SP	LS	SP	LS	SP	LS	% RLSOC*	SP	LS	% RLSOC	SP
Dimethomorph 50% WP	0.2	0.0	0	0.0	0	0.0	0	0.0	0	0.0	0	0.0	0	0.0	100.0	0	0.0	100.00	0
Cymoxanil 8% + mancozeb 64% WP	0.2	0.0	0	0.0	0	0.0	0	0.0	0	0.0	0	0.0	0	0.0	100.0	0	4.40	49.83	2.8
Mancozeb 75 % WP	0.25	0.0	0	0.0	0	0.0	0	0.0	0	0.0	0	0.0	0	0.0	100.0	0	4.16	52.57	3.4
Metalaxyl 8% + mancozeb 64 % WP	0.2	0.0	0	0.0	0	0.0	0	0.0	0	0.0	0	0.0	0	1.64	83.08	3.3	4.23	51.77	3.1
Control		8.80	6.2	5.78	5.0	6.43	5.2	8.38	5.2	7.08	5.8	6.41	4.6	9.69	–	4.7	8.77		4.8

DAS, days after spray application; LS, lesion size; SP, sporulation; RLSOC, reduction lesion size over control.

KEYWORDS

- **climate change**
- **CO$_2$**
- **late blight**
- **phytopthora**
- **potato**
- **temperature**

REFERENCES

Agrios, G. N. *Plant Pathology*. 5th edn, Elsevier Academic Press, 2006, p 267.

Agrios, G. N. *Plant Pathology*; Elsevier: London, 1997.

Ainsworth, E. A.; Long, S. P. What Have We Learned From 15 Years of Free-air CO$_2$ Enrichment (FACE)? A Meta-analytic Review of the Responses of Photosynthesis, Canopy Properties and Plant Production to Rising CO$_2$. *New Phytol.* **2005,** *165*, 351–371.

Annual Report CIP. International Potato Center. Annual Report, 2015, p 2.

APN Project Report. The Effects of Climate Change on Pests and Diseases of Major Food Crops in the Asia Pacific Region, 2005.

Anderson, P. K.; Cunnighum, A. A.; Patel, N. G.; Morales, F. J.; Epstein, P. R.; Daszak, P. Emerging Infectious Diseases of Plant: Pathogen Pollution, Climate Change and Agrotechnology Drivers. *Trends Ecol. Evol.* **2004,** *19* (10), 535–544.

Baker, K. M.; Kirk, W. W.; Andresen, J.; Stein, J. M. A Problem Case Study: Influence of Climatic Trends on Late Blight Epidemiology in Potatoes; Bertschinger, L., Anderson, J. D., Eds.; *Acta Horticult.* **2004,** *638*, 37–42.

Baker, K. M.; Kirk, W. W.; Stein, J. M.; Andresen, J. A. Climatic Trends and Potato Late Blight Risk in the Upper Great Lake Region. *Hort. Technol.* **2005,** *15* (3), 510–518.

Becktell, M. C.; Daughtrey, M. L.; Fry, W. E. Temperature and Leaf Wetness Requirements for Pathogen Establishment, Incubation Period, and Sporulation of *Phytophthora infestans* on *Petunia × hybrid. Plant Dis.* **2005,** *89*, 975–979.

Boland, G. J.; Melzer, M. S.; Hopkin, A.; Higgins, V.; Nassuth, A. Climate Change and Plant Diseases in Ontario. *Can. J. Plant Pathol.* **2004,** *26*, 335–350.

Boonekamp, P. M. Are Plant Diseases Too Much Ignored in the Climate Change Debate? *Eur. J. Plant Pathol.* **2012,** *133*, 291–294.

Bourgeois, G.; Bourque, A.; Deaudelin, G. Modeling the Impact of Climate Change on Disease Incidence: A Biloclimatic Challenge. *Can. J. Plant Pathol.* **2004,** *26* (3), 284–289.

Carter, T. R.; Saarikko, R. A.; Niemi, K. J. Assessing the Risk and Uncertainties of Regional Crop Potential Under a Changing Climate in Finland. *Agric. Food Sci. Finland* **1996,** *5* (3), 329–350.

Central Potato Research Institute. *Annual Report 2011–2012*. Central Potato Research Institute: Shimla, India, 2011.

Central Potato Research Institute. *Annual Report 2012–2013*. Central Potato Research Institute: Shimla, India, 2012.

Central Potato Research Institute. *Annual Report 2015–2016*. Central Potato Research Institute: Shimla, India, 2015.

Chakraborty, S.; Tiedemann, A.V.; Teng, P. S. Climate Change: Potential Impact on Plant Diseases. *Environ. Pollut.* **2000**, *108* (2000), 317–326.

Chen, C. C.; McCarl, B. A. An Investigation of the Relationship Between Pesticide Usage and Climate Change. *Clim. Change* **2001**, *50*, 475–487.

Coakley, S. M. Variation in Climate and Prediction of Disease in Plants. *Ann. Rev. Phytopathol.* **1998**, *26*, 163–181.

Coakley, S. M.; Scherm, H.; Chakraborty, S. Climate Change and Plant Disease Management. *Ann. Rev. Phytopathol.* **1999**, *37*, 399–426.

Collins, W. W. The Global Initiative On Late Blight – Alliance for the Future. In *Potato Global Research and Development*; Khurana, S. M. P., Shekhawat, G. S., Singh, B. P., Pandey, S. K., Eds; Indian Potato Association, CPRI: Shimla, India, 2003; Vol 1, pp 513–524.

Crosier, W. Studies in the Biology of *Phytophthora infestans* (Mont.) De Bary. Cornell University Agricultural Experiment Station Memoir, 155, 1934.

Dua, V. K.; Singh, B. P.; Ahmad, I.; Sharma, S. Potential Impact Of Climate Change On Late Blight Outbreak In Western Uttar Pradesh And Punjab Using JHULSACAST Model. *Potato J.* **2015**, *42* (1), 58–71.

Fowler, A. M.; Hennessey, K. J. Potential Impacts of Global Warming on the Frequency and Magnitude of Heavy Precipitation. *Nat. Hazards* **1995**, *11*, 283–303.

Fry, W. E.; Birch, P. R. J.; Judelson, H. S.; Grünwald, N. J.; Danies, G.; Everts, K. L.; Gevens, A. J.; Gugino, B. K.; Johnson, D. A.; Johnson, S.B.; McGrath, M. T.; Myers, K. L.; Ristaino, J. B.; Roberts, P. D.; Secor, G.; Smart, C. D. Five Reasons to Consider *Phytophthora infestans* a Re-emerging Pathogen. *Phytopathology* **2015**, 105, 966–981.

Garrett, K. A.; Dendy, S. P.; Frank, E. E.; Rouse, M. N.; Travers, S.E. Climate Change Effects on Plant Disease: Genomes to Ecosystems. *Annu. Rev. Phytopathol.* **2006**, *44*, 489–509.

Ghini, R.; Hamada, E.; Bettiol, W. Climate Change and Plant Diseases. *Scientia Agricola (Piracicaba, Braz.)* **2008**, *65* (special issue), 98–107.

Goodwin, S. B.; Scheinder, R. E.; Fry, W. E. Use of Cellulose Acetate Electrophoresis for Rapid Identifications of Allozymes Genotype of *P. infestans*. *Plant Dis.* **1995**, *79*, 1181–85.

Goodwin, S. B.; Cohen, B. A.; Fry, W. E. Panglobal Distribution of a Single Clonal Lineage of the Irish Potato Famine Fungus. *Proc. Natl. Acad. Sci. U.S.A.* **1994**, *91*, 11591–11595.

Gregory, Peter J.; Scott N. Johnson; Adrian C. Newton; John S. I. Ingram. Integrating Pests and Pathogens into the Climate Change/Food Security Debate. *J. Exp. Bot.* **2009**, *60* (10), 2827–2838.

Haas, B. J.; Kamoun, S.; Zody, M. C.; Jiang, R. H.; Handsaker, R. E.; Cano, L. M.; Grabherr, M.; Kodira, C. D.; Raffaele, S.; Torto-Alalibo, T.; Bozkurt, T. O.; Ah-Fong, A. M., Alvarado, L.; Anderson, V. L.; Armstrong, M. R.; Avrova, A.; Baxter, L,; Beynon, J.; Boevink, P. C.; Bollmann, S. R.; Bos, J. I.; Bulone, V.; Cai, G.; Cakir, C.; Carrington, J. C.; Chawner, M.; Conti, L.; Costanzo, S.; Ewan, R.; Fahlgren, N.; Fischbach, M. A.; Fugelstad, J.; Gilroy; E. M.; Gnerre, S.; Green, P. J.; Grenville-Briggs, L. J.; Griffith, J.; Grünwald, N. J.; Horn, K.; Horner, N. R.; Hu, C. H.; Huitema, E.; Jeong, D. H.; Jones, A. M.; Jones, J. D.; Jones, R. W.; Karlsson, E. K.; Kunjeti, S. G.; Lamour, K.; Liu, Z.; Ma, L.; Maclean, D.; Chibucos, M. C.; McDonald, H.; McWalters, J.; Meijer, H.J.; Morgan, W.; Morris, P. F.; Munro, C. A.; O'Neill, K.; Ospina-Giraldo, M.; Pinzón, A.; Pritchard, L.; Ramsahoye, B.; Ren, Q.; Restrepo, S.; Roy, S.; Sadanandom, A.; Savidor, A.; Schornack, S.; Schwartz, D. C.; Schumann, U. D., Schwessinger, B.; Seyer, L.; Sharpe, T.; Silvar, C.;

Song, J.; Studholme, D. J.; Sykes, S.; Thines, M.; van de Vondervoort, P. J.; Phuntumart, V.; Wawra, S.; Weide, R.; Win, J.; Young, C.; Zhou, S.; Fry, W.; Meyers, B. C.; van West, P.; Ristaino, J.; Govers, F.; Birch, P. R.; Whisson, S. C.; Judelson, H. S.; Nusbaum, C. Genome Sequence and Analysis of the Irish Potato Famine Pathogen *Phytophthora infestans*. *Lett. Nat.* **2009**, *461*, 393–398.

Halterman, D.; Guenthner, J.; Collinge, S.; Butler, N.; Douches, D. Biotech Potatoes in the 21st Century: 20 Years Since the First Biotech Potato. *Am. J. Potato Res.* **2016**, *93*, 1–20.

http://www.climateandforming.org.

http://www.sciencemag.org/news/2012/11/.

http://cpri.ernet.in.

IISR. *PhytoFura Consolidated Report 2009–2012*; Indian Institute of Spices Research: Kozhikode, Kerala, 2012; p 164.

IISR. *Phytofura Annual Report (2015)*. ICAR-IISR: Kozhikode, Kerala, 2014; p 22.

Intergovernmental Panel on Climate Change. (2007). *Climate Change: Impact, Adaptation and Vulnerability, Contribution of WG II to the Third Assessment Report of the Intergovernmental Panel on Climate Change*; IPCC Fourth Assessment Report 2007 www.ipcc-wg2.org; Cambridge University Press: Cambridge, UK, 2007.

Intergovernmental Panel on Climate Change. *Summary for Policymakers*. In Climate Change 2013: The Physical Science Basis. Contribution of Working Group I to the Fifth Assessment Report of the Intergovernmental Panel on Climate Change. Stocker, T.F.; D. Qin, G.-K.; Plattner, M. Tignor; Allen, S. K.; Boschung, J.; Nauels, A.; Xia, Y.; Bex, V.; Midgley, P. M., Eds. Cambridge University Press: Cambridge, United Kingdom and New York, NY, USA, 2013.

Johnson, D. A.; Alldredge, J. R.; Hamm, P. B.; Frazier, B. E Aerial Photography Used for Spatial Pattern Analysis of Late Blight Infection in Irrigated Potato Circles. *Phytopathology* **2003**, *93*: 805–812.

Johnson, D.A.; Cummings, T. F.; Fox, A. D. Accuracy of Rain Forecasts for Use in Scheduling Late Blight Management Tactics in the Columbia Basin of Washington and Oregon. *Plant Dis.* **2015**, 99.

Kamoun, S.; Furzer, O.; Jones, J. D. G.; Judelson, H. S.; Ali, G. S.; Dalio, R. J. D.; Roy, S. G.; Leonardo, C.; Antonios Zambounis; Franck Panabières; David Cahill; Michelina Ruocco; Andreia Figueiredo; Xiao-Ren Chen; Jon Hulvey; Remco Stam; Kurt Lamour; Mark Gijzen; Brett M. Tyler; Niklaus J. Grünwald; Shahid Mukhtar, M.; Daniel F.A; Me Mahmut Tör; Guido van den Ackerveken; John Mcdowell; Fouad Daayf; William E. Fry; Hannele Lindqvist-kreuze; Harold J. G; Meijer Benjamin Petre; Jean Ristaino; Kentaro Yoshida; Paul Birch R. J. The Top 10 Oomycete Pathogens in Molecular Plant Pathology. *Mol. Plant Pathol.* **2015**, *16*, 413–434.

Kapsa, J. The Influence of Climatic Conditioned on Fungicide Effectiveness in Control of Potato Late Blight (*Phytophthora infestans*). *Prog. Plant Protect.* **2008**, 48, 1063–1072.

Kaukoranta, T. Impact of Global Warming on Potato Late Blight: Risk, Yield Loss, and Control. *Agric. Food Sci. Finland* **1996**, *5*, 311–327.

Klopping, H. L.; Delp, C. J. 2-Cyano-N [(Ethylamino) Carbonyl]-2-(Methoxylmino) Acetamide, a New Fungicide. J. Agric. Food Chem. **1980**, *28*, 467–468.

Kocmánková, E.; Miroslav, T.; Jan, J.; Martin, D. D. S.; Martin, M.; Zdeněk, Ž. Impact of Climate Change on the Occurrence and Activity of Harmful Organisms. *Plant Prot. Sci.* **2009**, *45* (Special Issue), S48–S52.

Lal, M.; Yadav, S. In Efficacy of Fungicides Against *Phytophthora infestans* in the Era of Global Climate Change. Presented at the Global Conference on Technological Challenges and Human Resources for Climate Smart Horticulture-issues and Strategies at NAU Navsari, May 2014.

Lal, Mehi.; Sanjeev Sharma; Islam Ahmad; BP Singh; Saurabh Yadav (2014). Development of Yield Loss Assessment Model for Potato Late Blight Disease in Indo-Gangetic Plains. *Potato J.* **2014,** *41* (2), 130–136.

Mariette, N.; Androdias, A.; Mabon, R.; Andrion, D. Local Adaptation to Temperature in Populations and Clonal Lineages of the Irish Potato Famine Pathogen *Phytophthora infestans. Ecol. Evol.* **2016,** doi: 10.1002/ece3.2282.

Mayton, H.; Forbes, G. A.; Mizubuti, E. S. G.; Fry, W. E. The Roles of Three fungicides in the Epidemiology of Potato Late Blight. *Plant Dis.* **2001,** *85,* 1006–1012.

Mitchell, C. E.; Reich, P. B.; Tilman, D.; Groth, J. V. Effects of Elevated CO_2, Nitrogen Deposition, and Decreased Species Diversity on Foliar Fungal Plant Disease. *Global Change Biol.* **2003,** 9, 438–451.

Mizubuti, E. S. G.; Fry, W. E. Temperature Effects on Developmental Stages of Isolates From Three Clonal Lineages of *Phytophthora infestans. Phytopathology* **1998,** *88,* 837–843.

Osswald, W. F.; Fleischmann, F.; Heiser, I. Investigations On the Effect of Ozone, Elevated CO_2 and Nitrogen Fertilization on Host–Parasite Interactions. *Summa Phytopathol.* **2006,** *32S,* S111–113.

Parker, J. M., Thurstan, H. D.; Villarreal, G. M.; Fry, W. E. Stability of Disease Expression in the Potato Late Blight Pathosystem: A Preliminary Field Study. *Am. J. Potato* **1992,** *69,* 635–644.

Petzoldt, C.; Seaman A. *Climate Change Effects on Insects and Pathogens.* In Climate Change and Agriculture: Promoting Practical and Profitable Responses, 2005, p. III 1–16.

Peters, R. D.; Platt, H. W.; Hall, R. Hypotheses for the Inter-regional Movement of New Genotypes of *Phytophthora infestans* in Canada. *Can. J. Plant Pathol.* **1999,** 21, 132–136.

Plessl, M.; Elstner, E. F.; Rennenberg, H.; Habermeyer, J.; Heiser, I. Influence of Elevated CO_2 and Ozone Concentrations on Late Blight Resistance and Growth of Potato Plants. *Environ. Exp. Bot.* **2007,** *60,* 447–457.

Rani, R.; Sharma, V. K.; Kumar, P.; Mohan, C. (2013). Impact of Simulated Rainfall on Persistence of Fungicides Used Against Late Blight (*Phytophthora infestans*) of Tomato – in Poster Presentation. *Plant Dis. Res.* **2013,** *28* (2), 219.

Rubio-Covarrubias, O. A.; Douches, D. S.; Hammerschmidt, R.; Rocha, A. da.; Kirk, W. W. Effect of Photoperiod and Temperature On Resistance Against *Phytophthora infestans* in Susceptible and Resistant Potato Cultivars. *Am. J. Potato Res.* **2006,** *82* (2), 139–146.

Scehepers, H. T. A. M. Effect of Rain on Efficacy of Fungicides Deposits Potato Against *Phytophthora infestans. Potato Res.* **2006,** *39,* 541–550.

Schaap, B. F.; Blom-Zandstra, M.; Hermans, C. M. L.; Meerburg, B. G.; Verhagen, J. Impact Changes of Climatic Extremes on Arable Farming in the North of the Netherlands. *Reg. Environ. Change* **2011,** *11,* 731–741.

Seidl Johnson, A. C.; Gevens, A. J. Investigating the Host Range of the US-22, US-23, and US-24 Clonal Lineages of *Phytophthora infestans* on Solanaceous Cultivated Plants and Weeds. *Plant Dis.* **2014,** *98,* 754–760.

Singh, B. P., & Bhat, M. N. (2010). Impact Assessment of Climate Change on Potato Diseases and Insect Pests. In Challenges of Climate Change Indian Horticulture; Singh, H. P., Singh, J. P., Lal, S. S., Eds., Westville Publishing House: New Delhi, pp 178–184.

Singh, B. P.; Dua, V. K.; Govindakrishnan, P. M.; Sharma, S. *Impact on Climate Change on Potato*. In Climate-Resilient Horticulture: Adaptation and Mitigation Studies; Singh H. P., et al, Eds.; Springer: India, pp 125–135, doi: 10.1007/978-81-322-0974.4_12.

Skelsey, P.; Cook, D. L. E.; Lynnot, J.; Lee, A. K. Crop Connectivity Under Climate Change: Future Environmental and Geographic Risks of Potato Late Blight in Scotland. *Global Change Biol.* **2016**.

Staubli, B.; Wenger, R.; Wymann Von Dach, S. Potatoes and Climate Change. Zollikofen, Switzerland: Info Resources Focus No. 1/08. 2008, http://www.inforesources.ch/pdf/focus08_1_e.pdf.

Subbarao, K. V.; George, S. W.; Klosterman, S. J. Focus Issue Articles on Emerging and Re-emerging Plant Diseases. Phytopathology **2015**, *105*, 852–854.

Yadav, S.; Lal, M.; Singh, B. P.; Kaushik, S. K.; Sharma, S. Evaluation of Fungicides Against *Phytophthora infestans* at Elevated Temperature. *Potato J.* **2016**, *43* (1), 98–102.

Yang, X. B.; Scherm, H. El Niño and Infectious Disease. *Science* **1997**, *275*, 7399.

Ywa, N. S.; Walling, L.; McCool, P. M. Influence of Elevated CO_2 on Disease Development and Induction of PR Proteins in Tomato Roots By *Phytophthora parasitica*. *Plant Physiol.* **1995**, *85* (Suppl), 113.

Zargarzadeh, F.; Ghorbani, A.; Asghari, A.; Nouri-Ganbalani, G. The Effect of Climatic Factors on Potato Late Blight Disease in Aradabil Plain of Iran. *J. Food Agric. Environ.* **2008**, *6* (3/4), 200–205.

Zyabliskaya, E. Ya.; Parshikov, V. V. The Effect of Chronic UV-B Irradiation On the Productivity of Agricultural Crops and the Susceptibility of Plants to Phytopathogens Under Conditions of the Noncherozem Zone. *Russ. J. Ecol.* **1998**, *29* (3), 163–166.

CHAPTER 2

Pathosystem in Cowpea: An Overview

SANTOSH KUMAR[1*], ERAYYA[1], MD. NADEEM AKHTAR[2],
MAHESH KUMAR[3], and MEHI LAL[4]

[1]*Department of Plant Pathology, Bihar Agricultural University,
Sabour 813210, Bihar, India*

[2]*Plant Pathology Section, KVK, Agwanpur, Saharsa,
Bihar Agricultural University, Sabour 854105, Bihar, India*

[3]*Department of Molecular Biology and Genetic Engineering, Bihar
Agricultural University, Sabour 813210, Bihar, India*

[4]*Plant Protection Section, Central Potato Research Institute Campus,
Modipuram, Meerut 250110, Uttar Pradesh, India*

Corresponding author. E-mail: santosh35433@gmail.com

ABSTRACT

Cowpea (*Vigna unguiculata* (L.) Walp.) is an important legume vegetable crop grown in tropics and subtropics region across the globe. The crop is used mainly as vegetable which is a major source of protein and fiber. The crop is severely challenged by different pathogens, including fungi, bacteria, and viruses and is well adapted to stress and has excellent nutritional qualities. Diseases caused by pathogens such as fungi, bacteria, viruses, nematodes, and parasitic plants are major constraints to profitable cowpea production. Integrated management, such as host plant resistance, agronomic practices, biological control, judicious use of fungicides, pesticides for vector control, biopesticides for pathogen control, can be applied complementarily at farmers' field to provide them maximum economic return.

2.1 INTRODUCTION

Cowpea (*Vigna unguiculata* (L.) Walp.), commonly known as lobia is a short duration, annual versatile pulse crop in the livelihood of millions of people in the tropic and subtropic countries (Singh and Sharma, 1996). It is the most significant legume vegetable crop grown in India (Pande and Joshi, 1995). The crop is well adapted to stress and has excellent nutritional qualities (El-Ameen, 2008). It provides food, animal feed, and cash for the rural population. In India, cowpea is grown in an area of about 1.5 million hectares with productivity of 567 kg/ha and production of 0.5 million tonnes. The low productivity is attributed to biotic and abiotic constraints. Biotic constraints, such as fungi, bacteria, viruses, nematodes, and parasitic flowering plants, confined profitable production of cowpea in all agroecological zones (Hampton et al., 1997).

TABLE 2.1   Important Diseases of Cowpea and Their Causal Organism.

S. no.	Name of disease	Causal organism	Categories of pathogen
1	Anthracnose	*Colletotrichum lindemuthianum*	Fungus
2	Cercospora leaf spot	*Cercospora canescens* and *Pseudocercospora cruenta*	Fungus
3	Web blight	*Rhizoctonia solani*	Fungus
4	Dry root rot	*Macrophomina phaseolina*	Fungus
5	*Sclerotium*	*Sclerotium rolfsii*	Fungus
6	Rust	*Uromyces phaseoli typical*	Fungus
7	Wilt	*Fusarium oxysporum* f.sp. *tracheiphilum*	Fungus
8	Stem rot	*Phytophthora cactorum*	Fungus
9	Cowpea aphid-borne mosaic	Cowpea aphid-borne mosaic virus (CABMV)	Virus
10	Cowpea golden mosaic	Cowpea golden mosaic virus (CPGMV)	Virus
11	Southern Bean Mosaic	Southern Bean Mosaic virus (SBMV)	Virus
12	Bacterial blight	*Xanthomonas axonopodis* pv. *vignicola*	Bacteria

2.2 FUNGAL DISEASES OF COWPEA

Anthracnose, leaf spot, web blight, root rot, stem rot, wilt, and rust are the major economically important fungal diseases affecting all parts at all growth stages of the crop.

2.2.1 ANTHRACNOSE

Anthracnose is caused by *Colletotrichum lindemuthianum* (Sacc. & Magn.) Scriber; the perfect stage is *Glomerella lindemuthianum* (Sacc. & Magn.) Shear which causes soil and seed-borne disease. Characteristic symptoms of the disease are appearance of black, sunken, and crater-like cankers on the pods. Similar spots may be seen on the cotyledons and stems of young seedlings. Symptoms may also be seen on foliage as blacken dead lesions on lower surface of the leaves.

2.2.1.1 ECOLOGY OF PATHOGEN, SURVIVAL, DISPERSAL/ TRANSMISSION

The causal pathogen is soilborne and seed borne. The primary inoculum usually comes from seeds (40% seed transmission) or from diseased crop residues that lives in the soil. In seeds, the pathogen survives till the seeds are viable. The pathogen grows on cotyledons from where it spreads to other parts of the plant. The spores are embedded in gelatinous mass and water is required for their release. Usually, they are washed down by rains or dew through stem or new leaves. Movement of insect or wild animals or man also helps in dissemination.

2.2.1.2 EPIDEMIOLOGY OF DISEASES

Cool and rainy weather is conducive for the development of the disease. Dense plant population under field conditions create situation for pathogen infection. The disease is particularly severe in monocropped cowpeas in which it can cause up to 50% loss in yield.

2.2.2 CERCOSPORA LEAF SPOT

Cercospora leaf spot caused by *Cercospora canescens* and *Pseudocercospora cruenta* has been observed in all the cowpea-growing areas. Despite the fact that Cercospora leaf spot develops late in the season, disease spread is often rapid, and premature defoliation might be very severe. Cercospora leaf spot is characterized mostly by circular to irregular cherry red to reddish-brown lesions on both the surfaces of leaves. *Pseudocercospora* leaf spot is characterized by chlorotic or necrotic spots on the upper leaf surface. The fungus *Pseudocercospora* also damage the stems and pods with less prominent and significant. The pathogen survives the no-crop period on infected crop residue and in infected seed (Williams, 1975).

2.2.2.1 ECOLOGY OF PATHOGEN, SURVIVAL, DISPERSAL/ TRANSMISSION

The disease starts appearing about 30–40 days after planting. The pathogen survives on seeds, and plant debris in soil is considered as the primary source of inoculum, especially the infected cotyledons attached with the seedlings or lying in the soil surface (Shree Kumar, 1974). The secondary spread of the disease is mainly by airborne conidia and is influenced by several factors. After transmission, the further development of disease is greatly predisposed by environmental and cultural factors.

2.2.2.2 EPIDEMIOLOGY OF DISEASES

The disease is encountered during the rainy season of intermittent rains with relatively hot and high humid conditions (Poehlman, 1991). Dense plant population under field condition favors the climatic condition. Mean temperature of 22.5–23.5°C, relative humidity (RH) between 77% and 85%, sunshine of 5 h/day and more number of rainy days are more conducive to the disease development (Sud and Singh, 1984).

2.2.3 WEB BLIGHT OF COWPEA

Web blight disease of cowpea is caused by *Rhizoctonia solani* Kuhn; the perfect stage is *Thanatephorus cucumeris* (Frank) Donk, which is an ubiquitous soil fungus and most important constraint to cause heavy yield reduction. Initial symptoms appear on the leaves are small, reddish-brown spots which enlarge and become surrounded by irregular shaped water-soaked areas. Under humid conditions, the lesions develop rapidly and coalesces leading to extensive blighting and defoliation (Allen and Lenne, 1998). Dubey (1997) reported that 26–28°C temperature and 90–100% RH favored maximum web blight disease development. The leaves infected with mycelium appear as a spider web and thus recommended the name web blight disease (Saksena and Dwivedi, 1973). These diseases are often severe and destructive under localized, water-logged conditions in the hot humid regions of India (Verma and Mishra, 1989). The pathogen spreads as sclerotia, and these sclerotia are dispersed by means of wind, water, or soil movement.

2.2.3.1 ECOLOGY OF PATHOGEN, SURVIVAL, DISPERSAL/ TRANSMISSION

R. solani is a soilborne fungus and can survive for longer either as sclerotia or thick-walled brown hyphae in plant debris in soil and on plant tissues. Most of the sclerotia occur on the tap root of diseased plants and in the soil adjacent to the diseased root. Over 80% sclerotia occurred on the top 10 cm of the soil. The pathogen is disseminated as sclerotia by various agencies such as irrigated water, infected/infested or contaminated seeds, transplanted material, wind, rain splash, and farm equipments.

2.2.3.2 EPIDEMIOLOGY OF DISEASES

The pathogen is known to favor warm wet weather, and its outbreaks arise typically during the beginning of summer months. A combination of environmental factors has been coupled to the prevalence of the pathogen such as the presence of host plant, frequent rainfall/irrigation, and raising temperature and RH in spring and summer. In addition, soil type, reaction, moisture, and temperature are also known to create favorable environments. Higher aerial temperature (26–32°C), RH near 100%, and soil temperature

30–33°C favor the development of high disease severity. Optimal hyphal growth was measured at a temperature range of 20–30°C with an optimum at 25°C, when pH range was 5–8 (Kumar et al., 2014). In case of web blight of legumes, the plants are susceptible from seedling stage to maturity and the severity is maximum in 30–70 days old plants probably because of the dense canopy of the crop which facilitates the early spread of the pathogen through contact of plants and leaves, forming mycelial bridges. Kumar et al. (2013) reported that, maximum web blight severity was recorded in the plants inoculated at 60 days (68.8%) after sowing followed by 45 days (65.1%) and 30 days (61.5%).

2.2.4   ROOT ROT OF COWPEA

Root rot of cowpea is caused by *Macrophomna phaseolina* (Tassi) Goid. *Rhizotonia* (Taub) Butler is one of the most destructive diseases of cowpea in the tropic and subtropical climate (Chidamboram and Mathur, 1975; Reuveni et al., 1983). The pathogen causes severe losses to the crop from seedling to maturity stage. It attacks any part of the plant at any stage of growth from seed germination to harvest, thus causing damping off, seedling blight, leaf blight, collor rot, charcoal rot, stem rot, and root rot (Singh et al., 1990; Roy et al., 2008). The epidemic outbreak and yield losses due to charcoal and root rot of cowpea have been observed in many cowpea-growing areas. Sclerotium rot is another important rot disease of cowpea caused by *Sclerotium rolfsii*, which infects the basal portion of stems near the ground level, producing a fan of silky mycelium and large round sclerotia which are initially white and gradually darken. The infected plants usually wilt and die (Adejumo and Ikotun, 2003).

2.2.4.1   ECOLOGY OF PATHOGEN, SURVIVAL, DISPERSAL/ TRANSMISSION

The fungus is facultative parasite, seed and soilborne and may survive in the soil in the form of sclerotia for long time. Soilborne inoculum is more important in causing infection and disease development. The fungus spreads from plant to plant through irrigation water, tools and implements, and cultural operations. Seed, soil, plant residue, and collateral host plants are the sources of primary inoculum (Short et al., 1980; Reuveni et al., 1983). Fungus survives in upper layers of the soil and enters plant through stem.

2.2.4.2 EPIDEMIOLOGY OF DISEASES

Moisture stress condition and temperature above 28–30°C favors disease development.

2.2.5 RUST OF COWPEA

Cowpea rust is caused by *Uromyces phaseoli vignae* (Barel.) Arth, an autoecious macrocyclic rust and occurs wherever cowpea is grown. Rust sori are reddish-brown circular in shape and are most frequent on lower surface of the leaves. Severe infections result in complete defoliation and total loss of the crop; soilborne teliospore initiates the primary infection and germinates to form the sporidia, which infect the young leaves. The urediniospores can also oversummer and perpetuate the fungus and are disseminated by wind and through contact with animals. Cloudy, humid weather with heavy dews favor disease development and spread of the pathogen.

2.2.5.1 ECOLOGY OF PATHOGEN, SURVIVAL, DISPERSAL/ TRANSMISSION

The pathogen survives in the soil as teliospores and as uredospores in crop debris. Primary infection is by the teliospores which germinate to form the sporidia which infect the young leaves. Secondary spread is by wind-borne uredospores, which perpetuate the fungus and are disseminated by wind and through contact with animals. The fungus also survives on other legume hosts.

2.2.5.2 EPIDEMIOLOGY OF DISEASES

Cloudy humid weather with heavy dews and temperature of 21–26°C favor disease development and spread of the pathogen.

2.2.6 WILT

The disease is caused by a fungus, *Fusarium oxysporum* f. sp. *tracheiphilum*. The symptoms of this disease do not appear until the plants are about

6-weeks old. Initially, a few plants are noticed with pale green flaccid leaves which soon turn yellow. Growth is stunted; chlorosis, drooping, premature shedding, or withering of leaves with veinal necrosis often occurs, and finally plant dies within 5 days. Brownish, purple discoloration of the cortical area is seen, which often extends throughout the plant. The fungus produces falcate-shaped macroconidia which are 4–5 septate, thin walled, and hyaline. The microconidia are single celled, hyaline and oblong, or oval. The chlamydospores are also produced in abundance. The disease is favored by a temperature of 20–25°C and moist humid weather. The fungus survives in the infected stubbles in the field. The primary spread is through chlamydospores and seed contamination. The secondary spread is through conidia by irrigation water.

2.2.6.1   ECOLOGY OF PATHOGEN, SURVIVAL, DISPERSAL/ TRANSMISSION

The pathogen is a facultative parasite and lives saprophytically in the soil without host in the form of chlamydospores or macroconidia and microconidia for a longer duration. The fungus may also be seed borne in nature. The pathogen enters the host vascular system at root tips through wounds. The primary infection is through chlamydospores in soil, which remain viable up to next crop season. The weed hosts also serve as a source of inoculum. The secondary spread is through irrigation water, cultural operations, and implements.

1.2.6.2   EPIDEMIOLOGY OF DISEASES

The disease is favored by low soil moisture, high temperature above 25°C with high atmospheric humidity. The presence of weed hosts, such as *Cyperus rotundus*, *Tribulus terrestris*, and *Convolvulus arvensis*, regulates the infection cycle.

2.2.7   STEM ROT

The disease is caused by the fungus, *Phytophthora cactorum*. The symptoms of the disease include formation of dark brown or black brown sunken lesions, which appear at the base of stem or branches, extending to several centimeters. Initially, lesions are small and smooth, later enlarging and slightly depressed. Infected tissue becomes soft and whole plant wilts. In the adult plants, infection is mostly confined to basal portions of the stem. The infected bark becomes brown and the tissue softening cause the plant to collapse. The infected branches may break off in wind. The upper portions of the infected twigs eventually wilt and dry. Localized yellowing of leaves starts from the tip and margin and gradually extends toward the midrib. The center of the spots later turns brown and hard. The spots increase in size and cover a major portion of the lamina, leading to drying. Fungus produces hyaline, coenocytic mycelium. The sporangiophores are hypha like with a swelling on the tip bearing hyaline, ovate, or pyriform, non-papillate sporangia. Each sporangium produces 8–20 zoospores. Oospores are globose, light brown, smooth, and thick walled.

2.2.7.1 ECOLOGY OF PATHOGEN, SURVIVAL, DISPERSAL/ TRANSMISSION

The fungus survives in the soil and plant debris in the form of oospores. Primary infection is from oospores and secondary spread by zoospores from sporangia. Rain splash and irrigation water help the movement of zoospores.

2.2.7.2 EPIDEMIOLOGY OF DISEASES

The disease becomes more predominant in soils with poor drainage, heavy rain during the months of July–September, and high temperature (28–30°C).

2.3 VIRAL DISEASES OF COWPEA

Among viral diseases, cowpea golden mosaic (CPGM), cowpea aphid-borne mosaic (CABM), and Southern Bean Mosaic (SBM) virus diseases are the most important viral diseases of cowpea. CPGM virus belongs to the genus *Begomovirus* and is transmitted by whiteflies (*Bemicia* sp.). It produces intense yellow leaves which after sometime become distorted and blistered. CABM virus (CABMV) is economically significant seed-borne virus of cowpea. CABMV has a wide experimental host range and belongs to the genus *Potyvirus* which has filamentous particles measuring 750 nm in length and is transmitted by several common species of aphid in non-persistent manner (Bock, 1974). Its symptoms appear on primary leaves as mild mosaic with vein banding followed by irregular mosaic or yellow mottle, puckering, slight distortion, and arching of trifoliate leaves. The infected seed provides the original inoculums, and aphids are responsible for the secondary spread of the disease under field conditions. SBM virus has isometric particle, 28 nm in diameter, and belongs to the genus *Sobemovirus*. It makes chlorotic spots, systemic vein clearing or vein banding, mosaic, leaf distortion, and stunting of affected plants. The virus is mainly transmitted by beetles such as *Epilachna variestis* as semi-persistent manner and also transmitted by mechanical inoculation, grafting, and by seed and pollen.

2.4   BACTERIAL BLIGHT OF COWPEA

Bacterial blight of cowpea caused by *Xanthomonas axonopodis* pv. *vignicola* (Burkholder) Dye is the most economic disease which cause complete defoliation, when crop is susceptible (Ekpo, 1978; Prakash and Shivashanker, 1982), and preseedling and postseedling mortality (Kishun, 1989). The yield loss may vary from 2.7% to 92.2% and it varies according to susceptibility (Kishun, 1989). Reddish and wrinkled cotyledons of seed are the primary symptoms. Secondary infection on leaves appeared as light yellow circular spots with necrotic and brown in the center. This is a systemic disease and spreads through vascular system. Initially, growing tips are damaged leading to complete death of plants. The veins are red in color. On pods, deep green or water-soaked streaks are formed. Such pods become yellow, shrivel, and die. The infected seeds and probably diseased crop residues serve as the source of primary inoculum. Secondary spread is by rain, insects, and implements. The disease spreads rapidly during heavy rainfall and during overhead irrigation. The disease is serious in soils with poor drainage.

2.5   INTEGRATED APPROACH FOR MANAGEMENT OF COWPEA DISEASES

Adoption of integrated disease management (IDM) practices is essential for economical and effective control. IDM comprises the use of pathogen-free seed, sanitation and destruction, removal of infected plant debris, crop rotation, and deep summer plowing. Crop rotation helps in controlling the disease to a greater extent as a break in the infection cycle. Use of disease-resistant cultivars is most effective in IDM strategies. Biological management of the disease by the application of antagonistic microorganism as seed treatment and soil is a potential and sustainable method. Use of suitable fungicide as seed treatment and foliar application is highly effective in reducing infection.

2.6   MANAGEMENT OF FOLIAR DISEASES

Foliar diseases of cowpea can be effectively controlled by combining resistant varieties with chemical control, and cultural practices may help in reducing the incidence considerably. Cultural practices such as field sanitation, destruction of infected crop debris, and avoiding collateral hosts is considerable highly effective. Use of mancozeb, carbendazim, copper oxychloride, and benomyl post flowering are reported to reduce infection (Singh et al., 2001). Weekly spraying of benomyl and carbendazim beginning at 3 weeks after planting gave the best control of the diseases and the highest grain yield (Amadi, 1995). Sprays of mancozeb or maneb or zineb (2 kg/ha) at 10 days interval has been recommended as a preventive method.

2.7   MANAGEMENT OF WILT COMPLEX DISEASES

Seed treatment and soil drenching with suitable fungicides and biocontrol agent along with cultural practices are highly suitable for managing the wilt complex diseases. Treating the seeds with carbendazim or thiram (2 g/kg) or with *Trichoderma viride* (4 g/kg), and spot drench with carbendazim (0.5 g/L) is found to be highly effective. Soil drenching is done for 20 days, 40 days, and 60 days after planting. Drenching the affected plants and the infested soil with Bordeaux mixture may help in reducing the inoculums potential. Application of organic amendments effectively manages the disease. Deep summer ploughing (May–June), mixed/intercropping with linseed/mustard, and crop rotation with non-legume crop break the infection cycle.

2.8   MANAGEMENT OF VIRAL DISEASES

Cultural practices such as utilization of healthy seed and planting materials, rouging, alteration in sowing dates, use of early maturing cultivars, and growth-resistant varieties are effective in minimizing virus disease incidence. Remove the infected plants up to 30 days after sowing. Control of insect vector populations in the fields and alternate host will reduce the incidence of viruses. Rogue out the weed hosts periodically. Insecticide (Imidacloprid 1 mL/3 L of water) spray to control the insect vector will help to checkthe diseases. Host plant resistance (HPR) is the most acceptable component in virus control because it is ecofriendly, practical, and economically acceptable to farmers. A nucleic acid-based serological technique is the most sensitive technique for detection of viruses. Enzyme-linked immunosorbent assay is the most appropriate serological method for the detection of the virus in the seed or plant tissue/s for seed certification programs. Bioassays in which seeds are planted in the greenhouse are determined by assessing plants for visual symptoms of the virus can result in false-negative results.

2.9 CONCLUSION

Cowpea is an important dietary component of India and other tropical cuisines. Diseases are no doubt one of the major constraints to its efficient production. Managing the yield and grain quality and reducing the disease involve a number of strategies. In this chapter, major emphasis is given on economic importance, etiology, symptoms, ecology, disease cycle, epidemi-ology, and management of different diseases of cowpea (caused by fungi, bacteria, viruses which are having different kind of physiology and para-sitism) and outline current and suggested future research on issues related to disease management strategies. Management of the diseases for sustain-able cowpea production should consider an holistic approach; taking into account integrated approach since no control strategy is grossly effective at all times and conditions. Besides eco-disruption, use of standard fungicides may not be economically feasible in many low-input farming systems of India. Therefore, cowpea health management for sustainable grain produc-tion in such farming systems should emphasize IDM programs alternative to synthetic chemicals. HPR, fungicides, biofungicides, botanicals and agronomic practices such as crop rotation, inter/mixed cropping, change in date of sowing should be potential viable parts for IDM.

However, the efficacy of control strategy(ies) is/are dependent on proper identification of the causal agent of the disease. In cowpeas, identification of the causal agent has been largely done by morphological characteristics and assumption of host–partner specificity of the fungus. These have spelt far-reaching problems to sustainable resistant cultivar development. Managing the diseases for sustainable cowpea production in the overall, should be holistic; taking into account integrated approach since no one control strategy is grossly effective at all times and conditions. Besides ecological imbalance, use of standard fungicides may not be economically feasible in many low-input farming systems of India. Therefore, cowpea health management for sustainable crop production in such farming systems should emphasize IDM programs instead of chemicals.

In this review, existence of pathosystem of important diseases of cowpea has been discussed. Earlier research was mainly focused on resistant sources and chemical control of few diseases. Now the major emphasis is on identifying, evaluating, and integrating location-specific components of IDM. Future IDM strategies will also account effects of climate variability on pathogen dynamics that is contributing to emergence and resurgence of new strains/pathogens, necessitating the appropriate refinement for disease control. HPR, fungicides, natural plant products, biofungicides, botanicals, and agronomic practices will remain the potential novel options for IDM. In addition, the future modules will incorporate the latest advancements being made in the identification of biocontrol agents and products from biotechnological approaches such as marker-assisted selection, genetic engineering, and hybridization to develop cultivars with resistance to wide range of diseases. Emphasis will be given to use of information technology for disease modelling, developing decision support systems, and utilization of remote sensing to refine upscale and disseminate IDM technologies.

KEYWORDS

- cowpea
- pathosystem
- diseases
- integrated management
- epidemiology

REFERENCES

Adebitan, S. A.; Ikotun, T. Effect of Plant Spacing and Cropping Pattern on Anthracnose (*Colletotrichum lindemuthianum*) of Cowpea. *Fitopatol. Brasileira.* **1996**, *21*, 5–12.

Adejumo, T. O.; Ikotun, T. Effect of Planting Date on Incidence and Severity of Leaf Smut of Cowpea in Northern Nigeria. *Moor. J. Agric. Res.* **2003**, *4*, 106–110.

Allen, D. J.; Lenne, J. M. Diseases as Constraints to Production of Legumes in Agriculture. In *Pathology of Food and Pasture Legumes*; Allen, D. J., Lenne, J. M. Eds.; CAB International: Wallingford, UK, 1998; pp 1–61.

Amadi, J. E. Chemical Control of *Cercospora* Leaf Spot Disease of Cowpea (*Vigna unguiculata* (L.) Walp. *Agrosearch* **1995**, *1*, 101–107.

Bock, K. R.; Conti, M. *Cowpea Aphid-borne Mosaic Virus*. CMI/AAB. Description of Plant Viruses No. 134. Kew, Surrey, England, 1974.

Chidamboram, P.; Mathur, S. B. Production of Pycnidia by *Macrophomina phaseolina*. *Trans. Br. Mycol. Soc.* **1975**, *64*, 165–168.

Dubey, S. C. Influence of Age of Plants, Temperature and Humidity on Web Blight Development in Groundnut. *Indian Phytopath.* **1997**, *50* (1), 119–120.

Ekpo, E. J. A. Effect of *Xanthomonas vignicola* on the Yield of Cowpea (*Vigna unguiculata* (L.) Walp.). *Afr. J. Agric. Sci.* **1978**, *5* (1), 67–69.

El-Ameen, T. M. Genetic Components of Some Economic Traits in Cowpea, *Vigna unguiculata*. *J. Agric. Sci.* **2008**, *33*, 135–149.

Hampton, R.O.; Thottappilly, G.; Rossel, H. W. Viral Diseases of Cowpea and Their Control by Resistance Conferring Genes; In Singh, B. B., Mohan Raj, D. R., Dashiell, K. E., Jackai, L. E. N., Eds. *Advances in Cowpea Research*, Vol. 3, 1997; pp 159–175.

Kishun, R. Appraisal of Loss in Yield of Cowpea Due to *Xanthomonas campestris* pv. *vignicola*. *Indian Phytopathol.* **1989**, *42*, 241–246.

Kumar, S.; Kumar, A.; Chand, G.; Lal, M.; Kumar, R. Dynamics of Mycelial Growth and Sclerotia Production of *Rhizoctonia solani* Kuhn (AG1-IB) of Urdbean. *The Ecoscan* **2014**, *8*, 273–277.

Kumar, S., Lal, M., Garkoti, G. and Tripathi, H.S. Standardization of Inoculation Techniques, Plant Age and Host Range of *Rhizoctonia solani* Kuhn, the Incitant of Web Blight of Urdbean. *Plant Dis. Res.* **2013**, *28*, 45–48.

Pande, S.; Joshi, P. K. *Constraints and Prospectus of Legumes in the Rice-Wheat Based Cropping System in Terai Region of Nepal*; Trip Report 7 Dec–31 Dec 1995; ICRISAT: Patancheru 502 324, Andhra Pradesh, India, 1995, pp 5.

Pande, S.; Sharma, M.; Kumari, S.; Gaur, P.M.; Chen, W.; Kaur, L.; MacLeod, W.; Basandrai, A.; Basandrai, D.; Bakr, A.; Sandhu, J. S.; Tripathi, H.S.; Gowda, C. L. L. Integrated Foliar Diseases Management of Legumes, International Conference on Grain Legumes: Quality Improvement, Value Addition and Trade, February 14–16, 2009; Indian Society of Pulses Research and Development, Indian Institute of Pulses Research: Kanpur, India, 2009, pp 143–161.

Poehlman, J. M. *The Mungbean*. Oxford and IBH Publishing Co. Pvt. Ltd.: New Delhi, 1991; p 375.

Prakash, C.S., Shivashanker, G. Evaluation of Cowpea for Resistance to Bacterial Blight. *Trop. Pest Manage.* **1982**, *28* (2): 131–135.

Reuveni, R.; Nachmias, A.; Kikun, J. The Seed Borne Inoculum on the Development of *Macrophomina phaseolina* on Melon. *Plant Dis.* **1983**, *74*, 280–281.

Roy, A.; De, R. K.; Ghosh, S. K. Diseases of Bast Fibre Crops and Their Management in Jute and Allied Fibres. In: Karmakar, P. G., Hazara, S. K., Subramanian T. R., Mandal, R. K., Sinha, M. K., Sen, H. S; Eds.;. Updates Production Technology, Central Research Institute for Jute and Allied Fibres: Barrckpore, West Bengal, India, 2008, pp 327.

Saksena, H. K.; Dwivedi, R. P. Web Blight of Blackgram Caused by *Thanatephorus cucumeris*. *Indian J. Farm Sci.* **1973**, *1*, 58–61.

Saxena, M.; Saxena, D. R.; Bhale, M. S.; Khare, M. N. Diseases of Cowpea and Their Management. In Diseases of Field Crops and Their Management; Thind, T. S. Ed.; National Agricultural Technology Information Centre: Ludhiana, India, 1998, pp 239–252.

Short, G. E.; Wyllie, T. D.; Bristow, P. R. Survival of *Macrophomina phaseolina* in Soil and in Residue of Soybean. *Phytopathology* **1980**, *70*, 13–17.

Shree Kumar, K. Studies on Foliar Fungal Diseases of Urid Leading to Their Control. M.Sc. Ag. Thesis, JNKVV, Jabalpur, MP, India, 1974, p 84.

Singh, B.B. and Sharma, B. Restructuring Cowpea for Higher Yield. *Ind. J. Genet.* **1996**, *56*, 389–405.

Singh, R.A.; De, R. K.; Gurha, S.N.; Ghosh, A. *Management of Cercospora Leaf Spot of Mungbean.* In National Symposium on Pulses for Sustainable Agriculture and National Security, Nov 17–19, 2001; New Delhi, ISPRD, IIPR: Kanpur, India, 2001, p 146.

Singh, S.K., Nene, Y.L., Reddy, M.V. Influence of Cropping System on *M. phaseolina* Population in Soil. *Plant Dis.* **1990**, *74*, 814.

Sud, V. K.; Singh, B. M. Effect of Environmental Factors on the Development of the Leaf Spot (*Cercospora canescens*) in Urdbean. *Indian Phytopath.* **1984**, *37*, 511–515.

Verma, J. S.; Mishra, S. N. Evaluation of Improved Lines from IITA in Humid-Subtropical India. *Trop. Grain Legume Bull.* **1989**, *36*, 38–39.

Williams, R. J. Diseases of Cowpea (*Vignna aunguiculata* (L.)Walp.) in Nigeria. *PANS*, **1975**, *21*, 253–267.

CHAPTER 3

A Population Genetics Perspective on Ecology and Management of *Phytophthora* Spp. Affecting Potato, Tomato, and Pointed Gourd in India

SANJOY GUHA ROY*

Department of Botany, West Bengal State University, Barasat, Kolkata 700126, India

Corresponding author. E-mail: s_guharoy@yahoo.com

ABSTRACT

This chapter discusses the three pathoecosystems of potato, tomato, and pointed gourd affected by *Phytophthora* spp. in India, which is the second largest producer of vegetables and fruit in the world, and argues for a paradigm shift from traditional management practices to those that are linked to population genetics of the pathogen populations in real time. These management systems will incorporate the genetic diversity, field characteristics, and geospatial data to create region-specific phylogeographic big databases which can be used to prevent crop losses. Examples of practical applications of such big datasets in reducing late blight disease loses are already there in some countries and the first step toward that goal in India would be through assessment of the local genetic diversity in these pathoecosystems by annual surveillance programs some of which have been initiated by the author, results of which have been shared here.

3.1 INTRODUCTION

India ranks second in vegetable and fruit production in the world after China (http://apeda.gov.in), most of which is domestically consumed by its burgeoning 1.34 billion population, a sizeable percentage of which are vegetarians. The diverse agro meterological zones and geography allow cultivation of a wide selection of crops, but its predominantly tropical climate predisposes it to attack from a range of pests and pathogens. Of the various vegetable crops cultivated in India, this chapter will restrict the discussion to potato (*Solanum tuberosum* L., Solanaceae), tomato (*Solanum lycopersicum* L., Solanaceae), and pointed gourd (*Trichosanthes dioica* Roxb., Cucurbitaceae), as these are the major crops but not limited to those severely affected by Phytophthora diseases, but now thankfully localized information is also being generated about the population dynamics of these pathosystems.

Implementation of effective control strategies requires more knowledge about the genetic structure of population of plant pathogens (Wolfe and Caten, 1987), as control strategies must target a population instead of an individual if they are to be effective. Defining the genetic structure of a population is a logical first step in studies of fungal population genetics because the genetic structure of a population reflects its evolutionary history and its potential to evolve: aspects important for formulating disease management strategies. "Genetic structure" refers to the amount and distribution of genetic variation within and among populations (McDonald, 1997).

In oomycetes that undergo both asexual and sexual reproduction, it is necessary to differentiate between diversity at individual locus, "gene diversity" and diversity based on the number of genetically distinct individuals in a population, "genotype diversity." Taken together, gene and genotype diversity constitute genetic diversity (McDonald, 1997). It is, however, important to distinguish between studies of population diversity and population genetics; the former yield the raw data, to which the latter can be applied to answer questions on the fundamental mechanisms and process of genetic change in populations (Cooke and Lees, 2004).

Detection of diversity is usually done through use of phenotypic and genotypic markers that are selectively neutral, highly informative, reproducible and relatively easy, and inexpensive to assay. The choice of genetic marker can have a substantial impact on the analysis and interpretation of data. As Phytophthora reproduce mainly asexually, producing a population structure that is largely composed of clonal lineages, DNA fingerprint (a moderately repetitive DNA probe, RG57) was used earlier which has now been supplemented by 12-plexed single sequence repeats (SSRs) (for *Phytothora infestans*) to address various questions relating to roles played by population size, mating systems and gene flow, and also for questions relating to effects of selections.

Characterization of these pathogen populations both genotypically (mitochondrial haplotypes, SSR profiles, etc.) and phenotypically (which are important in the field like germination temperature, mating type, and fungicide sensitivity) will lead to creation of a databases linked with management practices suitable for the specific populations. So in case of future outbreaks, a quick diagnostic test can be run with any one or two characters which can be rapidly detected and these then can be matched to the database, therefore, allowing identification of the population responsible, which in turn will lead to better management of the disease and prevent it from taking epiphytotic proportions.

3.2   POTATO, TOMATO, AND POINTED GOURD

Potato, though originally exotic to India, is now the third most important staple crop after wheat and rice and is mainly cultivated in the Indo-Gangetic belt, mainly in West Bengal and in parts of Assam, Meghalaya and also toward the west in Uttar Pradesh and Punjab states, which combined accounts for almost 80% of the total yield and makes India the second largest producer of potato in the world. Similarly, tomato is another important vegetable crop in India, which is mainly grown in wide acreage in the southern states, for example, Karnataka and Andhra Pradesh as well as the western states, for example, Maharashtra and Gujarat, and also to a lesser extent in the eastern states, for example, Jharkhand and West Bengal of the country. In contrast to these, another vegetable, pointed gourd, is a tropical perennial vegetable crop with its primary center of origin in the Bengal–Assam area of the Indian subcontinent (Chowdhury, 1996) and is cultivated widely in several states of India like West Bengal, Bihar, Odisha, Andhra Pradesh, and some parts of Uttar Pradesh. This dioecious vegetable is commonly named as "parwal," "palwal," "potol," or "parmal" in different parts of the country. Understandably, the cropping season of these crops varies and so does the ecology. A careful understanding of these is prerequisite for understanding how the pathogen population structure might evolve and allow us to model the pathogen dynamics. This understanding is needed to prevent the losses caused which can go up to 100% in cases of severe infection which is also rapid, the latter a characteristic of Phytophthora diseases.

3.3   THE PHYTOPHTHORA POTATO AND TOMATO PATHOECOSYSTEMS

Potato is one of the major winter crops in most parts of India. Exceptions are in the northern and northeastern hills, where potato is a summer crop, and the southern regions, where potato is autumn crop. The late blight pathogen, *P. infestans*, is the major pathogen of potato and represents a significant threat to food security (Derevnina et al., 2016; Fisher et al., 2012). It is a recurring problem in the northern hills every year. In the Indo-Gangetic plains, where the major vegetable-growing regions of India are located, the disease is mild to sporadic each year. However, once every 2–3 years, the pathogen becomes epiphytotic, causing up to 75% loss (Bhattacharyya et al., 1990). In the plains of eastern India, crop rotation is a routine practice with potato grown between November and February in fields that are used for rice cultivation for the rest of the year and are flooded as a result of that. The annual mean temperature in this subtropical region remains around 30°C, and during summer, temperatures can reach up to 42°C. Since the pathogen population is reproducing asexually, oospores are not present for overwintering. In addition, sporangia cannot survive beyond 20 days in waterlogged conditions (Porter and Johnson, 2004). Due to these adverse environmental factors, each year there is almost complete elimination of both host and pathogen, and populations of *P. infestans* face a narrow genetic bottleneck. The pathogen survives only in infected stored seeds (tubers) kept by small shareholders and this impacts pathogen dispersal between cropping seasons. These host dynamics affect the pathogen population through repeated extinction and recolonization events, which may affect gene flow as well as evolutionary trajectories (Dey et al., 2018).

The study of Indian populations over the past 20–25 years indicates that different lineages of *P. infestans* have migrated to India from other parts of the world. Two recent late blight epidemics in 2009–2010 in Southern India (Chowdappa et al., 2013) and in 2014–2015 in eastern India (Fry, 2016) have brought into focus the vulnerability of food security in the country. In both of these cases, it was a European strain, 13_A2 lineage (also known as Blue_13), which caused the late blight epidemic and in the latter case along with it social upheaval too in the form of farmer suicides and policy changes. This again brings the scientific community back to the question of understanding the ecology of the pathogen in relation to the population dynamics for better management. Studies based on 12-plex SSR genotyping (Li et al., 2013) and other techniques in the author's laboratory have identified the different population structures of *P. infestans* in the country which is distinctly region specific with certain overlaps between regions having a large subclonal variation with differences in their aggressiveness, host specificity (Dey et al., 2018), thermal tolerance, and fungicide sensitivity (Guha Roy et al., 2018). The possible reason for such rapid and devastating spread of the 2014–2015 epidemics was this high subclonal variation shown in the population. Understandably, uniform control measures in the form of single fungicides sprays for the whole country or several regions will not work. Because pathogens dynamics and population change continuously spatially and temporally, so surveillance of the pathogen populations is also needed continuously to update management decisions. The available knowledge generated allows us to make some recommendations. Phenylamide fungicides like Metalaxyl and its isomer Mefenoxam is not to be used at all, whereas Mancozeb, Azoxystrobin have shown to be very promising for the eastern region. Stricter quarantine in movement of potatoes along the border regions of the country in eastern India bordering Nepal and Bangladesh is needed as the diversity of pathogen population deduced based on the diversity of alleles is the highest in these regions allowing us to hypothesize a possible influx through these regions. In these border regions, the source of new inoculum may be the import of potatoes that these neighboring countries resort to either directly or in the form of aid from Europe or elsewhere. As potato is cultivated in close proximity to the border where similar climatic conditions prevail on both sides, there is a high likelihood of these strains crossing over and spreading over to India.

Tomato was traditionally grown in the winter season in India. In the northern plains, three crops are harvested; but in frost-affected area, winter crop is not lucrative. In India, the rainy crop is transplanted in July, and the winter crop in October or November and the summer crop in February. In the southern plains, where no incidence of frost generally occurs, the first transplanting is done in December and January, the second one is done during June and July and the third transplanting in September and October depending on the irrigation and other facilities available in the respective area. This extended the growing season and has an effect on the ecology of the *Phytophthora* spp. found on tomato resulting in multiple species of *Phytophthora* affecting the crop. In winter months, it is mostly *P. infestans*, where as in other seasons they are *Phytophthora nicotiane* and (rarely) *Phytophthora capsici*. No studies have been done till date to discern whether the *P. nicotiane* or *P. capsici* found on tomato in India are host adapted. Recent population genetic analyses of *P. infestans* and aggressiveness tests on both potato and tomato (Dey et al., 2018) have shown that the populations of *P. infestans* on tomato in India have been found to have some sort of host specificity and interestingly enough usually, the *P. infestans* strains are less aggressive on tomato than on potato. However, when often enough in winter season both tomato and potato are grown in close proximity, genetically similar strains might infect both crops showing similar aggressiveness (Dey et al., 2018). Even though in Assam, surprisingly, Metalaxyl sensitive strains have still been found; however, neither Metalaxyl nor Mefenoxam is recommended for control.

3.4   THE PHYTOPHTHORA POINTED GOURD PATHOECOSYSTEMS

Pointed gourd is a perennial, dioecious vine usually considered to be a summer and rainy season crop and is available in the market for approximately 8 months (February–September) in a year; but of late, many farmers prefer to grow them earlier in the cooler months for getting higher market price. Though this changed cultivation practice is now picking up pace, it brings along with it a higher probability of infection because with increasingly warmer winters (as a result of climate change) when accompanied by two to three showers, it would favor fruit and vine rot of pointed gourd as well as others like Sclerotinia stem or collar rot. Fruit and vine rot of pointed gourd caused by *Phytophthora melonis* Katsura is a potential disease that occurs extensively in major pointed gourd growing regions of the country and often becomes the major yield-limiting factor causing yield losses of up to 90% and might even damage the entire crop under favorable conditions of rain associated with high temperature (Guha Roy et al., 2006) (Fig. 3.1). The way in which the crop is grown either in soil-bed system or on raised trellis affects the ecology of the disease and severity. In the soil-bed system, the whole plant is in contact with the ground. Since the growing season is involving wetter months and the disease is soilborne, disease incidence is high compared to when the plant is grown on raised trellis in the same season (Sangeetha et al., 2016). Since, the pointed gourd is usually propagated through vine cuttings and root suckers, once infected if not treated, it easily often becomes a source of new inoculum in the next season and being a perennial plant, the pathogen does not have to go through genetic bottlenecks like that in the potato/tomato-*Phytophthora* pathosystem. The lack of genetically well-defined cultivars, because the farmers propagate in-house promising cultivars and designate them randomly, becomes another constraint, though efforts are made on systematically screening of the cultivars at Indian Institute of Vegetable Research, Varanasi, in India.

Information generated from ongoing diversity studies on *P. melonis* in the laboratory of the author (at the Department of Botany, West Bengal State University) shows that both A_1 and A_2 mating types of *P. melonis* are present in infected samples in India. The ratio of the A_1 and A_2 mating types of the isolates were close to 1.25:2, suggesting the possibility for sexual hybridization and recombination in the population and thereby indicating a scope for increase in the pathogen diversity.

FIGURE 3.1   (See color insert.) Pointed gourd infected by vine and fruit rot disease caused by *Phytophthora melonis*. A worried farmer looks at his field which has the initial stages of the disease, which will soon overtake the whole field.

Genetic diversity studies such as those in *P. infestans* are lacking till date in this species primarily due to the lack of published SSR markers, but this work is also underway at this laboratory. The effective concentration (EC_{50}) values were calculated for fungicide sensitivity assays for five fungicides: Metalaxyl (Glazer35WS®), Dimethomorph (Acrobat®), Fosetyl-AL (Alitte®), Cymoxanil 8% + Mancozeb 64% (CurzateM8®), and Mandipropamid (Revus®), which showed that most of the isolates were sensitive to them. This was further borne out by the fact that till date we have not detected any point mutations from these populations for the carboxylic acid amide fungicides using specific primers as has been seen in China (Chen et al., 2012). But the sensitivity to fungicides varied among isolates and from region to region perhaps indicating a different history of usages. Interestingly, all isolates were sensitive to Metalaxyl in contrast to that observed for the potato/tomato-*P. infestans* pathosystems. Further studies are needed to determine whether regional structuring is found in the population and for defining the pathogen population on pointed gourd in India.

3.5   THE COSTS OF DOING NOTHING

The growing threat of crop pathogens to global food security has resulted in series of meta-analyses interrogating large data sets. It has revealed some stark facts about the growing threat of pathogens to crop security (Bebber and Gurr, 2015). Very importantly, a global analysis of pathogen distributions shows that a country's ability to monitor and report accurately its pathogen load not only increases with per capita gross domestic product (GDP), but also relates to a country's research capacity and expenditure (Bebber et al., 2014). This poses a major challenge to the development of pathogen control strategies in poorer nations. Countries which have more information on the pathogen population pose more opportunity to better manage known threats than those that remain unknown. Unfortunately, the integration of ecological, phenotypic, and genotypic diversity information is an aspect much neglected in India for all *Phytophthora* pathosystems, primarily due to the lack of information about local *Phytophthora* pathogen populations at a genetic level and less appreciation that a disease or an epidemic is not caused by a single strain of a pathogen but by populations differing not only in the genetic diversity but also in their phenotypic diversity. More funded studies are needed to gap these current lacunae. Under field conditions, these diversity parameters translate to very important information as to which particular strains infecting the crop are aggressive enough or host adapted, how sensitive it will be to the fungicides, even whether the prevailing temperature will allow further sporangial germination and spread of the pathogen. When this information is seeded and linked with a distinct *Phytophthora* population, thereby characterizing it, they become powerful tools. Because *Phytophthora* diseases spread very rapidly and in certain cases within 48–72 h, rapid management decisions are needed to control the outbreaks. A model has been presented (Fig. 3.2) which epitomizes the procedures to be followed without any further delay. The cost of doing nothing in terms of social and economic implications is too great to ignore especially when *Phytophthora* spp. are continuously co-evolving and remerging at a rapid pace (Dong et al., 2015; Fry et al., 2016), and accordingly their ever changing characteristics have to be taken into taken if control measures are to be successful in the future.

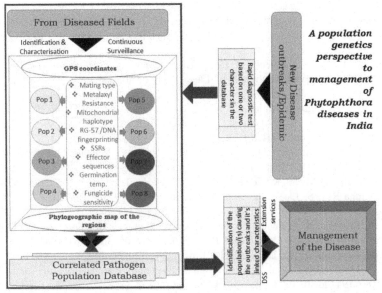

FIGURE 3.2   Model depicting how ecological and population genetics information can be integrated into disease management models for *Phytophthora* spp. in India. (Pop refers to the different pathogen populations of a *Phytophthora* spp. on a particular host.)

KEYWORDS

- *Phytophthora infestans*
- *Phytophthora nicotianae*
- *Phytophthora melonis*
- food security
- big data
- surveillance
- disease control

REFERENCES

Bebber, D. P.; Gurr, S. J. Crop-destroying Fungal and Oomycete Pathogens Challenge Food Security. *Fungal Genet. Biol.* **2015,** *74*, 62–64

Bebber, D. P.; Holmes, T.; Smith, D.; Gurr, S. J. Economic and Physical Determinants of the Global Distributions of Crop Pests and Pathogens. *New Phytol.* **2014,** *202*, 901–910.

Bhattacharyya, S. K.; Shekhawat, G. S.; Singh, B. P. Potato Late Blight. In *Technical Bulletin No. 27*. CPRI: Shimla, India, 1990.

Chen L.; Zhu S.; Lu X.; Pang Z.; Cai M.; Liu X. Assessing the Risk That *Phytophthora melonis* Can Develop a Point Mutation (V1109L) in CesA3 Conferring Resistance to Carboxylic Acid Amide Fungicides. *PLoS ONE* **2012**, 7 (7): e42069, https://doi.org/10.1371/journal. pone.0042069.

Chowdappa, P.; et al. Emergence of 13_A2 Blue Lineage of *Phytophthora infestans* Was Responsible for Severe Outbreaks of Late Blight on Tomato in South-west India. *J. Phytopathol.* **2013**, *161*, 49–58.

Chowdhury, B. *Vegetables*; National Book Trust: New Delhi, India, 1996, pp 173–175.

Cooke, D. E. L.; Lees, A. K. Markers, Old and New, for Examining *Phytophthora infestans* Diversity. *Plant Pathol.* **2004**, *53*, 692–704.

Derevnina, L; et al. Emerging Oomycete Threats to Plants and Animals. *Phil. Trans. R. Soc. B* **2016**, *371*, 20150459, http://dx.doi.org/10.1098/rstb.2015.0459

Dey, T.; Saville, A.; Myers, K.; Tewari, S.; Cooke, D. E. L.; Tripathy, S.; Fry, W. E.; Ristaino, J. B.; Guha Roy, S. Large Sub-clonal Variation in *Phytophthora infestans* from Recent Severe Late Blight Epidemics in India. *Sci. Rep.* **2018**, 8 (1), 4429.

Dong, S., Raffaele, S., Kamoun, S. The Two-speed Genomes of Filamentous Pathogens: Waltz With Plants. *Curr. Opin. Genet. Dev.* **2015**, *35*, 57–65, pmid:26451981.

Fisher, M. C.; Henk, D. A.; Briggs, C. J.; Brownstein, J. S.; Madoff, L. C.; McCraw, S. L.; Gurr, S. J. Emerging Fungal Threats to Animal, Plant and Ecosystem Health. *Nature* **2012**, *484*, 186–194, doi:10.1038/ nature10947.

Fry, W. E. *Phytophthora infestans*: New Tools (and Old Ones) Lead to New Understanding and Precision Management. *Ann. Rev. Phytopathol.* **2016**, *54*, 529–547.

Guha Roy, S.; Bhattacharyya, S.; Mukherjee, S. K.; Mondal, N.; Khatua, D. C. *Phytophthora melonis* Associated With Fruit and Vine Rot Disease of Pointed Gourd in India as Revealed By RFLP and Sequencing of ITS region. *J. Phytopathol.* **2006**, *154* (10), 612–615.

Guha Roy, S.; Dey, T.; Cooke, DEL. Fungicide Sensitivity of the Indian Sub-clonal Variants of the *Phytophthora infestans* 13_A2 Lineage. *Phytopathology* **2018**, *108* (10), 153–153S.

Li, Y.; Cooke, D. E.;L.; Jacobsen, E.; van der Lee, T. Efficient Multiplex Simple Sequence Repeat Genotyping of the Oomycete Plant Pathogen Phytophthora infestans. *J. Microbiol. Methods* **2013**, *92*, 316–322.

McDonald, B. A. The Population Genetics of Fungi: Tools and Techniques. Phytopathology **1997**, *87*, 448–453.

Porter, L. D.; Johnson, D. A. Survival of *Phytophthora infestans* in Surface Water. *Phytopathology* **2004**, *94*, 380–387.

Sangeetha, G.; Srinivas, P.; Singh, H. S.; Debasish, B.; Bharathi, L. K. Fruit and Vine Rot of Pointed Gourd (*Trichosanthes dioica* Roxb.) As Influenced By Planting Systems and Weather Parameters in East Coast Region of India. *J. Pure Appl. Microbiol.* **2016**, *10* (4), 2895–2900.

Wolfe, M. S.; Caten, C. E., Eds. Populations of Plant Pathogens: Their Dynamics and Genetics; Blackwell Scientific Publications: Oxford, UK.

CHAPTER 4

Phytoplasma in Vegetable Pathosystem: Ecology, Infection Biology, and Management

NAGENDRAN KRISHNAN[1*], SHWETA KUMARI[1],
KOSHLENDRA KUMAR PANDEY[1], AWADHESH BAHADUR RAI[1],
BIJENDRA SINGH[2], and KARTHIKEYAN GANDHI[3]

[1]Division of Crop Protection, ICAR – Indian Institute of Vegetable Research, Varanasi 221305, Uttar Pradesh, India

[2]Division of Crop Improvement, ICAR – Indian Institute of Vegetable Research, Varanasi 221305, Uttar Pradesh, India

[3]Department of Plant Pathology, Tamil Nadu Agricultural University, Coimbatore 641003, Tamil Nadu, India

*Corresponding author. E-mail: krishnagendra@gmail.com

ABSTRACT

Phytoplasmas are intracellular obligate pathogens, transmitted by phloem-feeding insect vectors. Phytoplasma diseases associated with vegetable crops occur all around the world and cause significant yield losses. Phytoplasma-infected plants show characteristic symptoms of yellowing, proliferation of shoot, reduction in leaf size, stunting, shortened internode, and phyllody. Different diagnostic methods such as electron microscopy and Dienes staining are developed for the detection of phytoplasmas; however, their detection mainly relies on molecular techniques like PCR, RFLP, nested PCR based on 16S rRNA gene, etc. Phytoplasma diseases were transmitted by phloem-feeding insects, grafting, and through dodder. Epidemic of phytoplasma disease outbreaks can be controlled either by the eradication of pathogens from the infected plant and alternative hosts or by controlling the vectors. Moreover, development of resistant cultivars against phytoplasma diseases is still far from complete. Therefore, the most effective method to prevent the outbreak is the use of clean propagating materials, altering the date of sowing and control of vectors. This chapter helps in understanding the symptomatology, transmission, distribution, ecology, detection, infection biology, and management of phytoplasma disease on vegetable crops.

4.1   INTRODUCTION

Phytoplasmas are important microorganisms causing diseases in both monocot and dicot plants, including vegetable, cereal, fruit, ornamental, and forest crops, all over the world (Lee et al, 2000). These specialized wall-less bacteria, diverged from gram-positive bacteria, obligate parasites harbor and perpetuate in the isotonic niche prevailing in the plant phloem tissues and insect hemolymph. These pleomorphic organisms with variable sizes (200–800 nm) possess a very small genome size of about 680–1600 kb (Bertaccini et al., 2014) as compared to their ancestors (walled bacteria *Bacillus/Clostridium*). They lack genes for the synthesis of several compounds essential for their survival (Bai et al., 2006), which are procured from their host plants and vectors.

In earlier days, there was a myth that plants exhibiting typical yellowing symptoms were thought to be infected by viruses because of symptomatology, their infective nature, and transmission by vectors. Doi et al. (1967) first demonstrated that the yellows-type diseases in plants could be caused by pleomorphic prokaryotes rather than viruses, which leads to a breakthrough in the discovery of a novel class of pathogens in the field of Plant Pathology. Initially, based on their ultrastructural and morphological similarity to mycoplasmas (Mollicutes), they were called as mycoplasma-like organisms (MLOs). However, these MLOs differ from mycoplasmas, which cause diseases in humans and animals. Also, these phytopathogenic MLOs failed to culture in vitro in cell-free media and were proposed to use the name "Phytoplasma" in the subcommittee on the taxonomy of Mollicutes in 1992 (Lee et al., 2000). Based on the ribosomal rDNA sequences, the phylogeny of these prokaryotes is within the class Mollicutes with the designation of new taxon "*Candidatus* Phytoplasma" for the plant pathogen belongs to the trivial name "phytoplasma," formerly MLOs (IRPCM, 2004). The term "*Candidatus*" should be dropped only when biological properties were used for their classification, that is, after attaining consistent success in culturing them to a pure culture in cell-free media (Bertaccini et al., 2010).

The association of phytoplasma with crops such as vegetable crops, fruit crops, ornamental plants, trees, and food crops was reported worldwide. For example, Tomato big bud, a disease caused by phytoplasma in tomato, has been reported to be from Australia (Davis et al., 1997), China (Xu et al., 2013), California (Shaw et al., 1993), Italy (Del Serrone et al., 2001), Egypt (Omar and Foissac, 2012), and Iran (Salehi et al., 2014), indicating its widespread distribution throughout the world. Phytoplasmas move within plants from insect stylet through sieve tube elements of phloem tissues from source to sink (Christensen et al., 2004). Inside the plant system, uneven distribution of phytoplasma has been described by several studies. In general, low titer was noticed in roots and moderate in stems (sink organ), whereas the highest titer accumulation of even ~40 times higher than that of roots was observed in matured leaves (source organs). Also, in young leaves, phytoplasma titer is very low, even below the detection limit. Moreover, the quantity of phytoplasma DNA may also varied among vegetatively propagated plantlets from the infected parent plant (Christensen et al., 2004). In some cases, phytoplasmas disappear during the winter and recolonize in aerial parts from the root system during onset of spring.

Different types of symptoms were reported in plants infected by phytoplasmas, namely, phyllody, virescence, witches' broom, yellowing, and stunting. Phyllody, a disease in which flowers are converted into green-leaf-like structures, is a common symptom caused by phytoplasmas. Flower deformities such as abnormal development of green pigmentation in floral parts, known as virescence, are also observed in phytoplasma-infected plants, and in some cases, it leads to plant sterility (Cousin and Abadie, 1982). In addition, yellowing of leaves is also associated with phytoplasma infection, which occurs due to the modification in synthesis and transportation of carbohydrates. Bertamini and Nedunchezhian (2001) had reported that chlorophyll and carotenoid contents were significantly reduced in phytoplasma-infected plants. This led to an inhibition of photosynthesis, particularly photosystem II activity. Hren et al. (2009) demonstrated the induced expression of alcohol dehydrogenase I and sucrose synthase genes responsible for alcoholic fermentation, a sign of hypoxic conditions, caused in phytoplasma-infected plants. There are some instances of virescence in the phytoplasma association (Lee et al., 2000). Also, flowers of the diseased plants were sterile sometimes.

The proliferation of axillary shoots gives the phytoplasma-infected plants a bushy or witches' broom appearance. This is mainly due to the variation in their normal growth patterns and loss of apical dominance. Besides these symptoms, phytoplasma infection causes other beneficial effects on the crop production. For example, phytoplasma association in poinsettia plants produces a more number of smaller sized flowers amenable to cultivate them in pots (Bertaccini et al., 1996; Lee et al., 1997). The association of little leaf phytoplasma in brinjal crop causes severe economic losses worldwide. Mitra (1993) recorded 40% yield losses in phytoplasma-infected eggplant. Furthermore, 60% total yield losses were recorded in tomato plants infected by stolbur phytoplasma (Navratil et al., 2009).

Insect vectors strongly decide the host range of phytoplasma and are spread mainly by phloem-feeding insect vectors of the order Hemiptera belonging to the families Cicadellidae (leafhoppers), Fulgoridae (planthoppers), Psyllidae (psyllids), Cixiidae, Cercopidae, Delphacidae, Derbidae, Menoplidae, and Flatidae from infected to healthy plants. Insect vectors truly act as the alternate host for the phytoplasma. Phytoplasma interaction with the vector varies with the species. For example, *Hishimonus sellatus* alone can be able to transmit six different types of phytoplasmas. In some cases, a single phytoplasma can also be transmitted by more than one species of insect (Weintraub and Beanland, 2006). Commonly, phytoplasmas are transmitted by the insect vector in a persistent manner (Tedeschi and Alma, 2004). During overwintering of phytoplasmas in the insect vector or in perennial plants, their interaction with insect hosts gets altered, that is, their fitness may get enhanced or reduced (Christensen et al., 2005). Weintraub and Beanland (2006) determined that exposure to a strain of aster yellows (AY) phytoplasma significantly increased the fecundity and longevity of females of *Macrotylus quadrilineatus*; however, exposure to another strain of the same phytoplasma increased the longevity but not the fecundity of the insects tested. In general, transmission of phytoplasmas occurs between the stages of the vector rather than transovarial transmission. But there are some instances of the transovarial transmission of phytoplasmas in insects such as AY by *Scaphoideus titanus* (Danielli et al., 1996; Alma et al., 1997) and mulberry dwarf by *Hishimonoides sellatiformis* and *Cacopsylla melanoneura* in apple but not reported in vegetables infected by phytoplasmas (Kawakita et al., 2000; Tedeschi et al., 2006).

Upon feeding plant sap, the vector acquires the phytoplasma through their stylets, which were absorbed into the hemolymph by passing through the intestine. Meanwhile, salivary glands were also being colonized. On the cell surface, phytoplasma possesses an antigenic protein, a key determinant for transmission, and infection was shown to interact with microfilament present in the intestine of the vector (Suzuki et al., 2006; Hoshi et al., 2007). Genus, sex, and age of the insect vector (nymphs or adults) play a major role in transmission specificity of the insect vectors. Females of *M. quadrilineatus* were more efficient than males at transmitting AY phytoplasma to lettuce (Beanland et al., 2000). On the other hand, high titers of phytoplasmas were accumulated in the salivary glands of young males rather than females of *Euscelidius variegatus* (Lefol et al., 1994). In the same way, fifth instar nymphs of *E. variegatus* are highly efficient in transmission, whereas recently emerged nymphs are not acquired (Palermo et al., 2004), and in some cases, phytoplasmas are transmitted more efficiently when they are acquired at the nymphal stage rather than adults (Moya-Raygoza and Nault, 1998; Murral et al., 1996). In addition, race of the phytoplasma and the environmental conditions also influence the transmissibility of leafhoppers (Murral et al., 1996).

Phytoplasmas were also transmitted from infected to healthy plants of same or different species through the parasitic plant dodder (*Cuscuta* sp.). Phytoplasmas were also efficiently spread by vegetative propagation, micropropagation practices, and any other method used to multiply plant material that avoids sexual reproduction (Bertacini et al., 2014).

In recent days, transmission of phytoplasmas through seeds has also been investigated in several crops. Cordova et al. (2003) detected phytoplasma DNA in embryos of coconut palms causing a lethal yellowing disease. Evidence of seed transmission of phytoplasma was witnessed on in vitro grown commercial seedlings of alfalfa (*Medicago sativa*) and in vitro seedlings grown from seeds of phytoplasma-infected plants of tomato, lime, corn, and oilseed rape under insect-proof conditions (Calari et al., 2011). Some of these seeds evidenced with seed transmission of phytoplasmas were belonging to 16SrI, 16SrXII, and 16SrII ribosomal groups (Khan et al., 2002; Botti and Bertaccini, 2006).

4.2 PHYTOPLASMA GROUPS INFECTING VEGETABLE CROPS

Utilization of reliable diagnostic techniques like serological and nucleic-acid-based assay brought new insights for the identification of phytoplasmas, which made them into several groups. PCR assay coupled with RFLP analysis along with sequence-based phylogenetic analysis was well established and standardized for the identification and taxonomic classification of phytoplasmas (Lee et al., 1998a; Seemuller et al., 1998). Due to its inability to grow in axenic culture, the taxonomy of phytoplasmas becomes complicated as other prokaryotes. RFLP analysis of PCR-amplified fragments of the 16S rDNA with a set of 17 restriction enzymes grouped with various phytoplasmas by their distinctive RFLP patterns is the widely adopted methodology for the taxonomy of phytoplasmas. Lee et al. (1998b) initially proposed 14 groups and 41 subgroups of phytoplasma strains.

Currently, phytoplasmas were grouped under 33 major groups consisting of 116 subgroups validated based on the phylogenetic analysis of near full-length 16S rDNA gene sequence. These frequently infecting vegetable crops were found to belong to 16SrI group: AY; 16SrII: Peanut witches' broom; 16SrVI: Clover proliferation; and 16SrXII: Stolbur phytoplasma. Following are the list of *Ca.* Phytoplasma infecting vegetable crops all over the world (Table 4.1).

TABLE 4.1   Major Groups and Subgroups of Phytoplasma Infecting Vegetable Crops.

16Sr subgroup	Strain (acronym)	Vegetable crops
16SrI: AY (America, Europe, Asia, and Africa)		Brinjal, soybean, garlic, *Cucurbita pepo*, broccoli
I-A	AY witches' broom	Potato
I-A	Tomato big bud (BB)	Tomato
I-B	Onion yellows mild strain (OY-M)	Potato, chilli, cabbage, lettuce, onion, prickly lettuce, chayote, pumpkin, *C. maxima*
I-C	Clover phyllody	Celery
16SrII: Peanut witches' broom (America, Africa, Europe, Asia, and Australia)		*C. pepo*, chayote, bitter gourd, pumpkin, *Luffa cylindrica*, *Sicana odorifera*
II-A	Peanut witches' broom (PnWB)	Tomato, cauliflower
II-C	Faba bean phyllody (FBP)	Tomato, soybean
II-D	Papaya mosaic (PpM) '*Ca.* P. australasia'	Chilli, tomato, faba bean, carrot, squash
16SrIII: X-disease (America, Europe, and Asia)		Muskmelon, brinjal, faba bean, *L. cylindrica,* chayote, broccoli
III-B	Clover yellow edge (CYE)	Potato
III-J	Chayote witches' broom (ChWBIII)	Cauliflower, pumpkin
16SrIV: Coconut lethal yellows (America and Africa)		Tomato
16SrV: Elm yellows (Europe, America, Asia, Africa)		Tomato
16SrVI: Clover proliferation (Europe, America, and Asia)		Cucumber, chilli
VI-A	Clover proliferation (CP) "*Ca.* P. trifolii"	Tomato, cabbage
VI-C	Illinois elm yellows (EY-IL1)	Potato, tomato
VI-D	Periwinkle little leaf (PLL-Bd)	Tomato

TABLE 4.1   *(Continued)*

16Sr subgroup	Strain (acronym)	Vegetable crops
16SrVIII: Loofah witches' broom (Asia)		Potato, *L. cylindrica*
16SrIX: Pigeon pea witches' broom (Europe, Asia, and America)		Potato
IX-C	Naxos periwinkle virescence (NAXOS)	Brinjal
16SrXII: Stolbur (Europe, Asia, America, Africa, and Australia)		Onion, lettuce, celery
XII-A	Stolbur (STOL11) "*Ca*. P. solani"	Potato, chilli, tomato
XII-E	Yellows diseased strawberry (StrawY) "*Ca*. P. fragariae"	Potato
16SrXIII: Mexican periwinkle virescence (America)		Broccoli
16SrXV: Hibiscus witches' broom (America)		
XV-A	Hibiscus witches' broom (HibWB) "*Ca*. P. brasiliense"	Cauliflower

4.3   PHYTOPLASMA DETECTION AND IDENTIFICATION

The identification of phytoplasma disease was cumbersome due to its unculturable nature in earlier days. Symptomatology, transmission study (insect/graft/dodder), and electron microscopy were also utilized for the detection of phytoplasma in infected plants. ELISA-based serological diagnostics were also used for the identification of phytoplasma during 1980s. Precise classification of phytoplasma into groups and subgroups was made possible only after the introduction of PCR RFLP analysis (Namba et al., 1993; Lee et al., 1993).

4.3.1   BIOLOGICAL DETECTION

First attempts made to classify and differentiate phytoplasmas were mainly based on the biological properties like symptomatology, host range, and transmission (graft/dodder/insect vector) (Jarausch et al., 2000; Pastore et al., 2001). Preliminary diagnosis of phytoplasma infection relied upon symptomatology shown by infected plants, such as yellowing, stunting, small leave shoot proliferation resulting in a witches' broom appearance, phyllody, virescence, floral sterility decline, and death (Lee and Davis, 1992).

Doi et al. (1967) and Lee and Davis (1992) described complete remission of symptoms shown by phytoplasma-infected plants upon treatment with antibiotic (i.e., tetracycline), which helps differentiate from virus symptoms. But discrimination of phytoplasma through biological properties is labor-intensive and time-consuming; also, in some cases, results were inconsistent (Errampalli et al., 1991; Chiykowski, 1991; Lee and Davis, 1992; Bertacini et al., 2014).

4.3.2 MICROSCOPIC TECHNIQUES

Detection of phytoplasma through a normal light microscope is not possible. Visualization of the localized phytoplasma in the infected tissue under a light microscope is possible using the DNA binding fluorochrome, 4'-6-diamidino-2-phenylindole stain, also called as Dienes reagent, which gives blue color (Dienes et al., 1948). Though this procedure is simple, quick, and less expensive, it is not specific for different phytoplasmas. Detection of phytoplasma using the technique is possible when the concentration is high in the infected tissues (Lee and Davis, 1992). Electron microscopic techniques like scanning electron microscopy (SEM) and transmission electron microscopy (TEM) are employed in the phytoplasma detection. Under SEM, phytoplasmas are visualized as short, branched, filamentous forms in sieve tubes of phloem of infected plants (Marcone and Ragozzino, 1996). Similarly, phytoplasma cells are localized in the sieve tubes as small and pleomorphic through TEM, and it is found to be most reliable and robust method for the phytoplasma detection (Poghosyan et al., 2004). Morphological and structural characteristics of phytoplasmas within the diseased host plant sieve tubes (Lee et al., 2000; Chapman et al., 2001) and plant--phytoplasma interaction (Chapman et al., 2001; Musetti et al., 2000) were also studied under TEM. In the phytoplasma detection, a confocal laser scanning microscope was used with membrane potential dyes and DNA-specific dyes, DiOC7 (3) (3, 3'-diheptyloxacarbocyanine iodide) and SYTO 13 (green fluorescent nucleic acid stain) respectively (Christensen et al., 2004). The fluorescence in situ hybridization (FISH) technique, involving nonradioactive oligonucleotides probes, allows in situ detection in both host plant tissues and insect vectors of specific phytoplasma taxonomic groups, subgroups, and/or strains, studying colonization patterns of phytoplasma, population dynamics of phytoplasma in response to the environment, and phytoplasma–endophyte interactions (Bulgari et al., 2011).

4.3.3   SEROLOGICAL DETECTION

Serological detection of phytoplasmas were successfully made using polyclonal and monoclonal antisera, and for some economically important phytoplasmas, antisera were available commercially (Lee et al., 1993; Chen et al., 1993). Several serological tools were employed in the detection of phytoplasmas from insect vectors and plants such as immunofluorescence, immunosorbent electron microscopy, dot blot or ELISA, and tissue blotting (Lherminier et al., 1990; Sinha and Benhamou, 1983; Boudon-Padieu et al., 1989; Lin and Chen, 1985). Antibodies raised against the recombinant immunodominant proteins in *Escherchia coli* were also utilized in the detection of phytoplasmas (Berg et al., 1999; Hong et al., 2001; Mergenthaler et al., 2001; Blomquist et al., 2001; Kakizawa et al., 2009).

4.3.4   MOLECULAR DETECTION

The phytoplasma diagnostic procedure should aim to be a quicker, more economic, robust, sensitive, and reliable methodology. This made possible only after the introduction of molecular techniques such as PCR, PCR-RFLP, nested PCR, immunocapture PCR, and real-time PCR. Universal primers, group specific primer, and species-specific primer pairs were utilized for the phytoplasma detection. Nucleic-acid-based detection and identification of phytoplasma associated in plants and vectors were initiated during the first cloning of phytoplasma DNA made by Kirkpatrick et al. (1987). Many nucleic-acid-based probes were designed from randomly cloned DNA of phytoplasmas for their detection from chromosomal and extrachromosomal region (Lee and Davis, 1988; Bertaccini et al., 1990a; Bonnet et al., 1990; Harrison et al., 1992). This probe-based detection provides the first evidence for prevalence of genetic differences as strains and species among the phytoplasmas infecting different host plants in different geographical locations (Lee et al., 1992; Bertaccini et al., 1990b, 1993).

In recent days, PCR-based phytoplasma diagnosis has been widely used as a robust and sensitive technique. In preliminary detection, universal-primer-based PCR assays are adopted. For routine detection of phytoplasmas, several universal and group-specific primers based on 16S ribosomal gene, ribosomal protein, *tuf*, *SecY* genes, and 16S-23S rRNA intergenic spacer region were being used (Deng and Hiruki, 1991a, 1991b; Ahrens and Seemuller, 1992; Namba et al., 1993; Davis and Lee, 1993; Lee et al., 1993, 1994, 1998a, 1998b, 2004a, 2004b, 2006; Lorenz et al., 1995; Smart et al., 1996; Schneider et al., 1997; Martini et al., 2002, 2007). Subgroup delineation was achieved through both RFLP analyses of 16S rDNA region and sequence analysis of *rp* gene. Validation of this subgroup identified by genome analysis made through dot blot and Southern hybridization using DNA probes based on the phytoplasma genome (Lee et al., 1992, 1998a; Gundersen et al., 1996; Martini et al., 2007).

In the phytoplasma-infected plant tissues, the phytoplasma DNA constitutes only less than 1% of the total DNA (Bertaccini, 2007). To enrich the phytoplasma DNA, nested PCR was employed for its sensitive and specific detection in the infected plant tissues, as demonstrated by many workers (Gundersen and Lee, 1994; Gundersen et al., 1994). Generally, nested PCR is performed with a universal primer pair for preliminary amplification followed by amplification by another universal primer pair or group-specific primer pair. Nested PCR is capable of detecting even more than one phytoplasma infection at a time (Lee et al., 1994).

Immunocapture PCR assay is also being used for sensitive phytoplasma detection, which is an alternative to nested-PCR assay. Initially, the phytoplasma of interest is captured on microtitter plates by specific antibody followed by amplification of the phytoplasma DNA using a specific or universal primer (Rajan and Clark, 1995; Heinrich et al., 2001). For further improving sensitivity, it can be coupled with nested PCR also. Immunocapture PCR results are more reliable and authentic since it involves both serological and molecular methods of detection. Molecular detection based on 16S rRNA gene, ribosomal protein gene operon, and *tuf* and *SecY* genes was a major breakthrough in the classification of phytoplasmas into different groups and subgroups (Gundersen et al., 1996; Schneider et al., 1997; Marcone et al., 2000; Martini et al., 2002; Wei et al., 2004; Martini et al., 2007). RFLP patterns are conserved for phytoplasmas, and they will be useful in identifying unknown phytoplasmas by comparing the patterns with already reported RFLP patterns (Lee, 1998a, 1998b; Wei et al., 2007, 2008a; Cai et al., 2008). According to the recommendation by the IRPCM Phytoplasma Taxonomy Group (IRPCM, 2004), a "*Candidatus* Phytoplasma" species should possess single unique 16S rRNA gene sequence (>1200 bp). A novel strain of "*Ca*. Phytoplasma" can be named if its 16S rRNA gene sequence has <97.5% similarity to that of any previously described species (Arocha et al., 2005, 2007; IRPCM, 2004). Because of the conserved nature of 16S rRNA gene, many distinct phytoplasma strains may fail in meeting requirement of <97.5% sequence similarity with existing "*Ca*. Phytoplasma." In such a case, additional properties such as antibody specificity, host range, and vector transmission, as well as other molecular criteria, may be employed for speciation (Seemuller and Schneider, 2004).

4.4   BIOLOGY AND GENOMICS

Phytoplasmas were introduced into sieve elements by the insect vector; there was systemic spread of phytoplasma throughout plants (Bertaccini et al., 2014). Among the phytopathogenic bacteria, phytoplasmas possess the smallest genome similar to those symbiotic bacteria associated with some insects. In spite of their smaller genome, phytoplasmas possess multiple copies of many basic housekeeping genes; also, they possess two rRNA operons showing sequence heterogeneity (Bertaccini et al., 2014). Gene duplication and redundancy were well studied in the phytoplasma genome. There was redundancy observed in the onion yellows (OY) phytoplasma genome, in which 18% of gene complement was shown by multiple redundant copies of only five genes (Oshima et al., 2004). Genomes of phytoplasmas contain several unique insertion and transposon sequences containing similar compositions and genes called as either variable mosaics (Jomantiene and Davis, 2006; Jomantiene et al., 2007; Wei et al., 2008b) or potential mobile units (Bai et al., 2006).

Plasmids, also known as extra-chromosomal DNA, of various sizes (1.7–7.4 kb) are found with AY (16SrI), stolbur (16SrXII), X-disease (16SrIII), and clover proliferation (16SrVI) phytoplasma groups (Kuboyama et al., 1998; Rekab et al., 1999; Nishigawa et al., 2001; Oshima et al., 2001; Nishigawa et al., 2002). There are some pieces of evidence suggesting that these plasmids play a role in insect vector interaction with phytoplasmas. Putative transmembrane protein ORF3 is present in the plasmids suggested to play a role in the interaction with the insect vector for its transmission. In plasmids of "Ca. P. asteris" (OY strain), ORF3 is missing in the noninsect transmissible line on plasmid pOYNIM, whereas it is present in the insect transmissible mildly pathogenic strain. Moreover, the expression of ORF3 protein is more specific on OY-M-infected insects than that of plants (Ishii et al., 2009). Also, gene-encoding malate, metal-ion, and amino acid transporters are present in multiple copies in the phytoplasma genome. It is evidenced that severe strain of Ca. P. asteris (OY-W) genome approximately had 30-kb duplicated region in the 80-kb total genomic DNA sequence (Oshima et al., 2007). It is also hypothesized that phytoplasma may import host metabolites for its metabolic process by disturbing the metabolic balance, which ultimately leads to the development of symptoms on the infected plants. In the OY-M phytolasma genome, lacking ATP synthase gene suggests that these phytoplasmas depend upon host ATP for its glycolysis process (Bertaccini et al., 2014).

4.5 HOST–PHYTOPLASMA INTERACTION

Proteins that are present on the cell surface of phytoplasmas are distinct to each species and are playing a key role in specific interaction with host plants. The three types of genes encoding these proteins are: (1) immunodominant membrane protein (Imp), detected in sweet potato witches' broom phytoplasma (Christensen et al., 2005); (2) immunodominant membrane protein A (IdpA), detected in western X-disease (WX) phytoplasma (Blomquist et al., 2001); and (3) antigenic membrane protein (Amp), detected in the phytoplasma associated with AY, clover phyllody, and OY diseases (Morton et al., 2003; Barbara et al., 2002). The OY-phytoplasma-associated Amp was reported to form a complex with microfilament of insects made of actin and myosin. This observation leads to a conclusion that this Amp is important in transmissibility of phytoplasmas by leafhoppers (Suzuki et al., 2006). In addition to IdpA gene on the WX phytoplasma genome, gene homologous to Imp was also present. This suggests that Imp might be a common ancestor for all phytoplasma (Liefting and Kirkpatrick, 2003). Detection of these membrane proteins is also possible in the plants infected with phytoplasmas at high concentration. Since the expressed membrane proteins are retained on the cell membrane of the phytoplasma cell, antibodies raised against these membrane proteins can be utilized for the detection of phytolasmas in the affected plants (Barbara et al., 2002; Arashida et al., 2008). Since phytoplasmas are intracellular parasites, around 33% of genes alter their expression to adapt between diverse environments in *Ca*. P. asteris' (OY-M). It is also shown that some proteins are expressed variably in insect and plant host (Oshima et al., 2011; Toruno et al., 2010; Hoshi et al., 2009).

4.6 VIRULENCE AND PATHOGENICITY

Next to no is thought about phytoplasma destructiveness. To fulfill their need of basic metabolic pathways, they may depend on the host cells for the assimilation of required materials by causing damage to the hosts. Oshima et al. (2007) correlated the differential symptom expression with the genome size of the phytoplasma. Mild strain of onion yellow (OY-M) phytoplasma having a chromosome size of 860 kbp produced mild proliferation and yellowing, whereas severe strain (OY-W) with a chromosome size of 1000 kbp produced yellowing, stunting, proliferations, and witches' broom, which are impelled by a severe strain (OY-W). Further investigation demonstrated that the phytoplasma concentration of OY-W was higher than that of OY-M in the infected plant tissue. Also, duplication of five glycolytic genes across the severe strain genome was also observed. In perspective of these observations, the higher carbon source utilization, accounting for the duplication of glycolytic genes, may promote the multiplication of OY-W phytoplasma, leading to expression of severe symptoms (Oshima et al., 2007).

Also, a virulence factor, "tengu-su" inducer (TENGU), was additionally demonstrated to induce witches' broom and stunted growth in transgenic *Nicotiana benthamiana* and *Arabidopsis thaliana* plants upon its expression. In microarray analysis, there was a downregulation of auxin-responsive genes in the "TENGU" protein expressed transgenic plants when compared with control plants (Hoshi et al., 2009).

Similarly, it is highly essential to study the genes of a secretary system in the phytoplasma genome. According to Bertaccini et al. (2014), secretary proteins and immunoproteins from the phytoplasma cell membrane may assume a critical role in virulence and/or pathogenecity when they are exposed to the cytoplasm of the host or vector cell. Phytoplasmas are known to have two secretary systems. A system for the integration of membrane proteins is called as "*YidC* system" and the other for both secretion and integration of secretary proteins into the host cell cytoplasm is called as "Sec system (*SecA, SecY,* and *SecE* genes)" (Kakizawa et al., 2001. 2004; Wei et al., 2004). Amp is accounted to be a substrate of the Sec framework and has a Sec sign arrangement at its N-end, which is cut in "*Ca.* P. asteris" OY-M. Hence, it is proposed that the phytoplasma Sec framework uses acknowledgment and cleavage of a sign succession, as in other bacterial Sec frameworks (Kakizawa et al., 2001; Barbara et al., 2002). Among the gene coding secretary proteins, SAP11 signals eukaryotic nuclear localization and accumulation in host cell nuclei (Bai et al., 2006). Also, transgenic plants expressing SAP11 gene exhibit leaf crinkling and stem proliferation. In addition, fecundity of insect vectors also increased upon feeding the transgenic plants expressing SAP11 rather than nontransformed plants (Sugio et al., 2011). There are some reports shown to cause a change in floral morphology in *A. thaliana* resembling typical phytoplasma symptoms upon expression of secretary proteins genes SAP11 and SAP54 belonging to AY phytoplasma (MacLean et al., 2011).

4.7 PHYTOPLASMA DISEASES IN VEGETABLE CROPS

4.7.1 SOLANACEOUS VEGETABLES

Infection of phytoplasma on solanaceous crops such as tomato, chilli, brinjal, and potato was reported worldwide with a considerable yield loss. Infection on brinjal causes severe symptoms, namely, size reduction and malformation of leaves (little leaf), conversion of floral parts into green-leaf-like structures (phyllody), excessive proliferation of auxiliary buds, infected plants devoid of any fruits, and looks taller than noninfected plants (Figs. 4.1 and 4.2) (Siddique et al., 2001; Omar and Foissac, 2012; Saranya and Umamaheswaran, 2015). Tomato plants infected by phytoplasmas exhibit a variety of symptoms including stunting, yellowing, or purpling of leaves, profuse proliferation of laterals buds (Fig. 4.3), formation of hypertrophic calixes (big bud) (Fig. 4.4), virescence of floral parts, absence of anther, and ovary formation and fusion of petals causing big bud symptoms. In general, infected plants lack leaves at their apex, and the youngest leaves become small, thick, and distorted (Omar and Foissac, 2012). Similar symptoms were reported on tomato plants infected by 16SrXII-A subgroup, stolbur phytoplasma (Pracros et al., 2006). Tomato infecting phytoplasmas were identified as 16SrI, II, III, IV, V, VI, and XII phylogenetic groups from Italy through 16S-rDNA-based PCR and RFLP analysis (Del Serrone et al., 2001). Tomato big bud, a phytoplasma disease, was found to be associated with 16SrII group phytoplasma in Australia (Davis et al., 1997) and China (Xu et al., 2013) and 16SrVI group in California (Shaw et al., 1993). Several workers reported "*Candidatus* Phytoplasma trifolii" causing diseases on tomato, chilli, and potato in North America (Deng and Hiruki 1991a; Lee et al., 1991; Choueiri et al., 2002; Jacobs et al., 2003). The same organism was found to cause infection on pepper in Spain and on tomato in Jordan (Castro and Romero 2002; Anfoka et al., 2003). Singh and Singh (2000) have reported phytoplasma causing little leaf disease on chilli for the first time in India. Chilli plants were reported to show yellowing of leaves, stunting of plant height, and fruit deformation followed by wilting and plant decline upon infection by stolbur phytoplasma (Navrátil et al., 2009) (Fig. 4.5).

FIGURE 4.1   Natural symptoms on eggplant showing phyllody.

FIGURE 4.2   Infected eggplant showing reduction of leaf size.

FIGURE 4.3   Stunted tomato plants with heavy proliferations of purplish leaves infected by 16SrII-D phytoplasma subgroup.
Source: Choueiri et al. (2007)

FIGURE 4.4   Stolbur phytoplasma infection causing big bud on tomato.
Source: Navrátil et al. (2009)

FIGURE 4.5 Chilli plants exhibiting symptoms of leaf yellowing caused by stolbur phytoplasma infection.
Source: Navrátil et al. (2009)

4.7.2 CUCURBITS

Phytoplasma diseases on cucurbits have been reported in *Cucurbita pepo* L., *Cucumis sativus* L., *Cucurbita maxima* Duchesne, *Cucurbita mixta* Pangalo, *Lagenaria leucantha* Rusby, *Lagenaria siceraria* (Molina) Standley, *Cucurbita moschata* Duchesne, *Luffa cylindrica* L., *Momordica charantia* L., *Sechium edule* (Jacquin) Swartz, and *Sicana odorifera* (Vellozo) Naudin (McCoy et al., 1989; Lee et al., 1993; Gundersen et al., 1994; Seemuller et al., 1998; Montano et al., 2000; Villalobos et al., 2002; Montano et al., 2006; Montano et al., 2007a, 2007b). Salehi et al. (2015) reported the phytoplasma disease causing phyllody on cucumber and squash from Iran, and the phytoplasmas belong to 16SrII group. Furthermore, 16S-rDNA-based RFLP analyses revealed the affiliation to subgroup 16SrII-D infecting squash and to a new subgroup 16SrII-M infecting cucumber. Omar and Foissac (2012) observed that phytoplasma-infected squash plants (*C. pepo*) showed stunting and virescence. Upon molecular characterization, the associated phyto-plasma was found to belong to 16SrII-D group with a disease incidence of about 1% across 20 different fields in Egypt (Figs. 4.6–4.9). In addition, several reports such as 16SrI group ("Ca. P. asteris") on *C. pepo* from Italy and in *S. edule* from Costa Rica; 16SrII group phytoplasmas on *C. pepo* from Australia, Egypt, and Iran and in *S. edule*, *M. charantia*, *S. odorifera*, *C. moschata*, and *L. cylindrica*, from Brazil; 16SrVIII group phytoplasmas on *L. cylindrica* from Taiwan were made on the occurrence of phytoplasma diseases on cucurbits (Lee et al., 1994; Davis et al., 1997; Seemüller et al., 1998; Montano et al., 2000; Villalobos et al., 2002; Montano et al., 2006; 2007a, 2007b; Dehghan et al., 2014) were presented. In addition, 16SrIII group phytoplasma on *L. cylindrica* and *S. edule* from Brazil; AY subgroup 16SrI-B phytoplasma on chayote from Costa Rica; *Ca.* Phytoplasma asteris on *M. charantia* L. from Myanmar; and *Ca.* Phytoplasma australiense on *C. maxima* and *C. moschata* from Australia (Montano et al., 2000; William et al., 2002; Streten et al., 2005; Montano et al., 2007c; Win et al., 2014) were also reported.

FIGURE 4.6   Reduction of leaf size, shortening of internode, virescence, and phyllody of flowers and witches' broom on cucumber.
Source: Salehi et al. (2015)

FIGURE 4.7   Stunting and virescence on squash plant.
Source: Omar and Foissac (2012)

FIGURE 4.8   Severe proliferation of stem, chlorosis, reduction of leaf size, shortening of internodes, and witches' broom on squash.
Source: Salehi et al. (2015)

FIGURE 4.9   Proliferation of abnormal flowers and phyllody on infected flowers (right) compared to a normal flower (left) on squash.
Source: Salehi et al. (2015)

4.7.3   CRUCIFERS

Infection of *"Ca.* Phytoplasma asteris" belonging to 16SrI group on genotypes of *Brassica oleracea* × *Brassica napus* shows failure of flower bud formation with well-developed green foliage in the second year of their growth in Poland (Kamińska and Berniak, 2012). In Iran, phytoplasma infection on cabbage causes Iranian cabbage yellows (ICY) with symptoms of yellowing, little leaves, stunting, opening of head, and proliferation of the buds at the base of the stem into wiches' broom. Based on PCR-RFLP analysis, *"Ca.* Phytoplasma trifolii" belonging to 16SrVI group is the cause for ICY and is transmitted by the leafhopper, *Circulifer haematoceps* (Salehi et al., 2007). In Brazil, broccoli stunt, a new disease on broccoli, showed characteristic symptoms such as stunting of plants, malformation of inflorescence, reddening of leaves, and necrosis of phloem tissues. Through nested-PCR assay, the association of phytoplasmas was confirmed and found to belong to 16SrI, III, and XIII groups (Eckstein et al., 2013).

4.7.4 LEGUMES

In Spain, Faba bean infected with phytoplasmas showed symptoms of shoe stringed leaves, phyllody, and flower abortion. Causal organism was found to belong to 16SrIII group, member of X-disease phytoplasma through PCR assay and phylogenetic analysis (Castro and Romero, 2004). Soybean plants showing typical symptoms of phytoplasma infection such as phyllody and witches' broom were observed in Malawi and Mozambique of South Africa with a yield loss of 100% in infected plants and is adversely affecting its production. It is found to be caused by 16Sr group II, subgroup C (cactus phytoplasma) through sequencing and RFLP analysis (Lava Kumar et al., 2011). Phyllody disease on lentil crop showed floral malformation, chlorosis, little leaf, and extensive proliferation of branches Pakistan. Akhtar et al. (2016) identified that phyloplasmas belong to 16SrII-C subgroup associated with the lentil phyllody disease and are transmitted by the leafhopper, *Orosius albicinctus*. In cowpea, flattened stem was reported to cause by the infection of *Ca*. P. cynodontis (16SrXIV-A) from southern India (Rao et al., 2017) (Fig. 4.10).

FIGURE 4.10 Flattening of stem and resetting of leaves on vegetable cowpea.

4.7.5 OTHER VEGETABLE CROPS

In Mexico, grain amaranth (*Amaranthus hypochondriacus* L. and *A. cruentus* L.) plants exhibited conspicuous symptoms such as excessive proliferation of stem and bud, mosaic, and unusual coloration. Sequence analysis of 16S rDNA gene revealed *Ca*. Phytoplasma aurantifolia as a causal agent for the disease (Ochoa-Sánchez et al., 2009). In north-east Italy, infection of stolbur group (16SrXII-A) phytoplasmas caused severe yellowing on celery plants (Carraro et al., 2008). Sichani et al. (2014) reported the infection of 16SrI-B group phytoplasmas on cabbage lettuce, onion, and prickly lettuce from Iran. Garlic plants and onion plants infected with 16SrI (AY) group phytoplasmas exhibited leaf discoloration varying from yellow to purple, stunting, necrosis of leaf tips, wilting, and death of plants in severe cases. Bulbs from infected plants were softer and smaller than the bulbs from normal-looking plants (Khadhair et al., 2002; Mollov et al., 2014). Several reports of infection of aster yellow phytoplasma on onion were reported from different parts of the world (Jomantiene et al., 2010; Sichani et al., 2014).

4.7.6   MANAGEMENT

Since phytoplasma are intracellular devastating prokaryotic pathogens found associated with wide range of host plants causing diseases. The durable resistance identification against phytoplasmas is the most effective and long lasting way for its management. Selection and screening of genotypes, which is resistant to either phytoplasma or insect vector, is the other possible choice for phytoplasma disease management. Achievement of this technique mainly depends on the efficient screening system. However, little information is available about the resistant vegetable cultivars against phytoplasma disease. Crop hygiene practices such as early roughing of infected plants, adjustment in date of sowing, use of clean propagating materials, and eradication of weed hosts that serve as a reservoir for phytoplasma and its vector and need-based spraying of insecticides may significantly help reduce the incidence of diseases. The most promising strategies to restrict the disease spread are prophylactic measures such as vector control, use of healthy planting materials, or eliminating the pathogens from infected plants through the use of meristem tip culture, antibiotics, or any other chemicals (Bertaccini, 2007). Although it is impossible to eliminate all the vectors, but well-managed control of vectors significantly reduces the chance of spread of diseases. Scott and Zimmerman (2001) suggested different control measures such as eradication of alternate hosts and application of antibiotic through injection for phytoplasma diseases. Temporary remission in symptoms was noticed in little leaf of the affected brinjal plant by tetracycline treatment (Anjaneyulu and Ramakrishnan, 1973; Raychaudhuri et al., 1970). In addition, several reports such as application of oxytetracycline in *Dendranthema grandiflorum* from

China (Chung and Choi, 2002), tetracycline in *Portulaca grandiflora* from India (Ajayakumar et al., 2007), and *in vitro* chemotherapy in *Catharanthus roseus* (Singh et al. 2007) were made for phytoplasma disease management.

Current research studies are focusing on the stimulation of plant defense through the application of biotic or abiotic elicitor molecules, which can be integrated with the other earlier discussion options of disease management. Some of the abiotic resistant inducers used against the phytoplasma are benzothiadiazole, phosetyl-aluminium, prohexadione calcium, indole-3-butyrric acid, indole-3-acetic acid, chitosan, salicylic acid, and mixture of glutathione and oligosaccharines (Jarausch and Torres, 2014). Upregulation of jasmonate gene has been demonstrated in apple trees recovered from apple proliferation disease (Musetti et al., 2013). In addition to the abiotic indicers, biotic inducers such as arbuscular mycorrhiza (AM) and endophytes are also playing a major role in dropping the phytoplasma disease. AM-treated tomato plants showed agglutinations and degeneration of phytoplasma cells, coupled to reduced symptom expression upon infection of stolbur phytoplasma (Jarausch and Torres, 2014). Recently, exploration on endophytic bacteria and fungus in the suppression of pathogens has been focused. Studying the endophytes associated with healthy and infected plants will identify some putative biocontrol agents for the phytoplasma disease management. Bianco et al. (2013) demonstrated reduction of symptom severity and phytoplasma titer in host tissues of *C. roseus* plant infected with *"Ca. P. mali"* upon pretreatment with endophytic strain of *Epicoccum nigrum* (Fig. 4.11).

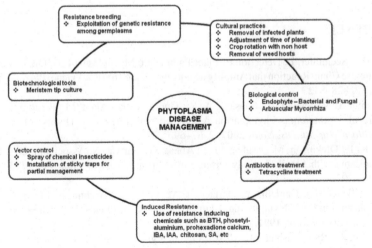

FIGURE 4.11   Strategies for the phytoplasma disease management in vegetable crops.

4.8   CONCLUSION

Similar to the viral diseases, insect-transmitted phytoplasma disease on many crops has increased in recent years. Also, it is extending its host range to various vegetable crops across the family and is focused as the emerging problem with increased disease incidence on vegetable crops in past few years. In addition to characterization of phytoplasmas, it is also essential to study their mode of survival on alternative host during the absence of the main host to prevent further escalation of the disease among vegetable crops for its efficient management. Management through a resistant source is durable; hence, future works should focus on the identification of the resistance sources.

KEYWORDS

- **phytoplasma**
- **phloem feeding**
- **symptoms**
- **detection**
- **vegetable**
- **management**

REFERENCE

Ahrens, U.; Seemüller, E. Detection of Plant Pathogenic Mycoplasma Like Organisms by a Polymerase Chain Reaction that Amplify a Sequence of the 16SrRNA Gene. *Phytopathology* **1992**, *82*, 828–832.

Ajayakumar, P. V.; Samad, A.; Shasany, A. K.; Gupta, M. K.; Alam, M.; Rastogi, S.,. First Record of a *Candidatus* Phytoplasma Associated with Little Leaf Disease of *Portulaca grandiflora. Aus. Pl. Dis. Notes* **2007**, *2*, 67–69.

Akhtar, K. P.; Dickinson, M.; Asghar, M. J.; Abbas, G.; Sarwar, N. Association of 16SrII-C Phytoplasma with Lentil Phyllody Disease in Pakistan. *Tropical Plant Pathol.* **2016**, *41* (3), 198–202.

Alma, A.; Bosco, D.; Danielli, A.; Bertaccini, A.; Vibio, M.; Arzone, A. Identification of Phytoplasmas in Eggs, Nymphs and Adults of Scaphoideus titanus Ball Reared on Healthy Plants. *Insect Mol. Biol.* **1997**, *6*, 115–121.

Anjaneyulu, A.; Ramakrishnan, K. Host Range of Eggplant Little Leaf Disease. *Mysore J. Agric. Sci.* **1973**, *7*, 568–579.

Anfoka, G. H. A.; Khalil, A. B.; Fattash, I. Detection and Molecular Characterization of a Phytoplasma Associated with Big Bud Disease of Tomatoes in Jordan. *J. Phytopathol.* **2003**, *151*, 223–227.

Arashida, R.; Kakizawa, S.; Ishii, Y.; Hoshi, A.; Jung, H. Y.; Kagiwada, S.; Yamaji, Y.; Oshima, K.; Namba, S. Cloning and Characterization of the Antigenic Membrane Protein (Amp) Gene and in Situ Detection of Amp from Malformed Flowers Infected with Japanese Hydrangea Phyllody Phytoplasma. *Phytopathology* **2008**, *98*, 769–775.

Arocha Y.; Antesana, O.; Montellano, E.; Franco, P.; Plata, G.; Jones, P. 'Candidatus Phytoplasma lycopersici', a Phytoplasma Associated with 'hoja de perejil' Disease in Bolivia. *Int. J. Syst. Evol. Microbiol.*, **2007**, *57*, 1704–1710.

Arocha, Y.; Lopez, M.; Pinol, B.; Fernandez, M. B.; Picornell, R.; Almeida, I.; Palenzuela, M. Wilson, R.; Jones, P. 'Candidatus Phytoplasma graminis' and 'Candidatus Phytoplasma caricae', Two Novel Phytoplasmas Associated with Diseases of Sugarcane, Weeds and Papaya in Cuba. *Int. J. Syst. Evolution. Microbiol.* **2005**, *55*, 2451–2463.

Bai, X.; Zhang, J.; Ewing, A.; Miller, S. A.; Radek, A. J.; Shevchenko, D. V.; Tsukerman, K.; Walunas, T.; Lapidus, A.; Campbell, J. W.; Hogenhout, S. A. Living with Genome Instability: The Adaptation of Phytoplasmas to Di- verse Environments of Their Insect and Plant Hosts. *J. Bacteriol.* **2006**, *188*, 3682–3696.

Barbara, D. J.; Morton, A.; Clark, M. F.; Davies, D. L. Immunodominant Membrane Proteins from Two Phytoplasmas in the Aster Yellows Clade (Chlorante Aster Yellows and Clover Phyllody) Are Highly Divergent in the Major Hydrophilic Region. Microbiology, **2002**, *148*, 157–167.

Beanland, L.; Hoy, C. W.; Miller, S. A.; Nault, L. R. Influence of aster yellows phytoplasma on the fitness of aster leafhopper (Homoptera: Cicadellidae). Annals of the Entomological Society of America, **2000**, *93* (2), 271–276.

Berg, M.; Davies, D. L.; Clark, M. F.; Vetten, J.; Maier, G.; Seemueller, E. Isolation of a Gene Encoding an Immunodominant Memebrane Protein Gene in the Apple Proliferation Phytoplasma and Expression and Characterization of the Gene Product. *Microbiology* **1999**, *145*, 1937–1943.

Bertaccini, A. Phytoplasmas: Diversity, Taxonomy, and Epidemiology. *Front. Biosci.* **2007**, *12*, 673–689.

Bertaccini, A.; Arzone, A.; Alma, A.; Bosco, D.; Vibio, M. Detection of Mycoplasmalike Organisms in *Scaphoideus titanus* Ball Reared on Flavescence dorée Infected Grapevine by Dot Hybridizations Using DNA Probes. *Phytopathologia Mediterranea* **1993**, *32*, 20–24.

Bertaccini, A.; Bellardi, M. G.; Vibio, M. Virus Diseases of Ornamental Shrubs. X. Euphorbia pulcherrima Willd. Infected by Viruses and Phytoplasmas. *Phytopathologia Mediterranea* **1996**, *35*, 129–132.

Bertaccini, A.; Contaldo, N.; Calari, A.; Paltrinieri, S.; Windsor, H. M.; Windsor, D. (2010). Preliminary Results of Axenic Growth of Phytoplasmas from Micropropagated Infected Periwinkle Shoots. (18th IOM 2010) Chianciano Terme, Italy, July, 11–16: 147–153.

Bertaccini, A.; Davis, R. E.; Lee, I.-M.; Conti, M.; Dally, E. L.; Douglas, S. M. Detection of Chrysanthemum yellows mycoplasmalike organism (MLO) by dot-hybridization and Southern blot analysis. Plant Disease, **1990a**, *74*, 40–43.

Bertaccini, A.; Davis, R. E.; Lee, I.-M. Distinctions among mycoplasmalike organisms (MLOs) in Gladiolus, Ranunculus, Brassica and Hydrangea Through Detection with Nonradioactive Cloned DNA Probes. *Phytopathologia Mediterrane* **1990b**, *29*, 107–113.

Bertaccini, A.; Duduk, B.; Paltrinieri, S.; Contaldo, N. Phytoplasmas and Phytoplasma Diseases: a Severe Threat to Agriculture. *Am. J. Plant Sci.* **2014**.

Bertamini, M.; Nedunchezhian, N. Effect of Phytoplasma, Stolbur-subgroup (Bois noir-BN)] of Photosynthetic Pigments, Saccarides, Ribulose-1,5-bisphosphate Carboxylase, Nitrate and Nitrite Reductases and Photosynthetic Activities in Field-grow Grapevine (Vitis vinifera L. cv Chardonnay) Leaves. *Photosynthetica* **2001**, *39*, 119–122.

Bianco, P. A.; Marzachì, C.; Musetti, R.; Naor, V. Perspectives of Endophytes as Biocontrolagents in the Management of Phytoplasma Diseases. *Phytopathogenic Mollicutes* **2013**, *3* (1), 56–59.

Blomquist C. L.; Barbara, D. J.; Davies, D. L.; Clark, M. F.; Kirkpatrick, B. C. An Immunodominant Membrane Protein Gene from the Western X-disease Phytoplasma is Distinct from Those of Other Phytoplasmas. *Microbiology* **2001**, *147*, 571–580.

Bonnet, F.; Saillard, C.; Kollar, A.; Seemüller, E.; Bové, J. M. Detection and Differentiation of the Mycoplasmalike Organism Associated with Apple Proliferation Disease Using Cloned DNA Probes. *Mol. Plant Microbe Interact.* **1990**, *3*, 438–443.

Botti, S.; Bertaccini, A. FD-Related Phytoplasmas and Their Association with Epidemic and Non Epidemic Situations in Tuscany (Italy). 15th *Meeting of the International Council for the Study of Virus and Virus-Like Diseases of the Grapevine* (ICVG), Stellenbosch, 2006, 163–164.

Boudon-Padieu, E.; Larrue, J.; Caudwell, A. ELISA and Dot-blot Detection of Flavescence dorée-MLO in Individual Leafhopper Vectors During Latency and Inoculative State. *Curr. Microbiol.* **1989**, *19* (6), 357–364.

Bulgari, D.; Casati, P.; Faoro, F. Fluorescence In Situ Hybridization for Phytoplasma and Endophytic Bacteria Localization in Plant Tissues. *J Microbiol Meth.* **2011**, *87*, 220–223.

Cai, H.; Wei, W.; Davis, R. E.; Chen, H.; Zhao, Y. Genetic Diversity Among Phytoplasmas Infecting Opuntia Species: Virtual RFLP Analysis Identifies New Subgroups in the Peanut Witches'-Broom Phytoplasma Group. *Int. J. Syst. Evol. Microbiol.* **2008**, *58*, 1448–1457.

Calari, A.; Paltrinieri, S.; Contaldo, N.; Sakalieva, D.; Mori, N.; Duduk, B.; Bertaccini, A. Molecular Evidence of Phytoplasmas in Winter Oilseed Rape, Tomato and Corn Seedlings. *Bull. Insectol.* **2011**, *64*, S157–S158.

Carraro, L.; Ferrini, F.; Martini, M.; Ermacora, P.; Loi, N. A Serious Epidemic of Stolbur on Celery. *J. Plant Pathol.* **2008**, *90*, 131–135.

Castro, S.; Romero, J. The Association of Clover Proliferation Phytoplasma with Stolbur Disease of Pepper in Spain. *J. Phytopathol.* **2002**, *150*, 25–29.

Castro, S.; Romero, J. First Detection of a Phytoplasma Infecting Faba Bean (Vicia faba L.) in Spain. *Span. J. Agric. Res.* **2004**, *2*, 253–256.

Chapman, G. B.; Buerkle, E. J.; Barrows, E. M.; Davis, R. E.; Dally, E. L. A Light and Transmission Electron Microscope Study of a Black Locust Tree, *Robinia pseudoacacia* (Fabaceae), Affected by Witches Broom, and Classification of the Associated Phytoplasma. *J. Phytopathol.* **2001**, *149*, 589–597.

Chen, K. H.; Guo, J. R.; Wu, X. Y.; Loi, N.; Carraro, L.; Guo, Y. H.; Chen, T. A. Comparison of Monoclonal Antibodies, DNA Probes, and PCR for Detection of the Grapevine Yellows Disease Agent. *Phytopathology* **1993**, *83*, 915–922.

Chiykowski, L. N. Vector–pathogen–host Plant Relationships of Clover Phyllody Mycoplasma Like Organism and the Vector Leafhopper *Paraphlepsius irroratus*. *Can. J. Plant Pathol.* **1991**, *13*, 11–18.

Choueiri, E.; Jreijiri, F.; El-Zammar, S.; Verdin, E.; Salar, P.; Danet, J. L.; Bove, J.; Garnier, M. First Report of Grapevine "Bois Noir" Disease and a New Phytoplasma Infecting Solanaceous Plants in Lebanon. *Plant Dis.* **2002,** *86,* 697.

Choueiri, E.; Salar, P.; Jreijiri, F.; El Zammar, S.; Massaad, R.; Abdul-Nour, H.; Bové, J. M.; Danet, J. L.; Foissac, X. Occurrence and Distribution of '*Candidatus* Phytoplasma trifolii' Associated with Diseases of Solanaceous Crops in Lebanon. *Eur. J. Plant Pathol.* **2007,** *118,* 411–416.

Christensen, N. M.; Axelsen, K. B.; Nicolaisen, M.; Schultz, A. Phytoplasmas and Their Interactions with Their Hosts. *Trend Plant Sci.* **2005,** *10,* 526–535.

Christensen, N. M.; Nicolaisen, M.; Hansen, M.; Schultz, A. Distribution of Phytoplasmas in Infected Plants as Revealed by Real Time PCR and Bioimaging. *Mol. Plant–Microbe Interact.* **2004,** *17,* 1175–1184.

Chung, B.; Choi, G. Elimination of Aster Yellows Phytoplasma from *Dendranthema grandiflorum* by Application of Oxytetracycline as a Foliar Spray. *Pl. Pathol.* **2002,** *18,* 93–97.

Cordova, I.; Jones, P. Harrison, N. A.; Oropeza, C. In Situ Detection of Phytoplasma DNA in Embryos from Coconut Palms with Lethal Yellowing Disease. *Mol. Plant Pathol.* **2003,** *4,* 99–108.

Cousin, M. T.; Abadie, M. Action des mycoplasmes sur l'anthère. Etude en microscopies photonique et électronique. *Revue de Cytologie, de Biologie Végétale et de Botanique,* **1982,** *5,* 41–57.

Danielli, A.; Bertaccini, A.; Alma, A.; Bosco, D.; Vibio, M.; Arzone, A. May Evidence of 16SrI-group-related Phytoplasmas in Eggs, Nymphs and Adults of *Scaphoideus titanus* Ball Suggest Their Transovarial Transmission? *IOM Lett.* **1996,** *4,* 190–191.

Davis, R. E.; Lee, I.-M. Cluster-specific Polymerase Chain Reaction Amplification of 16S rDNA Sequences for Detection and Identification of Mycoplasmalike Organisms. *Phytopathology* **1993,** *83,* 1008–1011.

Davis, R. I.; Schneider, B.; Gibb, K. S. Detection and Differentiation of Phytoplasmas in Australia. *Aust. J. Agric. Res.* **1997,** *48,* 535–544.

Dehghan, H.; Salehi, M.; Khanchezar, A.; Afshar, H. Biological and Molecular Characterization of a Phytoplasma Associated with Greenhouse Cucumber Phyllody in Fars Province. *Iran. J. Plant Path.* **2014,** *50,* 185–186.

Del Serrone, P.; Marzachi, C.; Bragaloni, M.; Galeffi, P. Phytoplasma Infection of Tomato in Central Italy. *Phytopathologia Mediterranea* **2001,** *40,* 137–142.

Deng, S.; Hiruki, C. Amplification of 16S rRNA Genes from Culturable and Nonculturable Mollicutes. *J. Microbiol. Methods* **1991a,** *14,* 53–61.

Deng, S. J.; Hiruki, C. Genetic Relatedness Between Two Nonculturable Mycoplasmalike Organisms Revealed by Nucleic Acid Hybridization and Polymerase Chain Reaction. *Phytopathology* **1991b,** *81,* 1475–1479.

Dienes, L.; Ropes, M. W.; Madoff, W. E.; Smith, S.; Bauer, W. The Role of Pleuropneumonia-like Organisms in Genitourinary and Joint Diseases. *New Engl. J. Med.* **1948,** *238,* 509–515.

Doi, Y.; Teranaka, M.; Yora, K.; Asuyama, H. Mycoplasma or PLT Grouplike Microrganisms Found in the Phloem Elements of Plants Infected with Mulberry Dwarf, Potato Witches' Broom, Aster Yellows or Pawlonia Witches' Broom. *Annal. Phytopathol. Soc. Japan* **1967,** *33,* 259–266.

Eckstein, B.; Barbosa, J. C.; Kreyci, P. F.; Canale, M. C.; Brunelli, K. R.; Bedendo, I. P. Broccoli Stunt, a New Disease in Broccoli Plants Associated with Three Distinct Phytoplasma Groups in Brazil. Journal of Phytopathology, **2013**, *161*, 442–444.

Errampalli, D.; Fletcher, J.; Claypool, P. L. Incidence of Yellows in Carrot and Lettuce and Characterization of Mycoplasmalike Organism Isolates in Oklahoma. *Plant Dis.* **1991,** *75*, 579–584.

Gundersen, D. E.; Lee, I.-M.; Rehner, S. A.; Davis R. E.; Kingsbury, D. T. Phylogeny of Mycoplasmalike Organisms (Phytoplasmas): a Basis for Their Classification. *J. Bacteriol.* **1994,** *176*, 5244–5254.

Gundersen, D. E.; Lee, I.-M. Ultrasensitive Detection of Phytoplasmas by Nested-PCR Assays Using Two Universal Primer Pairs. *Phytopathologia Mediterranea* **1996,** *35*, 144–151.

Gundersen, D. E.; Lee, I-M.; Schaff, D. A.; Harrison, N. A.; Chang, C. J.; Davis, R. E.; Kinsbury, D. T. Genomic Diversity Among Phytoplasma Strains in 16S rRNA Group I (Aster Yellows and Related Phytoplasmas) and III (X-Disease and Related Phytoplasmas). *Int. J. Syst. Bacteriol.* **1996,** *46*, 64–75.

Harrison, N. A.; Bourne, C. M.; Cox, R. L.; Tsai, J. H.; Richardson, P. A. DNA Probes for Detection of Mycoplasma like Organisms Associated with Lethal Yellowing Disease of Palms in Florida. *Phytopathology* **1992,** *82*, 216–224.

Heinrich, M.; Botti, S.; Caprara, L.; Arthofer, W.; Strommer, S.; Hanzer, V.; Katinger, H.; Laimer da Câmara Machado, M.; Bertaccini, A. Improved Detection Methods for Fruit Tree Phytoplasmas. *Plant Mol. Biol. Report.* **2001,** *19*, 169–179.

Hong, Y.; Davies, D. L.; Wezel, R. V, Ellerker, B. E.; Morton, A.; Barbara, D. Expression of the Immunodominant Membrane Protein of Chlorantie-aster Yellows Phytoplasma in Nicotiana Benthamiana from a Potato Virus X-based Vector. *Acta Horticulturae*, **2001,** *550*, 409–415.

Hoshi, A.; Ishii, Y.; Kakizawa, S. Oshima, K.; Namba, S. Hostparasite Interaction of Phytoplasmas from a Molecular Biological Perspective. *Bull. Insectol.* **2007,** *60*, 105–107.

Hoshi, A.; Oshima, K.; Kakizawa, S.; Ishii, Y.; Ozeki, J.; Hashimoto, M.; Komatsu, K.; Kagiwada, S.; Yamaji, Y.; Namba, S. *A Unique Virulence Factor for Proliferation and Dwarfism in Plants Identified from a Phytopathogenic Bacterium.* Proceedings of the National Academy of Sciences of the United States of America. **2009,** *106*, 6416–6421.

Hren, M.; Ravnikar, M.; Brzin, J.; Ermacora, P.; Carraro, L.; Bianco, P. A.; Casati, P.; Borgo, M.; Angelini, E.; Rotter, A.; Gruden, K. Induced Expression of Sucrose Synthase and Alcohol Dehydrogenase I Genes in Phytoplasma-infected Grapevine Plants Grown in the Field. *Plant Pathol.* **2009,** *58*, 170–180.

IRPCM. Candidatus Phytoplasma, a Taxon for the Wall-less, Non-helical Prokaryotes that Colonise Plant Phloem and Insects. *Int. J. Syst. Evol. Microbiol.* **2004,** *54*, 1243–1255.

Ishii, Y.; Kakizawa, S.; Hoshi, A.; Maejima, K.; Kagiwada, S.; Yamaji, Y.; Oshima, K.; Namba, S. In the Non-Insect-Transmissible Line of Onion Yellows Phytoplasma (OY-NIM), the Plasmid-Encoded Transmembrane Protein ORF3 Lacks the Major Promoter Region. *Microbiology* **2009,** *155*, 2058–2067.

Jacobs, K. A.; Lee, I. M.; Griffiths, H. M.; Miller, F. D.; Bottner, K. D. A New Member of the Clover Proliferation Phytoplasma Group (16SrVI) Associated with Elm Yellows in Illinois. *Plant Dis.* **2003,** *87*, 241–246.

Jarausch, W.; Torres, E. Management of Phytoplasma-associated Diseases. In *Phytoplasmas and Phytoplasma Disease Management: How to Reduce Their Economic Impact.* Ed. Bertaccini, A.; International Phytoplasmologist Working Group, Italy, 2014; pp 199–208.

Jarausch, W.; Lansac, M.; Portanier, C.; Davies, D. L.; Decroocq, V. In Vitro Grafting: a New Tool to Transmit Pome Fruit Phytoplasmas to Non-natural Fruit Tree Hosts. *Adv. Hortic. Sci.* **2000**, *14*, 32.

Jomantiene, R.; Davis, R. E. Clusters of Diverse Genes Existing as Multiple, Sequence-Variable Mosaics in a Phytoplasma Genome. *FEMS Microbiol. Lett.* **2006**, *255*, 59–65.

Jomantiene, R.; Davis, R. E.; Lee, I. M.; Zhao, Y.; Bottner-Parker, K.; Valiunas, D.; Petkauskaite, R. Onion is Host for Two Phytoplasma Lineages, Subgroups 16SrI-A and 16SrI-(B/L) L, in Lithuania: a HinfI Site Revealed a Snp Marking Divergent Branches of Evolution. *J. Plant Pathol.* **2010**, *92*, 461–470.

Jomantiene, R.; Zhao, Y.; Davis, R. E. Sequence-Variable Mosaics: Composites of Recurrent Transposition Characterizing the Genomes of Phylogenetically Diverse Phytoplasmas. DNA and Cell Biology, **2007**, *26*, 557–564.

Kakizawa, S.; Oshima, K.; Ishii, Y.; Hoshi, A.; Maejima, K.; Jung, H. Y.; Yamaji, Y.; Namba, S. Cloning of Immunodominant Membrane Protein Genes of Phytoplasmas and Their in Planta Expression. *FEMS Microbiol. Lett.* **2009**, *293*, 91–101.

Kakizawa, S.; Oshima, K.; Kuboyama, T.; Nishigawa, H.; Jung, H. Y.; Sawayanagi, T.; Tsuchizaki, T.; Miyata, S.; Ugaki, M.; Namba, S. Cloning and Expression Analysis of Phytoplasma Protein Translocation Genes. Molecular Plant-Microbe Interactions, **2001**, *14*, 1043–1050.

Kakizawa, S.; Oshima, K.; Nishigawa, H.; Jung, H. Y.; Wei, W.; Suzuki, S.; Tanaka, M.; Miyata, S.; Ugaki, M.; Namba, S. Secretion Oimmunodominant Membrane Protein from Onion Yellows Phytoplasma Through the Sec Protein-Translocation System in *Escherichia coli*. *Microbiology* **2004**, *150*, 135–142.

Kamińska, M.; Berniak, H. Failure of Flower Bud Formation in Brassica Plants Associated with Phytoplasma Infection. *J. Agric. Sci.* **2012**, *4*, 219–226.

Kawakita, H.; Saiki, T.; Wei, W.; Mitsuhasi, W.; Watanabe, K.; Sato, M. Identification of Mulberry Dwarf Phytoplasmas in Genital Organs and Eggs of the Leafhopper *Hishimonoides sellatiformis*. *Phytopathology* **2000**, *90*, 909–914.

Khadhair, A. H.; Evans, I. R.; Choban, B. Identification of Aster Yellows Phytoplasma in Garlic and Green Onion by PCR-based Methods. *Microbiol. Res.* **2002**, *157*, 161–167.

Khan, A. J.; Botti, S.; Paltrinieri, S.; Al-Subhi, A. M.; Bertaccini, A. Phytoplasmas in Alfalfa Seedlings: Infected or Contaminated Seeds? 13th *Congress IOM*, Vienna, 7–12 July 2002, p6.

Kirkpatrick, B. C.; Stenger, B. C.; Morris, T. J.; Purcell, A. H. Cloning and Detection of DNA from a Nonculturable Plant Pathogenic Mycoplasma-like Organism. *Science* **1987**, *238*, 197–200.

Kuboyama, T.; Huang, C.; Lu, X.; Sawayanagi, T.; Kanazawa, T.; Kagami, T.; Matsuda, I.; Tsuchizaki, T.; Namba, S. A Plasmid Isolated from Phytopathogenic Onion Yellows Phytoplasma and Its Heterogeneity in the Pathogenic Phytoplasma Mutant. *Mol. Plant-Microbe Interact.* **1998**, *11*, 1031–1037.

Lava Kumar, P.; Sharma, K.; Boahen, S.; Tefera, H.; Tamo. First Report of Soybean Witches'-Broom Disease Caused by Group 16SrII Phytoplasma in Soybean in Malawi and Mozambique. *Plant Dis.* **2011**, *95*, 492.

Lee, I. M.; Davis, R. E.; Hiruki, C. Genetic Interrelatedness Among Clover Proliferation Mycoplasmalike Organisms (MLOs) and Other MLOs Investigated by Nucleic Acid Hybridization and Restriction Fragment Length Polymorphism Analyses. *Appl. Environ. Microbiol.* **1991**, *57*, 3565–3569.

Lee, I.-M.; Davis, R. E. Mycoplasmas Which Infects Plants and Insects. In *Mycoplasmas: Molecular Biology and Pathogenesis*; Eds: Maniloff, McElhaney, Finch and Baseman. American Society of Microbiology, Washington, 1992; 379–390.

Lee, I.-M.; Davis, R. E.; Gundersen-Rindal, D. E. Phytoplasma: Phytopathogenic Mollicutes. *Ann. Rev. Microbiol.* **2000,** *54,* 221–255.

Lee, I.-M.; Davis, R. E.; Chen, T-A.; Chiykowski, L. N.; Fletcher, J.; Hiruki, C.; Schaff, D. A. A Genotype-base System for Identification and Classification of Mycoplasmalike Organisms (MLOs) in the Aster Yellows MLO Strain Cluster. *Phytopathology* **1992,** *82,* 977–986.

Lee, I.-M.; Gundersen, D. E.; Hammond, R. W.; Davis, R. E. Use of Mycoplasmalike Organism (MLO) Group-specific Oligonucleotide Primers for Nested-PCR Assays to Detect Mixed-MLO Infections in a Single Host Plant. *Phytopathology 84,* **1994,** 559–566.

Lee, I.-M.; Gundersen-Rindal, D.; Davis, R. E.; Bartoszyk, I. M. Revised Classification Scheme of Phytoplasmas Based on RFLP Analysis of 16SrRNA and Ribosomal Protein Gene Sequences. *Int. J. Syst. Bacteriol.* **1998a,** *48,* 1153–1169.

Lee, I.-M.; Gundersen-Rindal, D. E.; Bertaccini, A. Phytoplasma: Ecology and Genomic Diversity. *Phytopathology* **1998b,** *88,* 1359–1366.

Lee, I.-M.; Gundersen-Rindal, D.; Davis, R. E.; Bottner, K. D.; Marcone, C.; Seemüller, E. Candidatus Phytoplasma Asteris, a Novel Taxon Associated with Aster Yellows and Related Diseases. *Int. J. Syst. Bacteriol.* **2004a,** *54,* 1037–1048.

Lee, I.-M.; Hammond, R. W.; Davis, R. E.; Gundersen, D. E. Universal Amplification and Analysis of Pathogen 16SrDNA for Classification and Identification of Mycoplasma Like Organisms. *Phytopathology* **1993,** *83,* 834–842.

Lee, I.-M.; Klopmeyer, M.; Bartoszyk, I. M.; Gundersen-Rindal, D. E.; Chou, T.; Thomson, K. L.; Eisenreich, R. Phytoplasma Induced Free-branching in Commercial Poinsettia Cultivars. *Nat. Biotechnol.* **1997,** *15,* 178–182.

Lee, I.-M.; Martini, M.; Marcone, C.; Zhu, S. F. Classification of Phytoplasma Strains in the Elm Yellows Group (16SrV) and Proposal of 'Candidatus Phytoplasma ulmi' for the Phytoplasma Associated with Elm Yellows. *Int. J. Syst. Evol. Microbiol.* **2004b,** *54,* 337–347.

Lee, I.-M.; Davis, R. E. Detection and Investigation of Genetic Relatedness Among Aster Yellows and Other Mycoplasma Like Organism by Using Cloned DNA and RNA Probes. *Plant-Microbe Interact.* **1988,** *1,* 303–310.

Lee, I-M.; Bottner, K. D.; Secor, G.; Rivera Varas, V. 'Candidatus Phytoplasma americanum' a Phytoplasma Associated with a Potato Purple Top Wilt Disease Complex. *Int. J. Syst. Evol. Microbiol.* **2006,** *56,* 1593–1597.

Lefol, C.; Lherminier, J.; Boudon-Padieu, E.; Larrue, J.; Louis, C.; Caudwell, A. (1994). Propagation of Flavescence dorée MLO (Mycoplasma-like Organism) in the Leafhopper Vector Euscelidius variegatus Kbm. *J. Invertebrate Pathol. 63,* 285–293.

Lherminier, J. E.; Prensier, G. E.; Boudon-Padieu, E. L.; Caudwell, A. N. Immunolabeling of grapevine Flavescence dorée MLO in Salivary Glands of Euscelidius Variegatus: a Light and Electron Microscopy Study. *J. Histochem. Cytochem.* **1990,** *38,* 79–85.

Liefting, L. W.; Kirkpatrick, B. C. Cosmid Cloning and Sample Sequencing of the Genome of the Uncultivable Mollicute, Western X-Disease Phytoplasma, Using DNA Purified by Pulsed-field Gel Electrophoresis. *FEMS Microbiol. Lett.* **2003,** *221,* 203–211.

Lin, C. P.; Chen, T. A. Monoclonal Antibodies Against the Aster Yellows Agent. *Science* **1985,** *227,* 1233–1235.

Lorenz, K. H.; Schneider, B.; Ahrens, U.; Seemüller, E. Detection of the Apple Proliferation and Pear Decline Phytoplasmas by PCR Amplification of Ribosomal and Nonribosomal DNA. *Phytopathology* **1995,** *85,* 771–776.

MacLean, A. M.; Sugio, A.; Makarova, O. V.; Findlay, K. C.; Grieve, V. M.; Toth, R.; Nicolaisen, M.; Hogenhout, S. A. Phytoplasma Effector SAP54 Induces Indeterminate Leaf-Like Flower Development in Arabidopsis Plants. *Plant Physiol.* **2011,** *157,* 831–841.

Marcon, C.; Lee, I.-M.; Davis, R. E.; Ragozzino, A.; Seemüller, E. Classification of Aster Yellows-group Phytoplasmas Based on Combined Analyses of rRNA and Tuf Gene Sequences. *Int.l J. Syst. Evol. Microbiol.* **2000,** *50,* 1703–1713.

Marcone, C.; Ragozzino, A. Comparative Ultrastructural Studies on Genetically Differrent Phytoplasmas Using Scanning Electron Microscopy. *Petria* **1996,** *6,* 125–136.

Martini, M.; Botti, S.; Marcone, C.; Marzachì, C.; Casati, P.; Bianco, P. A.; Benedetti, R.; Bertaccini, A. Genetic Variability Among Flavescence dorée phytoplasmas from Different Origins in Italy and France. *Mol. Cellular Probes* **2002,** *16,* 197–208.

Martini, M.; Lee, I.-M.; Bottner, K. D.; Zhao, Y. Botti, S. Bertaccini, A.; Harrison, N. A.; Carraro, L.; Marcone, C.; Khan, J.; Osler, R. Ribosomal Protein Gene-based Phylogeny for Finer Differentiation and Classification of Phytoplasmas. *Int. J. Syst. Evol. Microbiol.* **2007,** *57,* 2037–2051.

Mccoy, R. E.; Caudwell, A.; Chang, C. J.; Chen, T. A.; Chiykowski, L. N.; Cousin, M. T.; Dale, J. L.; De Leeuw, G. T. N.; Golino, D. A.; Hackett, K. J.; Kirkpatrick, B. C.; Marwitz, R.; Petzold, H.; Sinha, R. C.; Sugiura, M.; Whitcomb, R. F.; Yang I. L.; Zhu, B. M.; Seemüller, E. Plant Diseases Associated with Mycoplasma-like Organisms. In *The Mycoplasmas,* Whitcomb, R., F., Tully, J. G., Eds.; Academic Press: San Diego, USA, 1989; Vol. 5, pp 545–640.

Mergenthaler, E.; Viczian, O.; Fodor, M.; Sule, S. Isolation and Expression of an Immunodominant Membrane Protein Gene of the ESFY Phytoplasma for Antiserum Production. *Acta Horticulturae* **2001,** *550,* 355–360.

Mitra, D. K. (1993). Little Leaf, a Serious Disease of Eggplant (*Solanum melongena*). In *Management of Plant Diseases Caused by Fastidious Prokaryotes*; Raychaudhuri, S. P., Teakle, D.S., Eds.; Associated Publishing Co.: New Delhi, India, pp 73–78.

Mollov, D.; Lockhart, B.; Saalau-Rojas, E.; Rosen, C. First Report of a 16SrI (Aster Yellows) Group Phytoplasma on Garlic (*Allium sativum*) in the United States. *Plant Dis.* **2014,** *98,* 419.

Montano, H. G.; Brioso, P. S. T.; Cunha, J. O.; Figueiredo, D. V.; Pimentel, J. P. First Report of Group 16SrIII Phytoplasma in Loofah (*Luffa cylindrica*). *Bull. Insectol. 60,* **2007c,** 277–278.

Montano, H. G.; Brioso, P. S. T.; Cunha, JUNIOR J. O.; Figueiredo, D. V.; Pimentel J. P. First Report of Group 16SrIII phytoplasma in loofah (Luffa cylindrica). *Bull. Insectol.* **2007a,** *60,* 277–278.

Montano, H. G.; Brioso, P. S. T.; Pereira, R. C.; Pimentel, J. P. *Sicana odorifera* (Cucurbitaceae) a New Phytoplasma Host. *Bull. Insectol.* **2007b,** *60,* 287–288.

Montano, H. G.; Brioso, P. S. T.; Pimentel, J. P.; Figueiredo, D. V.; Cunha, J. O. *Cucurbita moschata,* New Phytoplasma Host in Brazil. *J. Plant Pathol.* **2006,** *88,* 226.

Montano, H. G.; Davis, R. E.; Dally, E. L.; Pimentel, J. P.; Brioso, P. S. T. Identification and Phylogenetic Analysis of a New Phytoplasma from Diseased Chayote in Brazil. *Plant Dis.* **2000,** *84,* 429–436.

Morton, A.; Davies, D. L.; Blomquist, C. L.; Barbara, D. J. Characterization of Homologues of the Apple Proliferation Immunodominant Membrane Protein Gene from Three Related Phytoplasmas. *Mol. Plant Pathol.* **2003**, *4*, 109–114.

Moya-Raygoza, G.; Nault, L. R. Transmission Biology of Maize Bushy Stunt Phytoplasma by the Corn Leafhopper (Homoptera: Cicadellidae). *Ann. Entomol. Soc. Am.* **1998**, *91*, 668–676.

Murral, D. J.; Nault, L. R.; Hoy, C. W.; Madden, L. V.; Miller, S. A. Effects of Temperature and Vector Age on Transmission of Two Ohio Strains of Aster Yellows Phytoplasma by the Aster Leafhopper (Homoptera: Cicadellidae). *J. Econ. Entomol.* **1996**, *89*, 1223–1232.

Musetti, R.; Ermacora, P.; Martini, M.; Loi, N.; Osler, R. What Can We Learn from the Phenomenon of "Recovery"? *Phytopathogen. Mollicut.* **2013**, *3* (1), 63–65.

Musetti, R.; Favali, M. A.; Pressacco, L. Histopathology and Polyphenol Content in Plants Infected by Phytoplasmas. *Cytobios* **2000**, *102*, 133–147.

Namba, S.; Kato, S.; Iwanami, S.; Oyaizu, H.; Shiozawa, H.; Tsuchizaki, T. Detection and Differentiation of Plantpathogenic Mycoplasmalike Organisms Using Polymerase Chain Reaction. *Phytopathology* **1993**, *83*, 786–791.

Navratil, M.; Valova, P.; Fialova, R.; Lauterer, P.; afarova, D. S. Star, M. The Incidence of Stolbur Disease and Associated Yield Losses in Vegetable Crops in South Moravia (Czech Republic). *Crop Protect.* **2009**, *28* (10), 898–904.

Nishigawa, H.; Miyata, S.; Oshima, K.; Sawayanagi, T.; Komoto, A.; Kuboyama, A.; Matsuda, I.; Tsuchizaki, T.; Namba, S. Planta Expression of a Protein Encoded by the Extrachromosomal DNA of a Phytoplasma and Related to Geminivirus Replication Proteins. *Microbiology* **2001**, *147*, 507–513.

Nishigawa, H.; Oshima, K.; Kakizawa, S.; Jung, H.; Kuboyama, T.; Miyata, S.; Ugaki, M.; Namba, S. Evidence of Intermolecular Recombination Between Extrachromosomal DNAs in Phytoplasma: A Trigger for the Biological Diversity of Phytoplasma. *Microbiology* **2002**, *148*, 1389–1396.

Ochoa-Sánchez, J. C.; Parra-Cota, F. I.; Aviña-Padilla, K.; Délano-Frier, J.; Martínez-Soriano, J. P. *Amaranthus* spp.: a New Host of *Candidatus* Phytoplasma aurantifolia. Phytoparasitica, **2009**, *37*, 381–384.

Omar, A. F.; Foissac, X. Occurrence and Incidence of Phytoplasmas of the 16SrII-D Subgroup on Solanaceous and Cucurbit Crops in Egypt. *Eur. J. Plant Pathol.* **2012**, *133*, 353–360.

Oshima, K.; Ishii, Y.; Kakizawa, S.; Sugawara, K.; Neriya, Y.; Himeno, M.; Minato, N.; Miura, C.; Shiraishi, T.; Ya- maji, Y.; Namba, S. Dramatic Transcriptional Changes in an Intracellular Parasite Enable Host Switching Between Plant and Insect. *PLoS ONE* **2011**, *6*, Article ID: e23242.

Oshima, K.; Kakizawa, S.; Arashida, R.; Ishii, Y.; Hoshi, A.; Hayashi, Y.; Kawagida, S.; Namba, S. Presence of Two Glycolityc Gene Clusters in a Severe Pathogenic Line of 'Candidatus Phytoplasma asteris'. *Mol. Plant Pathol.* **2007**, *8*, 481–489.

Oshima, K.; Kakizawa, S.; Nishigawa, H.; Jung, H. Y.; Wei, W.; Suzuki, S.; Arashida, R.; Nakata, D.; Miyata, S.; Ugaki, M.; Namba, S. Reductive Evolution Suggested from the Complete Genome Sequence of a Plant-Pathogenic Phytoplasma. *Nature Genetics* **2004**, *36*, 27–29.

Oshima, K.; Kakizawa, S.; Nishigawa, H.; Kuboyama, T.; Miyata, S.; Ugaki, M.; Namba, S. A Plasmid of Phytoplasma Encodes a Unique Replication Protein Having both Plasmid and Virus-Like Domains: Clue to Viral Ancestry or Result of Virus/Plasmid Recombination. *Virology* **2001**, *285*, 270–277.

Palermo, S.; Elekes, M.; Botti, S.; Ember, I.; Alma, A.; Orosz, A.; Bertaccini, A.; Kölber, M. Presence of Stolbur Phytoplasma in Cixiidae in Hungarian Vineyards. *VITIS-J. Grapevine Res.* **2004**, *43*, 201–203.

Pastore, M.; Piccirillo, P.; Simeone, A. M.; Tian, J.; Paltrinieri, S.; Bertaccini, A. Transmission by Patch Grafting of ESFY Phytoplasma to Apricot (*Prunus armeniaca* L.) and Japanese Plum (*Prunus salicina* Lindl). *Acta Hortic.* **2001**, *550*, 339–344.

Poghosyan, A. V.; Lebsky, V. K.; Arce-Montoya, M.; Landa, L. Possible Phytoplasma Disease in Papaya (*Carica papaya* L.) from Baja California Sur: Diagnosis by Scanning Electron Microscopy. *J. Phytopathol.* **2004**, *152*, 376–380.

Pracros, P.; Renaudin, J.; Eveillard, S.; Mouras, A.; Hernould, M. Tomato Flower Abnormalities Induced by Stolbur Phytoplasma Infection Are Associated with Changes of Expression of Floral Development Genes. *Mol. Plant-Microbe Int.* **2006**, *19*, 62–68.

Rajan J.; Clark, M. F. Detection of Apple Proliferation and Other MLOs by Immunocapture PCR (IC-PCR). *Acta Hortic.* **1995**, *386*, 511–514.

Rao, G.P.; Madhupriya, Kumar, M.; Tomar, S.; Maya, B.; Singh, S. K.; Johnson, J. M. Detection and Identification of Four 16Sr Subgroups of Phytoplasmas Associated with Different Legume Crops in India. *Eur. J. Plant Pathol.* **2017**. DOI:10.1007/s10658-017-1278-6

Raychaudhuri, S. P.; Varma, A.; Chenulu, V. V.; Prakash, N. Singh, S. (1970). Association of Mycoplasma-like Bodies with Little Leaf of *Solanum melongena* L. *Xth Int. Cong. Microbiol. Mexico.* HIV-6 (Abstr).

Rekab, D.; Carraro, L.; Schneider, B.; Seemuller, E.; Chen, J.; Chang, C. J.; Locci, R.; Firrao, G. Geminivirus-Related Extrachromosomal DNAs of the X-Clade Phytoplasmas Share High Sequence Similarity. *Microbiology* **1999**, *145*, 1453–1459.

Salehi E.; Salehi M.; Taghavi S. M.; Izadpanah K. A 16SrII-D phytoplasma Strain Associated with Tomato Witches' Broom in Bushehr Province, Iran. *J. Crop Protect.* **2014**, *3*, 377–388.

Salehi, M.; Izadpanah, K.; Siampour, M. Characterization of a Phytoplasma Associated with Cabbage Yellows in Iran. *Plant Dis.* **2007**, *91*, 625–630.

Salehi, M.; Siampour, M.; Esmailzadeh Hosseini, S. A.; Bertaccini, A. Characterization and Vector Identification of Phytoplasmas Associated with Cucumber and Squash Phyllody in Iran. *Bull. Insectol.* **2015**, *68*, 311–319.

Saranya, S. S.; Umamaheswaran, K. Symptomatology, Transmission and Molecular Detection of Phytoplasma Infecting Brinjal (*Solanum melongena* L.). *Int. J. Appl. Pure Sci. Agric.* **2015**, *1*, 35–40.

Schneider, B.; Gibb, K. S.; Seeümller, E. Sequence and RFLP Analysis of the Elongation Factor Tu Gene Used in Differentiation and Classification of Phytoplasmas. *Microbiology* **1997**, *143*, 3381–3389.

Scott, S. W.; Zimmerman, M. T. Peach Rosette, Little Peach, and Red Suture Are Diseases Induced by a Phytoplasma Closely Related to Western X-disease. *Acta Hortic.* **2001**, *550*, 351–354.

Seemüller, E.; Schneider, B. *Candidatus* Phytoplasma mali, *Candidatus* Phytoplasma pyri and *Candidatus* Phytoplasma prunorum, the Causal Agents of Apple Proliferation, Pear Decline and European Stone Fruit Yellows, Respectively. *Int. J. Syst. Evol. Microbiol.* **2004**, *54*, 1217–1226.

Seemüller, E.; Marcone, C.; Lauer, U.; Ragozzino, A.; Göschl, M. Current Status of Molecular Classification of the Phytoplasmas. *J. Plant Pathol.* **1998**, *80*, 3–26.

Shaw, M. E.; Kirkpatrick, B. C.; Golino, D. A. (1993). The Beet Leafhopper-transmitted Virescence Agent Causes Tomato Big Bud Disease in California. *Plant Dis. 77*, 290–295.

Sichani, F. V.; Bahar, M.; Zirak, L. Characterization of Phytoplasmas Related to Aster Yellows Group Infecting Annual Plants in Iran, Based on the Studies of 16S rRNA and Rp Genes. *J. Plant Protect. Res.* **2014**, *54*, 1–8.

Siddique; Agrawal; Alam; Krishna Reddy. Electron Microscopy and Molecular Characterization of Phytoplasmas Associated with Little Leaf Disease of Brinjal (*Solanum melongena L.*) and Periwinkle (*Catharanthus roseus*) in Bangladesh. *J. Phytopathol.* **2001**, *149*, 237–244.

Singh, D.; Singh, S. J. Chilli Little Leaf-A New Phytoplasma Disease in India. *Indian Phytopathol.* **2000**, *53*, 309–310.

Singh, S. K.; Aminuddin, P.; Srivastava, B. R.; Singh, J.; Khan, A. Production of Phytoplasma-free Plants from Yellow Leaf Diseased *Catharanthus roseus* L. (G.) Don. Gewinnung phytoplasmenfreier Pflanzen aus Blattvergilbungs erkrankten *Catharanthus roseus* L. (G.) Don. *J. Plant Dis. Protect.* **2007**, *114*, 2–5.

Sinha, R. C.; Benhamou, N. Detection of Mycoplasmalike Organism Antigens from Aster Yellows-diseased Plants by Two Serological Procedures. *Phytopathology* **1983**, *73*, 1199–1202.

Smart, C. D.; Schneider, B.; Blomquist, C. L.; Guerra, L. J.; Harrison, N. A.; Ahrens, U.; Lorenz, K. H.; Seemüller, E.; Kirkpatrick, B. C. Phytoplasma-specific PCR Primers Based on Sequences of 16S rRNA Spacer Region. *Appl. Environ. Microbiol.* **1996**, *62*, 2988–3033.

Streten, C.; Conde, B.; Herrington, M.; Moulder, J.; Gibb, K. *Candidatus* Phytoplasma Australiense is Associated with Pumpkin Yellow Leaf Curl Disease in Queensland, Western Australia and the Northern Territory. *Aust. Plant Pathol.* **2005**, *34*, 103–105.

Sugio, A.; Kingdom, H. N.; MacLean, A. M.; Grieve, V. M.; Hogenhout, S. A. *Phytoplasma Protein Effector SAP11 Enhances Insect Vector Reproduction by Manipulating Plant Development and Defense Hormone Biosynthesis*. Proceedings of the National Academy of Sciences of the United States of America, 2011, 108, E1254–E1263.

Suzuki, S.; Oshima, K.; Kakizawa, S.; Arashida, R.; Jung, H. Y.; Yamaji, Y.; Nishigawa, H.; Ugaki, M.; Namba, S. *Interaction Between the Membrane Protein of a Pathogen and Insect Microfilament Complex Determines Insect-Vector Specificity*. Proceedings of the National Academy of Sciences of the United States of America, 2006, 103, 4252–4257.

Tedeschi, R.; Alma, A. Transmission of Apple Proliferation Phytoplasma by *Cacopsylla melanoneura* (Homoptera: Psyllidae). *J. Econ. Entomol.* **2004**, *97*, 8–13.

Tedeschi, R.; Ferrato, V.; Rossi, J.; Alma, A. Possible Phytoplasma Transovarial Transmission in the Psyllids *Cacopsylla melanoneura* and *Cacopsylla pruni*. *Plant Pathol.* **2006**, *55*, 18–24.

Toruno, T. Y.; Seruga Musić, M.; Simi, S.; Nicolaisen, M.; Hogenhout, S. A. Phytoplasma PMU1 Exists as Linear Chromosomal and Circular Extrachromosomal Elements and Has Enhanced Expression in Insect Vectors Compared with Plant Hosts. *Mol. Microbiol.* **2010**, *77*, 1406–1415.

Villalobos, W.; Moreira, L.; Rivera, C.; Bottner, K. D.; Lee, I.-M. First Report of an Aster Yellows Subgroup 16SrIB Phytoplasma Infecting Chayote in Costa Rica. *Plant Dis.* **2002**, *86*, 330.

Wei, W.; Davis R. E.; Lee, I.-M.; Zhao, Y. Computer-simulated RFLP Analysis of 16S rRNA Genes: Identification of Ten New Phytoplasma Groups. *Int. J. Syst. Evol. Microbiol.* **2007**, *57*, 1855–1867.

Wei, W.; Lee, I.-M.; Davis R. E.; Suo, X.; Zhao, Y. Automated RFLP Pattern Comparison and Similarity Coefficient Calculation for Rapid Delineation of New and Distinct Phytoplasma 16SrDNA Subgroup Lineages. *Int. J. Syst. Evol. Microbiol.* **2008a,** *58,* 2368–2377.

Wei, W.; Davis, R. E.; Jomantiene, R.; Zhao, Y. Ancient, Recurrent Phage Attacks and Recombination Shaped Dynamic Sequence-Variable Mosaics at the Root of Phytoplasma Genome Evolution. *Proc. Natl. Acad. Sci. U.S.A.* **2008b,** *105,* 11827–11832.

Wei, W.; Kakizawa, S.; Jung, H. Y.; Suzuki, S.; Tanaka, M.; Nishigawa, H.; Miyata, S.; Oshima, K.; Ugaki, M.; Hibi, T.; Namba, S. An Antibody Against the SecA Membrane Protein of one Phytoplasma Reacts with Those of Phylogenetically Different Phytoplasmas. *Phytopathology* **2004,** *94* (7), 683–686.

Weintraub P. G.; Beanland, L. A. Insect Vectors of Phytoplasmas. *Ann. Rev. Entomol.* **2006,** *51,* 91–111.

William, V. M.; Lisela, M. C.; Carmen, R. H.; Bottner, K. D.; Lee, I. M. First Report of an Aster Yellows Subgroup 16SrI-B Phytoplasma Infecting Chayote in Costa Rica. *Plant Dis.* **2002,** *86,* 330.

Win, N. K. Y, Young-Hwan, K.; Hee-Young, J. Bitter Gourd Little Leaf Disease Associated to Candidatus *Phytoplasma asteris. Trop. Plant Pathol.* **2014,** *39,* 82–88.

Xu, X.; Mou, H.; Zhu, S. F.; Liao, X. L.; Zhao, W. J. Detection and Characterization of Phytoplasma Associated with Big Bud Disease of Tomato in China. *J. Phytopathol.* **2013,** *161,* 430–433.

CHAPTER 5

Ralstonia solanacearum: Pathogen Biology, Host–Pathogen Interaction, and Management of Tomato Wilt Disease

BHUPENDRA S. KHARAYAT[1*] and YOGENDRA SINGH[2]

[1]Division of Plant Pathology, ICAR-Indian Agricultural Research Institute, Pusa, New Delhi 110012, India

[2]Department of Plant Pathology, G. B. Pant University of Agriculture and Technology, Pantnagar 263145, Udham Singh Nagar, Uttarakhand, India

*Corresponding author. E-mail: bhupendrakharayat@gmail.com

ABSTARCT

Ralstonia solanacearum causes wilt of tomato and huge economic loses worldwide. This disease causes very heavy loss varying from 2% to 90% in different climates and seasons in India. *R. solanacearum* is a soil-borne bacterium that enters in plants through roots, invades xylem vessels, and spreads rapidly to aerial parts of the plants through the vascular system where its high level of multiplication obstructs the xylem vessel and leads to wilting symptoms, ultimately, plant death. It grows well at 28–32°C strictly in an aerobic condition. Race 3 biovar 2 is most severe between 24 and 35°C in plants and however decreases in aggressiveness when temperatures exceed 35°C or fall below 16°C. *R. solanacearum* strains present an extensive genetic diversity and are divided into four phylotypes corresponding to the strains' geographic origin: Asia (phylotype I), America (II), Africa (III), and Indonesia (IV). Several quantitative virulence factors contribute to wilt disease development. The virulence factors for pathogenicity of *R. solanacearum* include production of extracellular polysaccharide (EPS), a consortium of plant cell wall-degrading enzymes, twitching and swimming motility, taxis, and several dozen effectors secreted by a type 3 secretion system. *hrp* genes, encoding type III secretion machinery, have been shown to be the key determinants for pathogenicity in the vascular phytopathogenic bacterium—*R. solanacearum*.

5.1   INTRODUCTION

Bacterial wilt of tomato (*Lycopersicon esculentum* Mill.) is caused by *Ralstonia solanacearum* (Smith, 1986; Yabuuchi et al., 1996) formerly called *Pseudomonas solanacearum* (Smith, 1986) and later renamed as *Burkholdeira solanacearum* (Smith, 1986; Yabuuchi et al., 1993) and finally as *Ralstonia solanacearum* (Smith, 1986; Yabuuchi et al., 1996). It belongs to the β-proteobacteria and considered as a "species complex." *R. solanacearum* is highly heterogeneous and complex bacterium. Different pathogenic races within the species, though, may show more restricted host range. *R. solanacearum* is a widespread phytopathogenic bacterium that causes a wilt disease, which deadly affects many agriculturally important crops and ornamentals plants. It has attacked almost 200 plant species in 33 different plant families and the disease results in very heavy economic losses (Chandrashekara and Prasannakumar, 2010). This is also considered as one of the largest known host ranges for any plant pathogenic bacterium which also offers an opportunity for the analysis of an array of virulence factors involved in the pathogenicity of this devastating bacterium (Genin and Boucher, 2004; Elphinstone, 2005; Denny, 2006; Mansfield et al., 2012).

   R. solanacearum is a soilborne bacterium that enters in plants through roots, invades xylem vessels, and spreads rapidly to aerial parts of the plants through the vascular system where its high level of multiplication obstructs and leads to wilting symptoms and, ultimately, death of plants (Genin, 2010). Invasion and highly multiplication of *R. solanacearum* in the xylem vessels exhibits a strong and tissue-specific tropism within the host (Salanoubat et al., 2002). In addition to its devastating nature, the ability of *R. solanacearum* to survive in soils for many years and to form latent infections within indigenous weeds is unique and it is difficult to eradicate and manage the bacterium (Hayward, 1991; Wenneker et al., 1999). Tomato crops can be infected by highly diverse race 1 lowland tropical strains of *R. solanacearum*, which are distributed in all four phylotypes. The race 3 highland temperate strains belonging to phylotype II (race 3-phylotype II), while being primarily adapted to potato (brown rot disease), are also pathogenic to tomato in natural environments. Because of the high susceptibility of tomato germplasm to race 3, the outbreaks reported in Europe (Carmeille et al., 2006, Prior et al.,

1998) and the recent fortuitous introduction of race 3 strains in the United States (Kim et al., 2003).

R. solanacearum strains present an extensive genetic diversity and are divided in four phylotypes corresponding roughly to the strains' geographic origin: Asia (phylotype I), the Americas (II), Africa (III), and Indonesia (IV). Phylotype II has two subclusters: IIA and IIB (Fegan and Prior, 2005) and only strains belonging to phylotype IIB are responsible for bacterial wilt of potato in cold and temperate regions (Janse et al., 2004). Several quantitative virulence factors contribute to bacterial wilt disease development. These virulence factors include production of extracellular polysaccharide (EPS), a consortium of plant cell wall-degrading enzymes (CWDE), twitching and swimming motility, taxis, and several dozen effectors secreted by a type 3 secretion system (TTSS; Denny, 2006). *R. solanacearum* has been studied intensively, both biochemically and genetically, and has long been recognized as a model system for the analysis of pathogenicity. It is well adapted to life in soil in the absence of host plants, thereby providing a good system to investigate functions governing adaptation to such an ecological niche (Salanoubat et al., 2002).

5.2 SIGN AND SYMPTOMS OF *RALSTONIA*

R. solanacearum race 3 biovar 2 causes brown rot (or bacterial wilt) of potato, Southern wilt of geranium, and bacterial wilt of tomato (Champoiseau et al., 2009). In case of bacterial wilt of tomato and potato caused by *R. solanacearum*, there is a sudden wilting (Fig. 5.1) of plants without showing any yellowing symptoms (Singh and Kharayat, 2013). At the early stages of the disease, the first visible symptoms of bacterial wilt are usually seen on the foliage of plants (Champoiseau et al., 2009). These symptoms consist of drooping of the youngest leaves at the ends of the branches during the hottest part of the day at this stage, only one or half a leaflet may wilt, and plants may appear to recover at night, after drop in the temperature. Slimy and sticky ooze forms tan-white to brownish beads where the vascular tissue is cut. When an infected stem is cut across and the cut ends held together for a few seconds, a thin thread of ooze can be seen as the cut ends are slowly separated. If one of the cut ends is suspended in clear container clean water, bacterial ooze will form a thread in the water.

FIGURE 5.1 Diagnostic wilt symptoms of *R. solanacearum* on tomato. (A) wilting of plants: drooping of leaves, (B) sudden and permanent wilt occurs, but without yellowing and necrosis, (C) brown discoloration of the vascular system, and (D) classical ooze test: ooze flow from a cut of infected stem.

Source: Bacterial Wilt management in Tomato, AVRDC.

5.3   HOST RANGE

Bacterial wilt is one of the major diseases of tomato and other solanaceous crop plants. The disease is known to occur in the wet tropics, sub-tropics, and some temperate regions of the world (Champoiseau et al., 2008). The bacterial wilt disease has been described, and the causal agent isolated, in more than 200 plant species belonging to 53 different botanical families (Elphinstone, 2005). According to host range, *R. solanacearum* strains have been classified into five races (Hayward, 1991). According to Elphinstone

(2005), the most frequently reported hosts for each of them have been identified, for example, race 1, which has the highest number of host species, solanaceous crops like potato, tomato and tobacco, chili and sweet pepper, eggplant; non-solanaceous crops like groundnut, bean and sunflower; ornamental plants like *Anthurium* spp., *Verbena* spp., *Zinnia* spp., *Dahlia* spp., *Hibiscus* spp., *Lilium* spp., marigold, *Heliconia* spp., palms, *Lesianthus* spp., *Pothos* spp., and *Strelitzia* spp.; trees like *Eucalyptus* and neem and fruit trees as black sapote and custard apple.

5.4 GEOGRAPHICAL DISTRIBUTIONS

The wilt disease caused by *R. solanacearum* has a worldwide distribution and strains of it have conventionally been classified as races and biovars. Bacterial wilt of tomato is caused by either race 1 or race 3 of *R. solanacearum* and, rarely, by race 2. Race 1 is present in the five continents, including Europe, with the exception of the member states of European Union (CABI, 2006). For race 2, plantain, cooking and dessert bananas, other *Musa* spp. and wild and ornamental *Heliconia* spp. This race occurs mainly in tropical areas of South America and also in the Philippines (EPPO/CABI, 2006). For race 3, tomato eggplant, potato, geranium, and *Capsicum* spp.; weeds like *S. nigrum* and *Solanum dulcamara*. The presence of race 3 in several European Union countries seems to be related to latent infections in imported potatoes from regions where the disease is endemic (Janse et al., 1998). For race 4, ginger and the related plant species patumma and mioga; race 4 occurs in Asia. For race 5, *Morus* spp; it is limited to China (Elphinstone, 2005). In India, *R. solanacearum* race 1 biovar 3 is dominated mostly in coastal and hilly and foothill areas including the states of Goa, Karnataka, Kerala, Maharashtra, Orissa, Jharkhand, West Bengal, and the states of Northeastern hills, like Himachal Pradesh, Jammu and Kashmir, and Uttarakhand (Devi and Menon, 1980).

5.5 ECONOMIC IMPORTANCE

Major economic losses caused by bacterial wilt epidemics have been produced in crop fields of infected tomato, potato, tobacco, ginger, banana, and groundnut plants (Elphinstone, 2005). In general, crop and yield losses depend on local climatic conditions, cropping practices, soil types, plant cultivar, choice of crop, and the virulent characteristics of the

R. solanacearum local strains (Elphinstone, 2005). Race 3 biovar 2 is an important crop pathogen in both developing and industrialized countries. It is endemic in most tomato-growing areas of Nigeria, causing 60–100% yield losses (Popoola, 2015). In Uganda, a survey in 17 districts showed that 88% of the tomato farms were affected by bacterial wilt (Katafiire et al., 2005). This disease causes very heavy loss varying from 2% to 90% in different climates and seasons in India (Singh et al., 2013; Mishra et al., 1995). Therefore, the level of damage is commonly expressed on a crop-by-crop basis and can range from minimal crop loss to very high economic damage. In India, a yield loss study with one cultivar of tomato showed 10–100% mortality of plants and 0–91% yield loss (Elphinstone, 2005). Integrated disease management strategies for *R. solanacearum* are complex because the bacterium is able to infect crops as a soil-borne, water-borne, or seed/tuber-borne organism.

5.6 DISEASE CYCLE AND EPIDEMIOLOGY

R. solanacearum primarily infects host plants through their roots, entering through wounds caused by cultivation, natural wounds during emergence of lateral roots, insect chewing or feeding damage, and nematode feeding (e.g., *Meloidogyne* spp. which causes the root-knot disease). The bacterium can also enter plants by way of stem injuries caused by insects, handling, and tools. Once the bacteria infect the roots or stems, they colonize the plant through the xylem in the vascular bundles. Race 3 biovar 2 is most severe on plants between 24°C and 35°C and decreases in aggressiveness when temperature exceeds 35°C or falls below 16°C. Active disease at temperature below 16°C is rare (Ciampi and Sequeira, 1980; Hayward, 1991).

Plant-to-plant infection can occur when bacteria shed from infected roots move to roots of nearby healthy plants. The pathogen can spread from infected to healthy fields by soil transfer on machinery and surface runoff water after irrigation or rainfall. It also can be disseminated from infested ponds or rivers to healthy fields by flooding or irrigation. *R. solanacearum* is primarily a soil-borne and water-borne pathogen. So far, no aerial spread of the pathogen has been reported. The bacterium also has an "exterior" phase (epiphyte) in which it can reside on the outer part of the plant. It is of minor importance in epidemiology of the pathogen since bacteria do not survive epiphytically for long periods of time when exposed to hot conditions or when relative humidity is below 95%.

Under favorable conditions, tomato plants infected with *R. solanacearum* may not show any disease symptoms. In this case, latently infected plants can play a major role in the spread of the bacterium (Champoiseau et al., 2009). Soil is the primary source of the disease. The bacterium can survive in soil for extended periods without a host plant. The bacterium can survive for days to years in infested water, wet soils or deep soil layers (> 75 cm), forming a reservoir of inoculum from which it can disperse. High temperatures (30–35°C) and high soil moisture favor disease development. Bacterial populations decrease rapidly at low temperatures (< 4°C), but race 3 biovar 2 can still survive, in a physiologically latent, viable, but non-culturable state (Champoiseau et al., 2009). In natural habitats, *R. solanacerum* race 3 biovar 2 can survive the winter in semi-aquatic weeds, in plant debris or in the rhizosphere of non-host plants that act as reservoirs for the pathogen. High soil moisture increases the survival of the pathogen, its rate of infection and development, and its spread through the soil.

Bacterial wilt is a greater problem in heavy soils and in low-lying areas that can retain soil moisture for long periods. When the diseased plant is removed from the field, the infected root pieces that remain in the soil provide bacteria for infection of new tomato roots. The dispersal of bacteria is by furrow irrigation or surface water, wounding, transplanting, cultivation, and pruning. Long-distance dispersal of the bacterium is through infested soil transported with seedlings or with farm implements or infected seedlings (Cerkauskas, 2004).

The bacterium can survive in diseased crop debris. The bacteria are released from the roots of the infected plant into the soil and can infect neighboring host plants and weeds. Colonization of weeds affects the degree of carryover of inoculum of *R. solanacearum*. Diverse biological (antagonistic microorganisms) and environmental (temperature, relative humidity, and soil moisture) factors can affect survival of *R. solanacearum* in soil and aquatic habitats (Champoiseau et al., 2009).

5.7 DETECTION, IDENTIFICATION, AND DIVERSITY ANALYSIS

R. solanacearum is a Gram-negative, rod-shaped bacterium measuring 0.5–0.7 × 1.5–2.0 μm in size. It grows well in the temperature range of 28–32°C and strictly in aerobic conditions (Schaad et al., 2001). Individual colonies of normal or virulent isolates are usually visible after 36–48 h, appearing as opaque white or cream-colored colonies that are irregular in shape and highly fluidal on a culture medium such as casamino acid-peptone-glucose-agar

(CPG). On tetrazolium chloride (TZC) medium, these colonies appear white with pink center (Fig. 5.2). Mutant or non-virulent type colonies of *R. solanacearum* are uniformly round and dark red, smaller in size and butyrous or dry on TZC. Strains of *R. solanacearum* have previously been grouped into five races based on susceptible host plants and biovar classification, which is determined by utilization of a set of five to eight carbohydrate substrates (Denny and Hayward, 2001).

R. solanacearum Race 3 biovar 2 can be identified following recovery from either symptomatic or asymptomatic plants and from water or soil samples by means of several microbiological and molecular methods (Priou et al., 2006). A combination of at least two different complementary tests is needed to unambiguously identify the species and biovar. Historically, the group was subdivided into "races" based loosely on host range (Buddenhagen and Kelman, 1964). The biovars of *R. solanacearum* are better defined than its races and are based on acidification of medium during metabolism of six carbohydrates (maltose, lactose, cellobiose, mannitol, sorbitol, and dulcitol; Hayward, 1994).

An entirely new phylogenetic classification system was proposed recently, consisting of four phylotypes, each further divided into sequevars (Fegan and Prior, 2005; Villa et al., 2005). Screening tests can facilitate early detection of *R. solanacearum* in plants or contaminated soil and water samples, but they cannot be used to identify the race or biovar. These screening tests include bacterial streaming, plating on a semi-selective medium such as modified SMSA, immunodiagnostic assays using species-specific antibodies, PCR with species-specific primers, and pathogenicity tests using susceptible hosts such as tomato seedlings (Weller et al., 2000; Schaad et al., 2001). Commercially available immunostrips can be used for rapid detection of *R. solanacearum* in the field or lab. Phylogenetic analyses based on different molecular approaches including RFLP sequence analysis of the 16S–23S rRNA intergenic spacer region (ITS), polygalacturonase and endoglucanase genes, and PCR-RFLP of the hrp genes region, revealed an extensive diversity and this group of organisms is now commonly called the *R. solanacearum* species complex (RSSC; Fegan and Prior, 2005; Genin and Denny, 2012; Wicker et al., 2012).

R. solanacearum is considered a species complex-a heterogeneous group of related but genetically distinct strains (Fegan and Prior, 2005). These strains are highly competent for genetic exchange in planta and show substantial pathogenic variability in host range and aggressiveness (Bertolla et al., 1999). This species complex includes strains with narrow and broad host ranges with different geographic origins. Recently, Fegan and Prior (2005) proposed a new hierarchical classification scheme, based on sequence analysis of the endoglucanase (*egl*) gene, *hrpB* gene and internal transcribed spacer (ITS) region, that subdivides *R. solanacearum* into phylotypes. Phylotypes are defined as "a monophyletic cluster of strains revealed by phylogenetic analysis of sequenced data" (Fegan and Prior, 2005, 2006). Twenty-three sequevars and four phylotypes were identified within the species that broadly reflects the ancestral relationships and geographical origins of the strains (Champoiseau et al., 2009). Phylotyping is based on DNA sequence analysis (Fig. 5.3), which divides the *R. solanacearum* species complex into four phylotypes (Table 5.1). Phylotypes can further be subdivided into sequevars based on the sequence of the endoglucanase (*egl*) gene (Fegan and Prior, 2005a,b). The endoglucanase gene encodes for the production of a cellulase (endoglucanase), a secreted virulence factor of *R. solanacearum* (Saile et al., 1997).

According to Champoiseau et al. (2009), race determination is not possible, because *R. solanacearum* strains do not have race-cultivar specificity on plant hosts and, with the exception of race 3 biovar 2, the old "races" do not have phylogenetic unity. Unequivocal identification must rely on at least two distinct methods, including the biovar test and one of the DNA-based tests that uses PCR to amplify a race 3 biovar 2-specific DNA fragment. *Ralstonia solanacearum* race 3 biovar 2 strains belong to phylotype II and sequevars 1 and 2 (Champoiseau et al., 2009). Phylotypes are not related with host preference as strains from all phylotypes are able to cause disease on potato, tomato, pepper, and eggplant (Cellier and Prior, 2010; Lebeau et al., 2011).

TABLE 5.1 Phylotype Group of *R. solanacearum* Strains and Their Geographical Origin.

Phylotype group	Geographical origin
Phylotype I strains	Asia
Phylotype II strains	America
Phylotype III strains	Africa
Phylotype IV strains	Indonesia

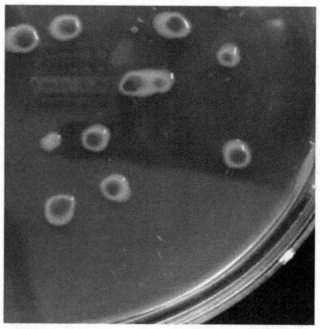

FIGURE 5.2 Typical growth of *R. solanacearum* on tetrazolium chloride medium: colonies are white with pink center.

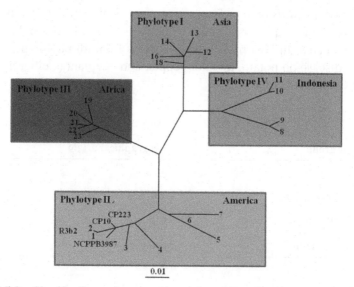

FIGURE 5.3 Classification and geographic origins of *R. solanacearum* strains based on sequence analysis of endoglucanase gene sequences. Numbers, 1–23, indicate sequevars.

Source: Fegan and Prior (2005)

5.8 GENE SEQUENCES AND CANDIDATE GENES RESPONSIBLE FOR PATHOGENESIS

After sequence analysis, the *R. solanacearum* genome assembled into two circular molecules: a large replicon of 3,716,413 bp and a smaller 2,094,509 base pair replicon, yielding a total genome size of 5,810,922 base pair (Salanoubat et al., 2002). A series of new genes putatively involved in pathogenicity were identified, apart from the virulence genes already described in *R. solanacearum*. These include genes coding for additional hydrolytic enzymes involved in the degradation of plant cell walls, and genes required for the production of the plant hormones auxin and ethylene or for the degradation of the plant signaling molecules salicylic acid-a mediator of the plant systemic acquired resistance (SAR) and ethylene (Salanoubat et al., 2002). Salanoubat et al. (2002) have described known and candidate genes responsible for pathogenesis in the genome of *R. solonacearum* strain GMI1000. They have found that the known pathogenicity genes in *R. solonacearum* strain GMI1000 genome are Type III secretion system and secreted effectors (31), global regulatory functions (11), exopolysaccharide biosynthesis (18), hydrolytic enzymes (4), hormone production (1) and candidate genes are Type III secretion-dependent effectors (51), Type III effectors based on homology (25), Type III effectors based on structural features (26), adhesion/surface proteins (93), hemagglutinin-related proteins (27), Type IV fimbrial biogenesis proteins (35), other pilli/fimbriae (24), hydrolytic enzymes/host cell wall degradation (5), toxins (13), resistance to oxidative stress (10), plant hormones and signaling molecules (7), and others (16).

5.9 HOST–PATHOGEN INTERACTION

In the presence of host plant cells, a regulatory cascade activates secretion of CWDE and the cluster of hypersensitive reaction (HR) and pathogenicity genes encoding components of a TTSS. Once bacterium enters into the plant tissues, high densities of the pathogen increase expression of virulence genes, and production of exopolysaccharide (EPS), the main pathogenicity determinant of *Ralstonia*. These genes are controlled by a density-dependent regulatory network taking part in a quorum sensing system. After destroying the plant, the bacterium can persist in the different environmental conditions through diverse survival forms until contact with a new host (Vasse et al., 1995).

It has been shown that bacterial motility, including both flagellum-driven swimming and, pilus-driven twitching is necessary for full virulence (Kang et al., 2002; Tans-Kersten et al., 2001). All of these virulence factors are controlled by an elaborate regulatory system that allows the bacterium to adjust the life in different ecological niches. The core of this regulatory network is the Phc (phenotype conversion) system (Schell, 2000), which specifically responds to 3-hydroxypalmitic acid methyl ester which is a quorum-sensing molecule (Clough et al., 1997). To sense specific environmental stimuli and move toward favorable conditions many bacteria use a form of motility called taxis which is directly involved in several plant–bacterium interactions (de Weert et al., 2002; Greer-Phillips et al., 2004). Virulence of *R. solanacearum* is complex and additive. Many factors that contribute to bacterial wilt disease have been identified; these include EPS, some type III-secreted effectors and several plant CWDE (Gonzalez and Allen, 2003; Liu et al., 2005).

Based on combinatory mutations in the CWDE, 15 different strains were produced and their virulence in tomato plants has been tested. Although the GMI-6 strain which was defective in all CWDE exhibited reduced virulence. For *R. solanacearum* virulence in tomato, cellulose rather than pectin degradation seemed to be more critical. More recently, it has been demonstrated that *R. solanacearum* endo- and exo-PG are inhibited by a tomato polygalacturonase-inhibiting protein (PGIP) activity which may contribute to plant resistance to the pathogen (Schacht et al., 2011).

According to Aziz et al. (2007), during plant–pathogen interactions, pectin degraded products such as (1–4)-a-linked oligogalacturonides (OG) elicit various plant defense responses with highly methylesterified pectins as the source of the most active host elicitor (Osorio et al., 2008). *R. solanacearum* secretes several extracellular plant CWDE via the type II secretion system (Poueymiro and Genin, 2009). It possesses several pectin-degrading enzymes including a pectin methylesterase (PME) which cleaves methoxyl groups from methylated pectin, thereby making them susceptible to the three polygalacturonases: PehA/PglA, PehB, and PehC (Tans-Kersten et al., 1998). Besides this suite of enzymes, bioinformatic searches indicated that *R. solanacearum* is not particularly rich in wall-degrading enzymes compared with other phytopathogenic bacteria (da Silva, 2002). It is possible that cell wall-derived pectic fragments of root cortical cells may induce programmed cell death (PCD) in the entire cortex, even at a distance from bacterial exposure.

5.9.1   INFECTION AND COLONIZATION OF TOMATO PLANTS

This soil and water-borne bacterium enters the plant through roots, multiplies in the xylem, collapses the host, and returns to the environment. It has effective pathogenicity determinants to invade and colonize host plants but, also exhibits successful strategies for survival in harsh conditions. The bacterium invades plant roots through wounds or natural openings such as lateral root emergence points. Once the bacteria enter a susceptible host, they colonize the intercellular spaces of the root cortex and vascular parenchyma, eventually entering the xylem vessels and spreading into the upper parts of the plant (Vasse et al., 1995).

The success of a given pathogen also depends on its capacity to degrade plant cell walls as it progresses within the host. Many pathogens possess enzymatic arsenals allowing the degradation of pectin including pectin lyases (PL), pectin methylesterases (PME), pectate lyases (PEL), and polygalacturonases (PG) and their enzyme efficiency is in some cases necessary for full virulence (Digonnet et al., 2012). *R. solanacearum* is also known to secrete several CWDE that are cellulolytic (Egl, CbhA, RSc0818, RSc1081) or pectinolytic (PglA/PehA, PehB, PehC, Pme; Huang and Allen, 1997; Gonzalez and Allen, 2003).

R. solanacearum GMI1000 has a large repertoire of 74 type III effectors; most of them remain functionally uncharacterized (Poueymiro and Genin, 2009). Some of these proteins including GALA7, an effector containing an F-box and a Leucine-rich repeat domain as well as AvrA, an HR (hypersensitive response) elicitor in tobacco, are required for host cell invasion (Turner et al., 2009). The infectious process of *R. solanacearum* has been described at the light microscopy level in tomato (Wydra and Beri, 2007). Upon infection, bacteria must penetrate the root surface.

In tomato seedlings, *R. solanacearum* requires wound sites or natural openings at the point of secondary root emergence or partially exfoliated cells of the outer parenchyma (Saile et al., 1997). Bacteria penetrate the root of tomato at the apex with no preferential point of entry, as well as via sites of secondary root emergence (Vasse et al., 2000). *R. solanacearum* moves toward the vascular cylinder via intercellular spaces of tomato plants (Vailleau et al., 2007). Digonnet et al. (2012) investigated the compatible interaction between the model plant—*Arabidopsis thaliana*, and the GMI1000 strain—*Ralstonia solanacearum*, in an in vitro pathosystem. At six days post-inoculation (dpi), bacteria tended to have a highly restricted localization with bacteria present in a few epidermal cells infected with a slightly broader colonization in cortical and endodermal cells (three to four cells). They described that the bacteria progresses through the root in a highly directed, centripetal manner to the xylem poles, without extensive multiplication in the intercellular spaces along its path.

Entry into the vascular cylinder was facilitated by cell collapse of the two pericycle cells located at the xylem poles. Once the bacteria reached the xylem vessels, they multiply profusely and move from vessel to vessel by digesting the pit membrane between adjacent vessels. Digonnet et al. (2012) also described that the degradation of the secondary walls of xylem vessels was not a prerequisite for vessel colonization xylem cell walls, even at very late stages in disease development.

5.9.2   *hrpB* AND *hrpG* GENES ARE REQUIRED DURING ROOT INFECTION

hrp genes, encoding type III secretion machinery, have been shown to be the key determinants for pathogenicity of *R. solanacearum* GMI1000 (Vasse et al., 2000). In *R. solanacearum,* the *hrp* gene cluster comprises more than 20 genes organized in at least seven transcriptional units (Arlat et al., 1992; van Gijsegem et al., 1995). The *hrpG* and the *hrpB* mutant strains are nonpathogenic toward tomato plants, while a *prhJ* mutant strain is hyperaggressive on this host (Genin et al., 1992; Brito et al., 1999). The recent identification of two additional transcriptional activators, PrhJ and HrpG, with distinct regulatory roles, and whose corresponding mutants display different pathogenic phenotypes on plants, prompted us to study in more detail the different roles played by PrhJ, HrpG, and HrpB during the interaction between *R. solanacearum* and tomato plants (Vasse et al., 2000). Vasse et al. (2000), in their experiment, took the phenotypes of *R. solanacearum* mutant strains which were disrupted in the *prhJ, hrpG,* or *hrpB* regulatory genes with respect to vascular colonization and root infection in tomato plants. An *hrpB* mutant shows reduced infection, colonization, and multiplication ability in planta, and induces a defense reaction similar to a vascular hypersensitive response at one protoxylem pole of invaded plants.

Vasse et al. (2000) showed that the phenotype of a *prhJ* mutant resembles that of the wild-type strain. In contrast, the *hrpG* mutant exhibited a wild-type level of infection at secondary root axils, but the ability of the infecting bacteria to penetrate into the vascular cylinder was significantly reduced. This indicates that bacterial multiplication at root infection sites and transit through the endodermis constitute critical stages in the infection process, in which *hrpB* and *hrpG* genes are involved. Moreover, these results suggested that the *hrpG* gene might control, in addition to *hrp* genes, other functions required for vascular colonization (Vasse et al., 2000).

5.9.3 ACTIVATION OF METABOLIC PATHWAYS DURING PATHOGENESIS

R. solanacearum typically enters hosts through root wounds and colonizes the xylem (water-transporting tissue), through which it spreads up into the plant stem. *R. solanacearum* infection reduces water transport, which leads to wilting and death of plant. Although xylem is considered a nutrient-limiting, low-oxygen environment. The bacterial wilt pathogen *R. solanacearum* is well adapted to plant xylem fluid, in spite of it is considered a nutrient-poor environment, and grows to 10^8 to 10^9 CFU/g in tomato stem (Jacobs et al., 2012). Several primary metabolic pathways highly express during pathogenesis. These pathways include sucrose uptake and catabolism, and components of these pathways are encoded by genes in the *scrABY* cluster. Significantly, a UW551 *scrA* mutant reduced in virulence on resistant and susceptible tomato.

According to Jacobs et al. (2012), genes encoding proteins involved in metabolism of sucrose were expressed by both strains in planta. Functional analysis of this conserved metabolic trait revealed that *R. solanacearum* requires sucrose for full virulence and to succeed in tomato xylem. Functional *scrA* contributes to pathogen competitive fitness during colonization of tomato xylem, which contained ~300 µM sucrose (Jacobs et al., 2012). Sucrose-induced *scrA* expression, but to a much greater degree by growth in planta. Unexpectedly, 45% of the genes directly regulated by HrpB, the transcriptional activator of the TTSS, were upregulated in planta at high cell densities. The active transcription of these genes in wilting plants suggests that TTSS has a biological role throughout the disease cycle.

5.10 PATHOGENICITY DETERMINANTS OF *R. solanacearum*

The virulence of *R. solanacearum* is quantitative and complex, with many contributing factors, such as a suite of type III-secreted effectors, bacterial EPS, several plant CWDE, motility (twitching and swimming), energy taxis and several dozen effectors secreted by type III secretion system (González, 2003; Yao and Allen, 2006; Denny, 2006). *R. solanacearum* has not only effective pathogenicity determinants to invade and colonize the host plants but also has successful strategies for survival in unfavorable environmental conditions. Thus, in the presence of host plant cells, a regulatory cascade activates pathogenicity genes, secretion of CWDE and the cluster of HR encoding components of a type III secretion system.

5.10.1   EXTRACELLULAR POLYSACCHARIDES

EPS play an important role in pathogenesis of many bacteria including Ralstonia, by both direct intervention with host cells functions and by providing resistance to oxidative stress. In the bacterial wilt of Solanaceous crops caused by *R. solanacearum*, EPS1 is the main virulence factor for the disease development. EPS1 is a polymer composed of a trimeric repeat unit consisting of N-acetyl galactosamine, deoxy-L-galacturonic acid, and trideoxy-D-glucose. At least 12 genes are involved in EPS1 biosynthesis. The bacterium produces EPS1 in massive amounts and makes up more than 90% of the total polysaccharide (Agrios, 2005). Once in the plant tissues, high densities of the pathogen increase expression of virulence genes and production of exopolysaccharide, the main pathogenicity determinant. These genes are controlled by a density-dependent regulatory network (Fig. 5.4) taking

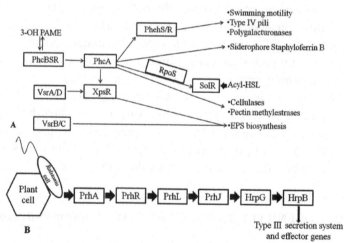

FIGURE 5.4   Simplified model of the network regulating virulence factors and motility in *R. solanacearum*. (A) The PhcA-regulated pathways. (B) The Hrp Type III secretion system regulatory pathway induced in response to the bacterium–plant cell contact. Lines with solid arrowheads or bars represent positive or negative control of gene expression, respectively. Lines with open arrowheads represent synthesis or sensing of the autoinducer molecule 3-OH PAME. PhcS/R, PehS/R, VsrA/D, and VsrB/C correspond to two-component regulator modules. Acyl-HSL: *N*-octanoyl and *N*-hexanoyl-homoserine lactone synthesized by *sol*I.

Source: (Fegan and Prior, 2005)

part in a quorum sensing system (Chandrashekara, 2012). At high cell densities (> 5×10^8 CFU/mL) in planta and in culture, accumulates 3-OH-PAME which triggers PhcA mediated activation of EPS biosynthesis (McGarvey et al., 1999). When the bacterium grows in culture, PhcA indirectly represses expression of *hrpB*, which encodes the transcriptional activator of the TTSS (Yoshimochi et al., 2009). It has been assumed from this in vitro result that the TTSS is active only in the early stages of host plant infection and is not needed at higher pathogen cell densities, after the development of wilt symptoms (Genin et al., 2005; Mole et al., 2007; Yoshimochi et al., 2009).

5.10.2   CELL-WALL-DEGRADING ENZYMES

R. solanacearum produces several plant CWDE, secreted via the type II secretion system (Huang and Allen, 1997; Tans-Kersten et al., 1998). These include one β-1,4-cellobiohydrolase (CbhA) and some pectinases whose activities have been identified respectively as one β-1,4-endoglucanase (Egl), two exopolygalacturonases (PehB and PehC), one pectin methyl esterase (Pme), and one endopolygalacturonase (PehA; Huang and Allen, 1997; Tans-Kersten et al., 1998; Liu et al., 2005). *R. solanacearum* Egl is a 43-kDa protein that has proved to be involved in pathogenicity. Inactivation of *pehA* or *pehB, egl*, genes revealed that each contributes to *R. solanacearum* virulence. A deficient mutant lacking the six enzymes wilted host plants more slowly than the wild-type (Liu et al., 2005).

5.10.3   THE TYPE THREE SECRETION SYSTEM

Pathogenicity of *R. solanacearum* relies on the Type III secretion system that injects Type III effector (T3E) proteins into plant cells. T3E collectively perturb host cell processes and functioning and modulates plant immunity to enable bacterial infection. As in several other plant pathogenic bacteria, effectors delivered to host cells by the Type III secretion system are essential to *R. solanacearum* pathogenicity (Deslandes and Genin, 2014). Among the virulence determinants of *R. solanacearum*, the TTSS, a molecular syringe whose structural and regulatory elements are encoded by *hrp* (hypersensitive response and pathogenicity) genes, is essential for pathogenicity (Vasse et al., 2000). Effector repertoires are diversified, with a core of 20–30 effectors present in most of the strains and the obtention of mutants lacking one or more effector genes revealed the functional overlap among this effector network (Deslandes and Genin, 2014). TTSS injects effectors proteins into plant cells to favor the bacterial infection by subverting and exploiting the host signaling pathways (Poueymiro and Genin, 2009). Effectors could promote nutrient leakage but mostly they are predicted to manipulate plant defenses (Goel et al., 2008; Deslandes and Rivas, 2012). The establishment of exhaustive effector repertoires in multiple *R. solanacearum* strains drew a first picture of the evolutionary dynamics of the pathogen effector suites.

Recently the transcriptome analysis of *R. solanacearum* cells collected from tomato xylem fluids revealed notable discrepancies in effector gene expression in planta: the half of them (32 of 71) had a twofold induction compared to expression in complete medium and interestingly this group comprises a majority of core effectors (Jacobs et al., 2012). Additional levels of control exist beyond gene expression and a study recently highlighted the key role of a chaperone-like protein in the TTSS-export process and the control of *R. solanacearum* effector translocation (Lohou et al., 2014). *R. solanacearum* virulence factors are regulated by a complex interlocking cascade involving quorum sensing and undefined plant signals (Mole et al., 2007; Schell, 2000). Two major virulence factors, EPS and TTSS, are controlled by a global regulator, PhcA, via the quorum-sensing molecule 3-OH-palmitic acid methyl ester (3-OH-PAME; Flavier et al., 1997; Genin et al., 2005).

5.10.4 MOTILITY

Most soil-borne plant-associated bacteria, including *R. solanacearum*, have swimming motility. Swimming motility allows *R. solanacearum* to efficiently invade and colonize host plants. However, the bacteria are essentially nonmotile once inside plant xylem vessels. Swimming motility makes an important quantitative contribution to bacterial wilt virulence in the early stages of host invasion and colonization. In soil-inoculated tomato plants, significant reduction in virulence of nonmotile flagellin (fliC) mutants of *R. solanacearum* has been observed (Tans-Kersten et al., 2004). Furthermore, taxis behavior is likely to play a role in the fitness of *R. solanacearum* during those parts of its life cycle when it is living in soil or water rather than inside a plant host.

Twitching motility is a flagella-independent form of bacterial translocation over moist surfaces. It occurs by the tethering, extension, and then retraction of polar type IV pili, which operate in a manner similar to a grappling hook (Mattick, 2002). Both swimming motility mediated by flagella and twitching motility mediated by type IV pili are required for full virulence in *R. solanacearum* (Liu et al., 2001; Tans-Kersten et al., 2001). However, injecting an aflagellate *fliC* (flagellin) mutant directly into host stems can rescue its wild-type virulence phenotype, suggesting that swimming motility makes its contribution to virulence early in disease development, perhaps enabling the pathogen to move toward the host rhizosphere, attach to host roots, and invade root tissue (Tans-Kersten et al., 2001). Motility also is an important virulence factor in several animal–host systems where it mediates adherence, invasion, or colonization (Giron et al., 2002; Ottemann and Lowenthal, 2002; Tomich et al., 2002). Moreover, biofilm formation mediated by motility or the flagella themselves is an integral part of certain infection processes.

5.10.5 TAXIS: AS A VIRULENCE FACTOR

5.10.5.1 AEROTAXIS

R. solanacearum needs taxis to locate and infect host plant roots. It is an ability of a bacterium to move toward more favorable conditions. Aerotaxis or energy taxis guides bacteria toward optimal intracellular energy levels. The genome of *R. solanacearum* encodes two putative aerotaxis transducers. Cloned *aer1* and *aer2* genes of *R. solanacearum* restored aerotaxis to an *E. coli aer* mutant, demonstrating that both genes encode heterologously functional aerotaxis transducers. Site-directed mutants lacking *aer1*, *aer2*, or both *aer1* and *aer2* were significantly less able to move up an oxygen gradient than the wild-type parent strain. In fact, the aerotaxis of the *aer* mutants was indistinguishable from that of a completely nonmotile strain. Slightly delayed wilt disease development n observed when tomato plants were inoculated with either the *aer2* or the *aer1/aer2* mutant. Furthermore, the *aer1/aer2* double mutant was significantly impaired in the ability to rapidly localize on tomato roots compared to its wild-type parent. Unexpectedly, all nonaerotactic mutants formed thicker biofilms on abiotic surfaces than the wild type. These results indicated that the ability of *R. solanacearum* to locate and effectively interact with its host plants, significantly contributed by energy taxis (Yao and Allen, 2007).

5.10.5.2   CHEMOTAXIS

To sense specific chemicals or environmental conditions and move toward attractants and away from repellants, many bacteria use a complex behavior called chemotaxis (Greer-Phillips et al., 2004; Terry et al., 2005). Taxis behavior is likely to play a role in the fitness of *R. solanacearum* during those parts of its life cycle when it is living in soil or water rather than inside a plant host (Yao and Allen, 2006). Qualitative and quantitative chemotaxis assays revealed that this bacterium is specifically attracted to diverse amino acids and organic acids, and especially to root exudates from the host plant tomato (Yao and Allen, 2006). Site-directed mutants of *R. solanacearum* lacking either CheA or CheW, which are core chemotaxis signal transduction proteins, were completely nonchemotactic but retained normal swimming motility (Yao and Allen, 2006). Both nonchemotactic mutants had significantly reduced virulence indistinguishable from that of a nonmotile mutant in biologically realistic soil soak virulence assays on tomato plants, demonstrating that directed motility, not simply random motion, is required for full virulence (Yao and Allen, 2006).

5.11   PROPHAGES AND GENOME EVOLUTION OF RALSTONIA

R. solanacearum has the capability of transferring blocks of genes among phylotypes, which is associated with the rapid adaptation to different environments and many hosts (Guidot et al., 2009). Recently, various temperate phages of *Myovirus* and *Inovirus* families were found infecting to *R. solanacearum* (Yamada et al., 2007; Murugaiyan et al., 2011). The infection of phage Φ RSS1 alters the pathogenic behavior of *Ralstonia*, increasing its virulence in host plants. The *R. solanacearum* GMI1000 presented three recognizable prophage regions, *Rasop1*, *Rasop2*, and ΦRSX, and at least seven phage remnants. The region *Rasop1* and *Rasop2* had no similarity with sequenced *Ralstonia* phages, whereas ΦRSX was inserted into a tRNA-Ser and was found closely related to temperate phage ΦRSA1 (Fujiwara et al., 2008). Addy et al. (2011) showed that this increase in virulence is related to an increased twitching motility, EPS, and earlier wilting of plants when compared with wild-type bacteria.

5.12 ROOT BORDER CELLS AS A DEFENSE SYSTEM

Root border cells separate from plant root tips and disperse into the soil environment. Each root tip in most plant species can produce thousands of metabolically active cells daily, with specialized patterns of gene expression (Hawes et al., 2016). Recent studies of Hawes et al. (2016) suggested that border cells operate in a manner similar to mammalian neutrophils: both cell types export a complex of extracellular DNA (exDNA) and antimicrobial proteins that neutralize threats by trapping pathogens and thereby preventing invasion of host tissues. Virulence and systemic spread of the microbes are promoted by extracellular DNases (exDNases) of pathogens. In plants, border cell extracellular traps and root tip resistance to infection eliminated by adding DNase I to root tips.

Mutation of genes encoding exDNase activity results in reduced virulence in plant-pathogenic bacteria (*Ralstonia solanacearum*) and fungi (*Cochliobolus heterostrophus*). The genome of *Ralstonia solanacearum*, encodes two putative secreted DNases. Two putative extracellular nucleases are expressed during pathogenesis of *R. solanacearum*, suggesting the hypothesis that the ability to degrade and escape from border cell traps contributes to virulence tested this hypothesis with *R. solanacearum* deletion mutants lacking one or both of these nucleases, named NucA and NucB (Hawes et al., 2016). Exposing tomato border cells to *R. solanacearum* triggered release of DNA-containing extracellular traps. This response occurred in a flagellin-dependent manner, suggesting that trap release called NETosis in animal systems (Brinkmann et al., 2013) could be an element of PAMP-triggered immunity (PTI; Pisetsky, 2012). These traps rapidly immobilized the pathogen, but the bacteria could be released by treatment with purified DNase. Taken together, these results indicate that extracellular DNase is a novel virulence factor that helps *R. solanacearum* successfully to infect plant roots and leads to wilt disease development (Hawes et al., 2016).

5.13 MANAGEMENT OPTIONS FOR WILT DISEASE

R. solanacearum has a wide host range, which makes it difficult to have a generalized estimate of the economic losses caused by the bacterial wilt disease. Direct yield losses vary widely according to host, cultivar, climate, soil type, cropping practices, and pathogen strain. There are no such chemicals available for effective management of this disease. Moreover, it causes soil and water pollution due to pesticide residues, which affects human health as well as development of resistant mutant by pathogen against pesticides. Hence, non-chemical methods including cultural methods, resistant cultivars and biocontrol with antagonistic bacterial agents have been made successfully to manage bacterial diseases of plants (Almoneafy et al., 2012). However, resistant cultivar is not completely effective due to lack of stability or durability (Boucher et al., 1992). In the absence of efficient strategies to eradicate *R. solanacearum* from infected soils and water, the use of resistant cultivars appears to be the best disease control strategy.

5.13.1.1   HOST RESISTANCE

Natural host resistance to disease is the preferred method for the efficient protection of crops since soil bacteria are notoriously difficult to eliminate in an eco-friendly way (Wang et al., 2000). Resistance to *R. solanacearum* has been described in different crop plants as polygenic and dependent upon environmental conditions (Hayward, 1991). Resistant varieties may tolerate the limited growth of the bacterial pathogen within their vascular systems (Grimault et al., 1995). To combat bacterial wilt, resistant cultivars have been recommended in areas of the world where the pathogen is endemic.

5.13.1.2   DEPLOYMENT OF IMMUNE SYSTEM RECEPTORS

Recent research program follows two main strategies to deploy the two classes of immune receptors. One is to transfer pattern recognition receptors (PRRs) that detect common microbial products into species that lack them. For example, the PRR EF-Tu receptor (EFR) of *Arabidopsis* recognizes the bacterial translation elongation factor EF-Tu (Dangl et al., 2013). Deployment of EFR into either *Nicotiana benthamiana* or *Solanum lycopersicum*, which cannot recognize EF-Tu, conferred resistance to a wide range of bacterial pathogens (Lacombe et al., 2010). Dangl et al. (2013) has been shown that the expression of EFR in tomato was especially effective against *R. solanacearum*

5.13.1.3   DEPLOYMENT OF EXECUTOR-MEDIATED DISEASE RESISTANCE

In contrast to PRRs and NLRs, another class of plant disease resistance genes has evolved to co-opt pathogen virulence functions and open a "trap door" that stops pathogen proliferation (Dangl et al., 2013). *Xanthomonas* and *Ralstonia* transcription activator-like (TAL) effectors are DNA-binding proteins delivered into plant cells, where they activate host gene expression to enhance pathogen virulence (Schornack et al ., 2013).

5.13.1.4 BREEDING FOR WILT DISEASE RESISTANCE

Resistance breeding to *R. solanacearum* in solanaceous crops appears to be regional or linked to climatic conditions and this limited success is due to all the constraints resistant cultivars must outsmart.

First, the breeding must combine durable resistance with desirable agronomic traits. Second, resistant cultivars must be able to face the diversity of agro-ecological zones where the bacteria proliferates and the high genetic variability of *R. solanacearum* strains. Third, breeding for highly resistant cultivars must be prioritized to avoid further *R. solanacearum* dissemination due to tolerant plants that shelter virulent bacteria without showing disease symptoms. In tomato, the polygenic resistance to bacterial wilt in the resistant cultivar Hawaii 7996 (Grimault et al., 1995) suggested to be strain specific (Wang et al., 2000) and, more recently, it was hypothesized that quantitative trait loci (QTLs) in Hawaii 7996 may deploy a phylotype-specific resistance (Carmeille et al., 2006). These results exemplify the difficulty of obtaining a worldwide resistance to *R. solanacearum*. Finally, the available sources of resistance have been found to be polygenic and, despite the identification of QTLs controlling resistance to bacterial wilt in tomato (Mangin et al., 1999; Wang et al., 2000, 2013; Carmeille et al., 2006).

5.13.1.5 QTL MAPPING

The use of genetic resistance appears to be the cheapest, most efficient, and environment-friendly method available to manage bacterial wilt. However, regarding race 3-phylotype II strains, only partial resistance sources have been identified, particularly in the *L. esculentum* cv. Hawaii 7996 (Carmeille et al., 2006). QTLs controlling resistance to race 1 strains (phylotypes I and II) have been identified in tomato progenies derived from different resistance sources (Wang et al., 2000). These studies revealed both shared and strain-specific QTLs. Recently, Carmeille et al. (2006) used a set of 76 molecular markers for QTL mapping using the single- and composite-interval mapping methods, as well as ANOVA. Four QTLs, named Bwr- followed by a number indicating their map location, were identified A major QTL, Bwr-6, and a minor one, Bwr-3, were detected in each season for all resistance criteria. Both QTLs showed stronger effects in the hot season than in the cool one. Their role in resistance to *R. solanacearum* race 3-phylotype II was subsequently confirmed in the RIL population derived from the same cross. Two other QTLs, Bwr-4 and Bwr-8, with intermediate and minor effects, respectively, were only detected in the hot season, demonstrating that environmental factors may strongly influence the expression of resistance against the race 3-phylotype II strain JT516 (Carmeille et al., 2006).

5.13.1.6   GRAFTING

Grafting of susceptible tomato cultivars onto resistant tomato or other Solanaceous rootstocks is effective against Asian strain of *R. solanacearum* and is used commercially in different location worldwide (Saddler, 2005). However, effectiveness of grafting against R3b2 has not been tested.

5.13.2.1   BIOLOGICAL CONTROL

Among bioagents, a number of soil bacteria and plant growth promoting rhizobacteria (PGPR) are currently being investigated for their role in the management of *R. solanacearum* R3bv2A (Champoiseau et al., 2010). However, these are currently unavailable commercially, and their efficacy is yet to be determined on a commercial scale (Champoiseau et al., 2010). Previous studies showed the potential value of some promising bioagents, which are dominantly avirulent or nonpathogenic strains of *R. solanacearum* and *Pseudomonas* spp., followed by *Bacillus* spp., *Streptomyces* spp., and other species, in controlling bacterial wilt.

5.13.2.2   NONPATHOGENIC STRAIN AS POTENTIAL BIOAGENT

Use biological control agent such as antagonistic bacteria or avirulent mutants of *R. solanacearum* is an alternative management strategy. *hrp* mutant strains of *R. solanacearum* is the most extensively studied avirulent mutants that no longer possess a functional TTSS (Frey et al., 1994). *hrp* mutant strains are still able to multiply, and, likely control bacterial wilt through competition with wild type strains for space and nutrients. Assays of protection by *hrp* mutants have been conducted on tomato and potato plants but, thus far, they did not help to reduce bacterial wilt in fields (Saddler, 2005; Denny, 2006). HrcV⁻ (formerly HrpO⁻) mutant strains are the most invasive and the most protective of all Hrp⁻ mutant strains of *R. solanacearum* so far tested (Etchebar et al., 1998). The biocontrol agent was first root inoculated and allowed to establish itself in the xylem vessels for a week before a challenge inoculation with the pathogenic strain was done, and the ratio of pathogenic strain to mutant strain in the inoculum was 0.1 (Trigalet and Trigalet-Demery, 1990).

According to Trigalet and Trigalet-Demery (1990), the protective ability of certain Hrp⁻ mutants against further invasion by a pathogenic strain has been correlated with their abilities to invade and survive within host plants; the more invasive the mutant strain the more protective it is. In microscopic studies of the colonization of the vascular tissues of tomato by an HrcV⁻ (formerly HrpO⁻) mutant strain of *R. solanacearum* after either root inoculation of the mutant strain alone or delayed challenge inoculation by a pathogenic strain shown that the population of the wild type strain is lower than in plants inoculated with the wild type alone. In contrast, growth of the HrcV⁻ mutant strain was significantly increased in the presence of the pathogenic strain (Etchebar et al., 1998).

Light microscopy of infection and colonization of tomato stems at the hypocotyl level: 2-mm-thick slices at lower part of hypocotyl near collar zone, shown that there is abundant colonization by the pathogenic strain of protoxylem, metaxylem, and secondary xylem vessels. Etchebar et al. (1998) shown that the entire stem section of a symptomless tomato plant that exhibits exclusive colonization of whole xylem vascular bundles by the Hrp⁻ mutant and entire stem section of a partially wilted tomato plant harbors both competitors, the HrcV⁻ mutant strain and the pathogenic strain.

Etchebar et al. (1998) suggested that prevention of invasion of the pathogenic strain by the mutant strain is probably not due to direct antagonism caused by a bacteriocin or a bacteriophage. Competition for space is strongly suggested by microscopic examination of roots or hypocotyls slices where invasion of the competitors always occurred in separate xylem vessels: the greater the number of vascular bundles invaded by the HrcV⁻ mutant strain the greater the prevention of the pathogenic strain.

5.13.2.3   USE OF BACTERIOPHAGES

The use of phages for disease control is a fast-expanding trend of plant protection with great potential to replace chemical measures. The filamentous φRSM-type phages (φRSM3 is a typical phage of this group) changed host bacterial cells to be avirulent after infection (Askora et al., 2009). The filamentous phage φRSM3 that infects *R. solanacearum* strains and inactivates virulence on plants is a potential agent for controlling bacterial wilt in tomato. Addy et al. (2012) demonstrated that inoculation of φRSM3-infected cells into tomato plants did not cause bacterial wilt. Instead, φRSM3-infected cells enhanced the expression of pathogenesis-related (*PR*) genes, including *PR-1a, PR-2b*, and *PR7*, in tomato plants. Moreover, pretreatment with φRSM-infected cells protects tomato plants from infection by virulent *R. solanacearum* strains. φRSM phages may serve as an efficient tool to control bacterial wilt in several crops including tomato.

5.13.3   SOIL AMENDMENT WITH AMINO ACIDS

Several approaches to improve the biocontrol efficacy of bacterial antagonists have been made time to time, for example, the use of nutritional amendments to enhance its colonization and population size (Davis et al., 1992). When *Pseudomonas putida* AP-1 was applied together with methionine than AP-1 alone, the suppression of Fusarium wilt of tomato was enhanced (Yamada and Ogiso, 1997). Thereby, the amendment of soil with organic matter or a substrate in the appropriate amount likely improves the rhizosphere competence and the survival of introduced or naturally occurring antagonists in the soil, resulting in effective and sustainable disease control. Posas and Toyota (2010) investigated the effect of four amino acids, unutilizable by *R. solanacearum*, on tomato bacterial wilt in three Japanese soils. Bacterial wilt was suppressed by lysine and serine, but not by tyrosine and valine. The number of the pathogen in non-rhizosphere soil, rhizosphere soil and the rhizoplane was markedly lower in the lysine and serine treatments than in the tyrosine and valine treatments, while the opposite result was obtained for the total bacterial population. They concluded, the lower disease incidence of TBW following lysine treatment is likely related to a specific bacterial community in the rhizoplane that developed on the addition of lysine.

Differential virulence between two race 3 biovar 2 strains of *R. solanacearum* was recently traced to the presence of methionone synthesis genes in the more virulent isolate, although methionine is not known to be limiting for growth in the host (Gonzalez et al., 2011).

5.13.4 CHEMICAL CONTROL

Chemical soil fumigation or application of phosphoric acid is effective for controlling bacterial wilt of tomato in the field (Ji et al., 2005). Similarly, soil treatments, including modification of soil pH, solarization and application of stable bleaching powder reduced bacterial populations and disease severity on a small scale (Saddler, 2005). Drawbacks of these methods may include environmental hazardous and pollution, cost and high labor input (Fajinmi and Fajinmi, 2010). Additionally, most of these methods still have to be tested against race 3 biovar 2 strain of *R. solanacearum.*

Recently, use of a Thymol derived volatile biochemical (currently not commercially available) reduced disease incidence and increase yield in field experiments in Florida (Ji et al., 2005). Chemical control should be integrated with other methods to reduce selection pressure for pathogen resistance (Fajinmi and Fajinmi, 2010). The chemical elicitor acibenzolar-*S*-methyl (ASM; Actigard 50 WG), which induces SAR, was investigated to determine the effect on bacterial wilt of tomato on moderately resistant cultivars enhanced resistance against bacterial wilt under greenhouse and field conditions (Pradhanang et al., 2005).

Chemical control, in addition to being potentially harmful to the environment, was not proved to be efficient to eradicate *R. solanacearum* (Denny, 2006). This can be explained by the bacterium localization in the deeper soil layers or sheltered in xylem vessels of infected plants and weeds (Wenneker et al., 1999). In addition, a soil dependent effect has been observed and therefore soil disinfection is not universally applicable (Saddler, 2005). The active portion of the potassium salts was found to be phosphorous acid (H_3PO_3). Phosphorous acid was found to inhibit in vitro growth of *R. solanacearum.* It is thought to be protecting plants from infection by acting as a bacteriostatic compound in the soil. The plants, however, are not protected from aboveground infection on wounded surfaces. Phosphorous acid drenches were shown to protect geranium plants from infection by either race 1 or 3 of *R. solanacearum* (Norman et al., 2006).

5.15.5   INDUCED RESISTANCE THROUGH SILICON (SI) AND CHITOSAN

Silicon (Si) and chitosan reduced the incidence of bacterial wilt through induced resistance (Kurabachew and Wydra, 2014). Wang et al. (2006) reported that Si-mediated resistance was associated with increase in the number of microorganisms in the soil as well as soil enzyme activity (urease and acid phosphatase). By soaking the seeds, in a low sodium chloride (NaCl) solution was previously found to increase seedling vigor and tolerance to *R. solanacearum* in the tomato (Nakaune et al., 2012). Some new innovative methods have been reported to suppress bacterial wilt in which live microbial cells of the pathogen are captured with 10 g/kg of coated sawdust with 1% of an equimolar polymer of Nbenzyl-4-vinylpyridinium chloride with styrene (PBVP-co-ST) (Kawabata et al., 2005a) or coagulated in the soil with 10 mg/ kg of a copolymer of methyl methacrylate with N-benzyl-4-vinylpyridinium chloride at a molar ratio 3:1 (PMMA-co-BVP; Kawabata et al., 2005b).

5.13.5   ESSENTIAL OILS

Natural plant products are important sources of new agrochemicals for the control of some plant diseases (Ji et al., 2005, Hassan et al., 2009, Regnault et al., 2005). Plant extracts are potentially environmentally safe alternatives and, thus, possible components of integrated pest and disease management programs. Several essential oils and their components have been reported to have a bactericidal effect against *R. solanacearum*. Plant-derived volatile essential oils such as thymol and palmarosa oil were effective biofumigants against *R. solanacearum* in greenhouse trials, but further field evaluation is necessary before their practical use for management of this disease is possible (Pradhanang et al., 2005). The use of plant essential oils as biofumigants has also been examined as a component in integrated disease management systems (IDM). Recently, palmarosa and lemongrass oils were shown to reduce the growth of *R. solanacearum* race 4 (phylotype I) (Paret et al, 2010).

5.13.6   ANTIMICROBIAL PEPTIDES

Peptides and small proteins exhibiting direct antimicrobial activity have been characterized by a vast number of organisms ranging from insects to humans (Zasloff, 2002). The general biological role of antimicrobial peptides (AMP) in defense against challenging microbes has already been recognized. AMP look like promising alternatives as novel therapeutics in combating the increasing incidence of antibiotic resistance in pathogenic microbes (Marcos et al., 2008). Plants have specific defense mechanisms that include small antimicrobial proteins or peptides with activity against phytopathogens (Van Loon et al., 2006). A recent study has suggested that these proteins might account for up to 2–3% of the genes predicted in plant genomes (Silverstein et al., 2007). AMPs are diverse and can be subdivided into several groups based on their origin, composition, and structure. However, they also share certain common structural characteristics such as amino acid composition: the residues most abundant in AMP are cationic (arginine and lysine) and hydrophobic (tryptophan, phenylalanine, leucine, and isoleucine) (Marcos et al., 2008).

The development of efficient methods for the synthesis of peptides and their analogs, peptide collections, and libraries, has resulted in substantial advances in the identification and characterization of bioactive peptides. A detailed knowledge of the mode of action of AMP is essential if the goal of applying them to plant protection is ever to be reached. For most of the cases in which activity was determined as cell-based growth inhibition assays of bacteria or fungi, the peptide mode of action needs further investigation (Marcos et al., 2008).

5.13.7 CULTURAL PRACTICES

These practices include crop rotation (3–4 years rotation) with non-host plants such as grasses, intercropping, control of weed and root-knot nematode population, planting at noninfested production sites, removal of weeds or crop residue where inoculum persists, selection of appropriate planting time to avoid heat deep ploughing of crop residues, satisfactory soil drainage and early- and late-season irrigation management (Hayward, 1991; Saddler, 2005). To avoid the dissemination of *R. solanacearum*, it is recommended to plant healthy seeds in pathogen-free soil combined with irrigation water known to be free of *R. solanacearum* (French, 1994). In infested soils, crop rotation (2–5 years), control of weed hosts and survey of water for irrigation can reduce the bacterial load (Lopez and Biosca, 2005).

5.13.7   SANITARY AND PHYTOSANITARY CONTROL MEASURES

The best strategy for controlling bacterial wilt in the field consists primarily of suitable sanitary and phytosanitary control measures (Fajinmi and Fajinmi, 2010). According to the IPPC, phytosanitary measures include any legislation, regulation or official procedure having the purpose of preventing the introduction and/or spread of plant pests and may normally only be applied to regulated pests, although the IPPC also provides for emergency measures to be taken against any pest that is a potential threat to its territories (Singh et al., 2013).

In regions where bacterial wilt of tomato is endemic or in locations where *R. solanacearm* is present but not yet established, these methods may be very effective (Fajinmi and Fajinmi, 2010). In locations where the pathogen is not present it is important to prevent the introduction and if inadvertently introduced subsequent movement of race 1 of *R. solanacearum* from infested to uninfested locations or fields (Fajinmi and Fajinmi, 2010). Sanitation efforts include planting only certified pathogen-free planting materials, disinfecting equipment before moving between fields, controlling floodwater flow and never using surface water for irrigation (Fajinmi and Fajinmi, 2010). A zero tolerance level includes reinforcement of quarantine regulation, exclusionary practices, sanitation protocols and inspection designed to prevent introduction of infected plants. Pest response guidelines for *R. Solanacearum* R3b2, presents current information for detection, control, containment and eradication of this pathogen (Floyd, 2008).

5.14   CONCLUSION

R. solanacearum is studied as a model organism because of pathogenicity in plants, survives in diverse environments and cause destructive disease on a wide range of plants by precisely controlling its gene expression through an elaborate regulatory network. The virulence factors of *R. solanacearum* include production of EPS, a consortium of plant CWDE, twitching and swimming motility, taxis, and several dozen effectors secreted by a TTSS. *R. solanacearum* has a wide host range, which makes it difficult to have a generalized estimate of the economic losses caused by the bacterial wilt disease. There are no such chemicals available for effective management of this disease. In the absence of efficient strategies to eradicate *R. solanacearum* from infected soils and water, the use of resistant cultivars appears to be the best disease control strategy.

KEYWORDS

- *Ralstonia*
- tomato
- wilt
- pathosystem
- diversity
- management

REFERENCES

Addy, H. S.; Askora, A.; Kawasaki, T.; Fujie, M.; Yamada, T. The Filamentous Phage Φrss1 Enhances Virulence of Phytopathogenic *Ralstonia Solanacearum* on Tomato. *Phytopathology* **2011**, *102* (3), 244–51.

Agrios, G. N. *Plant Pathology*, 5th Edition; Academic Press: San Diego, CA, 2005.

Almoneafy, A. A.; Xie, G. L.; Tian, W. X.; Xu, L. H.; Zhang, G. Q.; Muhammad Ibrahim. Characterization and Evaluation of Bacillus Isolates for Their Potential Plant Growth and Biocontrol Activities Against Tomato Bacterial Wilt. *Afr. J. Biotechnol.* **2012**, *11*, 7193–7201.

Arlat, M.; Gough, C. L.; Zischek, C.; Barberis, P. A.; Trigalet, A.; Boucher, C. A. Transcriptional Organization and Expression of the Large *hrp* Gene Cluster of *Pseudomonas solanacearum*. *Mol. Microbiol.* **1992**, *6*, 3065–3076.

Askora, A.; Kawasaki, T.; Usami, S.; Fujie, M.; Yamada, T. Host Recognition and Integration of Filamentous Phage ϕRSM in the Phytopathogen, *Ralstonia solanacearum*. *Virology* **2009**, *384*, 69–76.

Aziz, A.; Gauthier, A.; Bezier, A.; Poinssot, B.; Joubert, J. M., Pugin, A.; Heyraud, A.; Baillieul, F. Elicitor and Resistance-inducing Activities of Beta-1,4 Cellodextrins in Grapevine, Comparison with Beta-1,3 Glucans and Alpha-1,4 Oligogalacturonides. *J. Exp. Bot.* **2007**, *58*,1463–1472.

Bertolla, F.; Frostergard, A.; Brito, B.; Nesme, X.; Simonet, P. During Infection of its Host, the Plant Pathogen *Ralstonia solanacearum* Naturally Develops a State of Competence and Exchanges Genetic Material. *Mol. Plant Microbe Interact.* **1999**, 12, 467–472.

Boucher, C. A.; Gough, C. L.; Arlat, M. Molecular Genetics of Pathogenicity Determinants of *Pseudomonas solanacearum* with Special Emphasis on Hrp genes. *Ann. Rev. Phytopathol.* **1992**, 30 (1), 443–461.

Brinkmann, V.; Goosmann, C.; Kuhn, L. L.; Zychlinsky, A. Automatic Quantification of In vitro NET Formation. *Front. Immunol.* **2013**, *3*,1–8.

Brito, B.; Marenda, M.; Barberis, P.; Boucher, C.; Genin, S. *prhJ* and *hrpG*, Two New Components of the Plant Signaldependent Regulatory Cascade Controlled by PrhA in *Ralstonia solanacearum*. *Mol. Microbiol.* **1999**, *31*, 237–251.

Buddenhagen, I.; Kelman, A. Biological and Physiological Aspects of Bacterial Wilt Caused by *Pseudomonas solanacearum*. *Ann. Rev. Phytopathol.* **1964**, *2*, 203–230. DOI: 10.1146/annurev.py.02.090164.001223.

Carmeille, A.; Caranta, C.; Dintinger, J.; Prior, P.; Luisetti, J.; Besse, P. Identification of QTLs for *Ralstonia solanacearum* Race 3-Phylotype II Resistance in Tomato. *Theor. Appl. Genet.* **2006**, *113*, 110–121. DOI: 10.1007/s00122-006-0277-3.

Cellier, G.; Prior, P. Deciphering Phenotypic Diversity of *Ralstonia Solanacearum* Strains Pathogenic to Potato. *Phytopathology* **2010**, *100*, 1250–1261. DOI: 10.1094/PHYT O-02-10-0059.

Cerkauskas, R. Fact Sheet: Tomato Disease Bacterial Wilt; The World Vegetable Center. AVRDC Publication 04-611, 2004. http://www.avrdc.org.

Champoiseau, P. G. USDA NRI Project: *R. solanacearum* race 3 biovar 2: detection, exclusion and analysis of Select Agent Educational modules, 2008.

Champoiseau, P. G.; Jones, J. B.; Allen, C. *Ralstonia solanacearum* Race 3 Biovar 2 Causes Tropical Losses and Temperate Anxieties. *Plant Health Prog.* **2009**, DOI: 10.1094/PHP-2009-0313-01-RV.

Chandrashekara, K. N.; Mothukapalli, Krishnareddy P.; Manthirachalam, D.; Akella, V.; Abdul Nazir Ahmad, K. Prevalence of Races and Biotypes of *Ralstonia solanacearum* in India. *J. Plant Protect. Res.* **2012**, *52* (1), 53–58.

Chandrashekara, K. N.; Prasannakumar, M. K. New Host Plants for *Ralstonia solanacearum* from India. *Plant Pathol.* **2010**, *59* (6), 11–78. DOI:10.1111/J.1365-3059.2010.02358.X.

Ciampi, L.; Sequeira, L. Influence of Temperature on Virulence of Race 3 Strains of *Pseudomonas solanacearum*. *Am. Potato J.* **1980**, *57*, 307–317.

Clough, S.; Lee, K. E.; Schell, M.; Denny, T. A Two-component System in *Ralstonia* (*Pseudomonas*) *solanacearum* Modulates the Production of PhcA-regulated Virulence Factors in Response to 3-hydroxypalmitic Acid Methyl Ester. *J. Bacteriol.* **1997**, *179*, 3639–3648.

da Silva, A. C. et al. Comparison of the Genomes of Two *Xanthomonas* pathogens with Differing Host Specificities. *Nature* **2002**, *417*, 459–463.

Dangl, J. L.; Horvath, D. M.; Staskawicz, B. J. Pivoting the Plant Immune System From Dissection to Deployment. *Science* **2013**, *341* (6147). DOI:10.1126/science.1236011.

Davis, R. F.; Backman, P. A.; Rodriguez-Kabana, R.; Kokalis-Burelle, N. Biological Control of Apple Fruit Diseases by Chaetomium globosum Formulations Containing Cellulose. *Biol. Control* **1992**, *2*,118–123.

de Weert, S.; Vermeiren, H.; Mulders, I.; Kuiper, I.; Hendrickx, N.; Bloemberg, G.; Vanderleyden, J.; De Mot, R.; Lugtenberg, B. J. Flagelladriven Chemotaxis Towards Exudate Components is an Important Trait for Tomato Root Colonization by *Pseudomonas fluorescens*. *Mol. Plant Microbe Interact.* **2002**, *15*,1173–1180.

Denny, T. P. Plant Pathogenic *Ralstonia* Species. In *Plant-Associated Bacteria*; Gnanamanickam, S. S., Eds.; Springer: Dordrecht, 2006, pp 573–644.

Denny, T. P.; Hayward, A. C. *Ralstonia solanacearum*. In Laboratory guide for identification of plant pathogenic bacteria; Schaad, N. W., Jones, J. B., Chun, W., Eds;. APS Press: St. Paul, MN,2001, pp 151.

Deslandes, L.; Genin, S. Opening the *Ralstonia solanacearum* type III Effector Tool Box: Insights into Host Cell Subversion Mechanisms. *Curr. Opin. Plant Biol.* **2014**, *20*,110–117.

Deslandes, L.; Rivas, S. Catch Me if You Can: Bacterial Effectors and Plant Targets. *Trends Plant Sci.* **2012**, *17*, 644–655. DOI: 10.1016/j.tplants.2012.06.011.

Devi, R. L.; Menon, M. R. Seasonal Incidence of Bacterial Wilt of Tomato. *Ind. J. Microbiol.* **1980**, *20*, 13–15.

Digonnet, C.; Martinez, Y.; Denance, N.; Chasseray, M.; Dabos, P.; Ranocha, P.; Marco, Y.; Jauneau, A.; Goffner, D. Deciphering the Route of *Ralstonia solanacearum* Colonization in Arabidopsis thaliana Roots During a Compatible Interaction: Focus at the Plant Cell Wall. *Planta* **2012**, *236*,1419–1431.

Elphinstone, J. G. The Current Bacterial Wilt Situation: A Global View. In *Bacterial Wilt Disease and the Ralstonia solanacearum Species Complex*; Allen, C., Prior, P., Hayward, A. C., Eds.; APS Press: Saint Paul, MN, 2005, pp 9–28.

EPPO/CABI. Distribution Maps of Plant Diseases: *Ralstonia solanacearum* (2003–2006). http://www cabi org/DMPD 2006.

Etchebar, C.; Trigalet-Demery, D.; Gijsegem, F.; Vasse, J; Trigalet, A. Xylem Colonization by an HrcV⁻ Mutant of *Ralstonia solanacearum* is a Key Factor for the Efficient Biological Control of Tomato Bacterial Wilt. *Mol. Plant Microbe Interaction* **1998**, *11*, 9, 869–877.

Fajinmi, A. A.; Fajinmi, B. An Overview of Bacterial Wilt Disease of Tomato in Nigeria. *Agric. J.* **2010**, *5* (4), 242–247.

Fegan, M.; Prior, P. How Complex is the *Ralstonia solanacearum* Species Complex. In *Bacterial Wilt Disease*; Allen, C., Prior, P., Hayward, A. C., Eds.; APS Press: St Paul, MN, USA, 2005, pp 449–461.

Fegan, M.; Prior, P. Diverse Members of the *Ralstonia solanacearum* Species Complex Cause Bacterial Wilts of Banana. *Austr. Plant Pathol.* **2006**, *35*, 93–101.

Flavier, A. B.; Clough, S. J.; Schell, M. A.; Denny, T. P. Identification of 3-Hydroxypalmitic Acid Methyl Ester as a Novel Autoregulator Controlling Virulence in *Ralstonia solanacearum*. *Mol. Microbiol.* **1997**, *26*, 251–259.

Floyed, J. *New Pest Response Guidelines: Ralstonia solanacearum race 3 biovar 2. USDA-APHIS-PPQ-emergency and Domestic Programs*; Riverdale, MD, 2008.

French, E. R. Strategies for Integrated Control of Bacterial Wilt of Potatoes. In *Bacterial Wilt: The Disease and its Causative Agent, Pseudomonas solanacearum;* Hayward, A. C., Hartman G. L., Eds.; CAB International: Wallingford, 1994, pp 199–207.

Frey, P.; Prior, P.; Marie, C.; Kotoujansky, A.; Trigalet-Demery, D.; T rigalet, A. Hrp Mutants of *Pseudomonas solanacearum* as Potential Biocontrol Agents of Tomato Bacterial Wilt. *Appl. Environ. Microbiol.* **1994**, *60*, 3175–3181.

Fujiwara, A.; Kawasaki, T.; Usami, S.; Fujie, M.; Yamada, T. Genomic Characterization of *Ralstonia solanacearum* phageɸRSA1 and its Related Prophage (ɸRSX) in Strain GMI1000. *J. Bacteriol.* **2008**, *190* (1),143–56.

Genin, S.; Boucher, C. Lessons Learned From the Genome Analysis of *Ralstonia solanacearum. Ann. Rev. Phytopathol.* **2004**, *42*,107–34.

Genin, S.; Denny, T. P. Pathogenomics of the *Ralstonia solanacearum* species complex. *Ann. Rev. Phytopathol.* **2012**, *50*, 67–89.

Genin, S. Molecular Traits Controlling Host Range and Adaptation to Plants in Ralstonia solanacearum. *New Phytol.* **2010**, *187*,920–928.

Genin, S.; Brito, B.; Denny, T. P.; Boucher, C. Control of the *Ralstonia solanacearum* Type III Secretion System (Hrp) Genes by the Grobal Virulence Regulator PhcA. *Fed. Eur. Biochem. Soc.* **2005**, *579*, 2077–2081.

Genin, S.; Gough, C. L.; Zischek, C.; Boucher, C. A. Evidence that the *hrpB* Gene Encodes a Positive Regulator of Pathogenicity Genes from *Pseudomonas solanacearum. Mol. Microbiol.* **1992**, *6*, 3065–3076.

Giron, J. A.; Torres, A. G.; Freer, E.; Kaper, J. B. The Flagella of Enteropathic Escherichia coli Mediate Adherence to Epithelial Cells. *Mol. Microbiol.* **2002**, *44*, 361–379.

Goel, A. K.; Lundberg, D.; Torres, M. A.; Matthews, R.; Akimoto-Tomiyama, C.; Farmer, L., et al. The *Pseudomonas syringae* type III effector HopAM1 Enhances Virulence on Water-stressed plants. *Mol. Plant Microbe Interact.* **2008**, *21*, 361–370. DOI: 10.1094/MPMI-21-3-0361.

Gonzalez, A.; Plener, L.; Restrepo, S.; Boucher, C.; Genin, S. Detection and Functional Characterization of a Large Genomic Deletion Resulting in Decreased Pathogenicity in *Ralstonia solanacearum* race 3 biovar 2 strains. *Environ. Microbiol.* **2011**, *13*, 3172–85.

Gonzalez, E. T.; Allen, C. Characterization of a *Ralstonia solanacearum* Operon Required for Polygalacturonate Degradation and Uptake of Galacturonic Acid. *Mol. Plant-Microbe Interact.* **2003**, *16*, 536–544.

Greer-Phillips, S. E.; Stephens, B. B.; Alexandre, G. An Energy Taxis Transducer Promotes Root Colonization by *Azospirillum brasilense*. *J. Bacteriol.* **2004**, *186*, 6595–6604.

Grimault, V.; Prior, P.; Anais, G. A Monogenic Dominant Resistance of Tomato to Bacterial Wilt in Hawaii 7996 is Associated with Plant Colonization by *Pseudomonas solanacearum*. *J. Phytopathol.* **1995**, *143*, 349–352. 10.1111/j.1439-0434.1995.tb00274

Guidot, A.; Coupat, B.; Fall, S.; Prior, P.; Bertolla, F. Horizontal Gene Transfer Between *Ralstonia solanacearum* Strains Detected by Comparative Genomic Hybridization on Microarrays. *ISME J.* **2009**, *3*, 549–562.

Hassan, M. A. E.; Bereika, M. F. F.; Abo-Elnaga, H. I. G.; Sallam, M. A. A. Direct Antimicrobial Activity and Induction of Systemic Resistance in Potato Plants Against Bacterial Wilt Disease by Plant Extracts. *Plant Pathol. J.* **2009**, *25*, 352–360.

Hawes, M.; Allen, C.; Turgeon, G. B.; Curlango-Rivera, G.; Tran, T. M.; Huskey, D. A.; Xiong, Z. Root Border Cells and Their Role in Plant Defense. *Ann. Rev. Phytopathol.* **2016**, *54*, 5.1–5.19.

Hayward, A. C. Biology and Epidemiology of Bacterial Wilt Caused by *Pseudomonas solanacearum*. *Ann. Rev. Phytopathol.* **1991**, *29*, 65–87.

Hayward, A. C. The Hosts of *Pseudomonas solanacearum*. In *Bacterial Wilt: The Disease and its Causative Agent, Pseudomonas solanacearum*, Hayward, A. C., Hartman G. L., Eds.; CAB International: Wallingford, 1994.

Huang, Q.; Allen, C. An Exo-poly-alpha-D-galacturonosidase, PehB, is Required for Wild-type Virulence of *Ralstonia solanacearum*. *J. Bacteriol.* **1997**, *179*, 7369–7378.

Jacobs, J. M.; Babujee, L.; Meng, F.; Milling, A.; Allen, C.. The *in planta* Transcriptome of *Ralstonia solanacearum*: Conserved Physiological and Virulence Strategies During Bacterial Wilt of Tomato. *mBio* **2012**, *3* (4), e00114-12. doi:10.1128/mBio.00114-12.

Janse, J.; Beld, D.; van Den, H. Introduction to Europe of *Ralstonia solanacearum* biovar 2, race 3 in *Pelargonium zonale* Cuttings. *J. Plant Pathol.* **2004**, *86*, 147–155.

Janse, J. D.; Araluppan, F. A. X.; Schans, J.; Wenneker, M.; Westerhuis, W. Experiences with Bacterial Brown rot *Ralstonia solanacearum* biovar 2, race 3, in The Netherlands. In *Bacterial Wilt Disease. Molecular and Ecological Aspects;* Prior, P., Allen, C., Elphinstone, J., Eds.; Springer-Verlag: Berlin, 1998.

Ji, P.; Momol, M. T.; Olson S. M.; Pradhanang, P. M. Evaluation of Thymol as Biofumigant for Control of Bacterial Wilt of Tomato Under Field Conditions. *Plant Dis.* **2005**, 89, 497–500.

Kang, Y.; Liu, H.; Genin, S.; Schell, M. A.; Denny, T. P. *Ralstonia solanacearum* Requires Type 4 pili to Adhere to Multiple Surfaces and for Natural Transformation and Virulence. *Mol. Microbiol.* **2002**, 46, 427–437.

Kawabata, N.; Kishimoto, H.; Abe, T.; Ikawa, T.; Yamanaka, K.; Ikeuchi, H.; Kakimoto, C. Control of Tomato Bacterial Wilt Without Disinfection Using a New Function Polymer that Captures Microbial Cells Alive on the Surface and is Highly Biodegradable. *Biosci. Biotechnol. Biochem.* **2005a**, 69, 326–333.

Kawabata, N.; Sakakura, W.; Nishimura, Y. Static Suppression of Tomato Bacterial Wilt by Bacterial Coagulation Using a New Functional Polymer that Coagulates Bacterial Cells and is Highly Biodegradable. *Biosci. Biotechnol. Biochem.* **2005b**, 69, 537–543.

Kim, S. H.; Olson, T. N.; Schaad, N. W.; Moorman, G.W. *Ralstonia solanacearum* Race 3, biovar 2, the Causal Agent of Brown Rot of Potato, Identified in Geranium in Pennsylvania, Delaware, and Connecticut. *Plant Dis.* **2003**, 87, 450.

Kurabachew, H.; Wydra, K. Induction of Systemic Resistance and Defense-related Enzymes After Elicitation of Resistance by Rhizobacteria and Silicon Application Against *Ralstonia solanacearum* in Tomato (*Solanum lycopersicum*). *Crop Prot.* **2014**, 57, 1–7.

Lacombe, S.; Rougon-Cardoso, A.; Sherwood, E.; Peeters, N.; Dahlbeck, D.; van Esse, H. P.; Smoker, M.; Rallapalli, G.; Thomma, B. P.; Staskawicz, B.; Jones, J. D.; Zipfel, C. Interfamily Transfer of a Plant Pattern-recognition Receptor Confers Broad-spectrum Bacterial Resistance. *Nat. Biotechnol.* **2010**, 28, 365–369. DOI: 10.1038/nbt.1613.

Lebeau, A.; Daunay, M. C.; Frary, A.; Palloix, A.; Wang, J. F.; Dintinger, J.; Chiroleu, F.; Wicker, E.; Prior, P. Bacterial Wilt Resistance in Tomato, Pepper, and Eggplant: Genetic Resources Respond to Diverse Strains in the *Ralstonia solanacearum* Species Complex. *Phytopathology* **2011**, 101, 154–165. DOI: 10.1094/PHY T O-02-10-0048.

Liu, H.; Zhang, S.; Schell, M. A.; Denny, T. P. Pyramiding Unmarked Deletions in *Ralstonia solanacearum* Shows that Secreted Proteins in Addition to Plant Cell-wall-degrading Enzymes Contribute to Virulence. *Mol. Plant Microbe Interact.* **2005**, 18,1296–1305.

Liu, H. L.; Kang, Y. W.; Genin, S.; Schell, M. A.; Denny, T. P. Twitching Motility of *Ralstonia solanacearum* Requires a Type IV Pilus System. *Microbiology* **2001**, 147, 3215–3229.

Lohou, D.; Turner, M.; Lonjon, F.; Cazale, A. C.; Peeters, N.; Genin, S.; Vailleau, F. HpaP Modulates Type III Effector Secretion in *Ralstonia solanacearum* and Harbors a Substrate Specificity Switch Domain Essential for Virulence. *Mol. Plant Pathol.* **2014**, 15 (6), 601–14. DOI: 10.1111/mpp.12119. Epub: Feb 19, 2014.

Lopez, M. M.; & Biosca, E. G. Potato Bacterial Wilt Management: New Prospects for an Old Problem. In *Bacterial Wilt Disease and the Ralstonia solanacearum species complex*; Allen, C., Prior, P. Hayward, A. C., Eds.; APS Press: Saint Paul, MN, 2005, pp 205–224.

Mangin, B.; Thoquet, P.; Olivier, J.; Grimsley N. H. Temporal and Multiple Quantitative Trait Loci Analyses of Resistance to Bacterial Wilt in Tomato Permit the Resolution of Linked Loci. *Genetics* **1999**, 151, 1165–1172.

Mansfield, J.; Genin, S.; Magori, S.; Citovsky, V.; Sriariyanum, M.; Ronald, P.; Dow, M.; Verdier, V.; Beer, S. V.; Machado, M. A.; Toth, I.; Salmond, G.; Foster, G. D. Top 10 Plant Pathogenic Bacteria in Molecular Plant Pathology. *Mol. Plant Pathol.* **2012**, 13, 614–629. DOI: 10.1111/j.1364-3703.2012.00804.x.

Marcos, J. F.; Munoz, A.; Perez-Paya, E.; Misra, S.; Lopez-García, B. Identification and Rational Design of Novel Antimicrobial Peptides for Plant Protection. *Ann. Rev. Phytopathol.* **2008**, 46, 273–301.

Mattick, J. S. Type IV Pili and Twitching Motility. *Ann. Rev. Microbiol.* **2002**, 56, 289–314.

McGarvey, J. A; Denny, T. P.; Schell, M. A. Spatial-temporal and Quantitative Analysis of Growth and EPS I Production by *Ralstonia solanacearum* in Resistant and Susceptible Tomato Cultivars. *Phytopathology* **1999**, *89*, 1233–1239.

Mishra, A.; Mishra, S. K.; Karmakar, K.; Sarangi, C. R.; Sahu, G. S. Assessment of Yield Loss Due to Wilting and Some Popular Tomato Cultivars. *Enviorn. Ecol.* **1995**, *13*, 287–290.

Mole, B. M.; Baltrus, D. A.; Dangl, J. L.; Grant, S. R. Global Virulence Regulation Networks in Phytopathogenic Bacteria. *Trends Microbiol.* **2007**, *15*, 363–371.

Murugaiyan, S.; Bae, J. Y.; Wu, J.; Lee, S. D.; Um, H. Y.; Choi, H. K.; Chung, E.; Lee, J. H.; Lee, S. W. Characterization of Filamentous Bacteriophage PE226 Infecting *Ralstonia solanacearum* Strains. *J. Appl. Microbiol.* **2011**, *110* (1), 296–303.

Nakaune, M.; Tsukazawa, K.; Uga, H.; Asamizu, E.; Imanishi S.; Matsukura, C.; Ezura, H. Low Sodium Chloride Priming Increases Seedling Vigor and Stress Tolerance to *Ralstonia solanacearum* in Tomato. *Plant Biotech.* **2012**, *29*, 9–18.

Norman, D. J.; Chen, J.; Yuen, J. M. F.; Mangravita-Novo, A.; Byrne, D.; Walsh, L. Control of Bacterial Wilt of Geranium with Phosphorous Acid. *Plant Dis.* **2006**, *90*, 798–802.

Osorio, S.; Castillejo, C.; Quesada, M. A.; Medina-Escobar, N.; Brownsey, G. J.; Suau, R.; Heredia, A.; Botella, M. A.; Valpuesta, V. Partial Demethylation of Oligogalacturonides by Pectin Methyl Esterase 1 is Required for Eliciting Defence Responses in Wild Strawberry (*Fragaria vesca*). *Plant J.* **2008**, *54*, 43–55.

Ottemann, K. M.; Lowenthal, A. C. *Helicobacter pylori* Uses Motility for Initial Colonization and to Attain Robust Infection. *Infect. Immun.* **2002**, *70*, 1984–1990.

Paret, M. L.; Cabos, R.; Kratky, B. A.; Alvarez, A. M. Effect of Plant Essential Oil on *Ralstonia solancearum* race 4 and Bacterial Wilt of Edible Ginger. *Plant Dis.* **2010**, *94*, 521–527.

Pisetsky, D. S. The Origin and Properties of Extracellular DNA: From PAMP to DAMP. *Clin. Immunol.* **2012**, *144*, 32–40.

Popoola, A. R.; Ganiyu, S. A.; Enikuomehin, O. A.; Bodunde, J. G.; Adedibu, O. B.; Durosomo, H. A.; Karunwi, O. A. Isolation and Characterization of *Ralstonia solanacearum* Causing Bacterial Wilt of Tomato in Nigeria. *Nig. J. Biotech.* **2015**, *29*, 1–10.

Posas, M. B.; Toyota, K. Mechanism of Tomato Bacterial Wilt Suppression in Soil Amended with Lysine. *Microbes Environ.* **2010**, *25* (2), 83–94.

Poueymiro, M.; Genin, S. Secreted Proteins from *Ralstonia solanacearum*: A Hundred Tricks to Kill a Plant. *Curr. Opin. Microbiol.* **2009**, *12*, 44–52.

Pradhanang, P. M.; Ji, P.; Momol, M. T.; Olson, S. M.; Mayfield, J. L.; Jones, J. B. Application of Acibenzolar-*S*-methyl Enhances Host Resistance in Tomato Against *Ralstonia solanacearum*. *Plant Dis.* **2005**, *89*, 989–993.

Prior, P.; Fegan, M. Recent Developments in the Phylogeny and Classification of *Ralstonia solanacearum*. *Acta Hort.* **2005**, *695*, 127–136.

Prior, P.; Allen, C.; Elphinstone, J. *Bacterial Wilt Disease: Molecular and Ecological Aspects*; Springer Verlag: Berlin, Germany, 1998.

Priou, S.; Gutarra, L.; Aley, P. Sensitive Detection of *Ralstonia solanacearum* in Soil: A Comparison of Different Detection Techniques. *Plant Pathol.* **2006**, *49*, 414–422.

Remenant, B.; de Cambiaire, J. C.; Cellier, G.; Jacobs, J. M.; Mangenot, S.; Barbe, V.; Lajus, A.; Vallenet, D.; Medigue, C.; Fegan, M.; Allen, C.; Prior, P. *Ralstonia syzygii*, the Blood Disease Bacterium and Some Asian R. Solanacearum Strains Form a Single Genomic Species Despite Divergent Lifestyles. *PLoS One*, **2011**, *6*, e24356.

Saddler, G. S. Management of Bacterial Wilt Disease. In *Bacterial Wilt Disease and the Ralstonia solanacearum Species Complex*; Allen, C., Prior, P., Hayward, A. C., Eds.; APS Press: Saint Paul, MN, 2005, pp 121–132.

Saile, E.; McGarvey, J. A.; Schell, M. A.; Denny, T. P. Role of Extracellular Polysaccharide and Endoglucanase in Root Invasion and Colonization of Tomato Plants by *Ralstonia solanacearum*. *Phytopathology* **1997**, *87*, 1264–1271.

Salanoubat, M.; Genin, S.; Artiguenave, F.; Gouzy, J.; Mangenot, S.; Arlat, M.; Billault, A.; Brottier, P.; Camus, J. C.; Cattolico, L.; Chandler, M.; Choisne, N.; Claudel-Renard, C.; Cunnac, S.; Demange, N.; Gaspin, C.; Lavie, M.; Moisan, A.; Robert, C.; Saurin, W.; Schiex, T.; Siguier, P.; Thebault, P.; Whalen, M.; Wincker, P.; Levy, M.; Weissenbach, J.; Boucher, C. A. Genome Sequence of the Plant Pathogen *Ralstonia solanacearum*. *Nature* **2002**, *415*, 497–502.

Schaad, N. W.; Jones, J. B.; Chun, W. *Laboratory Guide for the Identification of Plant Pathogenic Bacteria*, 3rd Ed.; American Phytopathological Society: St. Paul, MN, 2001.

Schell, M. A. Control of Virulence and Pathogenicity Genes of *Ralstonia solanacearum* by an Elaborate Sensory Array. *Ann. Rev. Phytopathol.* **2000**, *38*, 263–292.

Schornack, S.; Moscou, M. J.; Ward, E. R.; Horvath, D. M. Engineering Plant Disease Resistance Based on TAL Effectors. *Ann. Rev. Phytopathol.* **2013**, *51*, 383–406. DOI: 10.1146/annurev-phyto-082712-102255.

Silverstein, K. A. T.; Moskal, W. A.; Wu, H. C.; Underwood, B. A.; Graham, M. A.; Town, C. D.; VandenBosch, K. A. Small Cysteine-rich Peptides Resembling Antimicrobial Peptides have been Underpredicted in Plants. *Plant J.* **2007**, *51*, 262–80.

Singh, D.; Yadav, D. K.; Sinha, S.; Mondal, K. K.; Singh, G.; Pandey, R. R.; Singh, R.. Genetic Diversity of Iturin Producing Strains of *Bacillus* Species Antagonistic to *Ralstonia solanacerarum* Causing Bacterial Wilt Disease in Tomato. *Afr. J. Microbiol. Res.* **2013**, *7*, 5459–5470.

Singh, Y.; Kharayat, B. S. *Plant Bacteriology*: Laboratory Manual; G.B.P.U.A.&T.: Pantnagar, 2013, pp 109.

Singh, Y.; Kharayat, B. S.; Gupta, A. K. Sanitary and Phytosanitary Measures for Controlling Plant Pathogens. In *Innovative Approaches in Plant Disease Management*;Singh, K. P., Prajapati, C.R., Gupta, A. K., Eds.; LAP Lambert Academic Publishing GmBH & CO.KG: Germany, 2013, pp 1065–1105.

Tans-Kersten, J.; Brown, D.; Allen, C. Swimming Motility, A Virulence Trait of *Ralstonia solanacearum*, is Regulated by FlhDC and the Plant Host Environment. *Mol. Plant Microbe Interact.* **2004**, *17*, 686–695.

Tans-Kersten, J.; Guan, Y.; Allen, C. *Ralstonia solanacearum* Pectin Methylesterase is Required for Growth on Methylated Pectin but not for Bacterial Wilt Virulence. *Appl. Environ. Microbiol.* **1998**, *64*, 4918–4923.

Tans-Kersten, J.; Huang, H.; Allen, C. *Ralstonia solanacearum* Needs Motility for Invasive Virulence on Tomato. *J. Bacteriol.* **2001**, *183*, 3597–3665.

Terry, K.; Williams, S. M.; Connolly, L.; Ottemann, K. M. Chemotaxis Plays Multiple Roles During *Helicobacter pylori* Animal Infection. *Infect. Immun.* **2005**, *73*, 803–811.

Tomich, M.; Herfst, C. A.; Golden, J. W.; Mohr, C. D. Role of Flagella in Host Cell Invasion by *Burkholderia cepacia*. *Infect. Immun.* **2002**, *70*, 1799–1806.

Trigalet, A.; Trigalet-Demery, D. Use of Avirulent Mutants of *Pseudomonas solanacearum* for the Biological Control of Bacterial Wilt of Tomato Plants. *Physiol. Mol. Plant Pathol.* **1990**, *36*, 27–38.

Turner, M.; Jauneau, A.; Genin, S.; Tavella, M. J.; Vailleau, F.; Gentzbittel, L.; Jardinaud, M. F. Dissection of Bacterial Wilt on *Medicago truncatula* Revealed Two Type III Secretion System Effectors Acting on Root Infection Process and Disease Development. *Plant Physiol.* **2009,** *150,* 1713–1722.

Vailleau, F.; Sartorel, E.; Jardinaud, M. F.; Chardon, F.; Genin, S.; Huguet, T.; Gentzbittel, L.; Petitprez, M. Characterization of the Interaction Between the Bacterial Wilt Pathogen *Ralstonia solanacearum* and the Model Legume Plant *Medicago truncatula. Mol. Plant Microbe Interact.* **2007,** *20,* 159–167.

Van Gijsegem, F.; Gough, C.; Zischek, C.; Niqueux, E.; Arlat, M.; Genin, S.; Barberis, P.; German, S.; Castello, P.; Boucher, C. The *hrp* Locus of *Pseudomonas solanacearum* that Controls the Production of a Type III Secretion System, Encodes Eight Proteins Related to Components of the Bacterial Flagellar Biogenesis Complex. *Mol. Microbiol.* **1995,** *15,* 1095–1114.

Van-Loon, L. C.; Rep, M.; Pieterse, C. M. J. Significance of Inducible Defense-related Proteins in Infected Plants. *Ann. Rev. Phytopathol.* **2006,** *44,* 135–62.

Vasse, J.; Frey, P.; Trigalet, A. Microscopic Studies of Intercellular Infection and Protoxylem Invasion of Tomato Roots by *Pseudomonas solanacearum. Mol. Plant Microbe Interact.* **1995,** *8,* 241–251.

Vasse, J.; Genin, S.; Frey, P.; Boucher, C.; Brito, B. The hrpB and hrpG Regulatory Genes of *Ralstonia solanacearum* are Required for Different Stages of the Tomato Root Infection Process. *Mol. Plant Microbe Interact.* **2000,** *13,* 259–267.

Villa, J.; Tsuchiya, K.; Horita, M.; Opina, N.; Hyakumachi, M. Phylogenetic Relationships of *Ralstonia solanacearum* Species Complex Strains from Asia and Other Continents Based on 16S rDNA, Endoglucanase, and *hrpB* Gene Sequences. *J. Gen. Plant Pathol.* **2005,** *71,* 39–46.

Wang, J. F.; Ho, F. I.; Truong, H. T. H.; Huang, S. M.; Balatero, C. H.; Dittapongpitch, V.; Hidayati, N. Identification of Major QTLs Asociated with Stable Resistance of Tomato Cultivar "Hawaii 7996" to *Ralstonia so*lanacearum. *Euphytica* **2013,** *190,* 241–252.

Wang, J. F.; Olivier, J.; Thoquet, P.; Mangin, B.; Sauviac, L.; Grimsley, N. H. Resistance of Tomato Line Hawaii7996 to *Ralstonia solanacearum* Pss4 in Taiwan is Controlled Mainly by a Major Strain-specific Locus. *Mol. Plant Microbe Interact.* **2000,** *13,* 6–13. DOI: 10.1094/MPMI.2000.13.1.6.

Weller, S. A.; Elphinstone, J. G.; Smith, N. C.; Boonham, N.; Stead, D. Detection of *Ralstonia solanacearum* Strains With a Quantitative, Multiplex, Real Time Fluorogenic PCR (TaqMan) Assay. *Appl. Environ. Microbiol.* **2000,** *66,* 2853–2858.

Wenneker, M.; Verdel, M.; Groeneveld, R.; Kempenaar, C.; van Beuningen, A.; Janse, J. *Ralstonia (Pseudomonas) solanacearum* race 3 (biovar 2) in Surface Water and Natural Weed Hosts: First Report on Stinging Nettle (*Urtica dioica*). *Eur. J. Plant Pathol.* **1999,** *105,* 307–315. DOI:10.1023/A:1008795417575.

Wicker, E.; Lefeuvre, P.; de Cambiaire, J. C.; Lemaire, C.; Poussier, S.; Prior, P. Contrasting Recombination Patterns and Demographic Histories of the Plant Pathogen *Ralstonia solanacearum* Inferred from MLSA. *ISME J.* **2012,** *6,* 961–974.

Wydra, K.; Beri, H. Immunohistochemical Changes in Methylester Distribution of Homogalacturonan and Side Chain Composition of Rhamnogalacturonan I as Possible Components of Basal Resistance in Tomato Inoculated with *Ralstonia solanacearum. Physiol. Mol. Plant Pathol.* **2007,** *70,* 13–24.

Yamada, M.; Ogiso, M. Control of Soil-borne Diseases Using Antagonistic Microorganisms. IV. Study on the Available Substrates for Antagonistic Bacterial Strains to Control Fusarium Wilt of Tomatoes. *Res. Bull. Aichi-Ken Agric. Res. Cent.* **1997,** *29,*141–144.

Yamada, T.; Kawasaki, T.; Nagata, S.; Fujiwara, A.; Usami, S.; Fujie, M. New Bacteriophages that Infect the Phytopathogen *Ralstonia solanacearum. Microbiology* **2007,** *153* (8), 2630–39.

Yao, J.; Allen, C. Chemotaxis is Required for Virulence and Competitive Fitness of the Bacterial Wilt Pathogen *Ralstonia solanacearum. J. Bacteriol.* **2006,** *188* (10), 3697–3708. DOI:10.1128/JB.188.10.3697–3708.2006.

Yao, J.; Allen, C. The Plant Pathogen *Ralstonia solanacearum* Needs Aerotaxis for Normal Biofilm Formation and Interactions with its Tomato Host. *J. Bacteriol.* **2007,** 6415–6424.

Yoshimochi, Y.; Hikichi, Y.; Kiba, A.; Ohnishi, K. The Global Virulence Regulator PhcA Negatively Controls the *Ralstonia solanacearum hrp* Regulatory Cascade by Repressing Expression of the PrhIR Signaling Proteins. *J. Bacteriol.* **2009,** *191,* 3424–3428.

Yuliar, Nion, Y. A.; Toyota, K. Recent Trends in Control Methods for Bacterial Wilt Diseases Caused by *Ralstonia solanacearum. Microbes Environ.* **2015,** *30* (1), 1–11.

Zasloff, M. Antimicrobial Peptides of Multicellular Organisms. *Nature* **2002,** 415, 389–395.

CHAPTER 6

Nematodes: Pest of Important Solanaceous Vegetable Crops and Their Management

AJAY KUMAR MARU*, TULIKA SINGH, and POONAM TAPRE

Department of Nematology, Anand Agricultural University, Anand 388110, Gujarat, India

Corresponding author. E-mail: maruajay@gmail.com

ABSTRACT

Plant parasitic nematodes adversely affect both agricultural production and productivity. Phytonematodes, in general, induce 12–13% loss in crop production. Due to its microscopic size, appearance of nutrition deficiency-like symptom is a challenge in front of farmers. They are also causing disease complex by interacting with other microorganisms and breaking plant resistance. Hence, they are known as "hidden enemy of farmers." Different management practices, such as sanitation and avoidance, host plant resistance, rotation with nonhosts crops, destruction of residual crop roots, and judicious use of nematicides have been attempted to control nematodes effectively. Among different nematode management strategies, chemical control has proved effective but that leads to polluting soil, underground water, and hazardous to all living beings. They also do not remain cost-effective to the end users. Therefore, in this chapter, we discuss about important nematodes pest of some important solanaceous vegetable crops and their effective, eco-friendly, and sustainable management practices to enhance crop productivity and quality as well.

6.1 INTRODUCTION

Plant parasitic nematodes are more virulent against vegetable crops. Different groups of the vegetable crop are prone to several species of nematodes, incurring significant damage/losses. Among the various group of plant pathogenic enemies (bacteria, virus, fungi, nematodes) plant-parasitic nematodes take a heavy toll of these crops. Some earlier milestones of report of plant-parasitic nematodes of vegetables are root-knot nematodes (*Meloidogyne* spp.), onion bloat/stem and bulb nematode (*Ditylenchus dipsaci*), Sugarbeet cyst nematode (*Heterodera schactii*), potato cyst nematode (*Globodera rostochiensis* and *G. pallida*), pea cyst nematode (*Heterodera goettingiana*), which are now the part of history, initiated the realization of the importance of nematodes on vegetable. Nematodes are affecting in almost all types of agricultural important crops, and the majority of nematodes are plant parasites and cause significant damage to crops. Among the phytonematodes, *Meloidogyne* species are one of the most important pests of most crop plants and cause root-knot disease.

Root-knot nematodes, *Meloidogyne* spp., are widely distributed in the tropics and subtropics and are common in temperate regions. Severe infestations cause total crop loss, while yield loss of 5–20% arise in some crops despite routine use of nematicides. More than 40 species of root-knot nematodes have been reported worldwide, out of them only four species, M. *arenaria, M. hapla, M. incognita,* and *M. javanica* are responsible for 95% of the damage. These species attack more than 2000 plant species, including most crop plants. Four species of root-knot nematodes, namely, *M. incognita, M. javanica, M. hapla,* and *M. arenaria* have been reported and all form conspicuous root galls. *M. incognita* and *M. javanica* are most widespread in distribution and have a wide host range among vegetables, whereas *M. hapla* attack potato and *M. arenaria* infect chilies.

Several phytonematodes have been found to be associated with horticultural crops in India, not much is known about their relative importance as potential pathogens of some have been consistently reported around plant roots and there is no information on economic loss caused by them individually and in concomitant association with each other. An estimated overall average annual yield loss of the world's major vegetable crops by nematode is 12.3%. Average losses for the 40 crops in developed countries were estimated to be 8.8% compared with 14.6% for developing countries. The root-knot nematode is commonly encountered causing damage to vegetable crops namely, tomato, brinjal, chili, okra, ginger, etc., leading to yield loss up to

31.7%. Economic losses in various solanaceous crops caused by nematodes particularly root-knot nematodes were reported as below:

Crop	Nematodes	Yield loss (%)	References
Tomato	*Meloidogyne incognita*	30.57–46.92	Bhatti and Jain, 1977; Reddy 1985; Darekar and Mahse, 1988
	Meloidogyne javanica	77.50	Anon, 1993
	Rotylenchulus reniformis	42.25–49.02	Subramaniyam et al. 1990
Brinjal	*Meloidogyne incognita*	27.30–48.55	Bhatti and Jain, 1977; Parvatha Reddy and Singh, 1981; Darekar and Mahse, 1988
Chili	*Meloidogyne* spp.	24.54–28.00	Singh et al. 2003
Potato	*Meloidogyne incognita*	42.50	Prasad, 1989
	Globodera rostochiensis	99.50	Prasad, 1989

6.2 NEMATODE ASSOCIATED WITH TOMATO CROP

6.2.1 ROOT-KNOT NEMATODE

The diagonstic symptoms exhibited above ground are stunted growth, yellowing of leaves, and wilting appearance in patchy areas. Medium to severe galling formed on roots is the typical below ground symptoms (Figs. 6.1, 6.6, and 6.7).

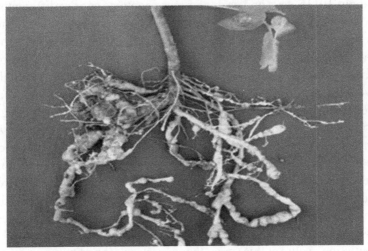

FIGURE 6.1 **(See color insert.)** Heavy galling on tomato roots.

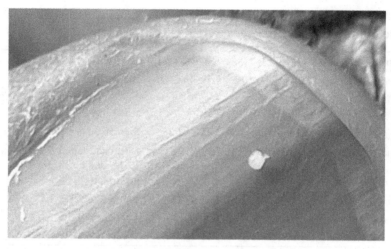

FIGURE 6.2   (See color insert.) White mature female of root-knot nematode on thumbnail.

6.2.1.1   LIFE CYCLE

It is a sedentary endoparasitic nematode. Female are saccate (Fig. 6.2), second-stage juvenile (J_2) is vermiform, and males are vermiform but nonparasitic. Reproduction is by parthenogenesis. Rectal gland of the female secretes gelatinous substance on the root surface in which eggs are deposited. Oviposition continues for 10–12 days, each female lays eggs about 200–400 eggs held together in an egg mass or egg sac. Embryogenesis takes about 10–15 days and each egg-mass contains eggs at different stages of development. First-stage juveniles (J_1) moult while still within the eggshell and become J_2. Root-knot nematode does not require any specific stimulus from the host and hatch freely in the water, depending on the availability of suitable temperature and moisture J_2 hatched out. Initially, the J_2 move in the soil randomly, but once in the vicinity of host roots, they are attracted toward them due to the presence of root exudates emanating from the roots. The J_2 penetrate root just behind the root tip (meristematic zone). Penetration is facilitated by repeated stylet thrust and/or enzymes secreted by the esophageal glands. The J_2 move through the root cell and position themselves with head located near the vascular tissues, while the rest of the body completely inside the cortex. All this stage, J_2 become sessile and initiates the development of feeding sites (giant cells). As the feeding process begins, J_2 start assuming swollen shape, now called parasitic J_2. Sex differentiation occurs at this stage, the juveniles destined to become females acquire V-shaped genital primordium, while in males it is I-shaped. Under optimum conditions, second moult occurs in about week and J_3 if formed. The third moult follows quickly and the J_3 convert to J_4. The J_3 and J_4 retain the old cuticles, the pointed tail of J_2 still visible, and hence are called "spike-tailed" stages. The body grows in width, genital primordial develop further, but these stages are nonfeeding as they lack stylet. At the last molt, the adult female becomes sac-like or pear shape. Stylet reappears, and the reproductive system gets fully developed with vulval opening making its appearance. Adult males are, however, vermiform, coiled inside J_4 cuticle, emerge out, and leave roots to come out into the soil. They are short-lived. Adverse environmental conditions after penetration may induce maleness in developing juveniles.

The whole life cycle is completed in about 25 days at 25–30°C, which is optimum for most species. During winter season under north India conditions, life cycle duration may be prolonged to 60–80 days depending on prevailing temperature. Thus, 7–8 overlapping generations are completed in a year depending on the host and environmental conditions (Fig. 6.3).

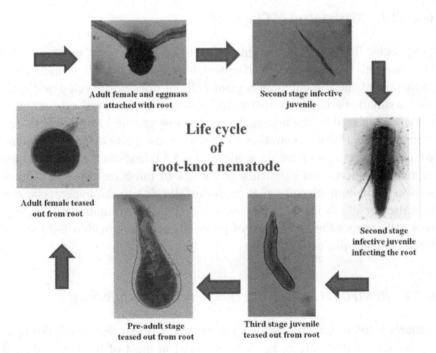

Adult female and eggmass
attached with root

Second stage infective
juvenile

**Life cycle
of
root-knot nematode**

Adult female teased
out from root

Second stage
infective juvenile
infecting the root

Pre-adult stage
teased out from root

Third stage juvenile
teased out from root

FIGURE 6.3   Life cycle of root-knot nematode.

6.2.1.2   GIANT CELLS

The free-living infective J_2 enter the plant roots and established themselves in the cortex while head lies near the vascular tissues. The enzyme secreted by oesophageal glands is injected by the stylet into the host cells and initiate a chain of reaction leading to the formation of feeding sites. Endodermal, pericycle, xylem, and phloem tissues in the root vicinity of nematode head undergo hypertrophy. Karyokinesis takes place without cytokinesis. Consequently, some 8–10 cells involving these tissues around the nematode head become enlarged, multinucleate with dense cytoplasm, showing hypermetabolism. These "giant cells" functions much like transfer cells or metabolic sinks where the nutrients absorbed by the roots are continuously pooled and diverted to nematode for its feeding, growth, and development. The disruption in the continuity of the conducting vessels obstructs the flow of nutrients and the water to the shoots and leaves, leading to reduced plant growth and yield.

6.2.1.2.1   Mechanism of Giant Cells

Giant cells formation and maintenance is very necessary for successful host–parasite relationship, and if a nematode unable or fails to induce the formation of these feeding sites/giant cells, it does not develop and dies. Such a situation arises in the incompatible host. Simultaneously, the protease enzymes released by the nematodes act on host protein breaking them into amino acids. The concentration of amino acids, particularly tryptophan, which is a precursor of indole acetic acid (IAA), leads to the accumulation of auxins or hormonal imbalance at the site of infection. Thus, instead of growing longitudinally, the root grows axially due to the hyperplasia and hypertrophy of cortical parenchyma cells. Thus, the formation of swelling root galls or root-knot at the site of juvenile penetration within 1–2 days of infection takes place.

6.2.2   RENIFORM NEMATODE (Rotylenchulus reniformis)

General stunted growth, yellowing of leaves, wilting, and deterioration in the quality of fruits are commonly observed in most of the host. Infected roots generally show necrosis and feeder roots may be destroyed.

6.2.2.1   LIFE CYCLE

Sedentary semiendoparasite of roots, preadult stage (immature female/J_4) is infective. The nematode head located near the stellar region, while rest of the body remains outside. Within a week's time, the posterior half of the body becomes swollen and reniform (kidney shaped). Eggs are deposited in an eggs mass surrounding the body on the root surface. The number of eggs laid by a single female range in between 40 and 80, although it varies with the host plants and environmental conditions. Second-stage juveniles hatch out from the eggs and develop to immature females without feeding through a series of three moults in the soil, a characteristic peculiar to this nematode. The life cycle is completed in about 25 days, and 25–30°C is the most optimum temperature for its development. Loamy and clay loam soils are considered more favorable for the reproduction of this nematode. Males are nonparasitic, but reproduction is by amphimixis in most of the population.

6.2.3   INTERACTION WITH OTHER PATHOGENS

6.2.3.1   BACTERIA

Pani and Das (1972) have reported the association of root-knot nematode with bacterial wilt of tomato.

Crop	Nematode	Pathogen	Reference
Tomato	*M. incognita*	*R. solanacearum*	Napiere and Quinio, 1980
	M. javanica	*R. solanacearum*	Libman et al. 1969
	Meloidogyne spp.	*Clavibacter michiganense*	Demoura et al. 1975
	H. nannus	*R. solanacearum*	Libman et al. 1969

6.2.3.2 FUNGI

M. incognita and *Rhizoctonia bataticola* or *Sclerotium rolfsii* when inoculated simultaneously in the soil reduced the germination of seeds in okra, brinjal, and tomato (Chhabra and Sharma, 1981; Shukla and Swarup, 1970). The incidence and severity of root rot of brinjal, tomato, and okra caused by the soilborne fungi such as *R. solani, R. bataticola,* and *Phomopsis vexans* was increased in the presence of *M. incognita* (Hazarika and Roy, 1974;

Chahal and Chhabra, 1984; Chhabra et al., 1977; Sharma et al., 1980). Jenkins and Coursen (1957) induced wilting in Fusarium wilt—resistant tomato variety Chesapeak only when root-knot nematode present along with fungal inoculum. Furthermore, when *M. hapla* was combined with the fungus, only 60% of the plants wilted, whereas *M. incognita acrita* promoted wilt in 100% of the plants. Bhagawati and Goswami (2000) reported the interaction of *M. incognita* and *F. oxysporum f. sp. lycopersici* on tomato (*Lycopersicon esculantum*). It was found that when either or both pathogens were inoculated simultaneously or nematode was inoculated 10 days prior to inoculation of the fungus, the symptoms were visible by 20 days as against no such symptoms in plants where the fungus was inoculated 10 days prior to nematodes. The intensity of wilt after 40–60 days of inoculation was significantly higher when simultaneously inoculated of fungus. The number of galls and egg masses and nematode population in soil was significantly reduced when simultaneously inoculated and prior inoculation. *M. javanica* increased the extent of damage by pre- and postemergence phases of damping off caused initially by *R. solani* and *Pythium debaryanum* in tomato (Nath et al., 1984).

Nematode fusarium wilt disease interaction		
Tomato	*M. incognita acrita and M. hapla*	*F. o. f. sp. lycopersici*	Jenkins and Coursen, 1957
.	*M. incognita and M. javanica*	*F. o. f. sp. lycopersici*	Bowman and Bloom, 1966
			Harrison and Young, 1941
			Chahal and Chhabra, 1984
			Hasan and Khan, 1985
	M. incognita	*F. o. f. sp. lycopersici*	Bhagawat and Goswami, 2000
	P. coffeae and P. penetrans	*F. o. f. sp. lycopersici*	Hirano and Kawamura, 1972
Nematode–Verticillium wilt disease interaction		
	P. penetrans	*V. alboutrum*	Conroy et al. 1972
	T. christiei	*V. alboutrum*	Conroy and Green, 1974
	G. tabacum	*V. alboutrum*	Miller, 1975
	T. capitatus	*V. alboutrum*	Overman and Jones, 1970
	M. incognita	*V. alboutrum*	Overman and Jones, 1970
	M. javanica	*V. dahlia*	Orion and Krikun, 1976
	M. incognita	*R. solani*	Chahal and Chhabra, 1984
	M. incognita	*Sclerotium rolfsii*	Shukla and Swarup, 1970

6.2.4 MANAGEMENT

6.2.4.1 PHYSICAL METHODS

Important pests controlled by greenhouse steaming are cyst nematode of potato attacking tomatoes and root-knot nematodes on tomatoes. Soil solarization or burning of paddy husk or sawdust on infected nursery proved to be effective.

Soil solarization of nursery beds with 100 gauge LLDPE clear plastic film during summer (in between April 15th and June 15th) for 15 days is recommended. It increased the number of transplantable seedlings by 615% giving a net profit of Rs. 7661 from 1000 sq.m. (ICBR 1:5.65) and decreased the root-knot disease and weeds by 66% and 93%, respectively, in Table 6.1.

Rubbing of root-knot infested tomato nursery with bajra husk at 7 kg/sq.m. a week prior to seeding is recommended for economic and effective management of root-knot disease (ICBR 1:5.12).

TABLE 6.1   Effect of Soil Solarization of Nursery Beds on Root Galling and Number of Transplantable Seedlings of Tomato in 1000 sq.m.

Treatments	No. of seedling	Weight of 100 seedlings	Root-knot index	Net profit	CBR
With plastic	260,400	36 kg	1.37	Rs. 7661	1:5.65
Without plastic	36,400	26 kg	3.97	-	-
% increase (+)/ decrease (−) over control	(+) 615	(+) 38	(−) 66	-	-

6.2.4.2 CULTURAL METHODS

Selection of nematode-free nursery area, removal and destruction of infested roots and plants debris after crop harvest, crop rotation with non-host crops (maize, wheat, sorghum) or antagonistic crops (marigold, mustard, sesame), soil solarization, flooding, fallowing, deep summer plowing, and harrowing either alone or in combination proved to be more effective and economical to reduce multiplication of root-knot, reniform, and lesion nematodes on the transplanted vegetable crops like tomato, brinjal, and capsicum. Deep summer plowing is the best way to disturbance and instability in nematode community and also causes their mortality by exposing them to direct solar heat and desiccation. Three summer plowings each at 10 days interval during

June leads to the reduction in *M. javanica* population while fallowing itself during the same period registered 44.5% reduction. (Jain and Bhatti, 1987).

Additional use of plastic sheets for covering soil either in nursery beds or in the field further enhance nematode reduction. Such an approach also helps in reducing the intensity of weeds, fungi, and bacteria in soil. Jain and Bhatti (1987) reported that planting of nematode-free seedlings in an infested field which received three summer plowings increased the yield of tomato by 55%.

The decrease in *M. incognita* population on tomato occurs when the field was left fallow or following crop rotation with marigold, spinach, and bottle gourd (Khan et al., 1975). To control *M. incognita* and *M. javanica*, a tomato crop can be followed by peanut without risk of damage to the peanuts. While the peanuts are growing, nematode cannot be produced. Instead many of the larvae in soil die or become noninfective because of starvations and the attacks of predators, fungi, and diseases. If the population is reduced sufficiently, tomatoes can be grown again after peanuts without serious injury (Taylor and Sasser, 1978). Tomato rotated with cotton-wheat-cotton resulted in low gall index due to *M. javanica* (RKI 1-2) and gave maximum yield (38% increases over continuous tomato).

Sesamum-tomato sequence produced the least reproduction rate (0.47), least root-knot index (1.33), and highest yield (86 q/ha) of tomato, which was at par with niger-tomato sequence that had reproduction rate of 0.56, root-knot index of 1.66, and highest yield of 75.5 q/ha of tomato (Sahoo et al., 2004). In cropping sequence studies for the management of *M. javanica* on tomato, nematode population decreased at low levels when carrot, capsicum, and onion were grown (Kanwar and Bhatti, 1992). The population of *M. incognita* was reduced consistently under groundnut-mustard-tomato cropping sequence (Sharma et al., 1980). Intercropping with marigold or mustard with tomato reduced the damage of root-knot and reniform nematodes. The root-knot development on tomato was low when interplanted with *Tagetes erecta*. The population of *Tylenchorhynchus, Helicotylenchus, Hoplolaimus, Rotylenchulus,* and *Pratylenchus* was also markedly reduced (Alam et al., 1977; Khan et al., 1971). Terthienyl is the active principle in *Tagetes* spp. which is toxic to these nematodes. Tomato nursery beds previously planted with trap crop (marigold) gave the maximum reduction in root-knot nematode population in soil, which increased the germination of tomato seeds and production of more healthy (nematode-free) seedlings (Rangaswamy et al., 1995). Marigold intercrop with tomato at 1:4 and 1:6 ratio and mustard at 1:2 ratio was found to be effective in reducing root galls, egg masses, and nematode population at harvest and gave cost:benefit ratio of 1:8.36, 1:7.88, and 1:3.31, respectively (Rangaswamy et al.,1999).

Interculture of onion with tomato reduced gall index due to *M. javanica* significantly and improved tomato fruit yield. Intercropping with two rows of onion with one row of tomato was the most effective treatment in reducing the nematode infection and population and enhancing tomato yield (Ram and Gupta, 2001). In pots having *Zinnia elegans* with tomato, the population of *M. incognita* and *R. reniformis* declined. The root-knot index on tomato was reduced to 1.08 compared to 3.75 in control. The nematode reproduction factor of *M. incognita* and *R. reniformis* was reduced to 0.88 and 0.80, respectively, compared to 3.86 and 2.43 in control (Tiyagi et al., 1986).

Cultivation of tomato with castor results in considerable reduction in the number of root-knot galls on tomato (Hackney and Dickerson, 1975). The inclusion of rice, maize, groundnut, and *Stylozanthes* in rotation sequences managed root-knot nematodes and increased yields of tomato. In crop rotation trial, it was found that a crop of groundnut or strawberry grown prior to tomato in soil heavily infested with root-knot nematodes greatly increased yields of tomato. Rotation of tomato with leek (*Allium porrus*), onion, groundnut, and sun hemp reduces the population of root-knot nematodes. Increased level of potash has significantly reduced the number of galls by *M. javanica* in tomato (Gupta and Mukhopadhyaya, 1971).

The growth of tomato increased (35%) and root galling due to *M. incognita* (root-knot index 1 as against 4 in control), nematode population in soil (80% reduction), and egg mass production reduced (root-knot index 1 as against 4 in control) in soil amended with 20% fly ash (Haq et al., 1985). The increase in plant growth may be ascribed to increased nutrient availability (N, K, Ca, Mg, Na, B, SO_4) and the reduction in nematode population might be due to toxic compounds (polycyclic aromatic hydrocarbon, dibenzofuran, and dibenzo-p-dioxin mixtures) present in fly ash. Row application of fly ash at 0.6 kg/sq.m enhanced the yield of tomato by 90.4% (Khan and Ghandipur, 2004). Red plastic mulch suppresses root-knot nematode damage in tomatoes by diverting resources away from the roots (and nematode) and into foliage and fruit (Adams, 1997). Soil application of neem and subabul leaves to nursery beds at 0.5 t/ha gave better seedlings growth and reduced root galling in tomato (Jain et al., 1988). Singh and Sitaramaiah (1966) applied finely divided oil cakes to root-knot infested soil and noted a reduction in disease incidence in tomato. The oil cakes (neem, groundnut, mustard, and castor) were effective in reducing parasitic nematode population (*Hoplolaimus, Tylenchorhynchus, Meloidogyne,* and *Helicotylenchus* species) and in increasing yields of tomato. Application of neem cake @ 15 g/spot or 100 g/m furrow, 3 weeks prior to transplanting of tomato and brinjal have been reported to give 36–450% enhanced yield with the corresponding reduction in gall index to 2.3 as against 4.3 in control (Anon., 1992). The combination of naffatia (*Ipomea fistulosa*) or water hyacinth (*Eichhornia crassipes*) or aak (*Calotropis procera*) or congress grasses (*Parthenium hysterophorus*) each @ 3 kg/m² in the soil before flowering and 15 days prior to seeding has been found effective for management of root-knot disease in nematode infested tomato nursery (Anon.,1992). Best result in respect of root-knot reduction due to *M. javanica* and increase in yield of tomato were obtained by mixing the soil with saw dust @ 2.5 t/ha 3 weeks before planting and then applying N through urea at 120 kg/ha (Singh and Sitaramaiaha, 1971). Rice hull ash at 2.5 t/ha increased the yield of tomato by 133–317% and reduced root-knot incidence by 46–100% (Sen and Dasgupta, 1981). Application of poultry manure at 2.0, 2.5, and 3.0 t/ha days prior to tomato seedlings is recommended for effective management of root-knot disease and production of more transplantable seedlings in tomato nursery giving ICBR of 1:4.54, 1:5.63, and 1:6.68, respectively. Kaplan and Noe (1993) tested five dosages of poultry litter (10–45 t/ha) and found an inverse relationship between dosage and both the total number of *M. arenaria* in tomato roots and the number of eggs in the soil. Green manuring with naffatia/besharami or neem leaves (ICBR 1:2.87) or water hyacinth (ICBR 1:2.527) or calotropis (ICBR 1:2.26) or congress grass (ICBR 1:2.09) cash at 3 kg/sq.m 15 days prior to seedling is recommended for economic and effective management of root-knot disease in tomato nursery. Incorporation of green materials of naffatia and neem leaves 2 kg/1.44 sq.m (ICBR 1:15.7) or congress grass and neem cake at 2 kg/1.44 sq.m (ICBR 1:15.6) or calotropis, naffatia, congress grass, and neem leaves each at 1 kg/1.44 m² (ICBR 1:14.7) 15 to 20 days prior to tomato seedling gave effective management of root-knot nematodes and production of more healthy seedlings tomato nursery in *kharif*. Gautam et al. (1995) studied integrated management of *M. incognita* on tomato using *Paecilomyces lilacinus, Baciilus subtillis,* and green manuring with *Eicchornia crassipes* alone and in combination for the management of *M. incognita* and revealed that combined used of these two biocontrol agents along with the green manuring of *E. crassipes* resulted in greatest plant growth of *M. incognita* inoculated plants. Soil amendment with fresh glyricidia leaves, fresh saw dust, and chicken manure controlled the root-knot and reniform nematode population on tomato (Castillo, 1985). The yield increases and the nematode suppressive effect of fresh chicken manure surpassed even phenamiphos.

6.2.4.3 BIOLOGICAL METHODS

Pasteuria penetrans and *Pseudomonas fluorescens* are the two most popular antagonistic bacteria for management of nematodes. *P. penetrans* effectively parasitized *M. incognita* in rotations that included tomato, eggplant, and beans or cabbage (Amer-Zareen et al., 2004), but its efficacy depended on cropping techniques and soil conditions. *P. fluorescens* also provide effective control of root-knot nematodes on vegetable crops (Haseeb and Kumar, 2006; Krishnaveni and Subramanian, 2004; Stalin et al., 2007).

Egg-parasitic fungi *Paecilomyces lilacinus* and *Pochonia chlamydosporia* are probably the most effective egg-parasites. *P. lilacinus* has been proven to successfully control root-knot nematodes, *M. javanica* and *M. incognita* on tomato, eggplant, and other vegetable crops (Verdejo-Lucas et al., 2003; Goswami and Mittal, 2004; van Damme et al., 2005; Goswami et al., 2006; Haseeb and Kumar, 2006; Kumar et al., 2009). Application of Royal 350 (*Arthrobotrys irregularis* culture on oat medium) @ 140 g/m^2 a month before transplanting of tomato resulted in good protection against root-knot nematodes. Among the egg parasitic, the efficacy of *Paecilomyces lilacinus* has been found to be comparatively higher in suppressing the population of *Meloidogyne* spp. and *R. reniformis* on tomato.

Bare root dip treatment of tomato seedlings in a spore suspension of *P. lilacinus* @ 4 × 10^5 cfu/mL reduced *Meloidogyne* infestation in roots by 50% and increases the growth of seedlings by 20%. Application of organic amendments increases the effectiveness of *P. lilacinus* against *M. incognita*.

Two bacteria *Bacillus licheniformis* and *Pseudomonas mindocina* and two fungi *Acrophialophora fusispora* and *Aspergillus flavus* were evaluated for their efficacy as biocontrol agent individually and in combination against *M. incognita* infecting tomato. Individually, *B. licheniformis* gave best management among although when all four agents were used, nematode multiplication was reduced by 82.06%. Application of *B. thuringiensis* causes 95% mortality of *M. javanica* juveniles due to β-exotoxin production and suppresses gall formation, egg mass production, and nematode production. Root material containing *P. penetrans* spores @ 212–600 mg powder/kg soil, when broadcasted in the soil, showed reduction in galling on tomato roots and final nematode population. Tomato yield increased by 20% over the untreated check, when *P. penetrans* was introduced to the nursery soil @ 1 × 10^5 spores/g soil. The persistence of *P. penetrans* in field soil was evident since about 50% of the final juvenile population was encumbered with bacterial spore (Walia and Dalal, 1994). *P. penetrans* is highly resistant to heat, desiccation, and the ability to survive for more than 2 years in soil qualifies as the most potential bicontrol agents against root-knot nematode infesting vegetable crops. Application *of Psedomonas fluorescence* at 10 g/m^2 in nursery bed gave good results to manage root-knot nematodes (Shanthi and Sivakumar, 1995).

AMF, *Glomus fasciculatum* was effective against root-knot nematode and *Glomus mosseae* against reniform nematode on tomato. Organic amendments (Oil cake, calotropis leaves) in combination with AMF enhance the colonization of AMF on tomato roots which further increases plant growth and reduced gall index. Sundaram and Arangarasam (1996) observed that inoculation of *G. fasciculatum* at 10 g/plant was effective in increasing the uptake of nutrients and yield and reduced galling followed by *G. mosseae, Gigaspora margarita,* and Acaulospora leaves. Application of *G. fasciculatum* at 50 g/m^2 (200 spore/g of soil) gave a minimum number of galls per root system (39.2) at harvest and recorded maximum yield (26.47 t/ha) with an increasing B:C ratio (1:3.34) and decrease nematode population (Kiran kumar et al., 2004).

Avermectins B1 and B2a (applied to soil through drip irrigation systems) at rates ranging from 0.093 to 0.34 kg a.i./ha applied as a single dose or 0.24 kg a.i./ha applied as three doses each at 0.08 kg on tomatoes against *M. incognita* were as effective as oxamyl and aldicarb at 3.36 kg a.i./ha (Garabedian and Van Gundy, 1983). The aqueous solution of avermectins (250 mL of 0.001%/m^2) significantly reduced root galls of tomato seedlings raised in root-knot-infested nursery beds, which yielded robust and healthy seedlings with no root infection (Parvatha Reddy and Nagesh, 2002). In root-knot infested microplot studies, avermectins at 0.015 kg/ha effectively controlled *M. incognita* on tomato and increased yield by 11% and 8% compared to untreated and carbofuran treated plots, respectively (Reddy and Nagesh, 2002). Seedling bare root-dip treatment in avermectin 100% gave highest reduction in root-knot index (1.3), soil nematode population (132.8), number of females/g root (12.8), egg masses/g root (5.4), and eggs/egg mass (125.2) in tomato (Jayakumar et al., 2005).

6.2.4.4   HOST RESISTANCE

Tomato cvs. Nematox, SL-120, NTR-1, SL-12, Patriot, VFN-8, VFN Bush, Piersol, Nemared, Ronita, Anahu, Bresch, Helani, Campbell-25, Punuui, Arka Vardan, Pelican, Hawaii-7746, Hawaii-7747, and Hisar lalit have been reported to be resistant to root-knot nematode.

In the field with resistance tomato (Hisar lalit) grown for a year, the yield was 27% higher than susceptible variety (HS-101) (Kanwar and Bhatti, 1990).

Balasubramanian and Ramakrishana (1983) observed that Kalyanpur Selection–I, Kalyanpur–III, CA-121 exhibited resistance reaction.

6.2.4.5 INTEGRATED METHODS

Nursery bed treatment with carbofuran @ 0.3 g a.i./m^2 + bare root dip treatment with carbosulfan (25 EC) @ 500 ppm for 1 h before transplanting. Seedlings may be raised in solarized nursery beds treated with carbofuran @ 0.3 g a.i./m^2 + neem cake @ 500 kg/ha in nematode-infested fields 10 days before transplanting in the field. The application of carbofuran @ 7 g/m^2 at sowing time in Haryana and the Bacillus macerans @ 25 g/m^2 along with 2% solution of formulated product of B. macerans is drenched at 7 DAS is advisable in vegetable crop nursery bed against root-knot nematodes in Kerala, respectively.

Nursery bed treatment with carbofuran @ 0.3 g a.i./m^2 reduced root-knot nematode. Seedlings of tomato raised with solarized bed treated with carbofuran 0.3 g a.i./m^2 and transplanted in the main field where neem cake is applied @ 500 kg/ha showed an appreciable reduction in nematode population in M.P., Rajasthan, and Maharashtra for management of root-knot nematode. Soil solarization with LLDPE sheet for 3–4 weeks during summer + application of P. fluorescence 10 g/m^2 in nursery area in Tamil Nadu for management of root-knot nematode.

In the nursery, integration of *Pasteuria penetrans* (28 × 10^4 spore/m^2), *P. lilacinus* (at 10 g/m^2 with 19 × 10^9 spores/g), and neem cake (at 0.5 kg/m^2) gave maximum plant growth and number of seedlings per bed. In the field, planting of tomato seedlings (raised in nursery beds amended with neem cake + *P. penetrans*) in pits incorporated with *P. lilacinus* (at 0.5 kg/ m^2) gave least root galling and nematode multiplication rate and increased fruit weight and yield of tomato (Parvatha Reddy et. al., 1997). *P. lilacinus* in combination with castor leaves reduced population up to 89% and increase plant growth and yield (Zaki and Bhatti, 1991).

The combined application of *P. penetrans* + *T. viride* + neem/castor cake at 1/3 dose was significantly superior compared to their individual applications in terms of increased plant growth and reduced root galling, egg mass production, and final populations of *M. incognita*. Similarly, total parasitization was higher in plants treated with bioagents in terms of number of juveniles encumbered and adult female infected with *P. penetrans* and egg parasitize by *T. viride* compared to the individual application (Rangaswamy et al., 1999).

Integration of soil solarization of nursery bed (with LLDPE transparent film of 25 μ for 15 days) with incorporation of Calotropis leaves at 4 kg/1.44 m² resulted in increased plant height (29.9 cm compared to 20 cm in control), transplantable seedlings (546/m² compared to 297 in control), fresh shoot weight (176.49 g compared to 100 g in control) with minimum root-knot index (1.83 compared to 3.85 in control (Patel et al., 2006). Soil solarization with clear LLDPE film (25 μ) for 15 days in hot summer in combination with poultry manure at 2 t/ha proved effective in the management of nematodes and higher production of transplants in tomato nursery with cost-benefit ratio of 1:3.10.

Combination of deep plowing (up to 20 cm) and nursery bed treatment with aldicarb at 0.4 g per m² and main field treatment with aldicarb/carbofuran at 1 kg a.i./ha proved effective in control of root-knot nematodes in tomato, which also registered maximum yield (Jain and Bhatti,1985).

In Tomato, application of aldicarb and carbofuran each at 1 kg a.i./ha in combination with neem cake and urea each at 10 kg N per ha at transplanting produced maximum yield with lowest gall index (2.5) and nematode population, 90 days after planting (Routaray and Sahoo, 1985).

The combination of treatment with *T. harzianum* + *T. viride* each at 50 g (4 × 10⁸ cfu/g) considerably increased plant growth, yield, and reduced the root galls and soil nematode population (Hassan and Sobita Devi, 2004). Integration of two bioagents, *P. lilacinus* and *P. chlamydosporia* resulted in combined and complementary effect for the successful management of *M. incognita* infecting tomato (Rao and Parvatha Reddy, 1992).

Maheshwari et al. (1987) reported that application of *P. penetrans* in combination with carbofuran, aldicarb, miral, sebufos, and phorate significantly improve plant growth of tomato by greatly reducing galling due to *M. javanica*. Nursery bed treatment with *P. fluorescens* (with 1 × 10⁹ spores/g) and *P. chlamydosporia* (with 4 × 10⁶ spores/g) each at 20 g/m² and field application of 5 t of enriched FYM with above bioagent each at 5 kg significantly reduced reniform nematodes in tomato by 72% overcheck. The yield increase was up to 21.7% with cost:benefit ratio of 1:4.9.

Application of neem cake in the nursery at 100 g/m² followed by carbofuran at 1 kg a.i./ha in the main field significantly reduced the soil population of *R. reniformis* and enhanced the fruit yield of tomato by 60% (Anitha and Subramanian, 1998). Integration of a bioagent *P. lilacinus* with carbofuran at 1 kg a.i./ha was found effective in the management of reniform nematode, *R. reniformis* infecting tomato (Parvatha Reddy and Khan, 1988). Solarization of nursery bed, treatment of carbofuran, and seed treatment with carbosulfan 25 DS @ 3% a.i. (w/w) were used for managing root-knot nematode in vegetable in U.P.

6.3 NEMATODE ASSOCIATED WITH BRINJAL CROP

There are two important nematodes that are infecting brinjal crop (a) root-knot nematode (*M. incognita and M. javanica*) and reniform nematode (*Rotylenchulus reniformis*). Both interact with some plant pathogenic bacteria and fungi.

6.3.1 BACTERIAL INTERACTION

The root-knot nematode, *M. incognita,* present along with *Ralstonia solanacearum* greatly increases the incidence of bacterial wilt of brinjal. The root-knot nematode is responsible for breaking bacterial wilt resistance Pusa Purple Cluster cultivar of brinjal (Parvatha Reddy et al., 1979). *M. incognita* interacts with *R. solanacearum* in increasing the speed of development and severity of wilt disease of eggplant. Maximum disease occurred when the bacterium and nematode were inoculated simultaneously. Inoculation with *R. solanacearum* alone resulted in less severe disease. It was concluded that root-knot nematodes modify the plant tissue facilitating bacterial colonization (Nayar et al., 1988).

6.3.2 FUNGAL INTERACTION

M. incognita and *Ozonium texanum var. parasiticum* together acted synergistically by reducing the germination of brinjal to 28% (Nath et al., 1976).

Eggplant	*M. incognita*	*F. oxysporum*	Smits and Noguera, 1982
	P. penetrans	*V. alboutrum*	Mckeen and Mountain, 1960
	M. incognita	*R. bataticola*	Chhabra and Sharma, 1981
		R. solani	Hazarika and Roy, 1974

6.3.3 MANAGEMENT

6.3.3.1 CULTURAL METHODS

Several cultural practices like selection of nematode-free nursery area, destruction of infested roots and plant debris after crop harvest, crop rotation with nonhost crops (maize, wheat, sorghum) or antagonistic crops (marigold, mustard, sesame), soil solarization, flooding, fallowing, deep summer plowing, and harrowing either alone or in combination proved to be effective and economical to reduce population density or multiplication of root-knot, reniform, and lesion nematodes on the transplanted vegetable crops like tomato, brinjal, and capsicum. Three summer plowings each at 10 days interval during June leads to the reduction in *M. javanica* population while fallowing itself during the same period registered 44.5% reduction (Jain and Bhatti, 1987). Additional use of plastic sheets for covering soil either in nursery beds or in the field further enhance nematode reduction. Such an approach also helps in reducing the intensity of weeds, fungi, and bacteria in soil. Bed treatment with neem cake at 200 g/m^2 gave a maximum reduction in nematode population giving the yield of 20.75 q/ha (Sheela and Nisha, 2004). The maximum reduction in *M. javanica* population and increase in fruit yield was recorded in brinjal by soil plowing + covering with polythene sheet, followed by soil plowing + exposure to the sun (Anon.,1989). The decrease in *M. incognita* population on brinjal occurs when the field was left fallow or following crop rotation with marigold, spinach, and bottle guard (Khan et al., 1975). Rotation of brinjal with sweet potato (cv. Shree Bhadra) reduced the root-knot nematode population by 47% and increased the fruit yield by 22% (Sheela et al., 2002). Crop rotation with sorghum, wheat, and chili reduced nematode population. The root-knot population is completely in marigold-marigold-fallow and Chilli-cauliflower-cauliflower. Garlic fallowing brinjal also brought about considerable reduction in larvae. The population levels of *M. incognita* on highly infested land dropped to zero after 18 months of continuous cultivation of *Panicum maximum*. Growing of *Panicum maximum* after was completely free from root-knot nematodes (Netscher, 1983). Intercropping brinjal with marigold, garlic, or resistant tomato (SL-120) improved plant growth and reduced number of galls and final soil population of nematode (Jain et al., 1990). Marigold (cv. Calcutta Yellow)-brinjal cropping sequence has been recommended to the farming community. Incorporation of neem cake 500 g/sq.m or 250 mL of 5% neem leaf extract/5% neem cake extract/sq.m in nursery beds of brinjal gave effective control of root-knot nematodes. Row application of fly ash at

0.6 kg/sq.m. enhances fruit yield by 27.7% and inhibited the reproduction of *M. incognita* (Khan and Ghadipur, 2004). Significant reduction in gall formation and less number of female nematodes in brinjal roots due to the addition of chopped leaves of *C. procera* and *Ricinus communis* was reported by Nandal and Bhatti (1990).

6.3.3.2 BIOLOGICAL METHODS

Application of Royal 350 (*Arthrobotrys irregularis* culture on oat medium) at 140 g/m² a month before transplanting of tomato resulted in good protection against root-knot nematodes. Among the egg parasitic, the efficacy of *Paecilomyces lilacinus* has been found to be comparatively higher in suppressing the population of *Meloidogyne* spp. and *R. reniformis* on tomato. Soil application of *P. lilacinus* at 20 g/sq.m to brinjal nursery reduced infection of *M. incognita* by 50% and improved growth of seedlings by 30%.

Application of *P. penetrans* in nursery soil of brinjal resulted in 31% and 36% suppression in root galling when seedlings were transplanted to sterilize and nematode infested soil, respectively. The results were indicative of easy dissemination of *P. penetrans* to the main field through infected seedlings (Walia et al., 1992). Nursery bed treatment with *Bacillus macerans* at 25 g/ m² + soil drench (2% solution) 10 days after sowing gave a maximum reduction in nematode population and maximum yield of 38.25 q/ha (Sheela and Nisha, 2004). Application of *P. chlamydosporia* at 50 g/m² in nursery beds significantly increased the plant growth parameters (42 and 50% increase in height and weight seedlings, respectively, over control) and root colonization by bioagent on transplants. When the above seedlings were transplanted in field, there was a reduction in root galling index (5.2 compared to 7.8 in control), number of egg mass in 5 g root (20 compared to 49 in control) number of juveniles in 5 g root (62 compared to control 162 in control) and number of juveniles in 100 cc of soil (67 compared to 152 in control) and increased plant growth and yield in eggplant at harvest (Naik, 2002). Rao et al. (1997) studied the integration of *Paecilomyces lilacinus* with neem leaf suspension for the management of root-knot nematodes on eggplant and revealed that the aqueous neem leaf suspension supported the growth of *P. lilacinus*. The number of spores per mL of neem leaf suspension of 5% and 10% was 17.4×10^6 and 22.7×10^6, respectively. Significant reduction in the root-knot index and the final population of *M. incognita* was observed in eggplant seedlings, which were given root-dip treatment in neem leaf suspension mixed with *P. lilacinus*. Rao et al. (1998) studied biointensive management of *M. incognita* on eggplant by integrating *P. lilacinus* (4×10^5 spore/mL) and *Glomus mosseae* (28–32 chlamydospores/g) and found that both the components of management did not affect each other's colonization on the roots resulting into an additive effect of both on management of *M. incognita* and also the final nematode population was significantly less in the treatments where *G. mosseae* and *P. lilacinus* were integrated. Avermectins effectively reduced root-knot nematode infection in brinjal under field condition (Reddy and Nagesh, 2002).

6.3.3.3   INTEGRATED METHODS

Integration of *P. chlamydosporia* (at 100 mL/seed pan containing 1.2×10^4 spores/mL) in castor cake (at 40 g/seed pan) amended soil was effective in increasing the seedling weight and colonization of roots with bioagents and also give least root galling and final nematode population. Borkakaty (1993) observed that inoculation of *P. lilacinus* at 4 g/kg of soil in combination with mustard oil cake at 0.5 and 1.0 t/ha increased plant growth with corresponding decrease in the number of galls, egg masses, and egg/egg mass of *M. incognita* in brinjal. Application of castor cake extract based formulation of *T. harzianum* (at 500 mL/m^2 containing 9.9×10^3 spores/mL) to nursery beds of brinjal was effective in producing vigorous seedling (with maximum seedling weight) with least root galling (Rao et al., 1998). Root-dip treatment of brinjal seedlings in neem cake extract based formulation of *P. lilacinus* (at 5×10^6 spores/mL) for min and planted in pots gave a significant increase in plant growth character and a drastic reduction in root galling, fecundity, and nematode population in soil and roots (Rao et al., 1998). Integration of a bioagent *P. lilacinus* with carbofuran at 1 kg a.i./ha was found effective in the management of reniform nematode in brinjal (Reddy and Khan, 1989). Split application of *T. harzianum* at 50 kg/ha (10^8 cfu/g) (before transplanting and 45 days after transplanting) + carbofuran at 16.5 kg/ha was effective in increasing fruit yield (Haseeb et al., 2004). The planting of marigold combined with the application of carbofuran at 1 kg a.i./ha controls *M. javanica* infestation in brinjal (Singh, 1991). Treatment of nursery bed with carbofuran at 3 kg a.i./ha along with summer plowing and covering of polythene sheets gave better yield and controlled root-knot nematode in brinjal (Anon, 1989c). Application of aldicarb at 1.0 kg/ha in nursery beds along with neem cake at 400 kg/ha increased yield and reduced the galling (Table 6.2) (Singh and Gill, 1998).

TABLE 6.2 Integrated Management of Root-knot Nematode in Brinjal with Neem Cake and Nematicide.

Treatments	% Reduction in gall index	% Increase in yield
Aldicarb (1.0 kg a.i/ha)+ neem cake (400 kg/ha)	92.90	58.70
Carbofuran (1.0 kg a.i/ha) + neem cake (400 kg/ha)	68.10	68.10

Soil solarization of nursery bed (using 100 gauge LLDPE clear film for 15 days) and application of neem cake 200 g/m^2 proved most effective for reduction in nematode population. Solarization of nursery bed for 15 days in summer and application of poultry manure at 200 kg/ha gave maximum yield. Summer plowing with soil solarization with polythethylene mulching effectively reduced *M. javanica* population in brinjal (Jain and Gupta, 1991).

6.4 NEMATODE ASSOCIATE WITH POTATO CROP

Root-knot and cyst nematodes are the economically important pest, and at present, 135 species belonging to 45 genera are reported to be associated with potato.

6.4.1 POTATO CYST NEMATODE (G. rostochinensis AND G. pallida)

Stunted growth with unhealthy foliage, premature yellowing, poor developments of the root system, and reduction in size and number of tubers are characteristic symptoms of heavy infestation by potato cyst nematodes. However, evenly distributed infestation levels, the nematode cause a gradual reduction in yield. Small patches of weakly growing plants may show in the field when soil population is sufficiently high. Temporary wilting of plants occurs during midday.

The cyst contains 200–500 eggs, which can remain viable for several years. Hatching commences with the planting of host crop, and potato root diffuses strongly stimulate the emergence of J2 from the cyst. Approximate 35% eggs hatch in a season, and the remaining eggs hatch over a period of 6–8 years. Life-cycle patterns are similar to other cyst-forming/heteroderid nematodes. Reproduction is sexual and is strongly favored by sex attractants. The life cycle is completed in about 5–7 weeks under Nillgiri conditions, where a second generation may also occur. But in European conditions only one generation completed in a year. In case of *Globodera rostochinensis*, the rounded white female turns to golden yellow or brown cyst directly, but in *G. pallida*, the females first turn creamish in color before shifting to the cyst. Feeding site comprises a "Syncytium," which is formed in the reaction to nematode enzymes. It involves endodermal, pericycle, xylem, or phloem cells surrounding the nematode cephalic region. Partial dissolution of cells walls occurs and the protoplasm coalesce to form a multinucleate mass with a common thickened cell wall, called syncytium. This is in contrast to giant cells formed by root-knot nematodes, where nuclear division leads to multinucleate condition and dissolution of cell walls does not occur.

The syncytium function like transfer cells is necessary for the growth and development of the nematode. The breakdown of conducting vessels leads to poor plant growth as the nutrient and water supply to the shoots is partially blocked.

6.4.2 ROOT-KNOT NEMATODES (MELOIDOGYNE Spp.)

The typical symptoms of infection of root-knot nematodes on the potato are small galls on the root system (Fig. 6.4). However, the infection on the tubers is expressed by the formation of pimple like the formation of galls on the tubers, which reduced the marketable value of tubers and storage life

*Photo curtsey by B. A. Patel

FIGURE 6.4   Root-knot nematode infected potato plant.

(Fig. 6.5). Heavily infested plants show sickly appearance and the foliage shows the nutrition deficiency symptoms. Such plants invariably produced small size tubers. When an infested tuber is cut across, the white glistening swollen females of the size of a pin head may be seen within potato tissue.

FIGURE 6.5   Root-knot nematode infected potato.

6.4.3   LESION NEMATODE

All life cycle stages are infective in lesion nematode. Zone of active growth is generally preferred for penetration. Nematode movement is both intercellular and intracellular within the cortical tissues. Eggs are laid scattered within the root cortex. The complete life cycle is done inside the root itself. Reproduction may be sexual or parthenogenetic. During the vegetative phase, root population increase enormously by the time of crop matures, nematodes leave roots, and again the soil population increase. In perennial crops, the infection spreads through feeder roots to the main roots as well, and nematode population fluctuations are correlated to season and availability of fresh roots. Species show different temperature optima for their biological activities, for example, *Pratylenchus penetrans* completes one life cycle in 4–5 weeks at 21°C on peach, while *P. minyus* takes same time at 38°C on tobacco. The nematode damaged is mostly confined to cortical parenchyma cells, where it causes extensive cavity formation due to migration and feeding. Lesion nematode was suspected to be the causal organism as it was constantly associated with diseased peach plants.

6.4.3.1   MECHANISM OF LESION FORMATION

Amygdaline (a carbohydrate) present in the peach roots is hydrolyzed by the nematode enzyme ß-glucosidase into sugar, benzaldehyde, and hydrogen cyanide. The latter two products are toxic to plant cells and cause cell death leading to lesion formation near the feeding site. The nematode being migratory in nature escapes the toxic effect of these chemicals and feed at a new site again causing necrosis.

6.4.3.2   INTERACTION WITH BACTERIA

Jatala et al. (1990) reported the interaction between *R. solanacearum* and nematodes, especially *Meloidogyne, Pratylenchus,* and *Globodera* spp., in potato in relation to breeding for combined resistance. The potato cyst nematode *R. solanacearum* causes wilt complex in potatoes.

6.4.3.3   INTERACTION WITH FUNGI

The potato cyst nematode *Globodera rostochiensis* in association with fungus *Rhizoctonia solani* caused root rot complex in potatoes.

Potato	*P. penetrans*	*V. dahliae*	Martin et al. 1982
	P. thornei	*V. dahliae*	Krikun and Orion, 1977
	P. neglectus/ Meloidogyne spp.	*V. dahliae*	Scholte and Jacob, 1989
	M. hapla	*V. alboutrum*	Jacobson et al. 1979
	G. rostociensis	*V. dahliae*	Corbett and Hide, 1971
	M. incognita	*R. solani*	Sharma and Gill, 1979

6.4.4   MANAGEMENT

6.4.4.1   REGULATORY METHODS

In India, there is a quarantine act against the cyst nematode of potato (*G. rostochiensis*) in Nilgiris, Tamil Nadu. Potatoes are not allowed to be transported from Nilgiris to other parts of India for seed and table purposes in order to prevent the spread of potato cyst nematode from Tamil Nadu to other states and Union Territories. Potato seed pieces free of the cyst nematode can be produced commercially by seed certification.

6.4.4.2   PHYSICAL METHODS

Fassuliotis and Sparrow (1955) demonstrated that irradiation of potato tuber with X-rays inhibited sprouting and also the development of *G. rostochiensis*. Cyst of this nematode exposed to 20,000 r contained only brown and dead eggs while at 40,000 r the eggs lost their contents completely. Careful washing of soil adhering to potato tubers can considerably reduce the risk of spreading cyst nematodes. Commercial potato tuber contents completely. Incubation of potato tubers at 45°C for 48 h has been to kill about 98.9% *M. incognita* without affecting tuber viability (Nirula and Bassi, 1965). Hot water treatment of potato tubers at 46–45.5°C for 120 min gave effective control of root-knot nematode infection. In small holding burning of trash in the field before taking up tuber planting not only help in reducing the nematode population but also enrich the soil.

6.4.4.3   CULTURAL METHODS

Potato cyst nematode (*G. rostochiensis*) can be eliminated by selecting nematode-free planting material. Early maturing varieties of potato like Irish Cobbler usually suffer less as compared with late maturing varieties. These varieties start growth in spring at soil temperature too low for much activity and may be harvested before many of the nematodes can reproduce. Continuous cropping with potatoes increases the cyst population of *G. rostochiensis*. More than 3-year rotation with wheat, strawberry, cabbage, cauliflower, pea, maize, and beans reduces nematode population to a safe level. Growing non host crop like radish, garlic, beetroot, French bean, cruciferous vegetables, turnip, or green manuring crops bring down the cyst nematode population by 50%. A crop rotation pattern (potato-French bean-peas-potato) wherein potato is grown once after 3–4 other crops, decreased nematode population by 98–99% and increase the yield of potato by 90%. In Nilgiris, crop rotation with cabbage and carrot gave effective control of cyst nematodes. Growing vegetables and resistant potato in long-term rotation trial brought down cyst propagule population up to 92% and was effectively used in nematode management at Nilgiris. Potato grown with white mustard in a pot of infested soil was less heavily attacked by root-knot nematodes than potatoes grown alone. Potato root spread was ineffective in the presence of leaching from roots of mustard seedlings. Mustard oils increased the yield of potatoes by reducing the severity of nematode attack. The active principle involved in mustard is allyl isothiocyanate, which is toxic to the nematodes. In a long season, warm climate regions, potatoes may be planted during the winter months and harvested before injury occurs in spring without visible infection. Frequently potatoes are planted in January–February and harvested in April–May before the second generation of nematodes attacks the tubers. Planting of potato during the third or fourth week of March in Shimla Hills would reduce the damage due to *M. incognita* (Prasad et al., 1983). In Shimla hills, early planting in the third or fourth week of March reduced both root and tuber infestation without affecting the yield (Ahuja, 1983). The yield was maximum, which was concomitant with lowest tuber infestation and lowest harvest population in soil at harvest. Late planting of autumn crop and early planting of spring crop in North Western Plains, while in hills, early planting of summer crop in the fourth week of March is ideal. Deep plowing and drying of soil in the summer months facilitate the drying and death of infective larvae thereby reducing initial inoculums in the soil. Dry fallow with 2 or 3 deep plowing during the hot summer month gives excellent control of root-knot nematodes. Fallowing 2 years crop rotational sequence of Maize-wheat-potato-wheat coupled with summer fallow after 2 or 3 deep plowings in northwestern hills, rotation potato with French beans, leafy vegetables, maize, and wheat helped in minimizing the nematode damage and increasing the productivity of crop (Raj and Nirula,1969). Crop rotation with nonhost crops like maize, wheat, barley, beans, etc., reduces nematode population. Growing of trap crop like French marigold, *Tagetes patula* in alternate rows with potato reduces nematode larval population in soil, root galling, and tuber infestation while yield increase up to 123% over control. Similarly, growing *Cassia* sp. and incorporation of sawdust @ 2.6 t/ha also helped in reducing tuber infestations. Intercropping of onion and maize reduced galling by *M. incognita* on potato roots. Oil cakes (neem, groundnut, mustard, and castor) were effective in reducing parasitic nematode population (*Hoplolaimus, Tylenchorhynchus, Meloidogyne,* and *Helicotylenchus* species) and increasing yield of potato. Application of eucalyptus leaf waste from oil distillation plants @ 2.5 t/ha helps in reducing the nematode population. The bacterium *P. penetrans* and Arbuscular mycorrhizal fungi such as *Glomus fasciculatum* and *G. mosseae* have shown promise in reducing the root-knot nematode infestation. Potato cv. Kufri Dewa was found resistant to root-knot nematode infestation.

6.4.4.4   BOTANICALS

Incorporation of eucalyptus litters @ 10 t/ha at the time of preparation of land reduces nematode population. Application of eucalyptus distillation @ 2.5t/ha also helps to reduce the cyst nematode population in the soil.

BIOLOGICAL METHOD

Application of *Pseudomonas fluorescence* at 10 g/m^2 gives good results for the management of nematodes.

6.4.4.5   INTEGRATED METHODS

The control of potato rot nematode was achieved by the combination of disease escape, hygiene, in the form of seed certification and crop rotation (Winslow and Willis, 1972).

6.5   POTATO ROT NEMATODE (*Ditylenchus destructor*) AND YAM NEMATODE (*Scutellonema bradys*): A NEW THREAT TO POTATO

Investigation in Nigeria revealed a relationship between the presence of the yam nematode *Scutellonema bradys* and potato tuber cracking and distortion, similar to the symptoms that the nematode causes on yam tubers (Coyne and Claudius-Cole 2009; Coyne et al., 2011). Other symptoms included scaly surface appearance and subsurface rot of tubers. Because potato is a major food crop in Europe, and production of ware and seed potato is increasing in Africa, the possible threats of *Scutellonema bradys* were investigated on potato production.

6.6   CHILI/CAPSICUM

6.6.1   ROOT-KNOT NEMATODE (MELOIDOGYNE Spp.)

Decrease in root-knot population on chilies occurs when the field was left fallow or following crop rotation with marigold, spinach, and bottlegourd

(Khan et al., 1975). Growing of chili along with marigold gave the highest degree of reduction in root-knot population followed by onion, garlic, and asparagus (Trivedi and Tiyagi, 1984). Row application of fly ash at 0.6 kg/m^2 enhanced the fruit yield by 21.3% and inhibited the reproduction of *M. incognita* (Khan and Ghadipur, 2004). Application of *P. lilacinus* reduces root-knot nematode population (Pandey and Trivedi, 2002). Avermectins effectively reduces root-knot nematode infection in chili under field condition (Parvatha Reddy and Nagesh, 2002). Chili cvs. Pusa Jwala, Wonder Hot, Tej, Utkal, Abha, and Utkal Rashami were found to resistance to root-knot nematode (Patnaik et al., 2004).

FIGURE 6.6   Root galling caused by root-knot nematodes showing by Chili grower.

*Photo curtsey by B. A. Patel

FIGURE 6.7   Capsicum infected by root-knot nematodes.

6.6.2   MANAGEMENT

Several cultural methods like selection of nematode-free nursery sites, destruction of infested roots after crop harvest, crop rotation with nonhost (maize, wheat, sorghum) or antagonistic crops (marigold, mustard, sesame), flooding, fallowing, deep summer plowing, and harrowing either alone or in combination proved to be reasonably effective and economical to check multiplication of root-knot, reniform, and lesion nematodes. Burning of paddy husk or sawdust on infested nursery proved to be reasonably effective. Plowing not only leads to disturbance and instability in nematode community but also causes their mortality by exposing them to solar heat and desiccation. Three summer plowings each at 10 days interval during June lead to the reduction in *M. javanica* population while fallowing itself during the same period registered 44.5% reduction (Jain and Bhatti, 1987). Additional use of plastic sheets for covering soil either in nursery beds or in the field further enhance nematode reduction. Such an approach also helps in reducing the intensity of weeds, fungi, and bacteria in soil. Application of Royal 350 (*Arthrobotrys irregularis*) at 140 g/m^2 a month before transplantation of capsicum resulted in good protection of root-knot nematode. Among the egg parasites, *P. lilacinus* has been found to be comparatively higher efficient to suppress the root-knot nematode and reniform nematode population.

Field application of *B. thuringiensis* strain to capsicum reduced root galling in roots due to root-knot and reniform nematode and increase yield (Zuckerman et al., 1993). Treatment of nursery bed with *P. chlamydosporia* at 50 g/m^2 was significantly effective in reducing galling index, number of nematodes in roots and soil, increasing the percent parasitization of eggs by bioagents, and also yields. Seed treatment with *P. chlamydosporia* and *P. fluorescence* were also effective (Naik, 2004). Capsicum cvs Mississipi-68, Santanka, Anaheim Chile, and Italian Pickling are reported to be resistant to root-knot nematode (Hare, 1951). In capsicum, the seedling stand was good where the combination of neem based *P. fluorescence* and soil application of *T. harzianum* (Naik, 2004).

Interaction with other pathogens

Plant	Nematode	Fungus	Reported by
Chili	*M. incognita*	*Pythium aphanidermatum/R. solani*	Hasan, 1985

6.7   MANAGEMENT OF ROOT-KNOT NEMATODE IN PROTECTED CULTIVATION IN SOLANACEOUS VEGETABLE CROPS

Effect of chemical nematicides (Carbofuran 3G, Phorate 10G, Trizophos 40EC) and neem seed powder was tested at IARI, New Delhi. The chemicals were applied @ 2 kg/2 L a.i./ha in two split doses 1 month apart; the first dose was applied 30 days after transplanting. It was concluded that the management of root-knot nematodes on tomato under polyhouse condition could be achieved with trizophos when applied at 2 L a.i./ha in two equal split dose (Sharma et al., 2008).

The nematicidal effect of *Pseudomonas fluorescens*, *Paecilomyces lilacinus*, *Pichia guilliermondii*, and *Calothrix parietina* individually or in combination were tested against root-knot nematode, *Meloidogyne incognita*. Treatments with *P. fluorescens* and *P. lilacinus* caused 45% and 30% mortality of *M. incognita* juveniles after 48 h of exposures, respectively, compared with control.

Under greenhouse conditions, all treatments reduced the disease severity and enhanced plant growth compared to untreated control. Application of *P. fluorescens*, *P. lilacinus*, and *P. guilliermondii* Moh 10 was more effective as compared to *C. parietina*. There was a negative interaction between *C. parietina* and either *P. lilacinus* or *P. guilliermondii* (Table 6.3).

TABLE 6.3   Effect of Different Biocontrol Agents on Nematode Population in Tomato Under Greenhouse.

Treatments	After 30 days		After 60 days	
	Root galls index	Nematode population (250 g soil)	Root galls index	Nematode population (250 g soil)
Nematode + Pseudomonas fluorescens (Pf)	2.5 d	1522 d	2.9 cd	2021 h
Nematode + Paecilomyces lilacinus (Pl)	2.8 c	1806 c	3.1 c	2010h
Nematode + Pichia gluilliermondii (Pg)	2.9 c	1800 c	3.2 cb	2551 g
Nematode + Calothrix parietina (Cp)	3 cb	2012 b	3.2 cb	2621 f
Nematode + Pf + Cp	1.5 f	1752 c	1.7 f	3010 d

TABLE 6.3   *(Continued)*

Treatments	After 30 days		After 60 days	
	Root galls index	Nematode population (250 g soil)	Root galls index	Nematode population (250 g soil)
Nematode + Pf + Pg	1 g	1280 e	1.5 f	2641 f
Nematode + Pf + Pl	2 e	1537 d	2.5 e	2011 h
Nematode + Cp + Pg	3 cb	2003 b	3.3 b	2912 c
Nematode + Cp + Pl	3.2 b	2001 b	3.5 b	3512 b
Nematode + Pg + Pl	2.7 c	1775 c	2.9 cd	2741 e
Nematode + Pf + Pl + Pg + Cp	0.9 g	1250 e	1.1 f	2211 h
Infested control (Nematode only)	4 a	2590 a	4.7 a	3800 a
Healthy control	NI	NI	NI	NI

NI, not inoculated with nematodes.
Values in the column followed by different letters indicate significant differences among treatments according to least significant difference test ($P > 0.05$).

Root-knot nematode infesting transplanted vegetable crops like tomato/ brinjal/capsicum can be successfully managed with increase in yield by application of *Paecilomyces lilacinus* (cfu 2×10^6) @ 50 g/m² in nursery bed (0.5 t/ha) + *P. lilacinus* (cfu 2×10^6) @ 5 kg along with 2.5t of FYM/ ha in main field prior to planting (Anon., 2013–2015). Furthermore, they enhanced the plant growth parameters compared with the control.

TABLE 6.4   Effect of Different Treatments on Root-knot Nematode in Tomato Under Polyhouse.

Center	% Reduction in FNP	% Reduction in galling	% Increase in yield
Kanpur—Tomato	19.1	60.8	5.72
Banglore—Tomato	24.6	62.8	84.3
Vellayani—Brinjal	77–88.93	27.8–95.5	20.63.7
Shivvamoga—Tomato	67.46	62.5	–
Jabalpur—Tomato	41.7	20	18
Anand—Tomato	73.8	32.56	21.36

These results proved that application of different biocontrol agents *(P. fluorescens, P. lilacinus,* and *P. guilliermondii)* not only has a fatal effect on the nematode, but also enhances the plant growth, supplying many nutritional elements and initiation the systemic resistance in plants. Presence of *C. parietina* as a soil inhabitant cyanobacterium could antagonize biocontrol agents leading to the reduction of their practical efficiency in the soil in the following table (Hashem et al., 2010).

KEYWORDS

- **root knot**
- **vegetable**
- **pathogen**
- **management**

REFERENCES

Adam, S. Seeing Red Colored Mulch Starves Nematodes. *Agric. Res.* **1997,** *45,* 18.

Ahuja, S. *Studies on the Variability in Infection of Root-knot Nematode Meloidogyne incognita on Potato.* Third Nematology Symposium, Himachal Pradesh Agriculture University, Solan, 1983; p 4.

Alam, M. M.; Saxena, S. K.; Khan, A. M. Influence of Inerculture of Marigold and Margosa with Some Vegetable Crops on Plant Growth and Nematodepopulation. *Act. Bot. Ind.* **1977,** *5,* 33–39.

Anita, B.; Subramanian, S. In *Management of Reniform Nematode Rotylenchulus reniformis in Tomato.* Proceedings of Nematology—Challenges and Opportunities in 21st Century, Sugarcane Breeding Institute: Coimbatore, Tamil Nadu, India, 1998; pp 249–250.

Anon. Bienniel Report (1987–1989) of AICRP on Plant Parasitic Nematodes with Integrated Approach for Their Control, Indian Agriculture Research Institute: New Delhi, India, 1989a, p 80.

Anon. Consolidated Bienniel Report (1987–1989) of AICRP on Plant Parasitic Nematodes with Integrated Approach for Their Control, Haryana Agriculture University: Hisar, India, 1989b, p 35.

Anon. QRT Report of AICRP on Nematodes (1989–1993), Deptartment of Nematology, Haryana Agriculture University, Hisar: India, 1993, p 149.

Anon. Consolidated Bienniel Report (2014–2015) of AICRP on plant parasitic nematodes with Integrated Approach for Their Control. UASH, Shiwamogga. 2015, 58–59.

Balasubramanian, P.; Ramakrishnan, G. *Nematol. Mediterr.* **1983,** *11,* 203–204.

Bhagawati, B.; Goswami, B. K. Interaction of *M. incognita* and *Fusarium oxysporum* f.sp. Lycopersici on Tomato. *Indian J. Nematol.* **2000,** *30,* 16–22.

Bhatti, D. S.; Jain, R. K. Estimation of Loss in Okra, Tomato and Brinjal Yield Due to *Meloidogyne incognita. Indian J. Nematol.* **1977,** *7,* 37–41.

Bowman, P.; Bloom, J. R. Breaking of the Resistance of Tomato Varieties to *Fusarium* Wilt by *Meloidogyne incognita. Phytopathology* **1966,** *56,* 871.

Chhabra, H. K.; Sharma, J. K. Combined Effect of *Meloidogyne incognita* and *Rhizoctonia bataticola* on Pre-emengence Damping Off of Okra and Brinjal. *Sci. Culture.* **1981,** *47,* 256–257.

Conroy, J. J.; Green, R. J.; Jr; Ferris, J. M. Interaction of *Verticillium alboutrum* and the Root Lesion Nematode, *Pratylenchus penetrans,* in Tomato Roots at Controlled Inoculum Densities. *Phytopathology* **1972,** *62,* 362–366.

Corbett, D. C. M; Hide, G. A. Interaction Between *Heterodera rostochiensis* Woll. and *Verticillium dahlia* Kleb. on Potatoes and the Effect of CCC on both. *Ann. Appl. Biol.* **1971,** *68,* 71–80.

De Moura, R. M.; Echandi, E; Powell, N. T. Interaction of *Corynebacterium michiganense* and *Meloidogyne incognita* on Tomato. *Phytopathology* **1975,** *65,* 1332–1335.

Fassuliotis, G.; Sparrow, A. H. Preliminary Report of X-ray Studies on Golden Nematode. *Pl. Dis. Reptr.* **1955,** *39,* 572.

Garabedian, S.; Van Gundy, S. D. Use of Avermectins for the Control of *Meloidogyne incognita* on Tomato. *J. Nematol.* **1983,** *15,* 503–510.

Gautam, A.; Siddiqui, Z. A.; Mahmood, I. Integrated Management of *Meloidogyne incognita* on Tomato. *Nematol. Meditt.* **1995,** *23,* 245–247.

Gupta D. C.; Mukhopadhyaya, M. C. Effect of N, P and K on the Root-knot Nematode, *Meloidogyne javanica* (Treub) Chitwood. *Sci. Culture.* **1971,** *37,* 246–247.

Hackney, R. W.; Dickerson, O. J. Marigold, Castorbean and Chrysanthemum as Controls of *Meloidogyne incognita* and *Pratylenchus alleni. J. Nematol.* **1975,** *7,* 84–90.

Hare, W. W. *Phytopathology* **1951,** *41,* 16.

Harrison, A. L.; Young, P. A. Effect of Root-knot Nematode on Tomato Wilt. *Phytopathology* **1941,** *31,* 749–752.

Hasan, A. Synergistic Interaction Between *Pythium aphanidermatum* and *Rhizoctonia solani* with *Meloidogyne incognita* on Chilli. *Nematologica.* **1985,** *31,* 216–217.

Hasan, A.; Khan, M. N. The Effect of *Rhizoctonia solani, Sclerotium rolfsii* and *Verticillium dahlia* on the Resistance of *Meloidogyne incognita. Nematol. Meditter.* **1985,** *13,* 133–136.

Hashem, A.; Abo-Elyousr, K. A. M. Management of the Root-knot Nematode *Meloidogyne incognita* on Tomato with Combinations of Different Biocontrol Organisms. *Crop Protection.* **2011,** *30,* 1251–1262.

Hazarika, B. P.; Roy, A. K. Effect of *Rhizoctonia solani* on the Reproduction of *Meloidogyne incognita* on Eggplant. *Indian J. Nematol.* **1974,** *4,* 246–248.

Hirano, K.; Kawamura, T. Incidence of Complex Disease Caused by *Pratylenchus penetrans* or *P. coffeae* and *Fusarium* Wilt Fungus in Tomato Seedlings. *Tech. Bull. Faculty Chaba Univ.* **1972,** *20,* 37–43.

Jacobsen, B. J. MacDonald, D. H.; Bissonnette, H. L. Interaction Between *Meloidogyne hapla* and Verticillium albo-utrum in Soil. *Phytopathology* **1979,** *69,* 971–981.

Jain, R. K.; Bhatti, D. S. Effect of Summer Ploughing Alone and in Combination with Some Other Effective Practices on the Incidence of Root-knot Nematode (*Meloidogyne javanica*) on Tomato (cultivar HS 101). *Indian J. Nematol.* **1985,** *15,* 262.

Jain, R. K.; Bhatti, D. S. Population Development of Root-knot Nematode (*Meloidogyne javanica*) and Tomato Yield as Influenced by Summer Ploughing. *Trop. Pest Manage.* **1987,** *33,* 122–125.

Jain, R. K.; Paruthi, I. J.; Gupta, D. C. Control of Root-knot Nematode (*Meloidogyne javanica*) in Tomato Through Nursery Bed Treatment Alone and in Combination with Application of Carbofuran at Transplanting. *Indian J. Nematol.* **1988,** *18,* 340–341.

Jain, R. K.; Paruthi, I. J.; Gupta, D. C. Influence of Intercropping on Incidence of Root-knot Nematode (*Meloidogyne javanica*) in Brinjal. *Indian J. Nematol.* **1990,** *20,* 236–238.

Jatala, P.; Martin, C.; Mendoza, H. A.; Hidalgo, O. A.; Rincon, R. H. (Eds.) Role of Nematodes in the Expression of *P. solanacearum* and Strategies for Screening and Breeding for Combined Resistance. In *Advances en el Mejoramiento Genetico de la papa en los Paises del cono sur;* 1990, pp 187–189.

Jayakumar, J.; Rajendra, G.; Ramkrishanan, S. Bio-efficacy of Avermectin Against Root-knot Nematode, *M. incognita* in Tomato. *Indian J. Nematol.* **2005,** *35,* 151–156.

Jenkins, W. R.; Coursen, B. W. The Effect of Root-knot Nematodes *Meloidogyne incognita* Acrita and *M. hapla* on Fusarium Wilt of Tomato. *Pl. Dis. Reptr.* **1957,** *41,* 182–186.

Kanwar R. S.; Bhatti, D. S. Yield Performance of Resistant and Susceptible Tomato Cultivars in Monoculture and Related Reproduction of *Meloidogyne javanica. Int. Nematol. Netw. Newslett.* **1990,** *7* (3), 16–18.

Khan, A. M.; Saxena, S. K.; Siddiqi, Z. A. Efficacy of Tagetes Erecta in Reducing Root Infesting Nematodes of Tomato and Okra. *Phytopathology* **1971,** *24,* 166–169.

Khan, A. M.; Saxena, S. K.; Siddiqi, Z. A.; Upadhyay, R. S. Control of Nematodes by Crop Rotation. *Indian J. Nematol.* **1975,** *5,* 214–221.

Khan, M. R.; Jain, P. K.; Singh, R. V., Pramanik, A. Economically Important Plant Parasitic Nematodes Distribution. *ATLAS;* 2010, pp 1–154.

Khan, M. R.; Gadhipur, M. H. In *Management of Root-knot Disease of Some Solanaceous Vegetables by Soil Applicationof Fly Ash,* National Symposium on Paradigm in Nematological Research for Biodynamic Farming. University of Agricultural Science: Banglore, India, 2004; p 8.

Kirankumar, K. C.; Krishanappa, K.; Karuna,K; Ravichandra, N. G.; Shreenivas, K. R. In Evaluation of Propose Management Package of Growth, Development and Yield of Tomato Infested by Root-knot Nematode. *National Symposium on Paradigm in Nematological Research for Biodynamic Farming.* University of Agricultural Science: Banglore, India, 2004; p 94.

Kirkun, J.; Orion, D. Studies of the Interaction of *Verticillium dahliae* and *Pratylenchus thornei* on Potato. *Phytoparasitica* **1977,** *5,* 67.

Libman, G.; Leach, J. G.; Adams, R. E. Role of Certain Plant Parasitic Nematodes in Infection of Tomatoes by *Pseudomonas solanacearum. Phytopathology* **1964,** *54,* 151–153.

Maheswari, T. U.; Mani, A.; Rao, P. K. Combined Efficacy of the Bacterial Spore Parasitic *Pasteuria penetrans* (Thorn, 1940) and Nematicides in the Control of *Meloidogyne javanica* on Tomato. *J. Biol. Control.* **1987,** *1,* 53–57.

Martin, M. J.; Riedel, R. M.; Row, R. C. *Verticillium dahlia* and *Pratylenchus penetrans*: Interaction in the Early Dying Complex of Potato in Ohio. *Phytopathology* **1982,** *72,* 640–644.

McKeen, C. D.; Moutain,W. B. Synergism Between *Pratylenchus penetrans* (cobb) and *Verticillium alboatrum* R. and B. in Eggplant Wilt. *Can. J. Botany* **1960,** *38,* 789–794.

Miller, P. M. Effect of Tobacco Cyst Nematode, *Heterodera tabaccum*, on Severity of Vericillium and Fusarium Wilts of Tomato. *Phytopathology* **1975**, *65*, 81–82.

Naik, D. Biotenological Approaches for Management of wilt Disease Complex in Capscicum (*Capscicum annum* L.) and Eggplant (*Solanum melongena*) with Special Emphasis on Biological Control. Ph.D. Thesis, Kuvempu University: Shiwamoga, Karnataka, 2004.

Nath, R.; Khan, M. N.; Wanshi, R. S. K.; Dwivedi, R. P. Influence of Root-knot Nematode, *Meloidogyne javanica* on Pre and Post Emergence Damping Off of Tomato. *Indian J. Nematol.* **1984**, *14*, 135–140.

Nath, R. P.; Sinha, B. K.; Haider, M. G. Studies on the Nematodes of Vegetables in Bihar–II. Combined Effect of *Meloidogyne incognita* and *Ozonium texacum var. parasiticum* on Germination of Eggplant. *Indian J. Nematol.* **1976**, *6*, 177–179.

Nayar, K.; Mathew, J.; Kurian, J; Gnanamanicum, S. S.; Mahadevan, A. (Eds) Role and Association of Root-knot Nematode (*Meloidogyne incognita*) in Bacterial Wilt of Brinjal Incited by *Pseudomonas solanacearum*. In *Advances in Research on Plant Pathogenic Bacteria*. Based on the Proceeding of the National Symposium on Phytobacteriology, University of Madras: Chennai, India, pp 45–48.

Netscher, C. Control of *Meloidogyne incognita* in Vegetable Production by Crop Rotation in Iovory Coast. *Acta Horticulture.* **1983**, *152*, 219–225.

Nirula, K. K; Bassi, K. K. Thermotherapy for Root-knot Nematode, *Meloidogyne incognita* in Potato Tubers. *Indian Potato J.* **1965**, *7*, 9–11.

Orion, D.; Krikum, J. Response of *Vericillium* Resistant and *Vericillium* Susceptible Tomato Varieties to Inoculation with the Nematode *Meloidogyne javanica* and *Verticillium dahlia. Phytoparasitica.* **1976**, *4*, 41–44.

Overman, A. J.; Jones, J. P. Effect of Stunt and Root-knot Nematodes on Verticillium Wilt of Tomato. *Phytopathology* **1970**, *60*, 1306.

Pandey, R. C.; Trivedi, P. C. Biological Control of Meloidogyne Incognita by *Paecilomyces lilacinus* in *Capscicum annum. Indian Phytopathol.* **1992**, *45*, 134–135.

Pani, A. K.; Das, S. N. Studies on Etiological Complexes in Plant Disease. I. Association of Root-knot Nematodes in Bacterial Wilt of Tomato. *J. Res. Orrisa Univ. Agriculture Technol.* **1972**, *2*, 54–59.

Parvatha Reddy, P.; Khan, M. R. Evaluation of *Paecilomyces lilacinus* for the Biological Control of *Rotylenchulus reniformis* Infecting Tomato as Compared with Carbofuran. *Nematol. Mediterr.* **1988**, *16*, 113–116.

Parvatha Reddy, P.; Nagesh, M. *Avermectins-Isolation, Fermentation, Preliminary Characterization and Screening for Nematicidal Activity.* Tech. Bull.17: Indian Institute of Hortilculture Research, Banglore, India, 2002; pp 28.

Parvatha Reddy, P.; Nagesh, M.; Devappa, V. Effect of Integration of *Pasteuria penetrans, Paecilomyces lilacinus* and Neem Cake for Management of Root-knot Nematodes Infecting Tomato. Pest Manage. *Horticultural Ecosyst.* **1997**, *3*, 100–104.

Parvatha Reddy, P.; Singh, D. B.; Ram Kishun. Effect of Root-knot Nematode on Susceptibility of Pusa Purple Cluster Brinjal to Bacterial Wilt. *Curr. Sci.* **1979**, *48*, 915–916.

Patel, S. K.; Patel, H. V., Patel, A. D. Integrated Management of Root-knot Nematode in Tomato Nursery Through Botanical Plant Material. *Indian J. Nematol.* **2006**, *36*, 307–308.

Patnaik, P. R.; Mohanty, K. C.; Mahapatra, S. N.; Shoo, S. In *Evaluation of Chilli Varieties Against Root-knot Nematode, Meloidogyne incognita.* National Symposium on Paradigm in Nematological Research for Biodynamic Farming. University of Agricultural Sciences: Banglore, India, 2004; p 100.

Prasad, K. S. K. Nematological Problems and Progress of Research on Potato in India. Fourth Group Meeting on Nematological Problems of Plantation Crops, University of Agricultural Sciences, Banglore, 1989.

Prasad, K. S. K.; Raj, D.; Sharma, R. K. In *Effect of Planting Time on Meloidogyne inognita Infesting Potato at Shimla Hills*. Third Nematological Symposium, HP, Solan. 1983.

Rao, M. S.; Parvatha Reddy, P. In *Innovative Approaches of Utilization of Biocontrol Agents – Paecilomyces lilacinus and Vericillium chlamydosporium Against Root-knot Nematode on Tomato (Abstr.)*. First Afro-Asian Nematology Symposium, Aligarh Muslim University, Aligarh. 1992pp 25–26.

Rao, M. S.; Parvatha Reddy, P.; Somasekhar, N.; Nagesh, M. Management of Root-knot Nematode, *Meloidogyne incognita* in Tomato Nursery by Integration of Ectomycorrhiza, *Glomus fasciculatum* with Castor. *Pest Manage. Hortic. Ecosyst.* **1997**, *3*, 31–35.

Rao, M. S.; Parvatha Reddy, P.; Sukhada, M.; Nagesh, M., Pankaj. Management of Root-knot Nematodes on Egg Plant by Integrating Endomycorrhiza (*Glomus fasciculatum*) and Castor (Ricinus communis) Cake. *Nematol. Meditt.* **1998**, *26*, 217–219.

Reddy, D. D. R. Analysis of Crop Losses in Tomato Due to *Meloidogyne incognita. Indian J. Nematol.* **1985**, *15*, 55–59.

Routaray, B. N.; Shoo, H. Integrated Control of Root-knot Nematode, *Meloidogyne incognita* with Neem Cake and Granular Nematicides on Tomato. *Indian J. Nematol.* **1985**, *15*, 261.

Sahoo, S.; Mohanty, K. C.; Sahoo, H. K.; Mohapatra, A. K. B.; Mishra, H. N. In *Effect of Different Cropping Sequences on Root-knot Nematode (Meloidogyne incognita) and Wilt Incidence in Tomato*. National Symposium on Paradigm in Nematological Research for Biodynamic Farming. University of Agricultural Science: Bangalore, India, 2004; p 37.

Sen, K.; Dasgupta, M. K. Effect of Rice Hull Ash on Root-knot Nematodes in Tomato. *Trop. Pest Manage.* **1981**, *27*, 518–519.

Shanti, A.; Sivakumar, C. V. Biocontrol Potential of Pseudomonas Fluorescence Against Root-knot Nematode, *Meloidogyne incognita* (Kofoid and White,1919) Chitwood, 1949 on Tomato. *J. Biol. Control.* **1995**, *9*, 113–115.

Sharma, J. K.; Singh, I.; Chhabra, H. K. Observations on the Influence of *Meloidogyne incognita* and *Rhizoctonia bataticola* on Okra. *Indian J. Nematol.* **1980**, *10*, 148–151.

Sharma, N. K.; Gill, J. S. Interaction Between *Meloidogyne incognita* and *Rhizoctonia solani* on Potato. *Indian Phytopathol.* **1979**, *32*, 293–298.

Sharma, H. K.; Pankaj; Jagan Lal. Effect of Chemical on Root-knot Nematode, *Meloidogyne incognita* Infecting Tomato Under Polyhouse Cultivation. *Indian J. Nematol.* **2008**, *38*, 124–125.

Sheela, M. S.; Nisha, M. S. In *Impact of Biocontrol Agents for the Management of Root-knot Nematode in Brinjal*. National Symposium on Paradigm in Nematological Research for Biodynamic Farming. University of Agricultural Science: Bangalore, India, 2004; pp 63–64.

Sheela, M. S.; Jiji, T; Nisha, M. S. In *Evaluation of Different Control Stratagies for the Management of Nematodes Associated With Vegetables (brinjal) (Abstr.)*. International Conference on Vegetables, Bangalore, India, 2002; p 268.

Shukla, V. N.; Swarup, G. Interrelationships of the Root-Knot Nematode *Meloidogyne incognita* (Kofoid and White, 1919) Chitwood, 1949 and *Sclerotium rofsii* Sacc. on the Emergence of Tomato Seedlings. *Bull. Indian Phytopathol. Soc.* **1970**, *6*, 52–54.

Singh, R. S.; Sitaramaiah, K. Incidence of Root-knot of Okra and Tomato in Oil Cake Amended Soil. *Pl. Dis. Reptr.* **1966**, *50*, 668–672.

Singh, R. S.; Sitaramaiah, K. Control of Root-knot Through Organic and Inorganic Amendments of Soil: Effect of Oil Cakes and Sawdust. *Indian J. Mycol. Plant Pathol.* **1971,** *1,* 20–29.

Smits, B. G.; Noguera, R. Effect of *Meloidogyne incognita* on the Pathogenicity of Different Isolates of *Fusarium oxysporum* on Brinjal (*Solanum melonena* L.). *Agronomia Trop.* **1982,** *32,* 284–290.

Taylor, A. L.; Sasser, J. N. *Biology, Identification and Control of Root-knot Nematodes (Meloidogyne spp.)*; North Carolina State University Graphics: Raleigh, North Carolina, USA, 1978; p 111.

Tiyagi, S. A.; Anvar, S.; Bano, M.; Siddiqi, M. A. Feasibility of Growing Zinnia as a Mix-crop Along with Tomato for the Control of Root-knot and Reniform Nematodes. *Int. Nematol. Netw. Newslett.* **1986,** *3* (3), 6–7.

Trivedi, P. C.; Tiyagi, A. Effect of Some Plants on *Meloidogyne incognita* Population in Chilli Soil. *Indian J. Botanical Sci.* **1984,** *63,* 416–419.

Walia, R. K.; Dalal, M. R. Efficacy of Bacterial Parasite (*Pasteuria penetrans*) Application as Nursery Soil Treatment in Controlling Root-knot Nematode, *Meloidogyne javanica* on Tomato. *Pest Manage. Econ. Zool.* **1994,** *2,* 19–21.

Walia, R. K.; Bansal, R. K.; Bhatti, D. S. In *Efficacy of Bacterium (Pasteuria penetrans) and Fungus (Paecilomuces lilacinus) Alone and in Combination Against Root-knot Nematode (Meloidogyne javanica) Infecting Brinjal.* Proceeding of the First Afro-Asian Nematology Symposium, Aligarh Muslim University, Aligarh, India, 1992; p 38.

Winslow, R. D.; Willis, R. J. *Nematode Diseases of Potatoes. In Economic Nematology*; Webster, J. M., Ed.; Academic Press: New York, 1968; pp 17–18.

Zaki, F. A.; Bhatii, D. S. Effect of Castor *Ricinus communis* and Biocontrol Fungus *Paecilomuces lilacinus* on *Meloidogyne javanica. Nematologica* **1991,** *36,* 114–122.

Zuckerman, D. M.; Dicklow, M. B.; Acosta, N. A Strain of *Bacillus thuringiensis* for the Control of Plant Parasitic Nematodes. *Biocontrol Sci. Technol.* **1993,** *3,* 41–46.

CHAPTER 7

Effect of Fungal Pathogen on Physiological Function of Vegetables

NISHANT PRAKASH[1*], ERAYYA[2], SRINIVASARAGHAVAN A.[3], and SUPRIYA GUPTA[4]

[1]Department of Plant Pathology, Krishi Vigyan Kendra, Lodipur Farm, Arwal, India

[2]Department of Plant Pathology, Dr. Kalam Agriculture College, Kishanganj, India

[3]Department of Plant Pathology, Bihar Agriculture University, Sabour, Bhagalpur, India

[4]Department of Plant Pathology, G.B. Pant University of Agriculture and Technology, Pantnagar, India

*Corresponding author. E-mail: gladiator.nishant@gmail.com

ABSTRACT

Vegetables are rich source of minerals and nutrition. In India, large quantity of vegetables are sold and purchased everyday. Ideal soil for cultivation of vegetables is sandy loam to clay. Moderate acidic and saline soils can also be preferred for vegetable cultivation. Various fungal and bacterial diseases are proved to be a threat for vegetable cultivation. Pathogens manifest their presence by releasing elicitors (mycelium, sclerotia, sporangia, conidia, etc.). The host plant surface bears a receptor which recognizes elicitors. The elicitor receptor interaction results in changes in plasma membrane permeability. Changes in plasma membrane permeability disturb Calcium-Proton influx and Potassium-chloride efflux. Oxidative burt is mediated by these ionic fluxes. Plant defense machinery mainly comprises reactive oxygen species (ROS) such as superoxide, hydrogen peroxide and hydroxyl free radical. During plant defense, reinforcement of cell wall is catalyzed by the

ROS which further signaled initiation of defense reactions. Large family of protein kinases (PK) carries defense signaling. PK include large number protein family called calmodulin like protein kinases (CDPK). Calcium signaling, mitogen activated protein kinase (MAPK) and ROS production is largely influenced by CDPK. Infection of fungal pathogen to plant enhance respiration, demand of carbon for energy generation, activity of plant growth hormone, namely, auxin, photosynthesis activity, carbohydrate content, etc. Fungal pathogens cause damage to the host by producing host specific toxins (HST) and non host specific toxins (non HST). HST cause damage to specific genotype while non HST does not have any specificity and attack a wide range of the host. Most of HST's are produced by *Cochliobolus* sp. and *Alternaria* sp. Non HST is produced by *Pseudomonas* sp., *Alternaria* sp., *Cercospora* sp., etc.

7.1   INTRODUCTION

Healthy vegetable crops serve as constant source of food for human population all over the world. Day by day, demand of good quality of vegetables is increasing and farmers are producing and supplying large volume of vegetables (Shehu et al., 2014d). Onion (*Allium cepa*), tomato (*Lycopersico esculentum*), cauliflower (*Brassica oleraceae* var. *botrytis*), carrot (*Daucus carota* subsp. *sativus*), etc., are among the most common vegetables cultivated by farmers in both the rainy and dry seasons (Salau et al., 2012; Shehu and Aliero, 2010). The cultivation of vegetables occurs in a varied range of soil type, namely, sandy loam, clay, moderately acidic, and saline soils. Nowadays, vegetable production is facing a great threat from fungal diseases. The fungal fruit rots in vegetables are caused by *Alternaria* sp. *Phytophthora* sp., *Anthracnose* sp., *Phoma* sp., *Fusarium* sp., etc. (Ibrahim, 2002; Iqbal et al., 2003; Patel et al., 2005; Ali et al., 2005; Salau et al., 2012). Purple blotch disease of onion caused by *Alternaria porri* was found to be endemic in sokoto state. It is an important disease of onion worldwide and is more prevalent in warm, humid environment. Purple blotch disease of onion is an endemic in sokoto state of Nigeria and cause significant yield loss of 25–40% annually ('Yar'adua, 2003; Shehu and Muhammad, 2011). This pathogen is soil borne and inoculums can remain viable in soils for many years (Shehu et al., 2014d).

A parasitic relationship exists between host plant and phytopathogenic fungi. Fungi thrive successfully host by securing its nourishment. Most of host parasite relationship studies are focussed on avirulence factors, toxins, cell wall degrading enzymes, and other components which participate in pathogenesis. Less information is available that how pathogen obatain nutrition, impact on photosynthesis, host respiration, hormonal balance, etc. An association between host and pathogen is established in such a way that pathogen utilizes the compounds produced by the host for its sustainability. Colonization of pathogen on host depends upon the successful establishment of this association. The contact of pathogen with the host sends a signal which initiates a seris of events of defense response. The resistance and susceptible response of host to pathogen depends upon the external environment factors. Likewise, the successful growth and multiplication of pathogen inside the host is determined by the internal environment. Invading pathogen causes mark changes in cell membrane permeability and disturb ionic balance, calcium ion channel, G-protein activity, photosynthesis and respiration (Gracia-Brugger et al., 2006). Apart from this fungus also produces toxin (both host specific and non host specific) which damage tissue of plant and cause death (Stergiopoulos et al., 2013; Bender et al., 1999). These toxins interfere in different metabolic process of plant and cause disease. It is difficult to assess that any event that arise after parasite invasion, is due to pathogen or host. Powdery mildew pathogen is suitable for making this distinction. In powdery mildew pathogen, mycelium infected tissues can be easily separated from mycelium-free tissue by microscopic observation and other means (Allen, 1938).

7.2   EARLY EVENTS AT CELL MEMBRANE DURING FUNGAL INFECTION

The release of elicitors indicates the presence of pathogen. The plasma membrane bears receptors which are located in plasmamembrane and cytosol. These receptors recognize elicitors released by the pathogen (Ebel and Scheel, 1997). Presence of fugnal pathogen on the host surface influences calcium ion channel (Jabs et al., 1997; Pugin et al., 1997). Interaction of elicitor to receptor led to the activation of signal transduction chain and results in defense response in the form of hypersensitive response (biotrophs). This host pathogen reaction does not initiate hypersensitive reaction in all cases (necrotrophs) (Ebel and Scheel, 1997).

7.2.1   CALCIUM AND ION CHANNELS

The response of presence of elicitors on the host cell surface was first observed in the form of changes in plasma membrane permeability. It results in change in ionic fluxes (calcium–proton influx and potassium–chloride efflux) (Ebel and Scheel, 1997). These ionic fluxes send signals for initiation of oxidative burst, activation of defene gene and production of phytoalexin (Jabs et al., 1997; Pugin et al., 1997). Receptor-elicitor binding increases the permeability of ion channel located at plasma membrane. It increases cytosolic calcium level, activate ion channels and pumps. It also activates other ion fluxes. Zimmerman et al. (1997) characterized a novel calcium inward channel in parsley, showing quick response to elicitor. Tomato-*Cladosporium fulvum* host pathogen system, calcium permeable channel activates in response to elicitor (Gelli et al., 1997). In tobacco cells various elicitors increase concentration of cytosolic calcium (Chandra et al., 1997). In resistant variety of cow pea, elevated calcium level is observed with appropriate race of rust fungu while in susceptible cow pea, no such change is observed (Xu and Heath, 1998).

Different calcium channel inhibitors, calcium chelators and omission of extracellular calcium hinder the initiation of hypersensitive response. It confirms the role of Calcium influx in initiating defense response. The currently available experimental evidence is also consistent with receptor-mediated release of calcium from internal stores as an immediate elicitor response, which triggers calcium influx.

7.2.2   PROTEIN KINASES

Signalling pathways in plants are also same as other living kingdoms with minor differences. PKs help in sensing Ca^{2+} in plants through their carboxy-terminal. Calmodulin (CaM)-like domain protein kinases (CDPK) comprise a large family of PKs of plants. These CDPKs are absent in animals (Hrabak et al., 2003). Five plants of PK classes have been defined with several subclasses (PlantsP: Functional Genomics of Plant Phosphorylation website). Many reports implicate PKs in plant defense reactions; most of them belong to the CDPK and MAPK families. After elicitor perception, PK activation may be the earliest induced event; this should be the case for those elicitors which are readily recognized by proteins such as receptor-like kinases (RLKs), which are cytoplasmic or plasma membrane localized.

PKs and calcium signalling relation: rapid increase in Ca^{2+} influx and intracellular Ca^{2+} observed after perception of elicitor by receptor (Fig. 7.1) (Kadota et al., 2004; Lecourieux et al., 2002; Tavernier et al., 1995). These signals, in turn, could activate Calcium dependent Protein Kinase (CDPKs). Under in vivo conditions, protein phosphorylation mostly depends upon calcium influx. In plant, defense reactions are coordinated by mitogen acti-vated protein kinase (MAPK) protein family. MAPKs are stimulated both under biotic stress like plant disease, and abiotic stresses such as wounding, salt, temperature and oxidative stress (Jonak et al., 2002). No direct correla-tion is established between Ca^{2+}, CDPKs and MAPKs however poor activa-tion of MAPK was observed when Ca^{2+} ion channel is blocked by chelators (Kurusu et al., 2005; Lebrun-Garcia et al., 1998; Romeis et al., 1999). Interestingly, overexpression of the rice two-pore channel one (OsTPC1), a putative voltage-gated Ca^{2+}-channel, is correlated with enhanced HR and activation of the rice MAPK OsMPK2 (Kurusu et al., 2005). Tomato cells on treatment with voltage pulse or elicitor preparation of *Fusarium oxysporum lycopersici*, showed enhance activity of MAPK but on treatment with nifedipine, a voltage-gated calcium channel inhibitor, suppress MAPK activity (Link et al., 2002). However, Ca^{2+} influx is not always a prerequisite for MAPK activation. In tobacco- *Pseudomonas syringae* pv. *phaseolicola* system, elicitor signal induce activation of MAPK system independent of Ca^{2+} influx (Lee et al., 2001; Vandelle et al., 2006).

Relationships between MAPK activation and ROS production: ROS may or may not participate in MAPKs activation. However, exogenous H_2O_2 is capable to activate MAPK. ROS can positively upregulate MAPK system. Hence, MAPK system can either reduce or increase ROS production (Moon et al., 2003; Ren et al., 2002). The role of exogenous H_2O_2 and endogenous H_2O_2 is still to be deciphered in activation of MAPK.

MAPK implications in host defense responses: study of several MAPK module indicates the role of MAPKs in positive transcriptional regulation of defense gene and disease resistance reaction. Several workers proved this in tobacco and arabidopsis (Asai et al., 2002; Ekengren et al., 2003; Jin et al., 2002, 2003; Kim and Zhang, 2004). Apart from upregulation of MAPKs, in some cases downregulation of MAPKs also activates defense reactions in host plant. In Arabidopsis, mpk4 and edr1 mutants showed that MAPK AtMPK4 and the MAPKKK EDR1 downregulate SA-dependent defense responses and increase the resistance to virulent pathogens (Frye et al., 2001; Petersen et al., 2000).

7.2.3 REACTIVE OXYGEN SPECIES

Calcium influx mediated by elicitor followed by rise in cytosolic calcium levels, are necessary for stimulation of oxidative burst (Jabs et al., 1997; Chandra et al., 1997). ROS is toxic to the pathogens (Baker and Orlandi, 1995). At initial stage, reinforcement of physical barriers is catalyzed by ROS (Tenhaken et al., 1995). This activates other defense reactions such as phytoalexin synthesis and defense gene activation (Jabs et al., 1996, 1997), programme cell death (Jabs et al., 1996, Levine et al., 1996). Production of ROS during oxidative burst is catalyzed by the NAD(P)H oxidase (located in plasma membrane). This phenomenon shows homology with mammalian oxidative burst phenomenon (Fig. 7.1). In many plant species (e.g., Parsley), elicitor stimulated oxidative burst was found to be necessary for phytoalexin production (Jabs et al., 1996, Apostol et al., 1989, Doke, 1983) while in soyabean and tobacco phytoalexin production is independent of ROS (Levine et al., 1996). Oxidative burst produces superoxide free radical. These free radicals donot cross the plasma membrane but interact with plasma membrane components, namely, fatty acid and proteins. These free radicals trigger programme cell death and PR gene expression (Jabs et al., 1996).

Upstream and downstream interaction of ROS with various signalling events, namely, membranes, NADPHoxidases, G-proteins, calcium, redox homeostasis, photosynthesis, MAPKs, plant hormones [such as salicylic acid (SA), jasmonic acid (JA), abscisic acid (ABA), and ethylene] and transcription factors. Gray arrows for direct ROS interactions with other signalling components, oragne arrows for expected indirect interactions.

Lesions produced by pathogen infection produce local and systemic lesions. Formation of systemic lesions and systemic acquired resistance (SAR) require hydrogen peroxide produced from oxidative burst. Unlike systemic lesion, both hydrogen peroxide and nitric oxide is required (Alvarez et al., 1997). Incompatible host pathogen reaction (avirulent strain) leads to the formation of hydrogen peroxide and nitric oxide which in turn generate sufficient defense response. Less production of hydrogen peroxide and nitric oxide and poor defense response are generated in compatible host pathogen reaction (virulent strain).

7.3   INCREASE IN NITROGEN AND AMINO ACID METABOLISM

The assimilation of nitrogen onto carbon skeletons has significant effects on plant development and yield (Lam et al., 1996). Inorganic nitrogen is assimilated in the form of amino acids glutamine, glutamate, asparagine, and aspartate. It serves as vital nitrogen-transport and storage molecules in crops (Lam et al., 1996). Host on infection with pathogen; generate strong demand of energy which leads utilization of amino acids into energy generating pathways like Tri carboxylic acid (TCA) cycle. Enzyme glutamate dehydrogenase (GDH) released amino nitrogen from amino acids and forms a keto acid and ammonia (NH_3). This further utilized in TCA cycle (Fig. 7.1) (Miflin and Habash, 2002). All the 20 amino acids metabolized in any seven intermediates of TCA, namely, α-ketoglutarate, acetoacetate, acetyl-CoA, fumarate, oxaloacetate, pyruvate, and succinyl-CoA and utilized for energy generation in plants. Apart from energy generation, nitrogen metabolism also plays role in plant defense. It mobilizes nitrogenous compound away from infection sites and starve pathogen (Tavernier et al., 2007). Example: root herbivore infection of *Agapeta zoegana, Centaurea maculosa* cause reduction of whole plant nitrogen uptake and reallocation of nitrogen from root to above ground tissues (Newingham et al., 2007). Genes normally involved in senescccence are reported to be upregulated during pathogen infection process (Pageau et al., 2006; Stephenson et al., 1997; Tavernier et al., 2007). For example, transcripts for *glutamine synthetase 1*(*GS1*), a marker for senescence, was induced within 2 h of inoculation (Pageau et al., 2006). Nitrogenous compounds also participate in defense process through production of reactive nitrogen species. Reactive nitrogen species such as nitric oxide (NO) are important component of several physiological processes and defense reaction (Lamotte et al., 2004).

NO is produced in plants using either a nitrate or nitrite-dependent pathway and an L-Arg-dependent pathway (Besson-Bard et al., 2008). NO acts as a toxicant to invading microbes, alter redox potential of cell, stimulate HR and defense reactions (Romero-Puertas et al., 2004; Zaninotto et al., 2006). Under nutrition and over nutrition both predispose host to disease. Under nutrition makes plant weak and invite certain pathogens for infection. Over fertilization makes plant suitable substrate for the growth of certain class of pathogen (Solomon et al., 2003). Nitrogen compounds also influence the expression of defense. Under in vitro condition these nitrogen compounds do not stimulate the expression of defense gene while enhance expression is observed under in planta condition. (Basse et al., 2000; Fudal et al., 2007; Guo et al., 1996; Khan and Straney, 1999; Thomma et al., 2006).

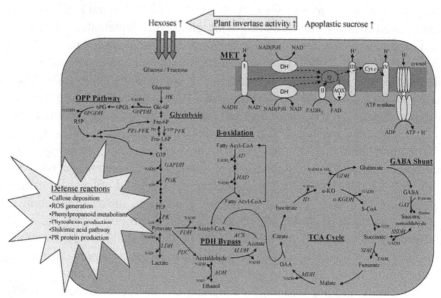

FIGURE 7.1   Primary metabolism pathways involved in plant defense.

Source: Bolton et al. 2009, MPMI Vol. 22, No. 5, pp 487–497.

The pathways of glycolysis, oxidative pentose phosphate (OPP), pyruvate metabolism including the pyruvate dehydrogenase (PDH) bypass, β-oxidation, TCA cycle, 4-aminobutyrate (γ-aminobutyrate) (GABA) shunt, and the mitochondrial electron transport (MET) are shown. Only enzymes that catalyze reactions that are related to energy production are shown. Electron donors (NAD(P)H and FADH2) from primary metabolism pathways can be used in the MET to produce ATP or be involved directly in various defense reaction pathways such as in reactive oxygen species (ROS) production. Abbreviations: HK, hexokinase; Glc-6P, glucose 6-phosphate; Fru-96P, fructose-6-phosphate; PPi-PFK pyrophosphate:fructose-6-phosphate 1-phosphotransferase; PFK, phosphofructokinase; Fru-1,6P, fructose-1,6-bisphosphate; G3P, glyceraldehyde-3-phosphate; GAPDH, glyceraldehyde-3-phosphate dehydrogenase; PGK, phosphoglycerate kinase; PEP, phosphoenolpyruvate; PK, pyruvate kinase; LDH, lactate dehydrogenase; PDH, pyruvate dehydrogenase; PDC, pyruvate decarboxylase; ALDH, acetaldehyde dehydrogenase; ACS, acetyl-CoA synthase; AD, acyl-CoA dehydrogenase; HAD, 3-L-hydroxyacyl-CoA dehydrogenase; ID, isocitrate dehydrogenase; α-KG, α-ketoglutarate; α-KGDH, α-ketoglutarate dehydrogenase; S-CoA, succinyl-CoA; OAA, oxaloacetate; GDH, glutamate dehydrogenase; GAT, 4-aminobutyrate aminotransferase; SSDH, succinic semialdehyde dehydrogenase; DH, internal/external NADH dehydrogenase; Q, ubiquinone; AOX, alternative oxidase; and Cyt *c*, cytochrome c.

7.4   RESPIRATION

During resistance reaction, rate of respiration is increased in host plant (Smedegaard-Petersen and Stolen, 1981). Respiration comprises three main pathways: glycolysis, the mitochondrial TCA cycle, and mitochondrial electron transport (Fernie et al., 2004). These pathways are interconnected. It functions to generate energy equivalents and carbon skeletons that may be used in the biosynthesis of various metabolites.

Glycolysis, a cytosolic pathway, converts glucose to pyruvate, resulting in a small net gain of ATP (Fig. 7.1). In aerobic conditions, phosphofructo-kinase (PFK) is the main regulator of glycolysis, catalyzing the irreversible conversion of fructose 6-phosphate to fructose 1, 6-bisphosphate, utilizing 1 mol of ATP in the process. Likewise, pyrophosphate: fructose-6-phosphate 1-phosphotransferase (PPi-dependent PFK [PPi-PFK]) catalyzes the same process but in an ATP-independent manner (Mertens et al., 1990). Transcripts for PPi-PFK are known to accumulate under anoxic conditions (Lasanthi-Kudahettige et al., 2007). Perhaps it suggests that this enzyme is used to conserve ATP under suboptimal conditions. Recently, transcripts for PFK and PPi-PFK were both found to be upregulated during the *Lr34*-mediated response to *Puccinia triticina* in wheat. It indicates that flux through glycol-ysis is an important aspect of the resistance response (Bolton et al., 2008b).

The oxidative metabolism of pyruvate by pyruvate dehydrogenase (PDH) forms acetyl-CoA that enters the TCA cycle (Fig. 7.1). PDH is thought to be a key regulatory point for pyruvate flux into the TCA cycle (Fernie et al., 2004). The TCA cycle is a central metabolic pathway for aerobic processes and is responsible for a major portion of carbohydrate, fatty acid, and amino acid oxidation that produces energy and reducing power (Fernie et al., 2004). In the TCA cycle, each acetyl-CoA produces one GTP, one FADH2 and three NADH. FADH2 and NADH are reoxidized through oxidative phosphorylation to form ATP, ultimately yielding 15 ATP equivalents per pyruvate molecule entered through the TCA cycle. The upregulation of citrate synthase or α-ketoglutarate dehydrogenase during a resistance response, both considered to be rate-limiting enzymes (Strumilo, 2005; Wiegand and Remington, 1986), suggests an elevated flux through the TCA cycle.

Another pathway involved with the TCA cycle is the 4-aminobutyrate (γ-aminobutyrate) (GABA) shunt (Fig. 7.1). This pathway produces succinate using either glutamate or α-ketoglutarate as substrates (Shelp et al., 1999). GABA is known to be an extracellular signal molecule (Shelp et al., 2006) and the GABA shunt may be involved in supporting the resistance response to pathogens (Bolton et al., 2008b). For example, under particularly energetically demanding conditions, pyruvate from glycolysis can be produced faster than PDH can convert it to acetyl CoA. By providing a second entry point for pyruvate, the GABA shunt provides a means to utilize excess pyruvate for energy production. In addition, the TCA enzymes aconitase, succinyl-CoA ligase, and α-ketoglutarate dehydrogenase are inactivated under oxidative stress conditions (Sweetlove et al., 2002; Tretter and Adam-Vizi, 2000). HR involves oxidative burst and in the course of oxidative burst, GABA shunt bypass oxidative stress sensitive enzymes and keeps NADH generation unhindered through TCA cycle.

Reducing equivalents are produced in the course of TCA cycle which were further used in ATP synthesis (Fig. 7.1). In Electron transporat chain, NADH and $FADH_2$ oxidised to ATP by transferring electron to O_2 via protein complexes embedded in inner membrane of mitochondria.

Under conditions of high TCA cycle flux, the alternative oxidase (AOX) pathway is often activated to divert electrons from the ubiquinone (Q) pool to form H_2O. It allows excess energy to be lost as heat.

ATP production through this pathway does not contribute much and it reduces ROS during programme cell death (Yip and Vanlerberghe, 2001).

Another pathway, that is, oxidative pentose pathway produced NADPH by the oxidation of glucose-6-phosphate (Fig. 7.1). The oxidative pentose phosphate (OPP) pathway is involved with the generation of NADPH by the oxidation of glucose-6-phosphate, an intermediate also shared with glycolysis (Fig. 7.1). Glucose-6-phosphate dehydrogenase, the rate-limiting step in the OPP pathway, generates NADPH and is also implicated in the proper localization of NPR1, a critical component of the SA defense pathway (Dong, 2004). With above discussion, we can draw an inference that NADPH oxidase enzyme are essential for ROS during HR, so large amount of NADPH is needed for generation of sufficient ROS.

In response to the elicitor cryptogein, glyceraldehydes-3-phosphate converts to fructose1,6-bisphosphate which in turn catalyzed by aldolase into fructose-6-phosphate and favours OPP pathway for generation of NADPH (Pugin et al., 1997). High stromal OPP pathways were found at infection site of tobacco by *Phytophthora nicotianae* (Scharte et al., 2005). Apart from TCA and OPP pathway, several other pathways are upregulated at the time of plant defense reaction. For example, NADP-malic enzyme mediated malate metabolism is enhanced during pathogen infection. It generates energy through production of pyruvate and NADPH (Casati et al., 1999; Schaaf et al., 1995; Widjaja et al., 2009; Zulak et al., 2009). The glyoxylate cycle involves the conversion of acetyl-CoA to succinate and play key role in *Pseudomonas syringae* pv. *tomato* in *Arabidopsis* (Scheideler et al., 2002). The resulting succinate can be transported from the glyoxysome to the mitochondrion, where it can be employed in the TCA cycle. The degradation of fatty acids during β-oxidation is another potential energy source during plant defense (Fig. 7.1). The fatty acid molecule oxidation produce significant amount of ATP molecule and fatty acid metabolism upregulated during host resistance response (Lynen, 1955; Bolton et al., 2008b; Schenk et al., 2003).

7.4.1   ACCUMULATION OF LIPIDS AND CAROTENOIDS

The site of pathogen infection showed a deposititon of significant amount of lipoidal material. *Uromyces betae* infection of sugarbeet leaves showed marked accumulation of lipoidal material. Similarly, accumulation of ß-carotene and γ-carotene in crab apple leaves occur in response to infection of pycnidia of *Gymnosporangium*.

7.6   CHANGES IN CONCENTRATION GROWTH FACTORS AND AUXINS

Fungal infection cause changes in concentration and distribution of auxin in plant. In response to fungal pathogen infection, concentration of auxin may increase and decrease and affect the growth and development of plant. Both the obligate and the facultative parasite are results in uneven quantities of auxin in plants. Indol acetic acid is a common if not universal product of fungus metabolism. Alteration in quantities of auxin cause abnormal growth and development of plant. These disorders also are accompanied by other symptoms depending upon the pathogen and host involved. It is often difficult to assign particular symptoms to the action of specifc auxins, since the changes in infected plants are influenced by other hormones, as well as other factors. Within these limitations, it is possible to relate some pathological changes to altered amounts of these hormones. Increase in auxin (Hyperauxiny) concentration is related to characteristic changes in plant growth and development. It results in formations of cankers, tumors, wart-like, or similar types of neoplastic growth stimulated by any of several pathogen bacteria and fungi. The auxin stimulates normal cell division. The elongations and/or enlargement of infected plant parts frequently are related to increased auxin levels. For example, abnormal hypocotyls elongation of plants of *Carthamus tinctorius* is symtomatic of infection by the rust fungus *Puccinia carthami*. Plants of *Euphorbia cyparissias* infected with the smut fungus *Uromyces pisi* are alongated plants. Increased growth of plants of Capsella bursa-pastoris infected with either *Albugo candida* or *Peronospora parasitica* and leaf elongation of plants of *Sempervivum tectorum* infected with the fungus *Endophyllum sempervivi*, as well as the enlargement and curling of peach leaves caused by *Taphrina deformans* all are examples of these types of changes. It is assumed that the increase in the size of plum fruits infected with *T. pruni* is induced by the same mechanism.

The formation of uneven quantities of auxin has been established in both obligate and facultative pathogenic fungi. The obligate pathogen, *Puccinia graminia tritici*, for example, causes a 24× increase in auxin content in cv. Little Club wheat plants. Otherwise, it is not know to cause hyperauxiny in cereal plants. Infection by *Erysiphe graminis* in cv. Atlas barley plants resulted in an increase of this hormone ranging from only slight to 5×. The obligate parasitic fungus *Gymnosporangium*, in its aecial stage, causes hyperauxiny in pear, similar to that caused by the fungus *Uromyces* in plants of *Ranunculus*. Hyperauxiny with up to 10× increases has been described on hypocotyls of *Carthamus tinctorius* infected with *Puccinia carthami*. Greater amounts of auxin were recorded in stems and leaves of *Brassica napus* infected with *Albugo candida* as well as in plants of *Capsella bursa-pastoris* infected with *A. candida* and *Peronospora parasitica*. The parenchyma tissue of tumors on potato tubers infected with *Synchytrium endobioticum* contained more indolauxin than uninfected ones. Higher levels of auxin were reported in smut galls on maize infected with *Ustilago maydis* than uninfected tissues.

7.6.1   PHOTOSYNTHESIS AND CARBOHYDRATE

Leaf is the site of metabolic activities and work as a photoautotrophic soruce of plant. This made it unfit for defense. Photosynthetic by-product, namely, carbohydrate is rapidly channelized to the site of infection while other ATP generating pathways viz glycolysis, OPPP, and shikmic acid pathway were slowed down. Early hour of pathogen infection is accompanied with ROS accumulation, deposition of callose, cessation of sucrose export and increase in apoplastic content. It indicates that pathogen infection precedes with the retention of carbohydrate at the infection site. *Phytophthora sojae* infection induces callose deposition which blocks the plasmodesmata (Enkerli et al., 1997). Depositions of callose at plasmodesmatal region adversely affect the intercellular transport of ions and carbohydrate (Olesen and Robards, 1990). Hence, we can draw a conclusion that callose depositied during plant defense reaction retard symplastic and apoplastic export.

Pathogen infection also accompanied significant increase in apoplastic invertase activity and sucrose content (Herbers et al., 2000). Simultaneous increase in sucrose activity along with callose helps in retention of carbohyadrate at infection site. Invertase activity results breakdown of sucrose into hexoses which again imported to the infected cells and serve as a fuel for metabolic pathways such as respiration and OPPP. These pathways are generally declined during defense reaction. During defense decline in photosynthesis and stimulation of G6PDH enzyme (a key enzyme of OPPP) as well as respiration takes place. Comparative analysis of photosynthesis induction images and photosynthesis capacity indicates that changes occur in photosynthesis and OPPP are in antiparallel fashion. At the infection site, level of stromal OPPP metabolite increased due to enhanced photosynthetic activity which further feed into the Calvin cycle. During this time period, photosynthetic flux goes down to zero. This declination takes place in two steps: (1) closure of stomata followed by (2) PET inhibition. Closure of stomata restricts CO_2 uptake. This creates ambient O_2 condition, high overall flux and leads to photorespiration at the cost of carbon fixation. This results in rise of OPPP metabolite and the glutamine-synthetase pathway and retardation of carboxylation reaction of RUBISCO. Further it causes reduction in the export of triose-phosphates from chloroplast to cytosol. Stomatal closure took place at late stage of plant-microbe interactions (Nogués et al., 2002) which further preceded by H_2O_2 accumulation and initiation of ROS. Calcium channel of guard cell is activated by ROS initiation which further involved in stomatal osmotic response (Pei et al., 2000).

In hypersensitive response of plant defense, photosynthesis has to be completely suppressed prior to initiation of cell death. In mesophyll cells, heterotrophic state of metabolism is the pre-requisite for cell death and cell wall fortification. At the primary stage of pathogen infection, high sugar cell formation transcript rbcS disappeared. Slow turnover is showed by Rubisco, which in turn results in steep depression in photosynthesis. Pathogen infection does not alter chlorophyll content of cell. It indicates during the collapse of photosynthetic flux, photosynthetic antennae remain intact. It causes rapid decline in electron transport chain followed by reduction in transfer of electron to oxidised PS-I reaction center (P-700). As electron transfer is restricted to PS-I, PS-I is not much affected and remain stable. But it cause moderate inhibition of PS-II. Electron donation shortage to PS-I restrict H_2O_2 release at PS-I and oxidized state of stroma. Oxidized state of stromal NADPH/NADP+ and thioredoxins make Calvin cycle inactive and activate key enzyme of the plastidic OPPP, that is, G6PDH. PEP (produced during glycolysis) and erythrose-4-phosphate (from stromal OPPP) drives shikimic acid pathway which is activated during plant defences. "High-OPPP state" and the decline in photosynthesis in stoma-dependent and -independent processes are localized in nature at initial infection site and distinctly separated from uninfected cells. Photosynthetically inactive and photosynthetically active cells may share common cell wall. Usually, rapid intercellular exchange of carbohydrates is observed in the mesophyll.

Sugar accumulation along with enhancing carbohydrate consuming reactions, it also upregulates gene expression. High sugar accumulation led to the suppression of assimilatory genes and activation of sink-specific genes (Rolland et al., 2002). PR proteins expression is induced by soluble hexoses (Herbers et al., 1996a, b, 2000). Hence, carbohydrate accumulation led to alteration of cellular state, metabolic regulation and in long term expression of gene. Hence, carbohydrate accumulation led to alteration of cellular state, metabolic regulation and in long-term expression of gene.

Metabolic and cellular changes preceding cell wall fortification and cell death during plant defence.

7.7　TOXIN

The role toxin in bringing about the symptoms of disease had been well established for facultative and obligate parasites. The violent response to initial penetration in resistant plants and the response of susceptible tissues free of fungus mycelium constitute convincing evidence for the participation of toxins. Response in healthy plants has been induced by extracts of diseased tissue but the substances responsible may be either of fungal or host origin. The pathogen, its part and pathogen originated substances played and important role in inducing resistance and susceptibility in host plant. Such substances must be assayed when the plant host cells respond to the entry of these pathogen parts like haustoria or hyphae as in a wheat plant showing hypersensitive reaction. If the same parasite fails to produce same response in another wheat plants whose cell it also enters (a susceptible genotype), the substances concerned must be either (1) donot have any effect upon the second cell, (2) neutralized by second cell, or (3) prevent from appearing in the second cell. All these three alternatives especially the second has been more or less explicitly proposed in case of resistant reaction. The hyphae growth of a fungal pathogen is not influenced by the resistance state of the plants upto 48 h post-inoculation. At later points, growth rate increased on susceptible plants, whereas it remains restricted on induced resistant plants. The interesting phenomenon observed is that there is no difference in nutrient uptake by haustoria on resistant and susceptible genotype.

Plant pathogenic fungi produce a range of toxic compounds (Phytotoxins), usually secondary metabolites, which affect the physiological functioning of higher plants. Phytotoxins often cause wilting, chlorosis and necrosis, although the importance and actual role in disease establishment is variable and, for some diseases, it is disputed. In general, toxins are low molecular weight compounds and it includes polypeptides, glycoproteins, phenolics, terpenoids, sterols, and quinines. Hormones, enzymes, genetic determinant, or compounds liberated during disruption of plant tissues are not considered as toxins.

Plant pathogenic microbes secrete phytotoxins which cause numerous devastating diseases of crop plants. Almost half a century ago, phytotoxins implicated in plant disease are placed in three major gropus, namely Phytotoxins, Pathotoxins and vivotoxins. Any compound produced by a microorganism which is toxic to plant is known as Phytotoxins are nonspecific, incite a few or none of the symptoms that are incited by the pathogen. It shows no relation between toxin production and pathogenicity. Pathotoxins play a crucial role in disease production and produce symptom characteristic of the disease in susceptible plants. These toxins may be produced by the pathogen, host or interaction between them. Vivotoxins are defined as a substance produced in the infected host by the pathogen or its host which function in the production of the disease but is not itself the initial inciting agent of disease. However, this definition was very much complicated and it led to the categorization of phytotoxin in simpler group namely Host-specific toxins and Non host-specific toxins.

7.7.1   HOST-SPECIFIC TOXINS

Tanaka first suggested about the presence of HST in black disease of Japanease pear in 1933. The disease was noticed only after the introduction of new pear cultivar Nijisseiki. Host specific toxins (HSTs) are defined as pathogen effectors that induce toxicity and promote disease only in the host species and only in genotypes of that host expressing a specific and often dominant susceptibility gene. Usually sensitivity is not shown by all the genotype of the host plant and in likewise toxin is also not produced by all the isolates of plant pathogen. Thus, the interaction between HSTs and plant genotype showed analogy with interaction between race specific effector (of biotrophic and hemibiotrophic pathogens) and corresponding resistance genes. The difference lies in the fact that HST specific allele in a host genotype makes plant susceptible in the presence of HST toxins, instead of resistance. As this concept is altogether contrary to classical gene-for-gene model (host and biotrophic pathogen interaction), so it is known as the "inverse gene-for-gene model" (Oliver and Solomon, 2010). Plant pathogens which produce HSTs include species of *Cochliobolus spp., Alternaria spp. and Pyrenophora spp.*.

Toxin secreted by *Cochliobolus* spp.: An array of structurally diverse HSTs are produced various species and races of *Cochliobolus* spp. These HSTs are considered as a primary virulence factor (Table 7.1). As *Cochliobolus* spp. mostly infect host plant of Gramineae family, so HSTs produced by *Cochliobolus* spp. are mostly pathogenic on grass host plant. The genetic basis of interaction between host plant and HSTs are governed by single Mendelian genes or genetic loci (Stergiopoulos et al., 2013).

TABLE 7.1   Different Host-specific Toxins Produced by *Cochliobolus* spp., *Periconia* spp., *Pyrenophora* spp., and *Stagonospora* spp.

Toxin	Species	Chemical structure	Mode of action	Plant target
HC-toxin	*Cochliobolus carbonum*	Cyclic tetrapeptide	Suppression of host defenses	HDACs
T-toxin	*Cochliobolus heterostrophus*	Linear polyketide	Disruption of mitochondrial activity	URF13: 13-kDa mitochondrial Protein

TABLE 7.1 *(Continued)*

Toxin	Species	Chemical structure	Mode of action	Plant target
PM-toxin	*Mycosphaerella zeae-maydis*	Linear polyketide	Disruption of mitochondrial activity	URF13: 13-kDa mitochondrial Protein
Victorin	*Cochliobolus victoriae*	Cyclic chlorinated pentapeptide	Induction of PCD	LOV1: CC-NB-LRR disease resistance protein
HS-toxin	*Bipolaris sacchari*	Sesquiterpene glucoside	Depolarization of plasma membrane	Unknown
PC-toxin	*Periconia circinata*	Peptidyl chlorinated polyketide	Induction of PCD	Unknown
PtrToxA	*Pyrenophora tritici-repentis*	13.2-kDa protein	Tsn1-mediated induction of PCD	Chloroplasts, ToxABP1
PtrToxB	*Pyrenophora tritici-repentis*	6.5-kDa protein	Induction of PCD	Probably chloroplasts
SnToxA	*Stagonospora nodorum*	13.2-kDa protein	Tsn1-mediated induction of PCD	Chloroplasts, ToxABP1
SnTox1	*Stagonospora nodorum*	10.3-kDa protein	Snn1-mediated induction of PCD	Probably chloroplasts
SnTox2	*Stagonospora nodorum*	7–10-kDa protein	Snn2-mediated induction of PCD	Probably chloroplasts
SnTox3	*Stagonospora nodorum*	25.8-kDa protein	Snn3-mediated induction of PCD	Unknown
SnTox4	*Stagonospora nodorum*	10.3-kDa protein	Snn4-mediated induction of PCD	Probably chloroplasts
SV-toxins I and II	*Stemphylium vesicarium*	Unknown	Induction of PCD	Unknown

HC-toxin: It is a cyclic tetrapeptide and secreted by a *Cochliobolus carborum* fungus. In maize, it causes Northern corn leaf spot and ear rot of maize (Walton, 2006). In 1960, HC-toxin was considered as a major determining factor of host specificity and virulence. Out of three races (races 1, 2, and 3) of *Cochliobolus carborum,* HC-toxin is produced by only race 1. In comparison to race 2 and 3 which are deficient in HC-toxin, race 1 produce larger lesions on susceptible maize cultivar (Walton et al., 1996). In the late 1930s, a new corn variety was introduced which contained HC-toxin susceptibility locus and paves the way for the discovery of HC-toxin (Ullstrup and Brunson, 1947). HC-toxin structure contains cyclo (D-Pro-L-Ala-D-Ala-L-Aeo). Here Aeo abbreviates 2aminoepoxyoxo decanoic acid. Aeo is essential for toxicity of HC-toxin on susceptible host (Fig. 7.1; Walton et al., 1982). HC-toxin inhibits the activity of an enzyme histone-deacetylases (HDACs) which in turn alter the activity of H_3 and H_4 and cause disease in maize (Brosch et al., 1995). Susceptibility of host to HC-toxin is also conferred due to the lack of HC-toxin detoxifying enzyme HC-toxin reductase (HCTR). HCTR reduce the side carbonyl group of HC-toxin and turn it inactive (Meeley and Walton, 1991; Walton, 2006). Hm1 a dominant nuclear gene of maize, shows resistance against HC-toxin. Hm1 was a first plant resistance gene cloned by Johal and Briggs (1992). Hm2 is another dominant resistance gene, which confers resistance against HC-toxin in the absence of Hm1 gene in adult plants.

T-toxin: Earlier it was known as HMT-toxin or BMT-toxin. It is produced by the southern corn leaf blight (SCLB) causing fungus *C. heterostrophus.* Maize cultivar carrying 'Texas male sterile cytoplasm' (T-cms), show high susceptibility to the race-T of *C. heterostrophus* and the toxin produced by the pathogen is termed as T-toxin. Race-O of *C. heterostrophus,* showed less virulence on T-cms genotype than race-T (Tegtmeier et al., 1982). In early 1970, race-T isolates of *C. heterostrophus,* were acknowledges widely after epidemic ravaging of corn field in US (corn plant carrying T-cms cytoplasm) by SCLB disease. Later SCLB epidemic was considered as one of the most economically devastating disease of 20th century (Ullstrup, 1972). This disease was the result of widely monoculture cultivation of maize genotype carrying T-cms cytoplasm. T-toxins are C41 derivative and a linear polyketide containing 35 to 49 carbons (Kono et al., 1985). T-toxin binds and cause conformational changes in the gene product of T-urf13 (a Texas male sterility gene). T-urf13 is a mitochondrial gene, encodes oligemeric protein URF13 of 13 kDa. URF13 protein assembled in the inner mitochondrial plasma membrane (Wise et al., 1987). T-toxin binds with URF-13 cause pore formation and rapid permeabilization of the inner mitochondrial membrane. This hampers the regular function of the inner mitochondrial membrane. This results in uncoupling of oxidative phosphorylation, leakage of small molecules, breakdown of electrochemical gradient andATP-production, and membrane swelling (Levings et al., 1995).

PM-toxin: It shows analogy with the T-toxin. *Mycosphaerella zeae-maydis* causing yellow leaf blight of corn, produce PM-toxin. Recently *Mycosphaerella zeae-maydis* is renamed as *Pyronellaea zeae-maydis* (Aveskamp et al., 2010). This fungus is also pathogenic on T-cms line of maize and produce PM-toxin. It is a polyketide with structure similar to T-toxin.

Victorin: It is secreted by the fungus causing Victoria blight of oats, *C. victoriae*. Victorin is a cyclic pentapeptide. During 1940s, US farmer's field were widely monocultured with oat variety carrying 'Victoria-type' of resistance gene against crown rust fungus *Puccinia coronata f. sp. avenae*, which gave rise an epidemic of Victoria blight of oat (Meehan and Murphy, 1946; Meehan, 1947). Pc-2 gene was resistant against crown rust but it had high susceptibility to the new pathogen *C. victoriae*. Virulence of *C. victoriae* on Pc-2 lines is largely decided by the production of victorin (Macko et al., 1985). A single dominant allele at Vb locus decides the sensitivity of oat to victorin. However, after extensive studies, it was suggested that Vb and Pc-2 locus are either same or closely linked (Mayama et al., 1995). Victorin toxin triggers host defense and initiates typical programme cell death (Navarre and Wolpert, 1999).

HS-toxin: It is produced on sugarcane infected with eyespot disease of sugarcane. The causal agent of this diseae is *Bipolaris sacchari* (also known as *Helminthosporium saccahri* or *Drechslera sacchari*). *Bipolaris sacchari* secretes HS-toxin. It has sesquiterpene core ($C_{15}H_{24}O_2$) which is asymmetrically linked with four ß-1,5-linked galactofuranose units at carbon positions 2 and 13 (Macko et al., 1983). It is also widely popular as Helminthosporoside. The sugarcane cultivars which are specifically show susceptibility to *B. sacchari*, show lesion development and leakage of electrolyte due to HS-toxin and thus it is considered as HST (Schroter et al., 1985). HS-toxin activities also cause membrane depolarization. Other than HS-toxin, low molecular weight toxin analogs (toxoids) can be produced by *B. sacchari*, which lacks one or more of the galactose residues. It is antagonistic in nature to HS-toxin. These toxoids protect sensitive tissues against HS-toxin (Livingston and Scheffer, 1984). It is not clearly understood about the biological relevance of the production of these toxoids.

Toxin produced by Alternaria spp.: Despite of identical spore morphology, different isolates of *A. alternata* secretes various HST with specific host range. A potential general aggressiveness is observed in all the isolates of *A. alternate* which is characterized by the ability to penetrate an artificial membrane (cellulose and polyvinyl membrane) through appressoria of germinate conidia (Nishimura et al., 1978; Nishimura, 1980). Culture filterates of low molecular alternaria HST are closely related compounds. As low as 10^{-8} to 10^{-9} M concentration of alternaria HSTs can cause necrosis of leaves of susceptible cultivar but no necrosis on resistant cultivar (Table 7.2) (Nishimura and Kohmoto, 1983; Kohmoto and Otani, 1991; Otani et al., 1995).

AM-toxins: It is produced by the pathogen *A. mali* which causes Alternaria blotch of apple. Its mode of action involves attack on plasma membrane and chloroplasts of apple cells (Park et al., 1981). In response to AM-toxin, plasma membrane undergoes invagination and loss of electrolyte. AM-toxin treatment to chloroplast can produce fragments of plasma membrane and vesicles within three hours. Fragmentation of grana causes chlorophyll disorganization and reduction of cholorophyll.

It leads to inhibition of photosynthetic process and assimilation of CO_2 in the leaves of susceptible apple cultivar (Kohmoto et al., 1982). These effects of AM-toxin are not observed in resistant cultivar of apple. The toxin showed marked specificity to the susceptible cultivar of apple. It was suggested that chloroplast is the primary site of action of AM-toxin (Otani et al., 1991). Genetically susceptibility of apple cultivar to AM-toxin is governed by single dominant gene (Saito and Takeda, 1984). Susceptible cultivar is heterozygous in nature while resistant cultivars are homozygous recessive.

AAL-toxin: Cell death of tomato susceptible cultivar takes place due to AAL-toxin (Akamatsu et al., 1997; Brandwagt et al., 2000; Yamagishi, et al., 2006). It is a chemically an aminopentol ester (Gilchrist and Grogan, 1976). Chemical structure of AAL-toxin is divided into TA and TB (minor). AAL-toxin shows analogy to sphingosine and sphinganine. Bezuidenhout et al. (1988) reported that Fumonisins (isolated from unrelated fungus Fusarium verticillioides) structurally identical AAL-toxin. Both the AAL-toxin and Fumonisins are toxic and inhibitory to ceramide synthase activity. Ceramide synthase is involved in sphingolipid biosynthesis (Gilchrist et al., 1994; Gilchrist, 1998). As AAL-toxins and fumonisins show structural and functional similarity, collectively they are known as sphinganine-analogue mycotoxins (SAMs) (Gilchrist et al., 1994). Sphinganine and phytosphingosine accumulation observed in the tissue of tomato plants on treatment susceptible tomato cultivar with SAMs. SAMs inhibit ceramide biosynthesis and induce programmed cell death both in the susceptible tomato cultivar and mammalian cell (Wang et al., 1996; Spassieva et al., 2002, 2006). Therefore, it can be concluded that ceramide imbalance is essential criteria for cell death. (Brandwagt et al., 2000; Markham and Hille, 2001).

ACR-toxin: The toxin is produced by the pathogen *A. alternata* pathotype rough lemon. The disease was first reported from South Africa in the year 1929. The pathogen does not restrict itself to rough lemon (*Citrus jambhiri*) host only but it also infects rangpur lime (*Citrus limonia*). Rangpur lime is the hybrid cross between acid mandarin and rough lemon. Study of oxidative phosphorylation process in physiologically active cells treated with ACR-toxin, indicates that target site of action is mitochondria (Kohmoto et al., 1984, Akimitsu et al., 1989). ACR-toxin cause uncoupling of oxidative phosphorylation and leakage of NAD+ cofactor from TCA (Akimitsu et al., 1989). ACR-toxin susceptible cultivar of citrus shows the susceptibility to the toxin while resistant cultivars are not affected by toxin. Structural similarity was shown between ACR-toxin and T-toxin (Kono and Daly, 1979; Gardner et al., 1985a, b; Nakatsuka et al., 1986b). ACR-toxin also has high host specificity.

AK-toxin, AF-toxin and ACT-toxin: Host plants show toxicity to all the three toxins. Minimum concentration of toxin which is toxic to host is 1–10 nM (Otani et al., 1985; Namiki et al., 1986a; Kohmoto et al., 1993). Resistant cultivar did not show any necrosis symptom. In strawberry, roots showed more susceptibility to AF-toxin than leaves (Yamamoto et al., 1985). The minimum concentration which is toxic to roots is 10- to 100-fold lower than the concentration which causes necrosis in leaf. In allelic pair, semidominant locus controls sensitivity of AK-toxin to Japanease pear and AF-toxin to strawberry (Kozaki, 1973; Yamamoto et al., 1985; Sanada, 1988). Dominant locus of homozygotes and heterozygotes show sensitivity to toxin but more sensistivity was observed in homozygous dominant locus. The cultivar which lacks this dominant locus become insensitive to toxin. Heterozygotes cultivars are Japanese pear cultivar Nijisseiki and strawberry cultivar Morioka-16. Plasma membrane is primarily affected by these toxins (Maekawa et al., 1984; Otani et al., 1985; Kohmoto et al., 1993; Park and Ikeda, 2008). The toxin treatment show immediate loss of K^+ ion and increase in K^+ efflux. The relationship between the amount of K^+ efflux and toxin treatment is linear. Electron micrographs study confirmed invagination of plasma membranes which was accompanied with fragmentation and vesiculation, desmotubules extension from plasmodesmata and Golgi vesicles fusion to the damaged plasma membrane. All these events are accomplished within one-three h after treatment of toxin (Park and Ikeda, 2008). Irreversible depolarization of plasma membrane of susceptible genotypes due to AK-toxin and AF-toxin, was observed by electrophysiological studies (Namiki et al., 1986b; Otani et al., 1989). Decrease in membrane potential gradient of susceptible cells observed on toxin treatment but it did not disappear completely. Depolarization mostly occurred in membrane potential which depends on respiration and supported by a H+ pump. Diffusion potential component was little effected by toxin treatment. In an isolated plasma membrane fraction of susceptible host cells, ATPase activity was not directly affected by AK-toxin and AF-toxin (Otani et al., 1991, 1995).

TABLE 7.2   Host-specific Toxins Produced by *Alternaria alternata.*

Disease	Pathogen		Host	
	Pathotype (previous name)	Toxin	Host range (susceptible cultivar)	Genetic background (dominance)
Alternaria blotch of apple	Apple pathotype (*A. mali*)	AM-toxin I, II and III	Apple (Red Gold, Starking)	Single (susceptible)
Black spot of strawberry	Strawberry pathotype	AF-toxin I, II and III	Strawberry (Morioka-16)	Single (susceptible)
Black spot of Japanese pear	Japanese pear pathotype (*A. kikuchiana*)	AK-toxin I and II	Japanese pear (Nijisseiki)	Single (susceptible)
Brown spot of tangerine	Tangerine pathotype (*A. citri*)	ACT-toxin I and II	Mandarins and Tangerines (Dancy, Emperor, Minneola)	Unknown
Leaf spot of rough lemon	Rough lemon pathotype (*A. citri*)	ACR-toxin I	Citrus rootstocks (Rough lemon, Rangpur)	Unknown
Alternaria stem canker of tomato	Tomato pathotype (*A. alternata f. sp. lycopersici*)	AAL-toxin Ta and Tb	Tomato (Earlypak 7, First)	Single (resistance)
Brown spot of tobacco	Tobacco pathotype (*A. longipes*)	AT-toxin	Tobacco	Unknown

7.7.2   NON–HOST-SPECIFIC TOXINS

Non-HSTs do not show specifity to the host and they are active against broad range of plant species (Berestetskiy, 2008). Unlike HSTs, non-HSTs are not solely responsible for disease development rather enhance the virulence of the pathogen. A wide range of non-HSTs involved in development of many disease, are produced by different species of *Pseudomonas spp., Alternaria spp., Fusarium spp.*, etc. In *Pseudomonas spp.*, different strain of *P.syringae* bacteria produce variety of bacterial toxin that are non host specific in nature (Bender et al. 1999; Yoder 1980.)

Tabtoxin: The toxin is produced by *P. syringae* pv. *tabaci*, coronafaciens, and garcae (Mitchell, 1991). It consists of monocyclic ß-lactam ring. The toxin is dipeptide in nature. Tabtoxinine-ß-lactam (TßL) of dipeptide toxin is linked to threonine by peptide bond. The production of Tabtoxin takes place as a primary intercellular metabolite (Durbin and Uchytil, 1984). The aminopeptidases secreted by plant or bacteria, cause cleavage of peptide bond of Tabtoxin and releases TßL, the toxic moiety. TßL rather not being an essential factor of the disease, is responsible for the chlorosis of leaves of plant, contribute in virulence and development of symptoms of wildfire disease on tobacco and halo blight of oats (Levi and Durbin, 1986; Uchytil and Durbin, 1980; Durbin, 1991). An enzyme glutamine synthetase is inhibited by TßL in irreversible manner which results in inhibition of glutamine synthesis in plants and accumulation of ammonia (as detoxification of ammonia does not takes in the absence of glutamine synthetase enzyme). Accumulatin of ammonia causes several harmful effects in plants like thylakoid membrane disruption, uncoupling of photophosphorylation, etc. (Durbin and Uchytil 1988, Turner and Debbage, 1982). Adenylation of glutamine synthetase protects it from toxicity of TßL (Knight et al., 1986). Another way of TßL detoxification is production of ß-lactamases which cause hydrolysis of ß-lactam ring of TßL and led to the release of nontoxic metabolite tabtoxinine (Coleman et al., 1996, Knight et al., 1987).

Phaseolotoxin: The toxin is produced by the bacteria *P. syringae* pv. *phaseolicola*. The toxin produce halo blight symptom on legumes plant and canker on kiwifruit (Mitchell, 1976; Swada et al., 1997). The structure of phaseolotoxin was deciphered by Mitchell (1976), which was further revised by Moore et al. (1984). The structure of phaseolotoxin consists of a sulfodiaminophosphinyl moiety having tripeptide linkage, that is, ornithine, alanine, and homoarginine. Patil et al. (1974) discovered the contribution of phaseolotoxin in virulence of *P. syringae* pv. *phaseolicola* and phaseolotoxin mutants were unable to move systemically in bean plants. Mutational complementation method has been used to isolate genes involved in phaseolotoxin production (Peet et al., 1986; Zhang et al., 1993). Phaseolotoxin inhibits the biosynthesis of ornithine carbamoyl transferase (OCTase) enzyme which helps in conversion of urea into ornithine and to citrulline.

Being a reversible inhibitor of OCTase, phaseolotoxin is hydrolysed in planta condition by peptidases producing N^δ-(N'sulfodiaminophosphinyl)-L-ornithine (also known as octicidine or PSorn). Octidine cause inhibition of OCTase in irreversible manner and present in tissue in predominant form (Mitchell and Bieleski, 1977). OCTase inhibition leads to the accumulation of ornithine and deficiency of intracellular arginine. It results in chlorosis. Two isozymes of OCTase is produced by *P. syringae* pv. *phaseolicola* due to toxic effect of phaseolotoxin namely ROCTase (resistant to the toxin) and SOCTase (sensitive to the toxin). Under condition, favorable for the production of phaseolotoxin, ROCTase isoform is produced by the *P. syringae* pv. *phaseolicola* (Peet and Panopoulos, 1987; Staskawicz et al., 1980; Templeton et al., 1986).

PSorn cause inhibition of OCTase (an enzyme which catalyse the conversion of ornithine and carbamoyl phosphate to citrulline in urea cycle). OCTase inhibition results in ornithine accumulation and arginine deficiency in intracellular pools which in turn cause chlorosis. Parallel lines in diagonal form are the indication of PSorn elicited blockage of urea cycle.

Coronatine: The toxin was produced by bacteria *Pseudomonas syringae*. Coronatine (COR) is an ethylcyclopropyl amino acid derived from isoleucine. The toxin has unusual structure contains two distinct components namely the polyketide coronafacic acid (CFA) and coronamic acid (CMA) (Ichihara et al., 1977; Mitchell, 1985; Parry et al., 1994).

COR is most toxic and predominant coronafacoyl compound secereted by bacteria. The toxin produces diffuse chlorosis of on a wide range of plant host (Gnanamaickam et al., 1982). *Arabidopsis thaliana's* reaction to COR is very unusual and interesting. Unlike chlorosis, accumulation of anthocyanins takes place at inoculation site and tissue turn in a strong purple hue (Bent et al., 1992). Hypertrophy, inhibition of root enlongation and enhanced ethylene production is a characteristic symptoms of COR (Ferguson and Mitchell, 1985; Kenyon and Turner, 1992, Sakai et al., 1979). COR shows structural and functional homologies with a plant growth regulator methyl jasmonate (MeJA) (a plant hormone activated during biological stress conditoin) (Sembdner and Parthier, 1993). COR is considered as molecular mimic of MeJA and have identical mode of actions (Feys et al. 1994, Greulich et al., 1995; Koda et al., 1996).

Tentoxin: The toxin is produced by a pathogenic *Alternaria* spp. It causes seedling albinism disease characterised by chlorosis of cotyledonary leaves. Later it was found that *A. tenuis* caused infections of cotton seed coat (Durbin and Uchytill, 1977) and of citrus seedlings (Schaeffer, 1976). The toxin causing chlrorosis was later known as Tentoxin (Earle, 1978). The death of seedlings takes place when >35% leaf area become chlorotic. It was found that tentoxin producing pathogen have considerably more virulence. Culturing of *A. mali* produce a product which reported to have tentoxin. Chlorosis is observed in cotyledons only if tentoxin treatment is done 48 h prior to commencement of germination (Walton et al., 1979). Instead of chlorophyll degradation, tentoxin disrupt chlorophyll synthesis process. Cotyledons of cucumber show high susceptibility to tentoxin. Tentoxin treatment to cucumber cotyledons does not show any interference in conversion of protochlorophyll to chlorophyll, although decline in chlorophyll synthesis was observed (Hanchey and Wheeler, 1979). Electron microscopic studies revealed that in cucumber, tentoxin hindered the development of prolamellae bodies and lamellae. It leads to deformation of plastids. Hence, it can be concluded that tentoxin disrupt plastid development and cause decline in chlorophyll synthesis in chloroplast.

Syringomycin: The toxin is produced by strains of bacteria *P. syringae* pv. *syringae*. It belongs to phytotoxin class known as cyclic lipodepsinonapeptide. The toxin consists of polar peptide head and a 3-hydroxy fatty acid tail (which is hydrophobic in nature) (Fukuchi et al., 1992a; Segre et al., 1989). Syringomycin is produced in three isoform which are distinguished from each other by the tail length, that is, 3-hydroxy fatty acid moiety. These three forms are decanoic (SRA1), dodecanoic (SRE), or tetradecanoic (SRG) acid. 3-hydroxy fatty acid is attached to N-terminal serine residue by amide bond, which is further linked to 4-chlorothreonine by an ester linkage at C-terminal end and form macrocyclic lactone ring. Three uncommon amino acids namely 2, 3-dehydroaminobutyric acid, 3-hydroxyaspartic acid, and 4-chlorothreonine form trio at the C terminal and disomers of serine and 2,4-diaminobutyric acid makes it distinct from others (Scaloni et al., 1994). Biological activity of syringomycin is activated by chlorination of syringomycin molecule (Goss, 1940). *P. syringae* pv. *syringae* which has wide range of plant species as a host produce syringomycin. Strains of citrus and lilac hosts produce syringotoxin and syringostatin, which belongs to lipodepsinonapeptides class of toxin respectively (Ballio et al. 1994; Fukuchi et al. 1992b). Primary target of syringomycin is plasma membrane of host cells and cause necrosis of plant tissue (Backman et al., 1971; Paynter et al., 1979). Syringomycin has lipopeptide structure which is amphipathic in nature. It facilitates the entry of toxin in lipid bilayers of cell membrane and forms the pore in membrane which allows free permeability of cations (Hutchison et al., 1995). The toxin enhances the transmembrane fluxes of toxic ions namely K^+, H^+, and Ca^{2+} (Bidwai et al., 1987; Mott and Takemoto, 1989; Takemoto, 1992). The phenomenon of pore formation in lipid bilayer of cell membrane indicates that syringomycin is required in nanomolar amount for its activity (Hutchison et al., 1995). Syringomycin is the first toxin produced by plant pathogenic bacteria which targets plasma membrane of the host cell, facilitate ion channel formation in lipid bilayers and cytolysis in cell (Hutchison et al., 1995, 1997).

Tagetitoxin: The toxin is produced by the strains of bacteria *Pseudomonas syringae* pv. *tagetis*. The toxin acts by inhibiting the activity of RNA polymerase of plastid which in turn inhibit RNS synthesis in chloroplast (Walton, 2006). It cause chlorosis and yellowing in plant. Tagetitoxin is also reported to show toxicity in animals by inhibiting RNA polymerase III (Lukens and Durbin, 1990) which in turn interfere in transcription and translation process in cell (Steinberg et al., 1990). The inhibition of RNA polymerase of plastid by tagetitoxin produces very specific and rapid response. On treatment of 100 µM tagetitoxin to isolated chloroplasts results in reduction in [^{3}H] uridine incorporation into protein but did not interfere [^{35}S] methionine incorporation. During RNA synthesis, that is, transcription process in the chloroplast, tagetitoxin cause inhibiton of [^{32}P] UTP incorporation into RNA. During DNA synthesis process, incorporation of [^{3}H] thymidine into DNA was little effected by the toxin. Other than plant and animal host, tagetitoxin also interfere in RNA synthesis in three different eubacteria namely *Escherichia coli*, Anabaena 7120, and *Pseudomonas syringae* pv. *tagetis* (Mathews, 1988).

Non host specific toxins are slow acting but major damage to plant metabolism. Fusicoccin is the toxin is produced by *Fusicoccum amygdale*. It causes wilting of peach and almond. Fusarium wilt causes plugging of xylem vessel by fungal mycelium, fungal spores, etc., and browning of vascular bundle. It impaired flow of water and food materials through vascular bundle.

7.8   CONCLUSION

Host surface recognizes fungal pathogens by receptors present on its surface. It involves influx of Calcium and potassium and efflux of potassium chloride. These ionic fluxes initiate the process of oxidative burst. Protein kinases (PK) protein family comprised calmodulin (CaM)-like domain protein kinases (CDPK), sense Ca^{2+} signalling and get activated. PKs further activate the Mitogen activated protein kinase (MAPK) and induce the production of reactive oxygen species (ROS) which initiates defense response in host plant. Assimilation of nitrogen is takes place in the form of amino acids like glutamine, asparagines and aspartate. During pathogen infection, energy demand rises very high which shuttle amino acids in energy generating pathways like tricarboxylic acid (TCA). Respiration in plant is greatly influenced by pathogen infection which involves increased activity of glycolysis, TCA and electron transport chain (ETC). As a by-product, NADPH is produced during glycolysis, TCA and OPP pathway. This NADPH serves as a substrate for generation of energy through ETC. Accumulation of lipids, carotenoid and alteration in the activity of growth hormone is observed due to pathogen infection of host plant. Pathogen infection induces stomatal closure resulting in decline in photosynthesis in plant. Stomatal closure also involves stomatal OPPP shikimmic acid pathway and cytosolic OPPP reaction. Phenolic compounds are produced by stomatal OPPP shikimmic acid pathway and ROS during cytosolic OPPP reaction. Cell wall fortification through callose deposition is induced by pathogen infection and cause hypersensitive death. Pathogen also produces toxins which facilitate disease in the host plant. This toxin can be host specific and non host specific. Host specific toxins are active against specific to the host while non host specific toxins can cause damage to broad range of host. Host-specific toxins are produced by *Cochliobolus* spp., *Alternaria* spp, *Periconia* spp., etc., while non-host specific by *Pseudomonas syringae, Fusarium* spp., etc.

KEYWORDS

- **fungi**
- **physiology**
- **pathogen**
- **ROS**
- **toxin**

REFERENCES

Akamatsu, H.; Itoh, Y.; Kodama, M.; Otani, H.; Kohmoto K. AAL-toxin Deficient Mutants of *Alternaria Alternata* Tomato Pathotype by Restriction Enzyme-mediated Integration. *Phytopathology* **1997**, *87*, 967–972.

Akimitsu, K.; Kohmoto, K.; Otani, H.; Nishimura, S. Host-specific Effect of Toxin from the Rough Lemon Pathotype of Alternaria Alternata on Mitochondria. *Plant Physiol.* **1989**, *89*, 925–931.

Allen, P. J. Physiological Aspects of Fungus Diseases of Plants. *Annu. Rev. Plant. Physiol.* **1954**, *5*, 225–248.

Alvarez, M. E.; Pennell, R. I.; Meijer, P-J.; Ishikawa, A.; Dixon, R. A.; Lamb, C. Reactive Oxygen Intermediates Mediate a Systemic Network in the Establishment of Plant Immunity. *Cell* **1998**, *92*, 773–784.

Apostol, I.; Heinstein, P. F.; Low, P. S. Rapid Stimulation of An Oxidative Burst During Elicitation of Cultured Plant Cells: Role in Defense and Signal Transduction. *Plant Physiol.* **1989**, *90*, 109–116.

Asai, T.; Tena, G.; Plotnikova, J.; Willmann, M. R.; Chiu, W.-L.; Gomez-Gomez, L.; Boller, T.; Ausubel, F. M.; Sheen, J. MAP Kinase Signaling Cascade in *Arabidopsis* Innate Immunity. *Nature* **2002**, *415*, 977–983.

Aveskamp, M. M.; de Gruyter, J.; Woudenberg, J. H. C.; Verkley, G. J. M.; Crous, P. W. Highhlights of the Didymellaceae: A Polyphasic Approach to Characterize Phoma and Related Pleosporalean Genera. *Stud. Mycol.* **2010**, *65*, 1–60.

Backman, P. A.; DeVay, J. E. Studies on the Mode of Action and Biogenesis of the Phytotoxin Syringomycin. *Physiol. Plant Pathol.* **1971**, *1*, 215–234.

Ballio, A.; Collina, A.; Di Nola, A.; Manetti, C.; Paci, M.; Segre, A. Determination of Structure and Conformation in Solution of Syringotoxin, a Lipodepsipeptide from *Pseudomonas syringae* pv. *syringae* by 2D NMR and Molecular Dynamics. *Struct. Chem.* **1994**, *5*, 43–50.

Basse, C. W.; Stumpferl, S.; Kahmann, R. Characterization of a *Ustilago maydis* Gene Specifically Induced During the Biotrophic Phase: Evidence for Negative as Well as Positive Regulation. *Mol. Cell. Biol.* **2000**, *20*, 329–339.

Bender C. L.; Alarco′N-Chaidez, F.; Gross, D. C. *Pseudomonas syringae* Phytotoxins: Mode of Action, Regulation, and Biosynthesis by Peptide and Polyketide Synthetases. *Microbiol. Mol. Biol. Rev.* **1999**, 266–292.

Bent, A. F.; Innes, R. W.; Ecker, J. R.; Staskawicz, B. J. Disease Development in Ethylene-insensitive *Arabidopsis thaliana* Infected with Virulent and Avirulent *Pseudomonas* and *Xanthomonas* Pathogens. *Mol. Plant-Microbe Interact.* **1992**, *5*, 372–378.

Berestetskiy, A. O. A Review of Fungal Phytotoxins: from Basic Studies to Practical Use. *Appl. Biochem. Microbiol.* **2008**, *44*, 453–465.

Besson-Bard, A. L.; Pugin, A.; Wendehenne, D. New Insights into Nitric Oxide Signaling in Plants. *Annu. Rev. Plant Biol.* **2008**, *59*, 21–39.

Bezuidenhout, S. C.; Gelderblom, W. C. A.; Gorst-Allman, C. P.; Horak, R. M.; Marasas, W. F. O.; Spiteller, G.; Vleggaar, R. Structure Elucidation of the Fumonisins, Mycotoxins from *Fusarium moniliforme*. *J. Chem. Soc. Chem. Commun.* **1988**, *11*, 743–745.

Bidwai, A. P.; Takemoto, J. Y. Bacterial Phytotoxin, Syringomycin, Induces a Protein Kinase-mediated Phosphorylation of Red Beet Plasma Membrane Polypeptides. *Proc. Natl. Acad. Sci. USA* **1987**, *84*, 6755–6759.

Bolton, M. D. Primary Metabolism and Plant Defense—Fuel for the Fire. *Mol. Plant-Microbe Interact.* **2009,** *22,* 487–497.

Bolton, M. D.; Thomma, B. P. H. J. The Complexity of Nitrogen Metabolism and Nitrogen-regulated Gene Expression in Plant Pathogenic Fungi. *Physiol. Mol. Plant Pathol.* **2008,** *72,* 104–110.

Bolton, M. D.; Kolmer, J. A.; Xu, W. W.; Garvin, D. F. *Lr34*-mediated Leaf Rust Resistance in Wheat: Transcript Profiling reveals a High Energetic Demand supported by Transient Recruitment of Multiple Metabolic Pathways. *Mol. Plant-Microbe Interact.* **2008b;** *21,* 1515–1527.

Brandwagt, B. F.; Mesbah, L. A.; Takken, F. L. W.; Laurent, P. L.; Kneppers, T. J. A.; Hille, J.; Nijkamp, H. J. J. A Longevity Assurance Gene Homolog of Tomato Mediates Resistance to *Alternaria alternata f. sp. lycopersici* Toxins and Fumonisin B1. *Proc. Natl. Acad. Sci. USA* **2000,** *97,* 4961–4966.

Casati, P.; Drincovich, M. F.; Edwards, G. E.; Andreo, C. S. Malate Metabolism by NADP-malic Enzyme in Plant Defense. *Photosynth. Res.* **1999,** *61,* 99–105.

Chandra, S.; Low, P. S. Measurement of Ca^{2+} Fluxes During Elicitation of the Oxidative Burst in Aequorin-transformed Tobacco Cells. *J. Biol. Chem.* **1997,** *272,* 28274–28280.

Coleman, R. H.; Shaffer, J.; True, H. Properties of ß-lactamase from *Pseudomonas syringae. Curr. Microbiol.* **1996,** *32,* 147–150.

Doke, N. Generation of Superoxide Anion by Potato Tuber Protoplasts During the Hypersensitive Response to Hyphal Cell Wall Components of *Phytophthora Infestans* and Specific Inhibition of the Reaction by Suppressors of Hypersensitivity. *Physiol. Plant Pathol.* **1996,** *23,* 359–367.

Doke, N. The Oxidative Burst: Roles in Signal Transduction and Plant Stress. In *Oxidative Stress and the Molecular Biology of Antioxidant Defenses,* Scandalios, J. G., Ed.; Cold Spring Harbor Laboratory Press: New York, 1997; pp 785–813.

Doke, N.; Miura, Y. *In Vitro* Activation of NADPH-dependent O_2^- Generating System in a Plasma Membrane-rich Fraction of Potato Tuber Tissues by Treatment with An Elicitor from *Phytophthora Infestans* or with Digitonin. *Physiol. Mol. Plant Pathol.* **1995,** *46,* 17–28.

Dong, X. NPR1, All Things Considered. *Curr. Opin. Plant Biol.* **2004,** *7,* 547–552.

Durbin, R. D. Bacterial Phytotoxins: Mechanism of Action. *Experientia* **1991,** *47,* 776–783.

Durbin, R. D.; Graniti, A. Possible Applications of Phytotoxins, In *Phytotoxins and Plant Pathogenesis,* Graniti, A., Durbin, R. D., Ballio, A., Eds.; Springer-Verlag: Berlin, 1989; pp 335–355.

Durbin, R. D.; Uchytil, T. F. The Role of Intercellular Fluid and Bacterial Isolate on the *In Vivo* Production of Tabtoxin and Tabtoxinine-blactam. *Physiol. Plant Pathol.* **1984,** *24,* 25–31.

Durbin, R. D.; Uchytil, T. F. The Mechanism for Self-protection Against Bacterial Phytotoxins. *Annu. Rev. Phytopathol.* **1988,** *26,* 313–329.

Durbin, R. D.; Uchytil, T. F. Cytoplasmic Inheritance of Chloroplast Coupling Factor 1 Subunits. *Biochem. Genet.* **1997,** *15* (11), 43–46.

Earle, E. D. Phytotoxin Studies with Plant Cells and Protoplasts. In *Frontiers of Plant Tissue Culture* 1978. Thorpe, T. A., Ed.; 1978; pp 363–72, Proc. 4th.

Ebel, J.; Scheel, D. Signals in Host–Parasite Interactions. In *The Mycota; Vol V, Plant Relationships, Part A.* G. C., Tudzynski, P., Eds.; Springer-Verlag: Berlin, 1997; pp 85–105.

Ekengren, S. K.; Liu, Y.; Schiff, M.; Dinesh-Kumar, S. P.; Martin, G. B. Two MAPK Cascades, NPR1, and TGA Transcription Factors Play a Role in Pto-mediated Disease Resistance in Tomato. *Plant J.* **2003,** *36,* 905–907.

Enkerli, K.; Hahn, M. G. Mims, C. W. Immunogold Localization of Callose and Other Plant Cell Wall Components in Soybean Roots Infected with the Plant Pathogenic Oomycete, *Can. J. of Bot.* **1997,** *75,* 1509–1517.

Ferguson, I. B.; Mitchell, R. E. Stimulation of Ethylene Production in Bean Leaf Discs by the Pseudomonad Phytotoxin Coronatine. *Plant Physiol.* **1985,** *77,* 969–973.

Fernie, A. R.; Carrari, F.; Sweetlove, L. J. Respiratory Metabolism: Glycolysis, the TCA Cycle and Mitochondrial Electron Transport. *Curr. Opin. Plant Biol.* **2004,** *7,* 254–261.

Feys, B. J. F.; Benedetti, C. E.; Penfold, C. N.; Turner, J. G. Arabidopsis Mutants Selected for Resistance to the Phytotoxin Coronatine Are Male Sterile, Insensitive to Methyl Jasmonate, and Resistant to a Bacterial Pathogen. *Plant Cell* **1994,** *6,* 751–759.

Frye, C. A.; Tang, D.; Innes, R. W. Negative Regulation of Defense Responses in Plants by a Conserved MAPKK Kinase. *Proc. Natl. Acad. Sci. U.S.A* **2001,** *98,* 373–378.

Fudal, I.; Ross, S.; Gout, L.; Blaise, F.; Kuhn, M. L.; Eckert, M. R.; Cattolico, L.; Bernard-Samain, S.; Balesdent, M. H.; Rouxel, T. Heterochromatin-like Regions as Ecological Niches for Avirulence Genes in the *Leptosphaeria Maculans* Genome: Map-based Cloning of *AvrLm6. Mol. Plant-Microbe. Interact.* **2007,** *20,* 459–470.

Fukuchi, N., Isogai, A.; Nakayama, J.; Takayama, S.; Yamashita, S.; Suyama, K.; Takemoto, J. Y.; Suzuki, A. Structure and Stereochemistry of Three Phytotoxins, Syringomycin, Syringotoxin and Syringostatin, Produced by *Pseudomonas Syringae* pv. *Syringae. J. Chem. Soc. Perkin Trans. I,* **1992a,** *9,* 1149–1157.

Fukuchi, N.; Isogai, A.; Nakayama, J.; Takayama, S.; Yamashita, S.; Suyama, K.; Suzuki, A. Isolation and Structural Elucidation of Syringostatins, Phytotoxins Produced by *Pseudomonas Syringae* pv. *Syringae* Lilac Isolate. *J. Chem. Soc. Perkin Trans. I* **1992b,** *7,* 875–880.

Gardner, J. M.; Kono, Y.; Tatum, J. H.; Suzuki, Y.; Takeuchi, S. Structure of Major Component of ACRL Toxins, Host-specific Phytotoxic Compound Produced by *Alternaria citri. Agric. Biol. Chem.* **1985a,** *49,* 1235–1238.

Gelli, A.; Higgins, V. J.; Blumwald, E. Activation of Plant Plasma Membrane Ca^{2+}-permeable Channels by Race-specific Fungal Elicitors. *Plant Physiol.* **1997,** *113,* 269–279.

Gilchrist, D. G.; Grogan, R. G. Production and Nature of a Host-specific Toxin *from Alternaria Alternata f. sp. lycopersici. Phytopathology* **1976,** *66,* 165–171.

Gilchrist, D. G. Programmed Cell Death in Plant Disease: the Purpose and Promise of Cellular Suicide. *Ann. Rev. Phytopathol.* **1998,** *36,* 393–414.

Gilchrist, D. G; Wang, H.; Bostock, R. M. Sphingosinerelated Mycotoxins in Plant and Animal Disease. *Can. J. Bot.* **1994,** *73,* S459–S467.

Gnanamaickam, S. S.; Starratt, A. N.; Ward, E. W. B. Coronatine Production in Vitro and in Vivo and Its Relation to Symptom Development in Bacterial Blight of Soybean. *Can. J. Bot.* **1982,** *60,* 645–650.

Goss, R. W. The Relation of Temperature to Common and Halo Blight of Beans. *Phytopathology* **1940,** *30,* 258–264.

Greulich, F.; Yoshihara, T.; Ichihara, A. Coronatine, a Bacterial Phytotoxin, Acts as a Stereospecific Analog of Jasmonate Type Signals in Tomato Cells and Potato Tissues. *J. Plant Physiol.* **1995,** *147,* 359–366.

Guo, W.; González-Candelas, L.; Kolattukudy, P. E. Identification of a Novel *pelD* Gene Expressed Uniquely in Planta by *Fusarium solani* f. sp. *pisi* (*Nectria haematococca*, Mating Type VI) and Characterization of its Protein Product as an Endo-pectate Lyase. *Arch. Biochem. Biophys.* **1996**, *332*, 305–312.

Hahlbrock, K.; Scheel, D.; Logemann, E.; Nu$rnberger, T.; Parniske, M.; Reinold, S.; Sacks, W. R.; Schmelzer, E. Oligopeptide Elicitor-mediated Defense Gene Activation in Cultured Parsley Cells. *Proc. Natl. Acad. Sci. U.S.A* **1995**, *92*, 4150–4157.

Hanchey, P.; Wheeler, H. The Role of Host Cell Membranes. See Ref. 26, 1979; pp 193–210

Herbers, K.; Meuwly, P.; Metraux, J.; Sonnewald, U. Salicylic Acid-dependent Induction of Pathogenesis-related Protein Transcripts by Sugars is Dependent on Leaf Developmental Stage. *FEBS Lett.* **1996b**, *397*, 239–244.

Herbers, K.; Takahata, Y.; Melzer, M.; Mock, H. P.; Hajirezaei, M.; Sonnewald, U. Regulation of Carbohydrate Partitioning During the Interaction of Potato Virus Y with Tobacco. *Mol. Plant Pathol.* **2000**, *1*, 51–59.

Hrabak, E. M.; Chan, C. W. M.; Gribskov, M.; Harper, J. F.; Choi, J. H.; Halford, N.; Kudla, J.; Luan, S.; Nimmo, H. G.; Sussman, M. R.; Thomas, M.; Walker-Simmons, K.; Zhu, J.-K.; Harmon, A. C. The *Arabidopsis* CDPK-SnRK Superfamily of Protein Kinases. *Plant Physiol.* **2003**, *132*, 660–680.

Hutchison, M. L.; Gross, D. C. Lipopeptide Phytotoxins Produced by *Pseudomonas Syringae* pv. *Syringae*: Comparison of the Biosurfactant and Ion Channel-forming Activities of Syringopeptin and Syringomycin. *Mol. Plant-Microbe Interact.* **1997**, *10*, 347–354.

Hutchison, M. L.; Tester, M. A.; Gross, D. C. Role of Biosurfac-tant and Ion Channel-rming Activities of Syringomycin in Transmembrane Ion Flux: A Model for the Mechanism of Action in the Plant-Pathogen Interaction. *Mol. Plant-Microbe Interact.* **1995**, *8*, 610–620.

Ibrahim, M. *Physiological Studies on Fungal Isolates Associated with Prehavest Rotten Tomato (Lycopersicum esculentus (Mill) Fruits in Some Aelected Local Government Areas of Sokoto State, Nigeria*, M.Sc. Thesis, Usmanu Danfodiyo University, Sokoto, Nigeria, 2002.

Ichihara, A.; Shiraishi, K.; Sato, H.; Sakamura, S.; Nishiyama, K.; Sakai, R.; Furusaki, A.; Matsumoto, T. The Structure of Coronatine. *J. Am. Chem. Soc.* **1977**, *99*, 636–637.

Iqbal, S. M.; Ghafoor, Z. A.; Haqqani . Pathogenicity and Fungicidal Efficacy for Sclerotinia Rot of Brinjal. *Int. J. Agric. Biol.* **2003**, *5* (4), 618–620.

Jabs, T.; Dietrich, R. A.; Dangl, J. L. Initiation of Runaway Cell Death in An *Arabidopsis* Mutant by Extracellular Superoxide. *Science* **1996**, *273*, 1853–1856.

Jabs, T.; Tschope, M.; Colling, C.; Hahlbrock, K.; Scheel, D. Elicitor Stimulated Ion Fluxes and O^{2-} from the Oxidative Burst are Essential Components in Triggering Defense Gene Activation and Phytoalexin Synthesis in Parsley. *Proc. Natl. Acad. Sci. USA* **1997**, *94*, 4800–4805.

Jin, H.; Axtell, M. J.; Dahlbeck, D.; Ekwanna, O.; Zhang, S.; Staskawicz, B.; Baker, B. NPK1, an MEKK1-like Mitogen-activated Protein Kinase Kinase Kinase, Regulates Innate Immunity and Development in Plants. *Dev. Cell* **2002**, *3*, 291–297.

Jin, H.; Liu, Y.; Yang, K. -Y.; Kim, C. Y.; Baker, B.; Zhang, S. Function of a Mitogen-activated Protein Kinase Pathway in *N*-gene Mediated Resistance in Tobacco. *Plant J.* **2003**, *33*, 719–731.

Johal, G. S.; Briggs, S. P. Reductase-activity Encoded by the Hm1 Disease Resistance Gene in Maize. *Science* **1992**, *258*, 985–987.

Jonak, C.; Okrész, L.; Bögre, L.; Hirt, H. Complexity, Cross-talk and Integration of Plant MAP Kinase Signaling. *Curr. Opin. Plant Biol.* **2002,** *5*, 415–424.

Kadota, Y.; Furuichi, T.; Ogasawara, Y.; Goh, T.; Higashi, K.; Muto, S.; Kuchitsu, K. Identification of Putative Voltage-dependent Ca^{2+}-permeable Channels involved in Cryptogein-induced Ca^{2+} Transients and Defense Responses in Tobacco BY-2 cells. *Biochem. Biophys. Res.* **2004,** *304* (17), 823–830.

Kenyon, J. S.; Turner, J. G. The Stimulation of Ethylene Synthesis in *Nicotiana tabacum* Leaves by the Phytotoxin Coronatine. *Plant Physiol.* **1992,** *100*, 219–224.

Khan, R.; Straney, D. C. Regulatory Signals Influencing Expression of the *PDA1* Gene of *Nectria Haematococca* MPVI in Culture and During Pathogenesis of Pea. *Mol. Plant-Microbe Interact.* **1999,** *12*, 733–742.

Kim, C.-Y.; Zhang, S. Activation of a Mitogen-activated Protein Kinase Cascade induces WRKY Family of Transcription Factors and Defense Genes in Tobacco. *Plant J.* **2004,** *38*, 142–151.

Kim, Y. M.; Bouras, N.; Kav, N. N.; Strelkov, S. E. Inhibition of Photosynthesis and Modification of the Wheat Leaf Proteome by Ptr ToxB: a Host-specific Toxin from the Fungal Pathogen *Pyrenophora tritici-repentis. Proteomics* **2010,** *10*, 2911–2926.

Kinscherf, T. G.; Coleman, R. H.; Barta, T. M.; Willis, D. K. Cloning and Expression of the Tabtoxin Biosynthetic Region from *Pseudomonas syringae. J. Bacteriol.* **1991,** *173*, 4124–4132.

Knight, T. J.; Durbin, R. D.; Langston-Unkefer, P. J. Role of Glutamine Synthetase Adenylation in the Self-protection of *Pseudomonas syringae* subsp. *"tabaci"* from Its Toxin, Tabtoxinine- ß-lactam. *J. Bacteriol.* **1986,** *166*, 224–229.

Knight, T. J.; Durbin, R. D.; Langston-Unkefer; P. J. Self-protection of *Pseudomonas syringae* pv. *"tabaci"* from its toxin, tabtoxinine- ß-lactam. *J. Bacteriol.* **1987,** *169*, 1954–1959.

Koda, Y.; Takahashi, K.; Kikuta, Y.; Greulich, F.; Toshima, H.; Ichihara, A. Similarities of the Biological Activities of Coronatine and Coronafacic Acid to those of Jasmonic Acid. *Phytochemistry* **1996,** *41*, 93–96.

Kohmoto, K.; Otani, H.; Nishimura, S. Action Sites of AM-toxins Produced by the Apple Pathotype of Alternaria alternata. In *Plant Infection: The Physiologocal and Biochemical Basis*, Asada, Y., Bushnell, W. R., Ouchi, S., Vance, C. P., Eds.; Springer-Verlag: Berlin, 1982; pp 81–136.

Kombrink, E.; Somssich, I. E. Defense Response of Plant to Pathogens. *Adv. Bot. Res.* **1995,** *21*, 1–34.

Kono, Y.; Daly, J. M. Characterization of the Hostspecific Pathotoxin Produced by *Helminthosporium maydis* Race T, Affecting Corn with Texas Male Sterile Cytoplasm. *Bioorg. Chem.* **1979,** *8*, 391.

Kozaki, I. Black Spot Disease Resistance in Japanese Pear. I. Heredity of the Disease Resistance. *Bull. Hort. Res. Stn. Jpn.* **1973,** A *12*, 17–27.

Kurusu, T.; Yagala, T.; Miyao, A.; Hirochika, H.; Kuchitsu, K. Identification of a Putative Voltage-gated Ca^{2+} Channel as a Key Regulator of Elicitor-induced Hypersensitive Cell Death and Mitogen-activated Protein Kinase Activation in Rice. *Plant J.* **2005,** *42*, 798–809.

Lam, H. M.; Coschigano, K. T.; Oliveira, I. C.; Melo-Oliveira, R.; Coruzzi, G. M. The Molecular-Genetics of Nitrogen Assimilation into Amino Acids in Higher Plants. *Ann. Rev. Plant Physiol. Plant Mol. Biol.* **1996,** *47*, 569–593.

Lamotte, O.; Gould, K.; Lecourieux, D.; Sequeira-Legrand, A.; Lebrun-Garcia, A.; Durner, J.; Pugin, A.; Wendehenne, D. Analysis of Nitric Oxide Signaling Functions in Tobacco Cells Challenged by the Elicitor Cryptogein. *Plant Physiol.* **2004,** *135,* 516–529.

Lasanthi-Kudahettige, R.; Magneschi, L.; Loreti, E.; Gonzali, S.; Licausi, F.; Novi, G.; Beretta, O.; Vitulli, F.; Alpi, A.; Perata, P. Transcript Profiling of the Anoxic Rice Coleoptile. *Plant Physiol.* **2007,** *144,* 218–231.

Lecourieux-Ouaked, F.; Pugin, A.; Lebrun-Garcia, A. Phosphoproteins Involved in the Signal Transduction of Cryptogein, an Elicitor of Defense Reactions in Tobacco. *Mol. Plant-Microbe Interact.* **2000,** *13,* 821–829.

Lee, J.; Klessig, D. F.; Nürnberger, T. A Harpin Binding Site in Tobacco Plasma Membranes Mediates Activation of the Pathogenesis-related Gene HIN1 Independent of Extracellular Calcium but Dependent on Mitogen-activated Protein Kinase Activity. *Plant Cell* **2001,** *13,* 1079–1093.

Levi, C.; Durbin R. D. The Isolation and Properties of a Tabtoxin Hydrolyzing Aminopeptidase from the Periplasm of *Pseudomonas syringae* pv. *tabaci. Physiol. Mol. Plant Pathol.* **1986,** *28,* 345–352.

Levine, A.; Pennell, R. I.; Alvarez, M. E.; Palmer, R.; Lamb, C. Calcium Mediated Ppoptosis in A Plant Hypersensitive Disease Resistance Response. *Curr. Biol.* **1996,** *6,* 427–437.

Levings, C. S.; Rhoads, D. M.; Siedow, J. N. Molecularinteractions of Bipolaris maydis T-toxin and Maize. *Can. J. Bot.* **1995,** *73,* S483–S489.

Link, V. L.; Hofmann, M.; Sinha, A. K.; Ehness, R.; Strnad, M.; Roitsch, T. Biochemical Evidence for the Activation of Distinct Subsets of Mitogen-activated Protein Kinases by Voltage and Defenserelated Stimuli. *Plant Physiol.* **2002,** *128,* 271–281.

Livingston, R. S.; Scheffer, R. P. Toxic and Protective Effects of Analogs of Helminthosporium Sacchari Toxin on Sugarcane Tissues. *Physiol. Plant Pathol.* **1984,** *24,* 133–142.

Lukens, J. H.; Durbin, R. D. Tagetitoxin Inhibits RNA Synthesis Directed by RNA Polymerases from Chloroplasts and *Escherichia coli. J. Biol. Chem.* **1990,** *265,* 493–498.

Lynen, F. Lipide Metabolism. *Annu. Rev. Biochem.* **1955,** *24,* 653–688.

Macko, V.; Acklin, W.; Hildenbrand, C.; Weibel, F.; Arigoni, D. Structure of Three Isomeric Host-specific Toxins from *Helminthosporium sacchari. Experientia* **1983,** *39,* 343–440.

Macko, V.; Wolpert, T. J.; Acklin, W.; Jaun, B.; Seibl, J.; Meili, J.; Arigoni, D. Characterization of Victorin-C, the Major Host-selective Toxin from *Cochliobolus victoriae*: Structure of Degradation Products. *Experientia* **1985,** *41,* 1366–1370.

Maekawa, N.; Yamamoto, M.; Nishimura, S.; Kohmoto, K.; Kuwada, K.; Watanabe, Y. Studies on Host-specific AF-toxins Produced by Alternaria Alternata Strawberry Pathotype Causing Alternaria Black Spot of Strawberry. (1) Production of host-specific Toxins and Their Biological Activities. *Ann. Phytopathol. Soc. Jpn.* **1984,** *50,* 600–609.

Markham, J. E.; Hille, J. Host-selective Toxins as Agents of Cell Death in Plant-fungus Interactions. *Mol. Plant Pathol.* **2001,** *2,* 229–240.

Mathews, D. E. *Mode of Action of Tagetitoxin.* Univ. of Wisconsin; Madison, WI (USA); University Microfilms, PO Box 1764, Ann Arbor, MI 48106, Order No.88-26,059; Ph.D.Thesis, 1988; p 147.

Mayama, S.; Bordin, A. P. A.; Morikawa, T.; Tanpo, H.; Kato, H. Association of Avenalumin Accumulation with Co-segregation of Victorin Sensitivity and Crown Rust Resistance in Oat Lines Carrying the Pc-2 Gene. *Physiol. Mol. Plant Pathol.* **1995,** *46,* 263–274.

Meehan, F. Murphy, H. C. A New Helminthosporium Blight of Oats. *Science* **1946,** *104,* 413.

Meehan F. Differential Phytotoxicity of Metabolic by Products of *Helminthosporium victoriae*. *Science* **1947**, *106*, 270–271.

Mertens, E.; Larondelle, Y.; Hers, H. -G. Induction of Pyrophosphate: Fructose 6-Phosphate 1-phosphotransferase by Anoxia in Rice Seedlings. *Plant Physiol.* **1990**, *93*, 584–587.

Miflin, B. J.; Habash, D. Z. The Role of Glutamine Synthetase and Glutamate Dehydrogenase in Nitrogen Assimilation and Possibilities for Improvement in the Nitrogen Utilization of Crops. *J. Exp. Bot.* **2002**, *53*, 979–987.

Mitchell, R. E. Isolation and Structure of a Chlorosis-inducing Toxin of *Pseudomonas phaseolicola*. *Phytochemistry* **1976**, *15*, 1941–1947.

Mitchell, R. E. The Relevance of Non-host-specific Toxins in the Expression of Virulence by Pathogens. *A. Rev. Phytopath.*, **1984**, *22*, 215–245.

Mitchell, R. E. Norcoronatine and *N*-coronafacoyl-L-valine, Phytotoxin Analogues of Coronatine Produced by a Strain of *Pseudomonas syringae* pv. *glycinea*. *Phytochemistry* **1985**, *24*, 1485–1488.

Mitchell, R. E. Implications of Toxins in the Ecology and Evolution of Plant Pathogenic Microorganisms: Bacteria. *Experientia* **1991**, *47*, 791–803.

Mitchell, R. E.; Frey, E. J. Production of *N*-coronafacoyl-Lamino Acid Analogues of Coronatine by *Pseudomonas syringae* pv. *atropurpurea* in Liquid Cultures Supplemented with L-amino Acids. *J. Gen. Microbiol.* **1985**, *132*, 1503–1507.

Mitchell, R. E.; Ford, K. L. Chlorosis-inducing Products from *Pseudomonas syringae* pathovars: new N-coronafacoyl Compounds. *Phytochemistry* **1998**, *49*, 1579–1583.

Mitchell, R. E.; Bieleski, R. L. Involvement of Phaseolotoxin in Halo Blight of Beans. Transport and Conversion to Functional Toxin. *Plant Physiol.* **1977**, *60*, 723–729.

Mitchell, R. E.; Durbin, R. D. Tagetitoxin, a Toxin Produced by *Pseudomonas-syringae* pv Tagetis: Purification and Partial Characterization. *Physiol. Plant Pathol.* **1981**, *18*, 157–168.

Moon, H.; Lee, B.; Choi, G.; Shin, D.; Prasad, D. T.; Lee, O.; Kwak, S.-S.; Kim, D. H.; Nam, J.; Hong, J. C.; Lee, S. Y.; Cho, M. J.; Lim, C. O.; Yun, D. -J. NDP Kinase 2 Interacts with Two Oxidative Stress-activated MAPKs to Regulate Cellular Redox State and Enhances Multiple Stress Tolerance in Transgenic Plants. *Proc. Natl. Acad. Sci. U.S.A.* **2003**, *100*, 358–363.

Moore, R. E., Niemczura, W. P.; Kwok, O. C. H.; Patil, S. S. Inhibitors of Ornithine Carbamoyltransferase from *Pseudomonas syringae* pv. *phaseolicola*. *Tetrahedron Lett.* **1984**, *25*, 3931–3934.

Mott, K. A.; Takemoto, J. Y. Syringomycin, a Bacterial Phytotoxin, Closes Stomata. *Plant Physiol.* **1989**, *90*, 1435–1439.

Nakatsuka, S.; Goto, T.; Kohmoto, K.; Nishimura, S. *Host-specific Phytotoxins. Natural Products and Biological Activities*, Imura, H., Goto, T., Murachi, T., Nakajima, T., Eds.; University of Tokyo Press: Tokyo, 1986; pp 11–18

Namiki, F.; Okamoto, H.; Katou, K.; Yamamoto, M.; Nishimura, S.; Nakatsuka, S.; Goto, T.; Kohmoto, K.; Otani, H.; Novacky, A. Studies on Host-specific Toxins Produced by *Alternaria alternata* Strawberry Pathotype Causing Alternaria Black Spot of Strawberry (5) Effect of Toxins on Membrane Potential of Susceptible Plants by Means of Electrophysiological Analysis. *Ann. Phytopathol. Soc. Jpn.* **1986**, *52*, 610–619.

Navarre, D. A.; Wolpert, T. J. Victorin Induction of an Apoptotic/Senescence-like Response in Oats. *Plant Cell* **1999**, *11*, 237–249.

Newingham, B. A.; Callaway, R. M.; BassiriRad, H. Allocating Nitrogen Away from a Herbivore: A Novel Compensatory Response to Root Herbivory. *Oecologia* **2007**, *153*, 913–920.

Nishimura, S.; Kohmoto, K. Host-specific Toxins and Chemical Structures from Alternaria Species. *Annu. Rev. Phytopathol.* **1983**, *21*, 87–116.

Nishimura, S.; Nakatsuka, S. *Trends in Host-selective Toxin Research in Japan. Host-Specific Toxins: Recognition and Specificity Factors in Plant Disease,* Kohmoto, K., Durbin, R. D., Eds.; Tottori University Press: Tottori, 1989; pp 19–31.

Nishimura, S.; Scheffer, R. P. Interactions Between *Helminthsporium Victoriae* Spores and Oat Tissue. *Phytopathology* **1965**, *55*, 629–634.

Nishimura, S. Host-specific Toxins from *Alternaria alternata*: Problems and Prospects. *Proc. Jpn. Acad.* **1980**, *56*B, 362–366.

Nishimura, S.; Sugihara, M.; Kohmoto, K.; Otani, H. Two Different Phases in Pathogenicity of the Alternaria Pathogen causing Black Spot Disease of Japanese Pear. *J. Fac. Agric. Tottori Univ.* **1978**, *13*, 1–10.

Nishimura, S.; Tatano, S.; Gomi, K.; Ohtani, K.; Fukumoto, T.; Akimitsu, K. Chloroplast-localized Nonspecific Lipid transfer Protein with Anti-fungal Activity from Rough Lemon. *Physiol. Mol. Plant Pathol.* **2008**, *72*, 134–140.

Nishimura, S.; Kohmoto, K. Host-specific Toxins and Chemical Structures from *Alternaria* Species. *A. Rev. Phytopath.* **1983**, *21*, 87–116.

Nogués, S.; Cotxarrera, L.; Alegre, L.; Trillas, M. I. Limitations to Photosynthesis in Tomato Leaves Induced by *Fusarium* wilt. *New Phytologist.* **2002**, *154*, 461–470.

Olesen, P.; Robards, A. W. The Neck Region of Plasmodesmata: General Architecture and Some Functional Aspects. In *Parallels in Cell to Cell Junctions in Plants and Animals,* Robards, A. W., Lucas, W. J., Pitts, J. D., Jongsma, H. J., Spray, D.C., Eds.;. Springer: Berlin, Germany, 1990; pp 145–170.

Oliver, R. P.; Solomon, P. S. New Developments in Pathogenicity and Virulence of Necrotrophs. *Curr. Opin. Plant Biol.* **2010**, *13*, 415–419.

Oliver, R. P.; Solomon, P. S. Recent Fungal Diseases of Crop Plants: is Lateral Gene Transfer a Common Theme? *Mol. Plant Microbe Interact.* **2008**, *21*, 287–293.

Otani, H.; Kohmoto, K.; Kodama, M. Alternaria Toxins and their Effects on Host Plants. *Can. J. Bot.* **1995**, *73*, S453–S458.

Otani, H.; Kohmoto, K.; Kodama, M.; Nishimura, S. Role of Host-specific Toxins in the Pathogenesis of *Alternaria alternata*. In *Molecular Strategies of Pathogens and Host Plants,* Patil, S. S., Ouchi, S., Mills, D., Vance, C., Eds.; Springer-Verlag: Berlin, 1991; pp 139–149.

Otani, H.; Kohmoto, K.; Nishimura, S.; Nakashima, T.; Ueno, T.; Fukami, H. Biological Activities of AK-toxins I and II, Host-specific Hoxins from Alternaria alternata Japanese Pear Pathotype. *Ann. Phytopathol. Soc. Jpn.* **1985**, *51*, 285–293.

Pageau, K.; Reisdorf-Cren, M.; Morot-Gaudry, J.-F.; Masclaux-Daubresse, C. The Two Senescence-related Markers, *GS1* (cytosolic glutamine synthetase) and *GDH* (glutamate dehydrogenase), Involved in Nitrogen Mobilization, Are Differentially Regulated During Pathogen Attack and by Stress Hormones and Reactive Oxygen Species in *Nicotiana tabacum* L. Leaves. *J. Exp. Bot.* **2006**, *57*, 547–557.

Park, P.; Ikeda, K. Ultrastructural Analysis of Responses of Host and Fungal Cells During Plant Infection. *J. Gen. Plant Pathol.* **2008**, *74*, 2–14.

Park, P.; Nishimura, S.; Kohmoto, K.; Otani, H.; Tsujimoto, K. Two Action Sites of AM-toxin I Produced by Apple Pathotype of *Alternaria alternate* in Host Cell: An Ultrastructural Study. *Can. J. Bot.* **1981,** *59,* 301–310.

Park, P. Effects of the Hostspecific Toxin and other Toxic Metabolites by *Alternaria Kikuchiana* on Ultrastructure of Leaf Cells of Japanese Pear. *Ann. Phytopathol. Soc. Jpn.* **1977,** *43,* 15–25.

Parry, R. J.; Mhaskar, S. V.; Lin, M. -T.; Walker, A. E.; Mafoti, R. Investigations of the Biosynthesis of the Phytotoxin Coronatine. *Can. J. Chem.* **1994,** *72,* 86–99.

Patil, S. S.; Hayward, A. C.; Emmons, R. An Ultraviolet-induced Nontoxigenic Mutant of *Pseudomonas phaseolicola* of Altered Pathogenicity. *Phytopathology* **1974,** *64,* 590–595.

Peet, R. C.; Panopoulos, N. J. Ornithine Carbamoyltransferase Genes and Phaseolotoxin Immunity in *Pseudomonas syringae* pv. *phaseolicola. EMBO J.* **1987,** *6,* 3585–3591.

Peet, R. C.; Lindgren, P. B.; Willis, D. K.; Panopoulos, N. J. Identification and Cloning of Genes Involved in Phaseolotoxin Production by *Pseudomonas syringae* pv. "phaseolicola." *J. Bacteriol.* **1986,** *166,* 1096–1105.

Pei, Z. M.; Murata, Y.; Benning, G.; Thomine, S.; Klusener, B.; Allen, G. J.; Grill, E.; Schroeder, J. I. Calcium Channels Activated by Hydrogen Peroxide Mediate Abscisic Acid Signalling in Guard Cells. *Nature* **2000,** *406,* 731–734.

Petersen, M.; Brodersen, P.; Naested, H.; Andreasson, E.; Lindhart, U.; Johansen, B.; Nielsen, H. B.; Lacy, M.; Austin, M. J.; Parker, J. E.; Sharma, S. B.; Klessig, D. F.; Martienssen, R.; Mattson, O.; Jensen, A. B.; Mundy, J. *Arabidopsis* MAP Kinase 4 Negatively Regulates Systemic Acquired Resistance. *Cell* **2000,** *103,* 1111–1120.

Phytophthora sojae. Can. J. Bot. **2000,** *75,* 1509–1517.

Pugin, A.; Frachisse, J. M.; Tavernier, E.; Bligny, R.; Gout, E.; Douce, R.; Guern, J. Early Events Induced by the Elicitor Cryptogein in Tobacco Cells: Involvement of a Plasma Membrane NADPH Oxidase and Activation of Glycolysis and the Pentose Phosphate Pathway. *Plant Cell* **1997,** *9,* 2077–2091.

Ren, D.; Yang, H.; Zhang, S. Cell Death Mediated by MAPK is Associated with Hydrogen Peroxide Production in *Arabidopsis. J. Biol.Chem.* **2002,** *277,* 559–565.

Rolland, F.; Moore, B.; Sheen, J. Sugar Sensing and Signalling in Plants. *Plant Cell* **2002,** *14,* 185–205.

Romero-Puertas, M. C.; Perazzolli, M.; Zago, E. D.; Delledonne, M. Nitric Oxide Signalling Functions in Plant-pathogen Interactions. *Cell. Microbiol.* **2004,** *6,* 795–803.

Saito, K.; Takeda, K. Genetic Analysis of Resistance to Alternaria Blotch (*Alternaria mali* Roberts) in Apple. (Studies on the breeding of the apple VIII). *Japan J. Breed.* **1984,** *34,* 197–209.

Sakai, R.; Nishiyama, K.; Ichihara, A.; Shiraishi, K.; Sakamura, S. The Relation Between Bacterial Toxic Action and Plant Growth Regulation. In *Recognition and Specificity in Plant Host-Parasite Interactions,* Daly, J. M., Uritani, I., Eds.; University Park Press: Baltimore, MD. 1979; pp 165–179.

Salau, I. A.; Kasarawa, A. B.; Tambuwal, N. I. *Fungi Associated with Vegetable Plant Diseases in the Fadama Land Areas of Sokoto Metropolis.* Proceeding of Agricultural Society of Nigeria (ASN) 46th Annual Conference ("Kano 2012"), 2012; pp 779–782.

Sanada, T. Selection of Resistant Mutants to Black Spot Disease of Japanese pear by using Host-specific Toxin. *Japan J. Breed.* **1988,** *38,* 198–204.

Sawada, H.; Takeuchi, T.; Matsuda, I. Comparative Analysis of *Pseudomonas syringae* pv. Actinidiae and pv. Phaseolicola Based on Phaseolotoxin-resistant Ornithine

Carbamoyltransferase Gene (*argK*) and 16S-23S rRNA Intergenic Spacer Sequences. *Appl. Environ. Microbiol.* **1997**, *63*, 282–288.

Scaloni, A.; Bachmann, R. C.; Takemoto, J. Y.; Barra, D.; Simmaco, M.; Ballio, A. Stereochemical Structure of Syringomycin, a Phytotoxic Metabolite of *Pseudomonas syringae* pv. *syringae*. *Nat. Prod. Lett.* **1994**, *4*, 159–164.

Schaaf, J.; Walter, M. H.; Hess, D. Primary Metabolism in Plant Defense (Regulation of a Bean Malic Enzyme Gene Promoter in Transgenic Tobacco by Developmental and Environmental Cues). *Plant Physiol.* **1995**, *108*, 949–960.

Scharte, J.; Schön, H.; Weis, E. Photosynthesis and Carbohydrate Metabolism in Tobacco Leaves During an Incompatible Interaction with *Phytophthora Nicotianae*. *Plant Cell Environ.* **2005**, *28*, 1421–1435.

Scheffer, R. P. Host-specific Toxins in Relation to Pathogenesis and Disease Resistance. See Ref. 20, **1976**, 247–69.

Scheideler, M.; Schlaich, N. L.; Fellenberg, K.; Beissbarth, T.; Hauser, N. C.; Vingron, M.; Slusarenko, A. J.; Hoheisel, J. D. Monitoring the Switch from Housekeeping to Pathogen Defense Metabolism in *Arabidopsis thaliana* Using cDNA Arrays. *J. Biol. Chem.* **2002**, *277*, 10555–10561.

Schenk, P. M.; Kazan, K.; Manners, J. M.; Anderson, J. P.; Simpson, R. S.; Wilson, I. W.; Somerville, S. C.; Maclean, D. J. Systemic Gene Expression in *Arabidopsis* During an Incompatible Interaction with *Alternaria brassicicola*. *Plant Physiol.* **2003**, *132*, 999–1010.

Schroter, H.; Novacky, A.; Macko, V. Effect of *Helminthosporium sacchari* Toxin on Cell-membrane Potential of Susceptible Sugarcane. *Physiol. Plant Pathol.* **1985**, *26*, 165–174.

Segre, A.; Bachmann, R. C.; Ballio, A.; Bosa, F.; Grgurina, I.; Iacobellis, N. S.; Marino, G.; Pucci, P.; Simmaco, M.; Takemoto, J. Y. The Structure of Syringomycins A1, E and G. *FEBS Lett.* **1989**, *255*, 27–31.

Sembdner, G.; B. Parthier. The Biochemistry and the Physiological and Molecular Actions of Jasmonates. *Annu. Rev. Plant Physiol. Plant Mol. Biol.* **1993**, *44*, 569–580.

Shehu, K.; Aliero, A. A. Effects of Purple Blotch Infection on the Proximate and Mineral Contents of Onion Leaf. *Int. J. Pharm. Sci. Res.* (IJPSR), **2010**, *1* (2), 131–133.

Shehu, K.; Muhammad, S.; Ezekiel, A.; Salau, I. A. Incidence and Control of Root-knot Nematodes (*Melodogyne incognita*) on Egg Plant (*Solanum melongene* Mill.) in Zamafara State. *Int. J. Adv. Res. Biol. Sci.* **2014d**; *1* (5), 76–83.

Shehu, K.; Muhammad, S.; Salau I. A. Fungi Associated with Fruit Rot of Tomato (*Solanum esculentum* Mill) in Sokoto. *American Journal of Research Communication*, **2014c**; 2014 www.usa-journals.com, ISSN: 2325–4076.

Shelp, B. J.; Bown, A. W.; Faure, D. Extracellular γ-aminobutyrate Mediates Communication Between Plants and Other Organisms. *Plant Physiol.* **2006**, *142*, 1350–1352.

Shelp, B. J.; Bown, A. W.; McLean, M. D. Metabolism and Functions of Gamma-aminobutyric Acid. *Trends Plant Sci.* **1999**, *4*, 446–452.

Smedegaard-Petersen, V.; Stolen, O. Effect of Energy Requiring Defense Reactions on Yield and Grain Guality in Powdery Mildew *Erysiphe graminis* sp. *hordei* resistant *Hordeum vulgare* cultivar Sultan. *Phytopathology* **1981**, *71*, 396–399.

Solomon, P. S.; Oliver, R. P. The Nitrogen Content of the Tomato Leaf Apoplast Increases During Infection by *Cladosporium fulvum*. *Planta* **2001**, *213*, 241–249.

Spassieva, S.; Seo, J. G.; Jiang, J. C.; Bielawski, J.; Alvarez-Vasquez, F.; Jazwinski, S. M.; Hannun, Y. A.; Obeid, L. M. Necessary Role for the Lag1p Motif in (dihydro) Ceramide Synthase Activity. *J. Biol. Chem.* **2006**, *281*, 33931–33938.

Spassieva, S. D.; Markham, J. E.; Hille, J. The Plant Disease Resistance Gene Asc-1 prevents Disruption of Sphingolipid Metabolism during AAL-toxin-induced Programmed Cell Death. *Plant J.* **2002**, *32*, 561–572.

Staskawicz, B. J.; Panopoulos, N. J.; Hoogenraad, N. J. Phaseolotoxin-insensitive Ornithine Carbamoyltransferase of *Pseudomonas syringae* pv. *phaseolicola*: Basis for Immunity to Phaseolotoxin. *J. Bacteriol.* **1980**, *142*, 720–723.

Steinberg, T. H.; Matthews, D. E.; Durbin, R. D.; Burgess, R. R. Tagetitoxin: A New Inhibitor of Eukaryotic Transcription by RNA Polymerase III. *J. Biol. Chem.* **1990**, *265*, 499–505.

Stergiopoulos, I.; Collemare, J.; Mehrabi, R.; De Wit, P. J. G. M. Phytotoxic Secondary Metabolites and Peptides Produced by Plant Pathogenic Dothideomycete Fungi. *FEMS Microbiol Rev.* **2013**, *37*, 67–93.

Strumilo, S. Short-term Regulation of the α-ketoglutarate Dehydrogenase Complex by Energy-linked and Some Other Effectors. *Biochem.* (Moscow), **2005**, *70*, 726–729.

Sweetlove, L. J.; Heazlewood, J. L.; Herald, V.; Holtzapffel, R.; Day, D. A.; Leaver, C. J.; Millar, A. H. The Impact of Oxidative Stress on *Arabidopsis* Mitochondria. *Plant J.* **2002**, *32*, 891–904.

Takemoto, J. Y. Bacterial Phytotoxin Syringomycin and Its Interaction with Host Membranes. *In Molecular Signals in Plant-microbe Communications*, Verma, D. P. S., Ed.; CRC Press Inc.: Boca Raton, Fla, 1992; pp 247–260.

Tavernier, V.; Cadiou, S.; Pageau, K.; Lauge, R.; Reisdorf-Cren, M.; Langin, T.; Masclaux-Daubresse, C. The Plant Nitrogen Mobilization Promoted by *Colletotrichum lindemuthianum* in *Phaseolus* Leaves Depends on Fungus Pathogenicity. *J. Exp. Bot.* **2007**, *58*, 3351–3360.

Tegtmeier, K. J.; Daly, J. M.; Yoder, O. C. T-toxin Production by Near-isogenic Isolates of *Cochliobolus heterostrophus* Race-T and Race-O. *Phytopathology* **1982**, *72*, 1492–1495.

Templeton, M. D.; Sullivan, P. A.; Shepherd, M. G. Phaseolotoxin-insensitive L-ornithine Transcarbamoylase from *Pseudomonas syringae* pv. *phaseolicola*. *Physiol. Mol. Plant Pathol.* **1986**, *29*, 393–403.

Tenhaken, R.; Levine, A.; Brisson, L. F.; Dixon, R. A.; Lamb, C. Function of the Oxidative Burst in Hypersensitive Disease Resistance. *Proc. Natl. Acad. Sci. USA* **1995**, *92*, 4158–4163.

Thomma, B. P.; Penninckx, I. A.; Broekaert, W. F.; Cammue, B. P. The Complexity of Disease Signaling in *Arabidopsis*. *Curr. Opin. Immunol.* **2001**, *13*, 63–68.

Tretter, L.; Adam-Vizi, V. Inhibition of Krebs Cycle Enzymes by Hydrogen Peroxide: A Key Role of α-ketoglutarate Dehydrogenase in limiting NADH Production under Oxidative Stress. *J. Neurosci.* **2000**, *20*, 8972–8979.

Turner, J. G.; Debbage, J. M. Tabtoxin-induced Symptoms Aren Associated with Accumulation of Ammonia Formed During Photorespiration. *Physiol. Plant Pathol.* **1982**, *20*, 223–233.

Turner, J. G.; Taha, R. R. Contribution of Tabtoxin to the Pathogenicity of *Pseudomonas syringae* pv. *tabaci*. *Physiol. Plant Pathol.* **1984**, *25*, 55–69.

Turner, N. C.; Graniti, A. Fusicoccin: A Fungal Toxin that Open Stomata. *Nature* **1969**, *223*, 1070–1071.

Uchytil, T. F.; Durbin, R. D. Hydrolysis of Tabtoxins by Plant and Bacterial Enzymes. *Experientia* **1980**, *36*, 301–302.

Ullstrup, A. J. The Impacts of the Southern Corn Leaf Blight Epidemics of 1970–1971. *Annu. Rev. Phytopal.* **1972**, *10*, 37–50.

Ullstrup, A. J.; Brunson, A. M. Linkage Relationships of a Gene in Corn Determining Susceptibility to a Helminthosporium Leaf Spot. *J. Am. Soc. Agron.* **1947**, *39*, 606–609.

Vandelle, E.; Poinssot, B.; Wendehenne, D.; Bentéjac, M.; Pugin, A. Integrated Signaling Network Involving Calcium, Nitric Oxide, Active Oxygen Species but Not Mitogen-activated Protein Kinases in BcPG1-elicited Grapevine Defenses. *Mol. Plant-Microbe Interact.* **2006**, *19*, 429–440.

Walton, J. D. Host-selective Toxins: Agents of Compatibility. *Plant Cell* **1996**, *8*, 1723–1733.

Walton, J. D. HC-toxin. *Phytochemistry* **2006**, *67*, 1406–1413.

Walton, J. D.; Earle, E. D.; Gibson, B. W. Purification and Structure of the Host-specific Toxin from *Helminthosporium carbonum* race-1. *Biochem. Biophys. Res. Commun.* **1982**, *107*, 785–794.

Walton, J. D.; Earle, E. D.; Yoder, O. C.; Spanswick, R. M. Reduction of Adenosine Triphosphate Levels in Susceptible Maize Mesophyll Protoplasts by Helminthosporium Maydis Race T Toxin. *Plant Physiol.* **1979**, *63*, 806–810.

Wang, H.; Li, J.; Bostock, R. M.; Gilchrist, D. G. Apoptosis: a Functional Paradigm for Programmed Plant Cell Death Induced by a Host-selective Phytotoxin and Invoked During Development. *Plant Cell* **1996**, *8*, 375–391.

Widjaja, I.; Naumann, K.; Roth, U.; Wolf, N.; Mackey, D.; Dangl, J. L.; Scheel, D.; Lee, J. Combining Subproteome Enrichment and Rubisco Depletion enables Identification of Low Abundance Proteins Differentially Regulated During Plant Defense. *Proteomics* **2009**, *9*, 138–147.

Wiegand, G.; Remington, S. J. Citrate Synthase: Structure, Control, and Mechanism. *Annu. Rev. Biophys. Biophys. Chem.* **1986**, *15*, 97–117.

Xu, H.; Heath, M. C. Role of Calcium in Signal Transduction During the Hypersensitive Response Caused by Basidiospore-derived Infection of the Cowpea Rust Fungus. *Plant Cell* **1998**, *10*, 585–597.

Yamagishi, D.; Akamatsu, H.; Otani, H.; Kodama, M. Pathological Evaluation of Host-specific AAL-toxins and Fumonisin Mycotoxins Produced by *Alternaria* and *Fusarium* Species. *J. Gen. Plant Pathol.* **2006**, *72*, 323–326.

Yamamoto, M.; Namiki, F.; Nishimura, S.; Kohmoto, K. Studies on Host-specific Toxins Produced by Alternaria alternata Strawberry Pathotype Causing Alternaria Black Spot of Strawberry (3) Use of Toxin for Determining Inheritance of Disease Reaction in Strawberry Cultivar Morioka-16. *Ann. Phytopathol. Soc. Jpn.* **1985**, *51*, 530–535.

Yar'adua, A. A. Some Studies on Epidemiology of Purple Blotch of Onion (*Allium cepa*) in Zaria, Nigeria. *Danmasani* **2003**, *1* (6–8), 98–106.

Yip, J. Y. H.; Vanlerberghe, G. C. Mitochondrial Alternative Oxidase Acts to Dampen the Generation of Active Oxygen Species During a Period of Rapid Respiration Induced to Support a High Rate of Nutrient Uptake. *Physiol. Plant.* **2001**, *112*, 327–333.

Yoder, O. C. Toxins in Pathogenesis. *Ann. Rev. Phytopathol.* **1980**, *18*, 103–29.

Zaninotto, F.; Camera, S. L.; Polverari, A.; Delledonne, M. Cross Talk Between Reactive Nitrogen and Oxygen Species During the Hypersensitive Disease Resistance Response. *Plant Physiol.* **2006**, *141*, 379–383.

Zhang, Y.; Rowley, K. B.; S. Patil, S. Genetic Organization of a Cluster of Genes Involved in the Production of Phaseolotoxin, a Toxin Produced by *Pseudomonas syringae* pv. phaseolicola. *J. Bacteriol.* **1993**, *175*, 6451–6464.

Zimmermann, S.; Nurnberger, T.; Frachisse, J -M.; Wirtz, W.; Guern, J.; Hedrich, R.; Scheel, D. Receptor-mediated Activation of a plant Ca^{2+}-permeable Ion Channel Involved in Pathogen Defense. *Phytophthora sojae. Can. J. Bot.* **1997**, *75*, 1509–1517.

Zulak, K. G.; Khan, M. F.; Alcantara, J.; Schriemer, D. C.; Facchini, P. J. Plant Defense Responses in Opium Poppy Cell Cultures revealed by Liquid Chromatography-tandem Mass Spectrometry Proteomics. *Mol. Cell. Proteom.* **2009**, *8*, 86–98.

CHAPTER 8

Foilar Fungal Pathogens of Cucurbits

ASHOK KUMAR MEENA*, SHANKAR LAL GODARA,
PRABHU NARAYAN MEENA, and ANAND KUMAR MEENA

Department of Plant Pathology, College of Agriculture, Swami Keshwanand Rajasthan Agricultural University, Bikaner, India

Corresponding author. E-mail: ak_patho@rediffmail.com

ABSTRACT

The major and minor cucurbits that are grown in India occupy about 5.6% of the total vegetable production. Cucumber (*Cucumis sativus* L), pumpkin (*Cucurbita maxima* Duch), melon (*C. melo* L.), watermelon (*Citrulluslanatus* (Thunb), squash (*C. pepo* L.) bottle gourd (*Lagenaria sciceraria* Standl. var. *hispida* Hara) which belong to the family of Cucurbitaceae are widely cultivated in India. They are used in different forms, that is, sweet (ash gourd, pointed gourd), salad (cucumber, gherkins, long melon), deserts (melons), and pickles (gherkins). These crops are often attacked by many foliar fungal pathogens which causes substantial yield loss. Cucurbits are susceptible to diseases such as downy mildew (*Pseudoperonospora cubensis*), powdery mildew (*Sphaerotheca fuliginea*, *Erysiphe cichoracearum*), fusarium wilt (*Fusarium* spp.), Cercospora leaf spot (*Cercospora citrullina*), bacterial wilt (*Erwinia tracheiphila*), Phytophthora blight (*Phytophthora* spp.) and anthracnose (*Colletotrichum* spp.). Researchers have utilized a number of cultivated and wild species to develop improved varieties and resistance parental lines. Approximately, 112 open pollinated varieties of cucurbits have been recommended for cultivation at national and state levels. Among them, 26 hybrids and seven disease resistant varieties of major cucurbits have also been developed against various pathogens. Similarly, 48 improved varieties in eight major cucurbits has been identified and recommended for cultivation. The main aim of research on cucurbitaceous vegetables in India is to improve production and productivity on sustainable basis through developing of biotic and abiotic resistant hybrids coupled with good quality attributes. The standardized agro-techniques and plant protection measures could be increased in the yield potential of cucurbits.

8.1   INTRODUCTION

Cucurbita belong to Latin word is a *genus* of herbaceous *vine* in the *family* of cucurbitaceae which is native to the Andes and Mesoamerica. Worldwide, five important species are grown for their edible fruit, variously known as squash, *pumpkin*, gourd depending on species, *variety*, and for their seeds. Among them cucurbits are most economically valuable vegetables cultivated throughout the world and stand next to the solanaceous group in terms of popularity and their presence in the international vegetable export market. *Cucurbita* plant bear male and female flower, which are either yellow or orange. The male flowers produce pollen and the female flowers produce the fruit. Snake–melon and muskmelon like other cucurbit plants, provide significant amount of minerals and vitamins namely; vitamins A, C, riboflavin, niacin, Fe, Na, K, P, and Ca. They are also good source of energy, protein, fat, and carbohydrates, as stated below (Watt and Merrill, 1963)

Crop	Energy (cal)	Protein (g)	Fat (g)	Carbohydrates (g)
Muskmelon	30	0.7	0.1	7.5
Snake–melon	15	0.9	0.1	3.4

The *Cucurbita* plants contain various toxins such as, cucurbitin, cucurbitacin, and cucurmosin because of existence of variety of plant type, morphology and fruits, then taxonomical classification and grouping is completed.

8.2   ECOLOGICAL IMPEDIMENTS

Cucumber is generally grown in summer and the plants thrive best at relatively high temperatures. The snake-melon could be grown during the slightly hot winter days, because it is short maturing and becomes very rewarding, and a highly profitable crop (Mohamed, 1997). It is presumed that all cucurbit vegetables are native to warm, humid and sometimes could be grown in arid regions in different parts of India as well as in world. Candolle (1882; c.f. Whitaker and Davis, 1962) reported that the cucumber has been cultivated in India, from there it has spread to westward. Cucurbitacin is a toxin compound present in plant that do not involve in defensive responses against viral, fungal, and bacterial leaf pathogens. Species in the genus *Cucurbita* are susceptible to some types of mosaic virus including: Papaya ring spot virus (PRSV) cucurbit strain, Cucumber mosaic virus (CMV), Tobacco ring spot virus (TRSV), Watermelon mosaic virus (WMV), Zucchini yellow mosaic virus (ZYMV), and Squash mosaic virus (SqMV). Among the viruses, PRSV is the only one that does not affect all cucurbits. SqMV and CMV are the most common viruses that infect all cucurbits. Symptoms produced by these viruses show a high degree of similarity, which is very difficult to differentiate without a laboratory investigation in affecting plants.

TABLE 8.1 Foliar fungal diseases of cucurbits.

S. no.	Name of disease	Pathogen
1	Downy mildew	*Pseudoperonospora cubensis*
2	Powdery mildew	*Sphaerotheca fuliginea, Erysiphe cichoracearum*
3	Anthracnose	*Colletotrichum orbicular, Colletotrichum lagenarium, Glomerel lalagenarium* [teleomorph]
4	Cercospora leaf spot	*Cercospora citrullina*
5	Alternaria leaf blight	*Alternaria cucumerina*

8.3 DOWNY MILDEW

8.3.1 DISTRIBUTION

Cucurbits downy mildew is a devastating foliar disease caused by the oomycete, *Pseudoperonospora cubensis* (Berk. et Curt.) Rostov is the most destructive pathogens of cucurbits (Lebeda and Cohen, 2011). In case of cucumber, downy mildew causes serious losses under favorable environmental conditions. It was first time reported in 1868 from Cuba, and even after 130 years it is considered as one of the serious problems. Bains and Jhooty, 1976b; Bains and V. Prakash, 1985 has reported that this is an important disease that affects many cucurbits such as ridge gourd, sponge gourd, musk melon and cucumber which are more severely affected than bitter gourd, bottle gourd and snake gourd. Pumpkin and vegetable marrow are less effective. According to Butler (1918), it is an interesting parasitic fungus that is described on what seem to have been a wild plant, in an isolated part of the world and considered of not economic importance, which by its gradual spread to other countries and the damage that it has caused on new hosts has won for itself a position amongst the major enemies of cultivated crop plants.

8.3.2   DISEASE SYMPTOMS

Affected leaves first show mosaic like mottling. The infected pale green areas are separated by dark green islands. The angular yellow colored spots developed on the upper surface restricted along with veins of leaves while, on lower side of leaf spot purplish downy growth of fungus appears. The affected leaves die quickly and middle leaves are infected first usually followed by other leaves. On infected plant, few small fruits with poor taste were developed. Sometime these symptoms and fungal growth can be confused with the bacterial angular leaf spot disease. Foliar symptoms vary depending on how quickly fungal infection occurs. The infected leaves show angular pale-green areas bounded by leaf veins on upper surfaces of leaves which may give the impression of viral mosaic. When the disease progress, the infected areas change to yellow angular spots and become necrotic. When leaves are infected by heavy spore showers, small individual necrotic and chlorotic flecks appear on the upper surface whereas in moist environment conditions it becomes sporulated on the lower leaf surface. The infected areas having active sporulation gives the undersurface a dark brown sooty appearance. Generally, the leaves near the center of row are infected first. The infected area spreads outward, causing stunted growth, defoliation, and poor fruit formation. The entire plant may eventually be killed (Bern-herdt et al. 1988). In both watermelon and melon foliar lesions are less defined than those on cucumber plant, and are not always bound by leaf veins (Thomas, 1996). As infection progresses, the chlorotic lesions expand and may become necrotic (Oerke et al., 2006), with necrosis occurring more quickly in hot, dry weather (Cohen and Rotem, 1971c). Leaves colonized by *Ps. cubensis* undergo various changes in temperature and transpiration rates, which mainly vary on the course of infection (Lindenthal et al., 2005; Oerke et al., 2006). Low temperatures can delay symptom development whilst still promoting colonization of the leaf tissue, whereas higher temperatures result in faster.

8.3.3   CAUSAL ORGANISM

The causal fungus *Pseudoperenospora cubensis* (Berk. and Curt.) is an obligate parasite that belongs to class Oomycetes, order Perenosporales and family Perenosporaceae. Other cucurbit crops are affected by downy mildew and previous infected crop trash. The fungus is easily carried by rain splash, wind currents, hands, farm implements, and clothes of farm workers. Hansen 2000 stated that it is favored by cool to moderately warm temperatures, but tolerates to hot days. The fungal mycelium consists of large size lemon-shaped spores often called sporangia that are borne on branched structures called sporangiophores. The sporangia are wind-borne and when air is moist it can survive for longer distance. They are also readily spread to adjoining plants by splashing rain. The sporangia are grayish to olivaceous purple, ovoid to elliptical, thin walled and with papilla at the distal end. They measure 21–39 × 14–23 microns. The germination of sporangia occurs by production of biflagellate zoospores which are 10–13 microns in diameter when in resting state. Oospores are not common in nature. In India these structures have been reported on certain cucurbits in M.P., Punjab, and Rajasthan (Mahrishi and Siradhana, 1980). The races of the pathogen differ in their ability to infect and sporulate on different hosts as well as in diameter of sporangiosphore and sporangia, number of zoospores per sporangium and in their response to metalaxyl (Bains and Sharma, 1986a). Thomas, et al. (1987) had identified five pathotypes that could be distinguished on the basis of occurrence of highly compatible reaction with specific cucurbits hosts. Thomas and Jourdain (1992) stated that the fungus originally isolated from a particular host but grown for several generations on a different cucurbit species, losses its affinity for the original host and sporulates better on the host on which it is being currently grown.

8.3.4 PATHOGEN BIOLOGY

The pathogen is survived only on living host as it is an obligate parasite. Downy mildew fungi are spread in cool, wet and relatively in warm weather. Presence of water on leaves is indispensable for infection. Lange et al. (1989) had observed that sporangia, which survived below freezing temperature (−18°C) for 3–4 month for able to serve as the over wintering structure in northern latitudes in Denmark in the absence of oospores. Cucurbit mildew pathogens require optimal temperature ranges between 18°C and 23°C. Under such conditions life cycle lasts 3–4 days only, and unprotected plants can be almost completely destroyed within several days. This fungus tolerates well even high temperatures above 30°C without loss of viability. The infected leaves usually serve as a source of inoculum for adjoining leaves and plants. Spores are spread through wind and water droplets of rain or overhead irrigation in the crop.

8.3.5   DISEASE CYCLE

The pathogen can survive from season to season and cause infection because of the wide diversity of climate and season in different parts of India (Butler 1918). The presence of suitable wild hosts would probably reservoir and ensures the continuous growth of the parasite. Bains and Jhooty (1976 a) studied this aspect in Punjab and observed that during winter season, the pathogen perpetuates in the form of active mycelium on cultivated sponge gourd. Where oospores are found, these can be an important source of survival and primary inoculum for disease initiation. Initially infected leaves serve as a source of inoculum for neighboring leaves and plants. In a crop, spores are spread by wind and water droplets during rain or overhead irrigation. Wet and relatively warm weather is favorable for the disease spread. The presence of water on leaves is essential for infection. Optimal temperature ranges between 18°C and 23°C. Under such conditions life cycle lasts only 3–4 days and unprotected plants can be almost completely destroyed within several days. This fungus tolerates well even high temperatures above 30°C without loss of viability. The varied severity of infection in different crops mainly depends on the presence of physiologic races in the pathogen. In India, five pathotypes infecting water melon, musk melon, bottle gourd, *Cucurbitapepo,* and *Luffaaegyptica* have been distinguished based on the high compatibility with specific hosts (Bains and Prakash, 1985).

8.4   DISEASE MANAGEMENT

8.4.1   CULTURAL PRACTICES

Downy mildew losses can be reduced by early sowing of crop. Used proper and balance nutrition (high N and P and low K). Destruction of weed hosts in the fields. Reducing the amount of moisture in the vines. Because the disease development depends on the outside inoculum, rotation cannot prevent the disease. Avoid overcrowding of plantings.

8.4.2   RESISTANT VARIETIES

According to Lebeda et al. (2002) and Cohen et al. (2003), Physiological specialization in *Pseudoperenospora cubensis* is a common phenomenon and must be considered seriously in the resistance breeding of cucurbitaceous vegetables. However, no specific resistant varieties were recommended in India. In Punjab, two lines of muskmelon namely, LC-8 and PPDMR-4 were identified as resistant to *cubensis*.

8.4.3 CHEMICAL CONTROL

Fungicidal seed treatment followed by foliar spray is a common practice to control downy mildews in pearl millet, sorghum, maize, mustard, and cucurbits in India (Thakur and Mathur, 2002). In order to control the disease in established infection, systemic fungicides, Ridomil MZ (0.25%) should be sprayed (Thind et al., 1991). Use a fungicide program that allows systemic and protective fungicides to rotate, which minimize the chance of fungicide resistance developing against the pathogens. Fungicide sprays effective for control on the susceptible varieties. Azoxystrobin (Amistar 25 SC) is a potent strobilurin fungicide with a novel biochemical mode of action (Hewitt, 1998). Its fungicidal activity results from the inhibition of mitochondrial respiration in fungi, which is achieved by the prevention of electron transfer between cytochrome b and cytochrome c (Becker et al., 1981). It is widely used against grapevine, cucumber and tomato foliar diseases. Foliar spraying of Dithane Z-78, Dithane M-45 and Tricop-50 are recommended for the disease management. Dithane Z-78 and Dithane M-45 protect the leaves for 9 days after spraying, while Tricop-50 gives protection for only 5 days (Bains and Jhooty, 1978). As tank mixture, Cymoxanil plus Mancozeb and Chlorothalonil also provide effective control (Robak, 1995). Khetmalas and Memane (2003) used three sprays of fosetyl-Al (@ 0.2%) at 15 days interval, starting with the initiation of the disease, and this was found to be effective in reducing the disease with increased yield.

8.5 POWDERY MILDEW

8.5.1 DISTRIBUTION

Powdery mildew is the most common fungal disease in cucurbits. Powdery mildew caused by *Erysiphe cichoracearum* (De Candolle) particularly, in cucumber (*Cucumis sativus* L.) is highly susceptible to this disease and suffers heavy losses in all localities of the Maharashtra state. The disease has been reported to be very widespread occurrence throughout the country and some work has been done on chemical control of this disease (Bandopadhyaya et al., 1980, Bhatia and Thakur 1989 ab, Sinha 1990, Iqbal et al.; 1994 and Gupta and Amita Gupta 2001. One of the species *Erysiphe cichoracearum,* is reported to attack potato and tobacco seedling, lettuce, sunflower, mango, castor, *Antirrhinum orontium, Hyssopus officinalis, Sedan* spurium, and Soncusasper (Stone, 1965). Powdery mildew has a negative impact on crop production, including reduced photosynthesis, impaired growth, premature senescence and yield loss. The yields are reduced due to the decline in size and number of fruits. Early death of leaves lowers market quality, because fruit gets sunburned doesn't store well and lose flavor. Powdery mildew adversely affected the fruit quality. Powdery mildew is caused by three compulsory biotrophicectoparasites, that is, *Golovinomyces cichoracearum* (Syn. *Erysiphe cichoracearum*), *Podosphaera xanthii* (Syn. *Sphaerotheca fuliginea*) and *Leveillula taurica* (Khan and Sharma, 1995 and Lebeda et al., 2010).

8.5.2   DISEASE SYMPTOMS

The disease starts on the older leaves of plants usually. It is the most widespread and serious foliar fungal disease of cucurbits (Zitter et al., 1996). Cucurbit leaves with severe infection lose their green color and becomes brown and withered. Due to the premature loss of leaves, cucurbit plants easily become stunted and die. The yield of the glasshouse cucumbers infected by powdery mildew, showed that the disease reduces yield and market quality greatly Dik (1999). Rhodes (1964) has studied that the powdery mildew on *Cucurbita moschata* appears on both the upper and lower surfaces of the leaf blades and later spreads to petiole and stem. First symptoms of powdery mildew are small, circular, white, and powdery fungal colonies consisting with the mycelia and conidia presence on upper leaf surfaces. White colonies of the fungus can be seen on the lower surface of infected leaves and stems occasionally, and white colonies merge together and cover most of the leaf surface. Until fruit initiation, plants in the field do not affected by powdery mildew. The disease affects all commercially grown cucurbit species, by causing premature leaf senescence, which exposes the fruit to sunscald. Normally, fruit produced from severely infected plants are of low quality in taste, texture and size and are considered to be low quality products on the market (Sitterly 1978). Severely infected leaves turn in yellow color, then wither, die and become dry and brittle. The pathogen does not infect cucurbit fruit directly. However, they may be malformed or gets sunburned due to early loss of foliage. Severe epidemics of disease can reduce both the size and number of fruits in infected plants.

8.5.3   CAUSAL ORGANISM

Powdery mildew caused by *Erysiphe* pathogen is a serious disease of cucurbits in India (Khan et al., 1971, 1972, 1974). *Erysiphe cichoracearum, Sphaerotheca fuliginea* are the pathogens, that causes this disease and the occurrence of both the pathogens on cucurbits in India is reported by Butler (1918). On the basis of conidial characteristics, Jhooty (1967) claimed that *Sphaeroth ecafuliginea* was responsible for causing of disease in Punjab state. However, Siradhana and Chaudhari (1972) reported that both the species occurs on cucurbits at Udaipur (Rajasthan) and *Erysiphe cichoracearum* is the main causal organism of the disease. Taxonomic evaluation of anamorphic characters for identification of fungi on cucurbits in India has been done by Khan and Sharma (1995). Pawar (2005) observed the conidiophores of *Sphaerotheca fuliginea* on cucurbits are long, straight simple 103–240 µm in length. It is cylindrical, hyline conidia are produced in chains on the conidiophores that arises from the surface mycelia growth. Bharat (2003) observed cliosthecia of *Sphaerotheca fuliginea* were globular, brown and 61.7–78.6 × 53.1–77.8 µm in size. Appendages were attached, variable in number, usually longer than the diameter of ascocarp. Mycelioid were brown in color. The ascocarp had a single ascus, which was elliptical and globose and measured 53–81 × 43–66 µm. Ascus had eight arcospores, which were ellipsoid to nearly spherical and measured 17–24 × 13–19 µm.

8.5.4   PATHOGEN BIOLOGY

The conidia on germination forked germ tube distinguished them from those of *Erysiph ecichoracearum,* which produce appressoria in germination (Zaracovits, 1965). The patches formed by this fungus are not white, but reddish brown. The clistothecia are different in structure from those of *Sphaerotheca.* Powdery mildew of cucurbits caused by *Sphaerotheca fuliginea* is a serious disease in northern India (Jhooty, 1967). Khan and Sharma (1993) has confirmed and reported that the three races of *Sphaerotheca fuliginea* Out of the three, race three is the most wide spread and infects most of the cucurbits in India. Kabitarani (1993) has reported the existence of heterothallism in the species and prevalence of two pathological races in Assam state. It is occurs in mid-summer on all cucurbits usually. The fungus requires high moisture and warm temperature and it often observed first in the low areas of a field with limited air flow. It does not need free water on the leaf surface to infect the host plants and favored by dense plantings, because the fungus likes shaded areas, such as the middle of the plant canopy. All leaves can be infected by the pathogen, but older leaves are more susceptible due to this infection pre-disposes plants to other diseases.

8.5.5   DISEASE CYCLE

The pathogens are obligate parasites and can persist either on the wild cucurbits or crop debris. The primary inoculum of the disease is presumed to be airborne conidia dispersed from southern states, from local areas that have greenhouse-grown cucurbits. The disease cycle is similar to the powdery mildew of cereals and pea. Perithecia and conidia are produced profusely in the white mold and dispersed by wind or air movement to adjacent leaves and plants, as well as to those farther away. Powdery mildew develops under favorable conditions quickly such as dense plant growth, low light intensity, warm temperatures, and high relative humidity. However, free water on leaf surfaces is not required for spore germination and disease development can occur in the presence or absence of dew. Unlike other fungal diseases, dry weather conditions will not necessarily stop the spread of this disease.

8.5.6   DISEASE MANAGEMENT

An effective management strategy for powdery mildew management should include proper cultural practices, resistant varieties, biorational products, applications of traditional fungicides, or biological agents.

8.5.7   CULTURAL PRACTICES

Plant cucurbits in a sunny location with good air circulation. Avoid planting new crops next to those that are already infected by powdery mildew. Remove old and heavily diseased leaves to improve air circulation and reduce inoculum. The use of crop rotation, removal of debris and alternative hosts breaks the disease cycle. Increasing ventilation reduces humidity, especially in glasshouses, and has given good control over powdery mildew on glasshouse grown cucurbits (Jhooty and McKeen, 1965; Reuveni and Rotem, 1974; Butt, 1978).

8.5.8   RESISTANT VARIETIES

The powdery mildew is effectively controlled by using genetically resistance varieties. However, some resistant cultivars may be susceptible to a specific fungal race, because there are several fungal races of powdery mildew are present. Resistant cucumber varieties are "Alibi," "Supremo," "Eureka," and "Marketmore." Resistant varieties of squash are "Sunray," "Soleil," "Sebring," "Payroll," "Royal Ace," and "Tay Bell PM." An extensive evaluation of zucchini, squash and cucumber varieties for resistance to powdery mildew was recently undertaken by Akem et al. (2011). However, the introduction of varieties with resistance to specific races increases the risk of development of new races. In addition, resistant varieties also require preventative spray programs to reduce the risk of the resistance failure.

8.5.9   FUNGICIDE APPLICATIONS

A scouting strategy is helpful to detect the beginning of the disease, so that a fungicide program can be initiated. Since most of the fungicides must be applied at very early stages of an epidemic to control powdery mildew. The fungicides that are registered for use on cucurbits to control powdery mildew, the systemic fungicides are azoxystrobin, trifloxystrobin, and myclobutanil, and the contact fungicide chlorothalonil. To avoid developing resistance to fungicides in pathogens and maintaining effectiveness, rotate fungicides in the application program. Dusting with Sulphur @ 20–25 kg/hectare and spray the crop with 0.05% Karathane (1000 L/ha). Systemic fungicides are such as triazoles, benzimidazoles, pyridine and pyrimidine carbinols, triforine and amino-pyrimidines (Bent et al., 1978; Sherald and Sister, 1979; Roy, 1973; Jhooty and Behar, 1972). Contact fungicides, such as quinomethionate, drazoxolon, ditalimofos and chloquinox were reported to control PMs (Bent et al., 1978). Waraitch et al. (1975) observed that Benomyl as a foliar spray is also quite effective in controlling powdery mildew of bottle guard. Bandopad-hyay et al., (1980) suggested that the powdery mildew of different cucurbits is significantly controlled by the treatment of karathane. Puzanova (1984) reported that extracted Ampelomycin from Ampelomyconartedmiside pre-

and post-inoculation spray of spore suspension of *Fusariumpallido-roseum* and *Fusarium moniliforme,* resulted in a 100% control of powdery mildew on detached leaves of cucumber. Gupta and Amita Gupta (2001) reported that the cucumber powdery mildew disease can be controlled by three sprays of Hexa-conazole (0.05%) at fortnightly interval were found highly effective followed by Carbendazim, Bitertanol, Difenconazole, Thiophanate methyl, and Triadimefon. Patel et al., (1990) reported the appearance of powdery mildews on bottle gourd during the rabbi season on *Lagenaria siceraria, Cucurbita moschata, Cucurbita maxima, Luffa cylindrica, Cucumis memo, Cucumis sativus,* and *Cocconia grandis* due to *Sphaerotheca fuliginea.* In the month of April to June, Branzanti and Brunelli (1992) reported powdery mildew on *Citrullus vulgaris* and *Cucumis sativus* due to *Sphaeroth ecafuliginea.*

8.6   ALTERNARIA LEAF BLIGHT

8.6.1   DISTRIBUTION

Alternaria blight, which favored warm and rainy weather, is caused by the fungus *Alternariacucumerina*s and is a common disease of most cucurbits. A similar cucurbit disease called Alternaria spot, caused by a related fungus (*Alternaria alternata*), is not known to occur in the high plains. It is a common disease of muskmelon, but it can also infect squash, watermelon, and cucumber. It occurs world-wide and infects most cucurbits. Leaf blight usually occurs from mid-season, can reduce late season fruit production, and can result in poor quality, if the disease is severe. Direct infection is primarily of the leaves and can lead to defoliation. In turn, this leads to a reduction in yield and may cause fruit to ripen prematurely. Fruit may also be damaged by exposure to the sun and wind. The reported losses have ranged up to 100%. Direct infection of the fruit also occurs (especially melons), but much less frequently. According to Bhargava and Singh (1985), yield losses due to leaf necrosis and foliage losses are variable according to the type of cucurbit crops and its susceptibility, reaching 80% on pumpkins and 88% on watermelons in India.

8.6.2   DISEASE SYMPTOMS

Lesions tend to appear first as small circular spots on the older leaves. Mature leaves near the crown of the plant are often infected first. The spots are light brown with a light center often with a yellow halo, and grow into irregular brown spots (up to 3/4") and form concentrated dark rings as they enlarge, thus naming the target spot. Lesions that form on the lower surface of the leaf tend to be more diffuse. Severely, infected leaves turn brown, curl upward, wither, and die. Another name for this disease is the target leaf spot, which refers to the appearance of the lesions on the upper surface of crown leaves as small spots 1 to 2 mm in diameter, although other lesions may be more than 10 mm in diameter (nearly one-half inch). Fruit infections start as sunken brown spots and may later develop a dark powdery appearance as the fungus sporulates. The infection may also begin at the blossom, end with the entire fruit eventually turning brown and shriveling.

8.6.3 CAUSAL ORGANISMS

Alternaria leaf blight of cucurbits is caused by *Alternaria cucumerina* and *Alternaria alternata*.The fungi survive in soil, on plant debris and seed may also be the source of new infections. Leaf blight over winters such as mycelium or chlamydospores in diseased plant debris and can probably survive for more than a year. Conidia are produced in the spring and act as the primary inoculum. Conidia produced on the infected plants provide inoculum for repeating secondary cycles during the season. A 2 year rotation out of cucurbits is the easiest way to break this cycle. There are few Alternaria-resistant varieties. Protective fungicides should be applied in mid–July, when vines run and fruit are set. Infection and disease development are favored by lengthy periods of high relative humidity (18 h) over a broad range of temperatures (68–90°F). *A. cucumerina* is favored by periods of warm moist weather. Especially, important are periods of leaf wetness, which allow spores enough time to germinate and penetrate the leaf cuticle. Temperatures of 70–90°F are optimal, along with 8 h continuous leaf wetness. Young plants and old plants are more susceptible than mid-season plants. Plants which have been weakened by poor nutrition, adverse growing conditions, other diseases, or heavy fruit set are also more susceptible.

8.6.4 DISEASE CYCLE

The Alternaria blight disease cycle begins when dormant mycelium in infested crop debris produces spores (conidia), which are disseminated by wind and splashing water to new plantings of susceptible cucurbit crops. Spores germinate in free moisture and penetrate their hosts through natural openings or wounds. Disease is the most severe during extended periods of leaf wetness (8–24 h) at moderate to warm (54–86°F) temperatures. Frequent rainfall, especially during the warm weather, is highly favorable to the pathogen and disease development. The pathogen survives between susceptible crops in infested crop debris up to 2 years.

8.7   DISEASE MANAGEMENT

8.7.1   CULTURAL PRACTICES

Practice a 3-year or longer crop rotation to non-hosts such as a millets, vegetables corn, etc. Deeply incorporate crop debris after harvest to reduce pathogen survival and primary inoculum that could incite future epidemics. Use disease-free seed. If possible, use drip irrigation instead of overhead sprinklers. Do not work in plants when wet. Rotate vegetables for three or more years before planting any member of the squash family in the same location. Fall plowing will bury the remaining waste and promote its breakdown. Provide adequate nutrition and appropriate growing crop conditions to reduce crop stress. Certain resistant varieties are available. Remove and destroy infected plants in small gardens at the end of the season. Remove crop refuse at the end of the season. Promote air movement within the canopy by planting rows parallel to the top revailing wind direc-tion, reducing plant population and planting on wider row spacing. These practices will reduce the duration of leaf and soil wetness, and can help reduce blight in semi-arid production areas. If possible, avoid irrigating near dusk and overhead irrigation to reduce periods of leaf wetness. Some melon types are highly resistant to Alternaria blight, but in most commercially acceptable cucurbits, little resistance is available. Leaf miner feeding can increase and should be controlled the incidence and severity of Alternaria blight. Provide adequate nutrition and appropriate growing crop conditions to reduce crop stress.

8.7.2   CHEMICAL CONTROL

Chemical controls are the most effective when integrated with sound cultural control practices. The fungicide 'Procure' is registered for powdery mildew control in cucurbits, but it may provide suppression or control of Alternaria blight as well. Fungicide treated disease-free seed may be helpful. Several fungicides against Alternaria leaf blight are registered for use. Laboratory studies indicated that the fungicide score was effectively inhibited the disease lesion due to *Alternaria cucumerina* (Batta, 2003). The use of fungicides will reduce the disease incidence when the applications are timed appropriately. Refer to the table below for some labeled fungicides.

8.8   ANTHRACNOSE

8.8.1   DISTRIBUTION

In 1867, anthracnose of cucurbits was recorded on gourds in Italy. Anthracnose is a fairly common problem on cucurbits in all temperate, subtropical and tropical regions and caused by *Colletotrichum lagenarium* (Sikora, 1997) was the most troublesome disease especially in "Sugarbaby" and the *Kakamega landrace*. 'Charleston Gray' which is reportedly resistant to most races of anthracnose (Wehner and Barrett, 2005), escaped infection in the first season but succumbed in the second season. In general, the severity of anthracnose in the second season was high compared to the first season. This disease can appear at anytime during the season, but most of the damage occurs late in the season after the fruit is set. It is caused by the fungus, *Colletotrichum orbiculare*. This pathogen can attack on all cucurbits, but it is most severe on cucumbers, water melons and muskmelons. The fungus is seed-borne and survives on infected plant residues during non-crop periods. Spores of the fungus are spread by wind, rain, machinery, and handling. Warm, wet weather is favorable for anthracnose infection and spread, with optimum temperatures of 26–32°C.

8.8.2   DISEASE SYMPTOMS

All parts of the above ground plant are severely affected. Leaf spots first appear as yellowish, water-soaked areas, turn brown and become brittle. Disease spots turn black on watermelon fruit. Diseases spots on petioles and stems are elongated and dark with a light center and can result in vine defoliation. Fruit becomes more susceptible to infection at the time of ripening. These are the most importunate noticeable symptoms on fruit. Cankers are circular, black, sunken lesions and often vary in size. Lesions expand rapidly in the field and during the transit or storage and may coalesce to form larger ones. The black center of the canker is filled with a gelatinous mass of pink spores in presence of moisture. Cankers do not penetrate the flesh of the fruit, but make the fruit to be bitter. Affected fruits are likely to be destroyed by soft rot organisms, which enter through the broken rind. Symptoms include fruit rot, leaf spots and seedling damage. The leaf lesions presence on watermelon is tend to be angular, but they are rounded on cucumber and muskmelon. Irregular brown leaf spots form on melon, squash and cucumber. The center of the leaf spot may drop out resulting in a ragged appearance. Cucumber leaf spots often have a yellow halo, while watermelon leaf spots are smaller and dark brown to black. Sunken elongate stem infections may occur on cucumber and melon, but do not appear on other cucurbits. Infections on melon often exude a reddish gum, while sunken black spots 1/4–1/2 inch across and 1/4 inch deep in fruit infections. Spots may have fluffy white cotton, such as mycelia and sticky salmon colored spores during wet weather. When humidity is high, the symptoms often become severe. Symptoms mainly develop about 4 days after infection. Two races of the anthracnose fungus (*Colletotrichum orbiculare*) are common on cucurbit crops. Race 1 causes lesions on cucumber, whereas race 2 is responsible for lesions on the watermelon. Muskmelon is more susceptible to race 1 than to race 2. Many watermelon cultivars are marketed as having resistance to anthracnose race 1, however, there are no commercial watermelon cultivars currently that have resistance to race 2 of the anthracnose pathogen.

8.8.3   CAUSAL ORGANISM

Colletotrichum orbiculare is widely distributed in cucurbits grown regions. The ascomycete state is rarely found in nature. Conidia are hyaline, non-septate and mostly oblong and measure 4–6 × 13–19 micrometers. A synonym for *Colletotrichum lagenarium* is *C. orbiculare* (Berk. and Mont.) Arx. (Some articles refer to it as the correct name and *C. lagenarium* as the synonym). *Gloeosporium lagenarium* is similar, if not identical, to *C. lagenarium*. The sexual or perfect stage (teleomorph) is *Glomerella lagenarium*. At any stage of growth, *C. lagenarium* can infect a plant successfully. When moisture is present, the spore germinates within 3 days and a penetration tube enters the plant. Invaded tissues die and form a canker, followed by the production of spores that repeat the infection process continuously. Acervulus produces conidia. Masses of conidia attain a pink or salmon color. Conidia are released from the acervuli and come into contact with susceptible plant or fruit hosts and germinate, when water is present and temperatures are optimal (20–32°C). The conidia germinate and penetrate the host tissues directly.

8.8.4 PATHOGEN BIOLOGY

The Anthracnose fungus can survive in infected plant debris, either inside or outside the seed. Spores are mostly produced on infected leaves and fruits and spread by splashing rain, irrigation, on workers' hands or equipment. The disease incidence is favored by warm, moist environmental conditions and severe in wet summer. Pathogen produces abundant spores in lesions under humid conditions and is favored by warm temperatures of 75°F and fungal spores disperse with splashing water. Sikora (1997) also reported that warm (24°C) and wet (frequent rains, poor drainage) conditions favor rapid development and spread of this disease. Symptoms were found to appear at the beginning of rainfall, which coincided with the start of reproductive phase in the second season and lasted up to maturity. The conidia are released and spread only when the acervuli are wet and generally spread by splashing water and blowing rain or by coming into contact with insects, other animals and humans and other tools. The funguses survive as overwinter mycelium on seeds and residues from diseased plants in and on the soil. The fungus also survives on important weeds of the cucurbit family.

8.8.5 DISEASE CYCLE

The fungus can live up to 2 years in the absence of a suitable host. The fungus is spread by seed and possesses a primary inoculums source. The spores depend on water for spread and infection. Warm and humid rainy weather at frequent intervals favors disease development. Spread by washing water on harvested fruit is important when cucumbers or melons are cleaned before packing. Spores can also be spread by cultivating equipment when the foliage is wet. The spotted cucumber beetle can spread the spores from plant to plant within a field and its adjoining fields. Seven (1–7) physiological races were differentiated in the fungus. Race 1 and 2 are the most prevalent and highly destructive *Agrobacterium tumefaciens*-mediated transformation (AtMT) has been suggested as a possible tool for tagging genes relevant to pathogenesity in the plant pathogenic fungus *C. lagenarium* (Tsuji et al. 2003).

8.9   DISEASE MANAGEMENT

8.9.1   CULTURAL PRACTICES

Crop rotation, the use of resistant varieties, good drainage, fall tillage and the destruction of wild cucurbits are important cultural practices. Purchase clean seed from a reputable source and do not collect seed from infected plants. If possible, use drip irrigation instead of overhead sprinklers. Do not work in plants when wet. At the end of the season, the infected vines could be removed and destroyed. Start transplants with disease free seeds and inspect transplants in the greenhouse frequently. Immunization is systemic and requires a 3–4 days lag period between inoculation and exposure to the pathogen, followed by a booster inoculation to maintain protection through fruiting. The immunized plant does not reduce conidial germination after inoculation of *C. lagenarium,* but reduces the penetration of the pathogen into plant tissues. Use fungicides to prevent disease at or before the time when the vines begin to touch within the row.

8.9.2   RESISTANCE

According to Wehner and Barrett (2005), most commercial watermelon cultivars, including "Crimson Sweet" have shown some resistance against anthracnose diseases. Now days, seven races of anthracnose pathogen have been reported. Races 4, 5, and 6 are virulent in watermelon and are more important. Races 1 and 3 are not virulent, and many varieties are resistant to them, while resistance to other races is sought (Wehner et al., 2001). Unfortunately, it was not possible to extract and identify specific race(s) of anthracnose pathogen that were present in the field.

8.9.3   CHEMICAL CONTROL

A fungicidal trial including six fungicides was tested in India (Prakash et al., 1974). Benlate (0.2%) provided the best control (applied at weekly intervals for 5 weeks, starting approximately 3 weeks after inoculation). Benlate treated plots also gave significantly higher yields than plots treated with other fungicides. Presently, several fungicides are registered against anthracnose diseases but they may not provide effective control if good coverage of fruit and leaves is not achieved.

8.10   CERCOSPORA LEAF SPOT

8.10.1   DISTRIBUTION

Cercospora leaf spot is caused by the fungus *Cercospora citrullina*. The disease is most damaging to watermelon, cucumber and other melons and in worldwide. The genus Cercospora was established by Fresenius in 1863, the name being derived from the Greek kerkos-a tail, and spora-a spore. However, Fresenius did not characterized the genus distinctly, but merely listed four species with descriptions, confining his remarks concerning the genus to a footnote to the description of C. apiiFresen., the type species. In the footnote he states: "DieserPilz, welchenich von HerrnFuckelmitgetheilterhielt, war unter den beschriebenenGattungennichtwohlunterzubringen. Ichbelegteihnda-herwegen der mitunterlangausgezogenenSchwanze der SporenmitobigenGat-tungsnamen, 'Cercospora'. CylindrischeSporenkommeneigentlichhiernichtvor und von den Fusisporienweichen die Cercosporen in der Beschaffenheit der sporentragendenFaden und der Sporenformenselbst ab." These are the only generalizations made by Fresenius. He states that it is not advisable to describe the genus further until there is a better understanding of this group of fungi. The limits of the genus are, therefore, not specifically defined, but are left for later interpretation. Cooke (1875) had described the genus Virgasporium, listing two species, *V. maculatum* Cooke and *V. clavatum* (Gerard) Cooke. Later, accepting Saccardo's statement that this genus was identical with Cercospora, Cooke (1875a) transferred the two species to that genus.

The disease Cercospora leaf spot was observed in this study has confined higher severity to the inbred cultivar 'Sugarbaby'. The fungus *Cercospo-racitrullina* causes this disease (Roberts and Kucharek, 2006). Its symptoms were also noted on the leaves of "Yellow Crimson," "Crimson Sweet" and the Kakamega landrace, but all three showed moderate resistance against the disease. "Charleston Gray" and the wild accession demonstrated good resistance to this disease. This disease was more troublesome in the first than second season, and it can be attributed to the different rainfall patterns that were observed in the two seasons. Cercospora was more severe in the early stages of plant development when the leaves were tender and more succulent.

8.10.2   DISEASE SYMPTOMS

Cercospora leaf spot symptoms were occur on foliage, petiole and stem lesions when conditions are highly favorable for disease development. The lesions are not confined to fruits. The older leaves contain small, circular to irregular circular spots with tan to light brown lesions. The number and size of lesions may increases, coalesce and cause the entire leaves to become diseased. The centers of lesions may grow thin and fall out on cucumber, squash, and melon. Lesion margins may appear dark purple and may have yellow halos surrounding them. Severely infected leaves turn yellow, senesce and fall off. In watermelon, the lesions may present on the younger rather than older foliage. It reduces fruit size and quality but economic losses are rarely severe. Spots enlarge rapidly and become circular to irregular, with pale brown and purple to almost black margins. Spots are coalescing and form blotches. Leaf may dry and die with a scorched appearance of the leaf. Stems and fruits are also attacked.

8.10.3   CAUSAL ORGANISMS

The specific causal agent of this disease is not known yet (Wehner et al., 2001), but it is thought to be caused by some bacteria that are present naturally in fruit (Roberts and Kucharek, 2006). Drought stress is also reported to predispose melons to rind necrosis (Warren et al., 1990). Some varieties resistant to this disease have been identified (Wehner et al., 2001), but there are no other control measures (Roberts and Kucharek, 2006). *Cercospora citrullina, Cercospora melonis,* and *Cercospora lagenarium* are common on watermelon, cucumber, and musk melon.

8.10.4   DISEASE CYCLE

This disease is caused by the fungus *Cercospora citrullina*. The Cerco-spora leaf spot disease is most damaging to watermelon, other melons, and cucumber. The disease cycle begins when conidia are deposited into leaves and petioles by wind and splashing water. Conidia germinate during moderate to warm (78–90°F) temperatures in the presence of moisture. New cycles of infection and sporulation occurs every seven to 10 days during warm or in wet weather. The pathogen is readily disseminated within and among fields by wind and splashing rain and irrigation water, and survives between cucurbit crops as a pathogen on weeds. Roberts and Kucharek (2006) reported that cercospora leaf spot is favored by wet conditions and warm temperatures of 27–32°C and that the spores are readily wind-borne and rain splashed. The first season received more rainfall during the vegetative phase of the plants thus creating more conducive conditions for growth and spread of the causal pathogen at an early stage. The second season received more rainfall during the reproductive phase running through to maturity, hence low disease severity in this season. There is no documented information regarding the resistant cultivars against Cercospora leaf spot. The fourth disease, rind necrosis, was observed in only two accessions namely 'Charleston Gray' and 'Sugarbaby.' The disease was limited to fruits only and was found to develop inside the fruit with no external symptoms. It could only be detected after harvesting and when the fruits were cut open. Development of resistant cultivars by breeding methods against this disease is highly desired on susceptible accessions, because this disease is very difficult to control by other methods.

8.11 DISEASE MANAGEMENT

8.11.1 CULTURAL PRACTICES

For this diseases management, the elimination of crop debris and cucurbit weed hosts is essential. Remove weeds from within and around the crop. Grow watermelons in full sun lights, not under the shade of coconuts or other trees. Practice a 2–3-year rotation to non-hosts. No resistance to Cercospora leaf spot resistance has been identified in commercial cucurbit varieties. Maintain good soil drainage and good aeration between veins.

8.11.2 CHEMICAL CONTROL

Foliar application of Mancozeb (Dithane M-45) 0.2% or Chlorothalonil (kavach) 0.2% or Copper hydroxide 0.2% as a protective application at a 7–10 days interval and Difenoconazole (Score) 0.5 mL/L 2–3 application per season is also effective for disease control. In bitter gourd, *Cercosporacitrullina* is an important disease under laboratory and screenhousetest, fungicides namely, benomyl, chlorothalonil, copper oxychloride, and mencozeb effectively suppressed the development of disease (Donayre and Minguez, 2014). Fungicide sprays are necessary for disease control in wet and humid weather, but are not required in the high plains in most years.

8.12 CONCLUSION

Cucurbita belongs to the Latin word is a genus of herbaceous vine in the cucurbitaceae family, which is native to the Andes and Mesoamerica.

Five species are grown worldwide for their edible fruit, variously known as squash, gourd, or pumpkin. Plants belonging to the genus *Cucurbita* are important food and oil sources for humans. The fruit contains a good source of various nutrients and vitamin A, vitamin C, niacin, folic acid, and iron and many of the plant products that are free from cholesterol. Nowadays, the species are accepted by different specialists, ranging from 13 to 30. The five domesticated species are belonging to *Cucurbita argyrosperma, C. ficifolia, C. maxima, C. moschata*, and *C. pepo*. All of these can be treated as winter squash, because the full-grown fruits can be stored for months; however, *C. pepo* includes some cultivars that are better used only as summer squash. Cucurbits are susceptible to diseases such as downy mildew (*Pseudoperonospora cubensis*), powdery mildew (*Sphaerotheca fuliginea, Erysiphe cichoracearum*), Cercospora leaf spot (*Cercospora citrullina*), bacterial wilt (*Erwinia tracheiphila*), anthracnose (*Colletotrichum* spp.), fusarium wilt (*Fusarium* spp.), and phytophthora blight (*Phytophthora* spp.). The main objective of this chapter was to identify and elestreted various foliar fungal pathogens and their management strategy to reduce the economic and aesthetic damage caused by plant diseases. Specific management programs against specific diseases are not intended, since these will often vary depending on circumstances of the crop, geographic location, disease severity, its regulations and other factors. Most of plant disease management practices mainly, rely on particular diseases development stages and attack vulnerable points in the disease cycle. Therefore, correct diagnosis of a disease and proper identification of the pathogen, which is the real target of any disease management program.

KEYWORDS

- **Cucurbitaceae**
- **disease**
- **management**
- *Pseudoperonospora*
- *Sphaerotheca*
- *Erysiphe*

REFERENCES

Akem, C.; Jovicich, E. Integrated Management of Foliar Diseases in Vegetable Crops. HAL Final Report VG07127. DEEDI: Queensland, 2011; p 205.

Bains, S.S.; Jhooty, J. S. Over Wintering of *Pseudoperenospora cubensis* Causing Downy Mildew of Muskmelon. *Indian Phytopathol.* **1976a,** *29,* 213–214.

Bains, S. S.; Sharma, N. K. Differential Response of Certain Cucurbits to Isolates of *Pseudoperonospora cubensis* and Characteristics of Identified Eaces. Phytoparasitica, **1986,** *18,* 31–33.

Bains, S. S.; Jhooty, J. S. Mode of Efficacy of Four Fungitoxicants Against *Pseudoperenospora cubensis*on Muskmelon. *Indian Phytopathol.* **1978,** *31,* 339–342.

Bains, S. S.; Vidyaprakash, C. Susceptibility of Different Cucurbits to *Psedoperenospora cubensis* Under Natural and Artificial Epiphytotic Conditions. *Indian Phytopathol.* **1985,** *38,* 377–388.

Bains, S. S.; Jhooty, J. S. Over Wintering of *Pseudoperonospora cubensis* Causing Downy Mildew of Muskmelon. *Indian Phytopathol.* **1976b,** *29,* 213–214.

Bandopadhyay, B. S.; Karmakar, A. K.; Chaudhary, Mukhopadhyay, S. Chemical Control of Powdery Mildew Caused by *Erysiphe cichoracearum* in Cucumber. *Pestiology* **1980,** *4,* 10–11.

Batta, Y. L. Alternaria Leaf Spot Disease on Cucumber: Susceptibility and Control using Leaf Disc Assay. *An-Najah Univ. J. Res. (N.Sc)* **2003,** *17* (2), 269–279.

Becker, W. F.; Von Jagow, G.; Anke, T.; Steglich, W. Strobilurin A, Strobilurin B and Myxothiazole: New Inhibitors of the bc 1 Segment of the Respiratory Chain with Emethoxyacrylate System as Common Structural Elements. *FEBS Let.* **1981,** *132,* 329–333.

Bent, K. J. Chemical Control of Powdery Mildew. In *The Powdery Mildew;* Spencer, D. M., Ed.; Academic Press: London, UK, 1978; pp 259–282.

Bernhaedt, E.; Dodson, J.; Waterson, J. *Cucurbit Diseases: a Practical Guide for Seeds Men, Growers and Agricultural Advisors;* Petoseed Co.: Saticoy, 1988.

Bharat, N. K. Morphological Characterization of Causal Organism of Sarda Melon (Cucumismela var. reticulatus) Powdery Mildew. *Pl. Dis. Rept.* **2003,** *18* (2), 201.

Bhargava, A. K.; Singh, R. D. Comparative Study of *Alternaria* Blight Losses and Caused Organism of Cucurbits in Rajasthan. *Indian J. Mycol. Plant Pathol.* **1985,** *15* (2), 150–154.

Bains S. S.; Jhooty J. S. Overwintering of *Pseudoperenospora cubensis* Causing Downy Mildew of Muskmelon. *Indian Phytopathol.* **1976,** *29,* 213–214.

Bhatia, J. N.; Thakur, D. P. Field Evaluation of Systemic and Non-systemic Fungicides Against Powdery Mildews of Different Economic Crops. *Indian Phytopathol.* **1989a,** *42,* 571–573.

Bhatia, J. N.; Thakur, D. P. *Pi. Dis. Res.* **1989b,** *4,* 146–147.

Branzanti, B.; Brunelli, A. Study of Etiology of Powdery Mildew of Cucurbits in EmilaRomanjna. *Informatore Eitopthologiea* **1992,** *42,* 37–44.

Butler, E. J. *Fungi and Diseases in Plants;* Thacker Spink and Co.: Calcutta, 1918; Vol. 6, pp 314–315.

Butt, D. J. Epidemiology of Powdery Mildews. In *the Powdery Mildews;* Spencer, D. M., Ed.; Academic Press Inc.: London, New York, 1978; pp 51–81.

Cohen, Y.; Meron, I.; Mor, N.; Zuriel, S. A New Pathotype of *Pseudoperenospora cubensis* Causing Downy Mildew in Cucurbits in Israel. *Phytoparasitica* **2003,** *31,* 458–466.

Cohen, Y.; Rotem, J. Rate of Lesion Development in Relation to Sporulating Potential of *Pseudoperonospora cubensis* in Cucumbers. *Phytopathology* **1971c**, *61*, 265–268.

Cooke, M. C. British Fungi (cont.). *Grevillea* **1875**, *4* (30), 66–69.

Cooke, M. C. (1875b). *Virgasporium* Cooke, *Grevillea* **3**:182 (1875) [Type Species: *V. maculatum* Cooke 1875].

Dik, A. J. *Damage Relationships for Powdery Mildews in Greenhouse Vegetables (Abstract)*. In Proceedings of the First International Powdery Mildew Conference.

Fresenius, J. G. B. W. (1863). Beiträgezur Mykologie. Frankfurt a. M.: H.L. Brönner

Gupta, S. K.; Gupta, A. Management of Cucumber Powdery Mildew Through Fungicides. *Pl. Dis. Res.* **2001**, *16*, 186–192.

Hansen, M. A. Downy Mildew of Cucurbits. Publication Number 450–707. Virginia Cooperative Extension. 2000. http://www.ext.vt.edu/pubs/plantdiseasefs/450-707/450-707.

Hewitt, H. G. *Strobilurins: Fungicides in Crop Protection*; CAB International: New York, 1998.

Jhooty, J.; McKeen, W. Studies on Powdery Mildew of Strawberry Caused by *Sphaerothecamucalaris*. *Phytopathology* **1965**, *55*, 281–285.

Jhooty, J. S. Identity of Powdery Mildew of Cucurbits in India. *Pl. Dis. Reptr.* **1967**, *51*, 263–269.

Kabitarani, A. Sexuality and Pathological Specialization in *Spaerothecafuliginea*. *Indian Phytopath.* **1993**, *46* (1), 50–53.

Khan, M. W.; Khan, A.; Akram, M.; Khan, A. M. Powdery Mildew Flora of Aligarh. *Proc. Indian Sci. Congress* **1972**, 569.

Khan, M. W.; Akram, M.; Khan, A. M. Status of Cucurbit Powdery Mildew and Perithecial Production in Cucurbit Powdery Mildew in Northern India. *Indian Phytopath.* **1971**, *23*, 497–502.

Khan, M. W.; Khan, A. M.; Akram, M. Studies on the Cucurbit Powdery Mildew III. Intensity and Identity of Cucurbit Powdery Mildew in Kashmir. *Indian Phytopath.* **1974**, *27*, 93–96.

Khan, M. W.; Sharma, G. K. Taxonomic Evaluation of Anamorph Characters in Identification of Powdery Mildew Fungi on Cucurbits. *Indian Phytopath.* **1995**, *48*, 314–324.

Khetmalas, M. B.; Memane, S. A. Management of Downy Mildew Disease of Cucumber During Rainy Season. *J. Maharashtra Agric. Univ.* **2003**, *28*, 281–282.

Lange, L.; Eden, U.; Olson, L. W. Zoosporongenesis in *Pseudoperonospora cubensis* the Causal Agent of Cucurbits Downey Mildew. *Nordic J. Bot.* **1989**, *8*, 497–504.

Lebeda, A.; Gadasova, V.; Nishimura, S.; Ezura, H.; Matsuda, T.; Tazuke, A. *Pathogenic Variation of Pseudoperonospora cubensis in the Czech Republic and Some Other European Countries*. Proceedings of the Second International Symposium on Cucurbits, Tsukuba, Japan, Acta Horticulturae, 2002; *588*: 137–141.

Lebeda, A.; Sedlakova, B.; Pejchar, M.; Jerabkova, H. Variation for Fungicide Resistance Among Cucurbit Powdery Mildew Population in the Czech Republic. *Acta Hort.* **2010**, *871*, 465–475.

Lebeda, A.; Cohen, Y. Cucurbit Downy Mildew (*Pseudoperonospora cubensis*)-Biology, Ecology, Epidemiology, Host-pathogen Interaction and Control. *Eur. J. Plant Pathol.* **2011**, *129*, 157–192.

Lindenthal, M.; Steiner, U.; Dehne, H. W.; Oerke, E. C. Effect of Downy Mildew Development on Transpiration of Cucumber Leaves Visualized by Digital Infrared Thermography. *Phytopathology* **2005**, *95*, 233–240.

Mahrshi, R. P, Siradhana, B. S. Some New Hosts of *Qidiopsistaurica* from Rajasthan. *Indian Phytopath.* **1980**, *33*, 513.

Mohamed, S. A. Comparison of Host Range of Powdery Mildew and Attempt of Control Using Potassium and Phosphate Salts, M.Sc. Thesis, University of Gezira, 1997.

Oerke, E. C.; Steiner, U.; Dehne, H. W.; Lindenthal, M. Thermal Imaging of Cucumber Leaves Affected by Downy Mildew and Environmental Conditions. *J. Exp. Bot.* **2006**, *57*, 2121–2132.

Patel, J. G.; Prajapti, K. J.; Patel, S. T. Effected of Fungicides Against Powdery Mildew of Bottlegourd. *Plant Pathol.* **1990**, *43*, 247.

Pawar, S. M. Studies on the Occurrence of Powdery Mildew Diseases on Crops and Wild Plants in Marathwada. Ph.D. Thesis, 2005.

Prakash, O.; Sohi, H. S.; Sokhi, S. S. Studies on Anthracnose Disease of Cucurbits Caused by *Colletotrichumlagenarium* (Pass.) Ell. & Halst. and Its Control. *Indian J. Hortic.* **1974**, *31*, 278–282.

Puzannova, L. A. Hyperparasite of the Genus *Ampelomyces.* Exschlecht and the Possibility of Their Use in the Biological Control of the Pathogen of Powdery Mildew of Plant. *Mikologia- i-Fitopathlogica* **1984**, *18*, 333–338.

Reuveni, R.; Rotem, J. Effect of Humidity on Epidemiological Patterns of the Powdery Mildew *Sphaerothecafuliginea* on Squash. *Phytoparasiticia* **1974**, *2*, 25–33.

Rhodes, A. M. Inheritance of Powdery mildew Resistance in the Genus Cucurbita. *Plant Dis. Reporter* **1964**, *48* (1), 54–55.

Robak, L. Epidemiology and Control of Cucumber Downey Mildew *Pseudoperonospora cubensis. Biuletyn Warzywniczy* **1995**, *43*, 5–18.

Roberts, P.; Kucharek, T. Watermelon; Specific Common Diseases. Florida Plant Disease Management Guide: Florida Cooperative Extension Service, Institute of Food and Agricultural Sciences: University of Florida, 2006; Vol. 3 (55), pp 1–4.

Sherald, J. L.; Sister, H. D. Antifungal Mode of Action of Triforine. *Pesticides Biochem. Physiol.* **1979**, *5*, 477–488.

Sikora, E. J. Common Diseases of Cucurbits. The Alabama Cooperative Extension System, Alabama A&M and Auburn Universities. UPS, 7.5M36, ANR-809, 1997.

Siradhana, B. S.; Chaudhary, S. L. Occurrence of *Erysiphecichoracearum and Sphaerothercafuligineaat* Udaipur, Rajasthan (India) *Indian J. Mycol. Pl. Path.* **1972**, *2*, 76–79.

Sitterly, W. R. Powdery Mildews of Cucurbits. In *The Powdery Mildews*, Spencer, D. M., Ed.; Academic Press: London, New York, 1978; pp 359–379.

Thakur, R. P.; Mathur, K. Review Downy Mildews of India. *Crop Protect.* **2002**, *21*, 333–345

Thind, T. S.; Singh, P. P.; Sokhi, S. S.; Grewal, R. K. Application Timing and Choice of Fungicides for Control of Downy Mildew of Muskmelon. *Plant Dis. Res.* **1991**, *6*, 49–53.

Thomas, C. E, Jourdain EL. Host Effect on Selection of Virulence Factors affecting Sporulation by *Pseudoperonosporacubensis.* Plant Disease, **1992**, *76*, 905–907.

Thomas, C. E, Inaba, T.; Cohen Y. Physiological Specialization in *Pseudoperonospora cubensis. Phytopathology* **1987**, *77*, 1621–1624.

Thomas, C. E. Downy Mildew. In *Compendium of Cucurbit Diseases*, Zitter, T. A., Ed.; Cornell University Press: Ithaca, NY, 1996; pp 25–27.

Tsuji, G.; Fujii, S.; Fujihara, N.; Hirose, C.; Tsuge, S.; Shiraishi, T.; Kubo, Y. *Agrobacterium tumefaciens*-mediated Transformation for Random Insertional Mutagenesis in *Colletotrichum langenarium. J. General Plant Pathol.* **2003**, *69*, 230–239.

Waraitch, K. S.; Munshi, C. D.; Jhooty, J. S. Fungicidal Control of Powdery Mildew of Bottle Gourd. *Indian Phytopath.* **1975,** *28,* 556–557.

Warren, R.; Motes, J.; Damicone, J.; Duthie, J.; Edelson, J. Watermelon Production. In *Division of Agricultural Sciences and Natural Resources*; Oklahoma State University: Oklahoma Cooperative Extension Fact Sheets, F-6236, 1990; pp 1–4.

Watt, B. K.; Merrill, A. L. *Composition of Food*; U.S. Dept. of Agr., Handbook No. 1.8. 190 P, 1963.

Wehner, T. C.; Barrett, C. Watermelon Cultivars. In, *Cucurbit Breeding,* Wehner, T. C. Ed.; Horticultural Science, North Carolina State University, 2005; Online Publication: http://www.cuke.hort.ncsu.edu/cucurbit (accessed Mar 24, 2011).

Wehner, T. C.; Shetty, N. V.; Elmstrom, G. W. Breeding and Seed Production. In *Watermelons: Characteristics, Production, and Marketing*, Maynard, D. N., Ed.; ASHS Press: Alexandria, Va., 2001; pp 27–73.

Whitaker, T. W.; Davis, G. N. *Cucurbits, World Crop Series*, Nicholas, P., Ed.; Interscience Publishers. Inc.: New York, 1962; pp 1–154.

Zarcovitis, C. Attempts to Identify Powdery Mildew Fungi by Conidial Characters. *Trans. Brit. Mycol. Soc.* **1965,** *48,* 553–558.

Zitter, T. A.; Hopkins, D. L.; Thomas, C. E. *Compendium of Cucurbit Diseases*. APS Press: St. Paul, USA, 1996.

CHAPTER 9

Yellow Vein Mosaic of Okra: A Challenge in the Indian Subcontinent

AMIT KUMAR[1*], R. B. VERMA[1], RANDHIR KUMAR[1], and CHANDAN ROY[2]

[1]Department of Horticulture (Vegetable and Floriculture), Bihar Agricultural University, Sabour, Bhagalpur, India

[2]Department of Plant Breeding and Genetics, Bihar Agricultural University, Sabour, Bhagalpur, India

*Corresponding author. E-mail: amit.koon@gmail.com

ABSTARCT

Yellow vein mosaic is one of the devastating diseases of okra found in Indian subcontinent. Mosaic is caused by mono and bipartite begomoviruses and their satellites. This disease mainly reduces the quality of fruit and is greatly affected in severe case of infection. Management of disease by the application of chemicals is quite difficult and hazardous to the environment. The development of resistant cultivar is only practical approach to control the disease. Several researchers work on genetics of disease resistance and inheritance of resistance to yellow mosaic disease. Different source of resistant in wild species of okra had been identified. For the improvement of cultivated okra species, it is necessary to understand and appreciate the studies related to the virus, transmission of virus, alternative host for virus, and wild relative of okra, cross abilities barrier between the species, favorable conditions for disease development, screening methods, and breeding strategies. Under this chapter, we have tried to incorporate all related aspects that directly or indirectly help in the development of resistant cultivar of okra.

9.1 INTRODUCTION

Okra (*Abelmoschus esculentus* L. Moench) is cultivated in Indian and all over the world for its green, tender pod with five to seven ridges, finger like fruit vegetable (Seth et al., 2017). It belongs to the cotton family, Malvaceae (Sanwal et al., 2016). Locally in India, it is called as bhindi (Shetty et al., 2013).The origin of okra remains unclear but centers of genetic diversity include West Africa, India and Southeast Asia (Charrier 1984; Hamon and Van Sloten 1989). It is supposed to be originated from tropical Africa, confirmed by Purseglove (1987). This tender vegetable is an amphidiploid of nature between *Abelmoschus tuberculatus* (2n=58) and an unknown species (2n= 72) (Datta and Naug 1968). The okra diploid chromosome number is 130 (Shetty et al., 2013). There were about 50 species reported in okra; eight are mostly accepted (Borssum, 1966). Ten species of *Abelmoschus* has been found in India are believed to be of Asiatic origin (Table 9.1). The only *Abelmoschus esculentus* is of Indian origin (Seikh et al., 2013).

TABLE 9.1   Wild Relative of Okra and Their Occurrence in India.

Sl. no.	Wild species	Occurrence in India
1.	*A. ficulneus* and *A. tuberculatus*	Semi-arid region of north and northwestern India
2.	*A. crinitus* and *A. manihot*	Terai region of lower Himalayas
3.	*A. manihot* (*tetraphyllus* types), *A. angulosus*, and *A. moschatus*	Western and Eastern Ghats
4.	*A. crinitus* and *A. manihot* (mostly *pungens* types)	North-east region

Okra is consumed as fresh and canned food (Saurabh et al., 2016). Okra tender fruit are consumed as boiled vegetable and fried (Guddadamath et al., 2011). The fibrous fruits of okra are harvested at green and immature stage. India contributes maximum in terms of production of okra in the world with a total area of 0.53 million ha, yielded 6.36 million tones and productivity of 11.9 tones/ha (Anonymous, 2015). However, okra is susceptible to many biotic and abiotic factors that cause heavy losses in yield. Among them, okra yellow vein mosaic disease (YVMD) is one of the major constraints.

The severity of YVMD depends on certain locations and specific seasons in India. In north India, the rainy season is the most conducive season for the occurrence and spread of YVMD. These locations include Karnal, Terai region of Uttarakhand, Nadia district of West Bengal and Varanasi area of Uttar Pradesh. In central and south India, the disease is pronounced in several locations in summer season, the prominent ones being Guntur in Andhra Pradesh, Jalgaon in Maharashtra, Surat in Gujarat and Coimbatore in Tamil Nadu. However, in western region like Maharashtra, summer season is favorable for YVMD than the rainy season (Prabu et al., 2007 and Deshmukh et al., 2011).The loss in marketable yield was estimated at 50–94%. It depends on the crop growth and stages at which the infection occurs (Sastry and Singh 1974, Pun and Doraiswamy 1999). The maximum losses (94%) were observed when infection occurred within 20 days after germination, resulting to retarded growth, little number of leaves and fruits (Sanwal et al., 2016). However, if infection occurs at late in okra, the damages are reduced. Plants infected 50 and 65 days after germination suffer a loss of 84% and 49%, respectively (Nath and Sakia, 1992; Ali et al., 2000).

9.2   YELLOW VIEN MOSAIC DISEASE

Okra is susceptible to a large number of plant viruses (19 plant viruses) and okra yellow vein mosaic virus (YVMV) is one of those virus which causes heavy damage to it (Fajinmi and Fajinmi, 2010). It was first reported by Kulkarni (1924) from Bombay, and later studied by Capoor and Verma (1950) and Verma (1952). However, Uppal et al. (1942) established the viral origin of YVMD and coined the name yellow vein mosaic (YVM).

9.3   PATHOGEN/VIRAL BIOLOGY

In India, Okra YVMD is caused by yellow vein mosaic virus (OYVMV) and an additional single-stranded DNA (ss-DNA) is called betasatellite (Jose and Usha, 2003). It is a single stranded DNA virus (Begomovirus) and monopartite in nature (Pun and Doraiswamy, 1999). It is the largest genus of the family *Geminiviridae* (Rishishwar et al., 2015). Geminiviruses are a family of small, circular, ss-DNA viruses that cause severe diseases in major crop plants worldwide. The YVMD does not occur only in India in the neighboring countries like Bangladesh, Pakistan (Ali et al., 2000), Srlanka (Samarajeewa and Rathnanayaka, 2004). A survey on begomoviruses associated with okra in India shown that the incidence of YVMD in terms of percentage ranged from 23.0 to 67.67% in Karnataka, 45.89 to 56.78% in Andhra, 23 to 75.64% in Tamilnadu, 42.45 to 75.64% in Kerala, 23 to 85.64% in Maharashtra, 24.85 to 65.78% in Haryana, 35.76 to 57% in Uttar

Pradesh, 45.45% in Delhi, 67.78% in Chandigarh and 45.89 to 66.78% in Rajasthan (Venkataravanappa 2008). Begomovirus can be bipartite ((having two DNA molecules, DNA-A and DNA-B) or monopartite (having a single DNA molecule, resembling DNA-A) (Stanley et al., 2005; Fondong, 2013). The genomes size of bipartite begomovirus is 2.6 kb (Venkataravanappa et al., 2012). The similarity between both DNA -A and DNA -B are due to common region (CR) made up of 200 nucleotides (Fauquet and Mayo, 2001). CR plays an important role in replication and transcription. It is the site of both first and second strand synthesis (Arguello-Astorga et al., 1994). The genes on DNA A encodes several multifunctional proteins like Rep, a replication-associated protein which is essential for DNA replication of virus and a transcription activator protein (TrAP) that controls late gene expression (Jose and Usha, 2003). The DNA A encodes seven open reading frames (ORFs), while DNA B encodes for a single ORF βC1 (Priyavathi et al., 2016). The DNA B encodes two proteins namely nuclear shuttle protein (NSP) and a movement protein (MP), which are essential for systemic infection of plants (Gafni and Epel, 2002). The genes present in the DNA-A molecules are involved in encapsidation and replication, whereas the genes in the B component are involved in viral systemic movement, host range determination, and symptom expression (Rojas et al., 2005; Venkataravanappa et al., 2012). In addition to DNA-A and DNA-B, several small circular molecules, approximately half of the size begomoviruses, have also been confirmed to be found in the plants infected with begomoviruses (Shanmugam et al., 2016). These are known as alpha and Beta satellite DNA molecules. The beta satellites of a class of ss DNA molecules are about half of the size 1350 bp of the DNA-A (Briddon et al., 2003). The Beta satellites helps in symptom induction, host range determination and overcoming host defenses with the help of the helper virus. Beta satellites encodes a single protein (βC1), which suppresses RNA-interference, a plant defense reaction against viruses (Saunders et al., 2004). In addition to beta satellites, the begomoviruses can also be associated with small ssDNA molecules, of ca 1.4 kDa known as alpha satellites (also known as DNA_1) (Venkataravanappa et al., 2013).The alpha satellite encodes a protein likely to Rep protein of begomoviruses. It has the capacity to auto-replicate. The role of alpha satellite in the advance of disease process is not clearly known, but Nawaz-ul-Rehman et al. (2010) recently suggested that it may plays role in overcoming host defense. In addition to monopartite, two recent bipartite begmovirus infecting okra crop has been reported by Venkataravanappa and his colleague in 2012 and

2013 like Okra yellow vein Delhi virus (OYVDV) and Okra Yellow vein Bhubneshwar virus (OYVBV) respectively (Table 9.4). By using specific primers to begomovirus, a PCR amplification was performed by Jose and Usha (2000), concluded that the presence of a begomovirus element equal to DNA A in infected okra plants. However, the second genomic component (DNA B) was not amplified. It suggested that okra YVMV is a monopartite begomovirus.

9.4   ROLE OF VECTOR IN DISEASE SPREADING

Plant viruses cannot enter the host body by itself. They can enter either by wound or mechanical damage in plant or through insect and soil borne fungal species. These species that helps in carrying the virus to the host is called "vectors." Actually, these insect penetrates the virus inside the plant during the course of feeding. Among the insects, aphids predominate in temperate regions whereas whitefly and leafhopper predominate in tropical and subtropical areas. White fly (*Bamisia tabaci*) is a very complex species that act as a vector of YVMV. It consists of 34 morphologically indistinguishable species. The Q biotype and B biotype are the most widely distributed biotypes of the species (Tiwari et al., 2013). Begomoviruses are transmitted exclusively by whitefly in persistent and circulative manner. Persistent transmission requires hours for acquisition, with a retention time in the hemolymph of days to the entire life of the insect. Persistent transmitted viruses give propagative circulative transmission and replicate in the insect (Tiwari et al., 2013). The whiteflies could secure the virus from diseased plants after feeding for 1 h and the viruliferous insects could transmit the pathogen to disease free plants after feeding on them for only 30 min. It found that a single whitefly can be able to spread the virus to new plant. The minimum number of insect was about 10 that cause 100% infection to plant. The male and female flies both can able to spread this virus; however the female flies can transmit better than the male flies. The white flies which had been feed on the infected plants for 12–24 h remained infected throughout their lives (Bose et al., 2003).   The virus could not be transmitted through sap or seed. It may also be transmitted by the beetles of Podagrica spp. (Lana and Taylor, 1975; Atiri, 1984, 1990; Alegbejo, 2001a,b).

9.4.1   ALTERNATIVE HOST

In Indian subcontinent, the available alternative hosts for YVMV are *Croton sparsiflora, Malvastrum tricuspidatum, Abelmoschus manihot, Althea rosea, Hibiscus tetraphyllu*s, and *Ageratum* sp. (Gupta and Thind 2006). Roy et al. (2015) have been stated first time that the dicotyledonous plant *L. apetala* and *L. sebifera* act as natural and alternative hosts of *BYVMV* in India.

9.5   SYMPTOMS AND ECONOMIC IMPORTANCE OF DISEASE

FIGURE 9.1   (See color insert.) Severe symptom of YVMD of okra, covering >90% plant.

When infected to YVMV, okra plant shows persistent symptoms of yellowing and vein clearing. The initial symptom on tender leaves is a diffuse, mottled appearance and on older leaves irregular yellow areas which are interveinal. Clearing of the small veins starts near the leaf margins, at various points, about 15–20 days after the infection. Thereafter, the vein clearing develops into a vein chlorosis. Initially, the leaves exhibit only yellow colored veins but under the severe infection, the leaves become chlorotic completely and turn yellow. The newly developed leaves exhibit an interlinked network of yellow vein, which enclose the green patches of the leaf (Fig. 9.1). The size of leaves and fruits becomes reduced, resulting significant reduce in the production of the okra (Palanisamy et al 2009) by 30 to 70 % (Duzyaman, 1997). YVMD causes a lowering of chorophyll content of leaf and plants become stunted with small-sized pale yellow fruits (Gupta and Paul, 2001). Sharma (1995) observed a sharp reduction of total chlorophyll, reducing sugar, phosphorus and potassium was observed whereas total phenol, total sugars, non –reducing sugars, nitrogen, and protein content increased. Palanisamy et al. (2009) had agreed that all the pathological change in the okra are closely related to reduced rates of Photosynthesis as well effects on chloroplast number, ultra structure, chlorophyll metabolism, deterioration of chloroplast structure, pigment composition, and electron transport. These transformations appeared due to mild damage caused mostly in PS II by infection (Balachandran et al., 1997). Over several report on primary photosynthetic reaction, it was found accumulation of virus coat protein in the membranes of chloroplast and thylakoids of the infected plants leads to damage to electron transport of PS II (Rahoutei et al., 2000). Palanisamy et al (2009) further concluded that the infection of okra YVMV leads to reduction of photosynthetic pigments, soluble proteins, ribulose-1, 5-bisphosphate carboxylase activity, and nitratereductase activity. Virus infection caused the marked inhibition of PS II activity. He concluded that the infection of YVMV inactivates the donor side of PS II.

Under the field condition, when this virus infected to the okra young plants, they will induce three types of clear symptoms. First, infection at very early stage of plants in the young leaves leads to complete yellow and later on turns brown and finally dry up. Secondly, infection starts after the flowering, the younger leaves and the all flowering parts show vein clearing symptoms and the fruits, they produce are of bad quality like becomes hard and yellow at harvesting stage. In the third type of infection, the plants continues to grow in a healthy with normal fruiting up to some extent of time but, at the end of season, a very few older leaves and shoots showed typical YVMV symptoms. Although in such kind of plants yield were as good as symptoms less plants (Venkataravanappa et al., 2012).

9.6   INFLUENCE OF CLIMATIC CONDITION ON VECTOR POPULATION AND VIRAL INCIDENCE

During the rainy season, the temperature and relative humidity is high enough to support disease development. A fall in temperature further may lead to a decline in vector population at the end of rainy season. It results into the reduction of disease expression (Sanwal et al., 2016). In north India, the crop sown in month of June, the pods reaching to marketable stage in month of July–August were least susceptible to YVMD (4.1 %). However, most of the infection were found 92.3 % approximately, when the crops were sown late month of July and produces the fruit in month of August–September (Roychaudhary et al., 1997). At Kalyani (West Bengal), the population dynamics of whitefly was examined during the seasons and it was observed that it was remarkably found low during February to 1st fortnight of April and reached its peak in the month of August (Chattopadhyay et al., 2011). Ali et al. (2005) has found disease occurrence that increased with the increase in minimum temperature. While, the whitefly population decreases with increase in the relative humidity. The bright sunshine hours associated significantly positive while the minimum temperature associated significantly negative to YVMD disease incidence (Dhankhar et al., 2012). In an investigation, Chaudhary et al., (2016) stated that among all environmental factors, two factors, that is, wind speed and rainfall shows non-significant correlation with okra YVMD incidence and whitefly population. They observed that, with increase in minimum temperature, the disease incidence and whitefly population also increased. Similarly, with decrease in relative humidity, the disease incidence and whitefly population increased. Spatial study of OYVMV and whitefly shows that, with increase in time and distance, the disease incidence and whitefly population is also increased.

9.6.1   GENETICS OF YELLOW VEIN MOSAIC DISEASE

Arumugam and Muthukrishnan (1978) evaluated different varieties of *A. esculentus*. After examination, he came into the conclusion that, none of the present varieties showed resistance reaction to YVMV and there is need to search new resistance source in the present cultivated species as well as related wild species (Prabhu et al., 2007; Deshmukh et al., 2011). Number of scientist had been worked on resistant breeding for YVMV but success was meager (Prabhu et al., 2007). Deshmukh et al. (2011) has reported that the disease resistant depends upon the environment where the cultivar had been grown. They screened the 35 new okra line and seen that a line, which was found in no sign of any symptoms of YVMV in one season, but in subsequent later season they found appearance of disease. He concluded that disease infection depend upon the climatic conditions especially temperature and humidity, which is influenced the vector (white fly) population directly (Samarjeewa and Rathnayaka, 2004). Among the wild genotypes *A. angulosus* was found resistant under Rahuri dist of Maharashtra (Prabu et al., 2007; Prabu and Warade 2009), but it was a contrasting result as these were susceptible to YVMV at some other places (Premnath 1975; Rajmony et al., 1995). There are large numbers of wild genotypes available, as a source of resistant to YVMV (Prabhu et. al., 2007). Among them the wild *A. manihot* ssp. *manihot* follows the dominance gene action to YVMV and a single dominant gene is responsible for resistant to YVMV (Jambhale and Nerkar 1981). *A. caillei* (the West African okra) is also found as a potential donor of resistant gene to *A. esculentus* species (Kumar et al., 2010). *A. caillei* is a photosensitive wild cultivated mainly in dry season have better adaptation to humid zone and tolerant to biotic stress.

From the grafting test, it was confirmed by the Ali and his colleague in 2000 that, the tolerance developed in the genotype IPSA Okra one was not due to the escape, rather it was due to the resistant gene. Further he confirmed that in the variety IPSA Okra one, the tolerance to YVMV governs by the dominant gene. Earlier, Jambhale and Nerkar (1981) had reported that a single dominant gene may be responsible for resistance to disease, however, two recessive genes for the same observed by Singh et al. (1962). Although the dominant complementary genes are responsible for disease control (Thakur 1976; Sharma and Dhillon, 1983; Sharma and Sharma, 1984). The Pullaiah et al. (1998) studied the genetics pattern of resistance. He was also agreed with the Thakur (1976), and that two complementary dominant genes controlled the disease in susceptible × susceptible (S × S) and susceptible × resistant (S × R) crosses but in resistant × resistant (R × R) crosses two duplicate dominant genes controlled the disease. These findings are also confirmed by Seth et al. (2017). Vashisht et al. (2001) studied the genetics of disease resistance to YVMV in okra. He observed that additive gene action controlled the resistance to YVMV. Further, the two complementary dominant genes controlled the inheritance of resistance to YVMV following by Mendelian segregation (Dhankar et al., 2005). Arora et al. (2008) further examined genetics of resistance to YVMV. He used two YVMV resistant variety (Punjab-8 and Parbhani Kranti) and two susceptible cultivars (Pusa Sawani and Pusa Makhmali) in his study. They studied all the segregating generations like F2 and back cross. The qualitative analysis for segregants showed that genes leading to resistance in different resistant parent were different. When these genes were brought together in the F1, their effect was duplicated. Single dominant genes control the disease resistance, when a hybrid made with one resistant and one susceptible parent. Sindhumole and Manju (2015) studied the gene action of for disease resistance to major diseases (YVM) under Kerala condition and found that the duplicate gene action may operate for resistance to YVMV, it further may causes problem in crop improvement by simple selection. Therefore, he suggested reciprocal recurrent selection might be useful for the effective utilization of both types of additive and non-additive gene effects simultaneously. Finally, none of the workers came on a same opinion for the genetics to YVMV resisance, it need further verification.

Certain bio-molecules like Phenols and their related enzymes plays an important role in imparting either a resistant or susceptible reaction in the host (Prabhu and warade, 2009). According to Bajaj (1981), the biochemical analysis revealed that the parent which showed resistant to YVMV contains higher moisture, phenol, orthodihydroxy phenols and total chlorophyll content while in susceptible cultivars it has been found lower. Expression of all the traits except, total chlorophyll content have controlled by the dominance gene action in this study. Nazeer et al., (1994) had also confirmed that the virus free cultivars resistant to okra YVM begomovirus contains high amount of total phenols, orthodihydroxy phenols, flavonols, total proteins, and soluble proteins. While the enzymes peroxidase and polyphenol oxidase showed non-significant differences between virus-free susceptible and resistant cultivars. However, changes in these constituents were induced after viral inoculation. Further after virus infection to the resistant and susceptible cultivars, he observed that total proteins increased to a greater extent in the latter one. After inoculation, total phenols, orthodi-hydroxy phenols and flavonols decreased in resistant lines accompanied by an increase in peroxidase and polyphenol oxidase activity, whereas this was almost reversed in susceptible lines (Prabu and warade, 2009). It was found that the virus multiplication may be reduced due to the higher amounts of phenols and their oxidation products such as quinones, formed by increased peroxidase and polyphenol oxidase. The higher amount of soluble sugar was observed in leaves of susceptible cultivar then the resistant cultivar (Bhagat and Yadav, 1997). Hossain et al. (1998) reported that the total sugar, reducing or non-reducing sugar and total chlorophyll were low in YVMV infected leaves than the healthy one, whereas total phenol, carotene and ortho-hydroxy phenol contents were high in infected leaves. Kousalya (2005) had also reported maximum peroxidase and polyphenoloxidase activity in resistant wild *A. caillei* while minimum in susceptible *A. esculentus*. In the resistant wild *Abelmoschus* species and their interspecific hybrids after infection with the OYVM, showed lower phenolic compound while in the susceptible cultivars these content becomes increased (Prabhu and warade, 2009). He also observed that the total nitrogen content was lower in the resistant wild okra species and their interspecific hybrid as compared to susceptible *A. esculentus* cultivars. In the healthy plants of resistant wild okra and their interspecific hybrids on an average exhibited more polyphenol oxidase and peroxidaseactivity than the susceptible cultivated *Abelmoschus esculentus* cultivars. It is, therefore, concluded that the initial higher total phenols and their subsequent decrease accompanied by an increase in peroxidase and polyphenol oxidase activity after infection in the resistant lines, as compared to the susceptible okra cultivars confirms that the higher enzymatic activity is important firstly in the biosynthesis of orthodihydroxy phenols from monophenols, and secondly in the oxidation of phenols to more toxic quinones. These phenols and their oxidative develop the resistant against the YVMV either by inhibiting the virus activity or by reducing their rate of multiplication (Bhaktavatsalam et al., (1983). Prabu et al. (2008) has studied the influence of biochemical factors on the incidence of major diseases including YVMD and different pests of okra. They found that, there were positive correlation between mean whitefly population per leaf and YVMV coefficient of infection with total N, reducing sugar, and total sugar contents. The correlation of total phenol and seed soluble protein contents with the number of days to the first appearance of YMV and YMV coefficient of infection was highly negative. In addition to this, Total N, reducing and non-reducing sugar, and total sugar contents were found positively associated with mean whitefly adult population per leaf and days to the first appearance of YVMV.

9.7 SOURCE OF RESISTANCE

As none of the present cultivar of *Abelmoschus esculentus* species is available as a stable source of resistance to this disease Sanwal et al. (2014). However, few wild species of okra have reported as stable and reliable sources of resistance to YVMV (Table 9.2). Out of eight species in India, the only one *Abelmoschus esculentus* is cultivated species (Singh et al., 2007). The rest are available in the wild form (Sanwal et al., 2016). Nariani and

Seth (1958) has reported that *A. tuberculatus* (2n = 58), *A. manihot* (2n = 66) and *A. angulosus* (2n = 138) carry the virus without exhibiting symptoms. The species *A. manihot ssp. manihot* also carry the virus, but the plant was free from symptoms (Thakur 1976). Samarajeewa and Rathnayaka (2004) reported *Abelmoschus angulosus* as a resistance source to YVMV while the *A. ficulneus* and *A. moschatus* as a susceptible source virus infection. The various accessions from the species *A. angulosus, A. crinitus, A. manihot, A. vitifolius, A. tuberculatus, A. panduraeformis, A. pungens,* and *A. tetraphyllus* were reported against the resistant to YVMV (Dhankar and Mishra 2004, Seth et al., 2017). Keeping this view, an interspecific hybrid was made between *A. esculentus* × *A. tetraphyllus*, namely Arka Anamika and Arka Abhay (resistant varieties). *A. manihot* ssp. *manihot* is good source of resistance to YVMV and is being used to develop resistant variety (Nerkar and Jambhale 1985). However, due to sterility problems, it is not so easy or useful to transfer the resistant gene from wild species to the cultivated one. The partial fertile hybrid was made between Pusa Sawani and *A. manihot* ssp. *manihot* (Jambhale and Nerkar 1981). Joshi et al. (1960) developed the variety Pusa Sawani by incorporating the resistance from the West Bengal strain I.C. 1542 (a symptomless carrier). But, nowadays the field resistance has been broken in pusa sawani, may be due to new biotype development of virus (Table 9.3). Batra and Singh (2000) had been examined the eight open pollinated and six hybrids of okra against YVMV for two successive years during the rainy season. He confirmed four lines namely okra no. six, LORM-1 VRO-6 and p-7 were free from disease incidence and VRO-4 showed moderate resistant.

Rashid et al (2002) screened twelve okra genotypes for resistance to okra YVMV under field conditions. Two variety OK-292 and OK-286 were resistant to YVMV in both seasons and OK-315, OK-316, and OK-317 were found tolerant. Panda and Singh (2003), screened eighty-three genotypes of okra during summer and rainy season including twenty parent line and rest were hybrids. Among the twenty-two parents seven (HRB-55, HRB-92, KS 404, Parbhani Kranti, P-7, IC-10272, and Arka anamika) were found highly resistant during both the seasons. Among the 61 crosses, twenty-six crosses during summer and 35 crosses in rainy were rated as highly resistant. Abdul et al. (2004) observed three accessions, IC 218887,IC 69286 and EC 305619 resistant and 43 lines moderately resistant to YVMV.*A. manihor* ssp *manihot* has been successfully used in the development of several varieties of okra, viz; 'Parbhani Kranti', 'P 7' and 'Punjab Padmini. Aparna et al. (2012) used ten homozygous varieties of okra to make F1 hybrids in diallel fashion without reciprocal cross. six out of the top ten high yielding cross genotypes, were marked to exert highly resistant (Arka Abhay × Hissar Unnat), resistant (Hissar Unnat × Lam-1), moderately resistant (VRO-6 × Pusa Makhmali, Parbhani Kranti × VRO-6, and Parbhani Kranti × Arka Abhay) and highly sensitive (Parbhani Kranti × HissarUnnat) reactions in both rainy and late rainy seasons. Kumar et al. (2015) reported that the out of thirty okra genotypes two (IIHR 129, IIHR 123) were highly resistant to YVMV and the genotypes IIHR 120, IIHR 53, IIHR 113 and Arka Anamika showed resistant reaction to YVMV. Meena et al (2015) screened 98 lines of okra accessions on different regions, and reported that out of 98 lines six (AO: 109, AO:118, AO:133, AO:151, AO:189, and AO:204) were completely free from OYVMV. Kumar et al. (2016) recorded no incidence of OYVMV disease in the genotypes DOV-12 and DOV-66 even after 90 days of sowing. Several varieties of okra have been developed using conventional breeding technique. But none of the resistance variety is stable to yvmv. Frequent breakdown of resistance has been observed in developed varieties so that there is an urgent need to adopt non-conventional methods breeding technique in combination to biotechnological tools for development of pre-breeding lines resistant to biotic stresses (Singh et al., 2007).

TABLE 9.2 Wild Source of Resistance to YVMV.

Wild source	Gene action	Reference
A. manihot ssp. manihot	Dominant genes	Sharma and Dhillon (1983)
	Complimentary dominant genes	Sharma and Sharma (1984a)
	Recessive genes	Singh et al. (1962)
A. tuberculatus		Nariani and Seth (1958)
A. callei		Sergius and Esther (2014)

TABLE 9.3 Diverse Begomoviruses have been Associated with Yellow Vein Mosaic Disease Transmitted by *Bemisia tabaci*.

Genome	Virus	Reference
Monopartite	Okra yellow vein mosaic virus	Kulkarni (1924)
	Okra yellow vein Madurai virus	
	Okra yellow vein Haryana virus	Venkataravanappa et al. (2008)
	Okra yellow vein mosaic virus	Fauquet et al. (2008)

TABLE 9.3 *(Continued)*

Genome	Virus	Reference
	Cotton leaf curl Allahabad virus, (CLCuAlV)	Venkataravanappa et al. (2013)
	Cotton leaf curl Bangaluru virus, (CLCuBaV)	
	Okra yellow vein Bhubaneswar virus (BYVBhV)	
	Okra yellow vein Maharashtra virus (BYVMaV)	Brown et al. 2012
	Okra enation leaf curl virus (OELCuV)	
	Radish leaf curl virus	Kumar et al. 2012
Bipartite	Okra yellow vein Delhi virus (BYVDV)	Venkataravanappa et al. 2012
	Tomato leaf curl New Delhi virus	Venkataravanappa et al. 2008

9.7.1 CROP MANAGEMENT PRACTICES

As whitefly is the main vector responsible for transmission of virus that spread form infected to healthy plant. The management of this disease turns around the control of this vector. The population of these vector influence by the temperature and rainfall (Horowitz et al., 1984). Chakraborti et al. (2014) considered that main limiting abiotic factor was rainfall but significant positive correlation with increasing temperature, closing of the canopy and repeated irrigation. A high of nitrogen application and low application potassium lead to accumulation of amino acid in crop which are highly attack by vectors like whitefly. Not only the population of whitefly need to control, however it is different host plant where the whitefly feed also need to remove from the all corners of the field. This versatile host range facilitates easy population development and smooth carryover of the pest from one crop to another.

Several approaches have been attempted to control virus. Pest can be control by the application of chemicals. Alam et al. (2010) has used different eco-friendly management agents like Oil @0.5% mixed with a 0.5% washing soap, Marigold as a trap crops and planted in between rows of okra and Admire (Imidacloprid) @ 0.05% to check the disease. He finally concluded that most effective one was admire spray on okra followed by neem oil and mustard oil. Imdiachloprid 17.8% SL applied twice and one seed treatment significantly reduce the pest population (Jambhulkar et al 2013) up to 90.2%. It is biological product like Azadirachtin spray at an interval of 15 days reduces the white fly population up to the 79.2%.Chemical spray has hazardous impact on the nature need to use less frequently. Plant growth promoting rhizobacteria (PGPR) has been promoted as an alternative approach for disease management which is ecological friendly and safe (Patil et al., 2011). Rhizobacteria control the viruses through systemic resistance defense mechanism by activating the genes encoding chitinase, beta-1, three glucans, peroxidase, PALase, and other enzymes (Srinivasan et al., 2005). According to Srinivasan et al. (2005), strain of fluorescent Pseudomonas was the most effective strain. It reduces the incidence of bhindi Yellow Vein Mosaic to the maximum extent (up to 86.6%) through induced systemic resistance by triggering defense molecules. They also observed that the biosynthesis of peroxidase and PALase activity becomes enriched in the plant by 79% and 47%, respectively over the disease control. Development of resistant variety is the safest and permanent nonhazardous technique (Dhankher et al., 2005). Bhyan et al. (2007) studied the effect of different phytopesticidal treatment on the incidence of okra mosaic virus and its severity on yield and nutrition of okra. He observed that among the different phytopesticide like an extract of neem (*Azadiracta indica*) fruits, garlic (*Allium sativum*) bulbs, Karamja (*Pongamia pinnata*) leaves and mehogani (*Swietenia macro-phylla*) seeds, Karamja extract treated plants had minimal rate of incidence of YVMV and produce maximum fruits formation and yield. Biswas et al. (2008) studied the efficacy of different plant oils against OYVMV. The plant oils were extracted from crozophera (Crozophera plicata), palmrosa (*Cymbopogon martini*), citronella (*Cymbopogon winterianus*), lemongrass (*Cymbopogon citratus*), coronza (*Pongamia glabra*) and neem (*Azadirachta indica*). Greater fruit yield of okra, and reduction in disease incidence and whitefly population were obtained with application of crozophera oil at 1.0 mL/L, followed by palmrosa oil at 1.0 mL/L. Gowdar et al. (2007) suggested agrochemicals like acetamiprid, imidacloprid, trizophos, and monocrotophos were gives positive result toward controlling the YVMV. Two spray of acetamiprid 20SP @40 g a.i/ha. In reducing the incidence of YVMV, whitefly population subsequently increasing the yield of okra was effective. Fajinmi and Fajinmi (2010) concluded that the easiest and method of reducing Okra mosaic disease of okra is planting of resistant varieties against this disease. He observed that the protection of plant up to the age of 28 days after germination reduced the spread of okra YVMV by checking its vector. Khajuria et al. (2015) suggested destruction of infected plants along with the spray with azadirachtin (0.03 %) in the form of neem oil followed by second spray with dimethoate (0.03 %) and timely monitoring of whitefly by installing yellow stick traps @ 10 traps/ha. Ansar et al. (2014) suggested seed treatment with Imidacloprid and sowing of two rows of maize border with spraying of imidacloprid +Neem oil spray until fruit formation showed least incidence (15.47%) of disease. Kumar et al. (2015) suggested that the use of mixed insecticide imidachlorpid and carbofuran at 60 days after sowing (DAS), was found statistically superior and recorded minimum disease intensity (1.94%) over the profenofos + carbofuran (3.61%), metasystox+ carbofuran (4.72%) and imidacloprid (5.00%) as compared to control treatment (39.17%). Virus infection in okra plants at growth stages earlier than 4 weeks has more severe effect on the physiological performance of okra plant and subsequent reduction in growth performance and yield of okra. Therefore, some effective control measure is very necessary at early growth stages of okra plant.

9.8   BREEDING FOR YVMV RESISTANT

The disease cannot be controlled by simply spraying the chemical to control the vectors. Most problematic situations rise as the varieties which showed resistance against virus earlier becomes susceptible in next 2–3 years (Dhanker et al., 2005). This breakdown in resistance probably happens due to development of new strains of begomovirus (Venkataravanappa et al., 2012). The breeding for germplasm collection and varietal improvement had been started under the supervision of Singh et al. (1950). Consequently, Pusa Makhmali was developed from the collection from west Bengal in 1955 and released for cultivation. Later, Joshi and his colleague developed a variety Pusa Sawani from an intervarietal cross between IC 1542 (symptomless carrier for YVMV from West Bengal) and Pusa Makhmali. After that by the introducing a line from Ghana (highly resistant to YVMV) by the NBPGR several varieties had been developed. These are G-2 and G-2-4 from NBPGR, Punjab Padmini, Punjab-7 (PAU), Parbhani Kranti (MAU), IIHR Sel-4, IIHR Sel-10,Sel-2, Varash Uphar, Hisar Unnat (CCSHAU), PusaA-4 (IARI), Kashi Vibhuti, Kashi Pragati, Kashi Sathdhariand Kashi Kranti (IIVR, Varanasi Table 9.5). But reduction in the production of okra in India was appears to be due to several parameter, such as loss of varietal resistance to YVMV (Borah et al., 1992), appearance of diverse viruses or strains (Venkataravanappa et al., 2012) evolution of new biotypes of whitefly vectors (Sanwal et al., 2014) and showing of moderate to strong resistance by vector to insecticides applied (Rashida et al., 2005). Two prominent varieties of okra namely Hisar Unnat and Varsha Uphar identified and released at National Level in the year of 1992 and 1996, respectively had wide adaptation all over the country. But no further resistant for YVMV exist in Hisar Unnat and Varsha Uphar. Therefore, Dhankar (2012) focused efforts had been taken to improve Hisar Unnat in respect to its tolerance to YVMV using wild relatives *A. manihot* ssp. *manihot.* Varsha Uphar was poorly compatible with A. mannihot spp. manihot. He made a cross between the Hisar Unnat and A. manihot ssp. Manihot. The F_1 was partial fertile, found free from YVMV disease throughout the season but emerging fruit, which was intermediate for most of the fruit traits. The 30% of the obtained seed from BC_1 plants were viable. He further crossed the F_1 with tolerant cultivar of cultivated species like US7109. US7109 is identified as a source of tolerant to YVMV with dark green fruit. Such crosses were made to remove all the intermediate traits in F_1 and just for improving the fruit shape and color characters. The segregating generation studied for the various morphological and fruit traits, found stable and uniform and further isolate line 10, 15 and 25 (0–5% disease incidence). All the three lines having dark green color pod with smooth surface were resistant to YVMV. It is not necessary that resistance to YVMV once developed in a cultivar is stable and permanent in the cultivated species but it breakdown may observed in developed varieties (Singh et al., 2007). The Interspecific hybridization technique is an effective method available now for developing YVMV resistant varieties (Reddy, 2015) due to availability of crossable species. The success rate in terms of fruit setting percentage in both way of crossing between different species of *Abelmoschus* has been given in the (Table 9.4). Keeping view of nature of cross-ability among the different species of *Abelmoschus*, Reddy (2015) perform an experiment for improvement of an inbred line RNOYR-19 for YVMV. This inbred was found horticulturally superior for all traits, but susceptible to yvmv. He made a cross between RNOYR-19 (female) and A. manihot subsp. tetraphyllus (male). It produces normal fruit set and seed set. But the crossability was 90% between two species. A F1 hybrid obtained from A.esculentus and A. manihot subsp. tetraphyllus was completely sterile. Colchicine application helps to restore fertility in F_1 hybrid plants. There was zero mortality rate observed upon application of colchicine to the interspecific F_1 seedlings at two leaf stage with normal fruit set (100%) and partial seed set (53.12%). On selfing of this F_1 resulted into production of fully fertile stabilized colchiploids. Shriveled and non-viable seeds obtained from the interspecific crosses between *A. esculentus* and *A. moschatus*. It was observed due to post zygotic incompatibility (Rajamony et al., 2006). Now it will be very fruitful to adopt the non-conventional method like interspecific hybridization technique with combination biotechnological approaches for development of pre-breeding lines resistant to YVMV.

TABLE 9.4   Cross-ability and Fruit Setting % Between Interspecific Crosses.

Female parent	Male parent	Fruit setting (%)	Under reciprocal fruit setting (%)
A. esculentus	A. tetraphyllus var. tetraphyllus	92.31	19.23
A. esculentus	A. moschatus subsp. moschatus	57.14	11.54
A. esculentus	A. caillei	38.89	25.00
A. esculentus	A. ficulneus	35.48	0.00
A. esculentus	A. tetraphyllus var. pungens	100.00	0.00
A. esculentus	A. tuberculatus	30.00	85.71

9.9   MUTATION BREEDING FOR YVMV

Phadvibulya et al. (2009) performed an experiment for development of resistant mutant line from the cultivated variety "Annie and Okura" in Thailand. Seeds of these varieties were irradiated using Gamma rays @ 400 and 600 Gy. He found that one M_4 plant of okra (B-21) was highly resistant, but none of Annie seed when the light intensity was 400 Gy. Under field condition as well as green- house condition, all tested mutant lines showed resistance up to a month. However, only a few mutant lines showed resistant throughout the whole growth periods. Boonsirichai et al. (2009) studied the genetics of the radiation induced YVMD resistance mutation in okra. The YVMD- resistant B4610 mutant generated through gamma irradiation of the Okura variety of okra. A BC_1F_1 and an F2 mapping population were generated from the cross between B4610 and Pichit 03, an YVMD-susceptible variety. Analysis of F1 and F2 progeny revealed the semi-dominant nature of the resistance which appeared to be caused by a single-locus mutation. From this experiment it cannot be stated that whether the YVMD mutation involves a loss or a gain of gene function. Dalve et al. (2012) confirmed disease intensity become less on higher doses of mutagenic. A dose of 40kR gamma rays + 0.1% EMS and 30kR gamma rays + 0.1% EMS had shown resistance against YVMV. Singh and Singh (2000) evaluated okra genotypes using gamma rays and EMS as a mutagens. A high dose of mutagens 45 and 60 kR gamma rays and 0.75 and 1.0% EMS had been given highly resistant to resistant plants in both M2 and M3 generations. Henceforth, incidence of YVMV is directly proportional to high dose of mutagens and vice versa.

9.10   RNAi STRATEGY AGAINST YMVD OF OKRA

YVMD of okra is major viral disease of okra throughout India, affecting all plants parts from seedling stages to young plant, resulting in deterioration in quality fruits. We have discussed the different cause of this disease, but the situation is not under control and day by day more number recombinant viruses arises which causes YVMV. To control this dreaded situation, it needs to rely on more advance technique like Gene silencing, recent biotechnological tools. This technique mainly suppresses either the pre transcription process or alters the post transcription process known as Transcriptional Gene Silencing (TGS) or through mRNA degradation, termed post transcriptional gene silencing respectively. RNAi is a promising tool to knock down or silence a gene expression because it can target multiple gene family members by same RNAi inducing transgene. Attempts are being made for incorporation of specific genes such as CP (coat protein) gene and antisense RNA gene for elevated viral resistance in okra (Sanwal et al., 2016).

TABLE 9.5   Resistant Varieties of Okra.

Resistant to disease	Variety of okra	Released from
YVMV and OLCV	Hybrid- Kashi Bhairav	Indian Institute of Vegetable Science (IIVR, Varanasi, Uttar Pradesh, India)
YVMV	Kashi Mohini	IIVR, Varanasi
YVMV and OLCV	Kashi Mangali	IIVR, Varanasi
YVMV and OLCV	Kashi Vibhuti	IIVR, Varanasi
YVMV and OLCV	Kashi Pragati	IIVR, Varanasi
YVMV	Kashi Satdhari	IIVR, Varanasi
YVMV and OLCV	Shitla Jyoti	IIVR, Varanasi
YVMV and OLCV	Kashi Mahima	IIVR, Varanasi
YVMV	Pusa A4	IARI, New Delhi

9.11 CONCLUSION

Okra is susceptible to large number of begomovirus, associated with YVMD in Indian subcontinents. Whitefly is a major transmitting mediator for this disease. Whitefly is a polyphagous insect feed on different host in absent of okra plant is one of major difficult to control vector of virus. Host genetic resistance to viruses is one of the most practical, economical and environment-friendly strategy to manage the disease. The resistant genes from the source could be possible to transfer in the commercial cultivar by the different breeding method. It will helps in development of resistance/ tolerance varieties of okra to YVMD.

KEYWORDS

- Okra virus genome
- resistance source
- economic importance
- genetic of resistant
- advance breeding technique
- resistant cultivar

REFERENCES

Abdul, N. M.; Joseph, J. K.; Karuppaiyan, R. Evaluation of Okra Germplasm for Fruit Yield, Quality and Field Resistance to Yellow Vein Mosaic Virus. *Indian J. Plant Genetic Resour.* **2004,** *17*, 241–244.

Alam, M. M.; Hoque, M. Z.; Khalequzzaman, K. M.; Humayun, M. R.; Akter, R. Eco-friendly Management Agents of Okra Yellow Vein Clearing Mosaic Virus of Okra (*Abelmoschus esculentus* L. Moench). *Bangla. J. Agric.* **2010,** *35* (1), 11–16.

Ali, S.; Khan, M. A.; Habib, A.; Rasheed, S.; Iftikhar, Y. Correlation of Environmental Conditions with Okra Yellow Vein Mosaic Virus and *Bemisia tabaci* Population Density. *Int. J. Agric. Biol.* **2005,** *7* (1), 142–144.

Anonymous. Indian Horticulture Database. National Horticulture Board, Ministry of Agriculture, Government of India, Gurgaon, 2015.

Ansar, M.; Saha, T.; Sarkhel, S.; Bhagat, A. P. Epidemiology of Okra Yellow Vein Mosaic Disease and Its Interaction with Insecticide Modules. *Trends Biosci.* **2014,** *7* (24), 4157–4160

Aparna, J.; Srivastava, K.; Singh, P. K. Screening of Okra Genotypes to Disease Reactions of Yellow Vein Mosaic Virus under Natural Conditions. *Vegetos.* **2012,** *25* (1), 326–328.

Arguello-Astorga, G. R.; Guevara-Gonzalez, R. G.; Herrera-Estrella, L. R.; Rivera-Bustamante, R. F. Gemini Virus Replication Has a Group Specific Organization of Iterative Elements: A Model for Replication. *Virology* **1994,** *203,* 90–100.

Bajaj, K. L. Biochemical Basis of Resistance to Yellow Vein Mosaic Virus in Okra (*Abelmoschus esculentus*). *Genetica Agraria* **1981,** *35,* 121–130.

Balachandran, S.; Hurry, V. M.; Kdley, S. E.; Osmond C. B.; Robinson, S. A.; Robozinski, J. Concepts of Plant Biotic Stress; Some Insights into Stress Physiology of Virus-infected plants, from the Perspective of Photosynthesis. *Physiologia Plantarum* 1997, *100,* 203–213.

Banerjee, M. K.; Kalloo, G. Sources and Inheritance of Resistance to Leaf Curl Virus in *Lycopersicon* spp. *Theor. Appl. Genet.* **1987,** *73,* 707–710.

Bhagat, A. P.; Yadav, B. P. Biochemical Changes in OYVMV Infected Leaves of Bhindi. *J. Mycol. Plant Pathol.* **1997,** *27,* 94–95.

Bhaktavatsalam, G.; Nene, Y. L.; Beniwal, S. P. S. Influence of Certain Physio-chemical Factors on the Infectivity and Stability of Urd Bean Leaf Crickle Virus. *Indian Phytopathol.* **1983,** *36,* 489–493.

Bhyan, S. B.; Alam, M. M.; Ali, M. S. Effect of Plant Extract on Okra Mosaic Virus Incidence and Yield Related Parameters of Okra. *Asian J. Agric. Res.* **2007,** *1* (3), 112–118.

Biswas, N. K.; Nath, P. S.; Srikanta, D.; De, B. K. Management of Yellow Vein Mosaic Virus Disease of Okra (*Abelmoschus esculentus* L. Moench) Through Different Plant Oils. *Res. Crops* **2008,** *9* (2), 345–347.

Borah, G. C.; Saikia, A. K.; Shadeque, A. Screening of Okra Genotypes for Resistance to Yellow Vein Mosaic Virus Disease. *Indian J. Virol.* **1992,** *8,* 55–57.

Bose, T. K.; Kabir, J.; Maity T. K.; Parthasarthy, V. A.; Som, M. G. Vegetable Crops, **2003,** *3,* 224–225.

Borssum, W.; Van, I. Malesian Malvaceae Revised. *Blumea* **1966,** *14,* 1–251.

Briddon, R. W.; Bull, S. E.; Amin, I.; Idris, A. M.; Mansoor, S.; Bedford, I. D.; et al. Diversity of DNA Beta: a Satellite Molecule Associated with Some Monopartite Begomoviruses *Virology* **2003,** *312,* 106–121.

Brown, J. K, Fauquet, C. M.; Briddon R. W.; Zerbini, M.; Moriones, E.; Navas-Castillo. J. Geminiviridae. In *Virus Taxonomy: Ninth Report of the International Committee on Taxonomy of Viruses,* King, A. M. Q., Adams, M. J., Carstens, E. B., Lefkowitz, E. J., Eds.; Associated Press: London, UK, **2012;** pp 351–373.

Brunt, A.; Crabtree, K.; Gibbs, A. *Viruses of Tropical Plants.* CAB International: Wallingford, 1990.

Chakraborti, S.; Pijush, K. S.; Chakraborty, A. Assessing the Impacts of Safer Management Strategies for Okra Yellow Vein Mosaic Virus (OYVMV) and its White Fly Vector with Emphasis on Understory Repellent Crop. *J. Ent. Res.* **2014,** *38* (1), 7–15.

Chattopadhyay, A.; Dutta, S.; Shatterjee, S. Seed Yield and Quality of Okra as Influenced by Sowing Dates. *Af. J. Biotechnol.* **2011,** *28* (5), 461–467.

Chaudhary, A.; Khan, M. A.; Riaz, K. Spatio-temporal Pattern of Okra Yellow Vein Mosaic Virus and Its Vector in Relation to Epidemiological Factors. *J. Plant Pathol. Microbiol.* **2016,** *7,* 360. DOI:10.4172/2157-7471.1000360.

Dalve, P. D.; Musmade, A. M.; Patil, R. S.; Bhalekar, M. N.; Kute, N. S. Selection for Resistance to Yellow Vein Mosaic Virus Disease of Okra by Induced Mutation. *Bioinfolet* **2012,** *9* (4b): 822–823.

Datta, P. C.; Naug, A. A Few Strains of *Abelmoschus esculentus* (L.) Moench, Their Karyological Study in Relation to Phylogeny and Organ Development. *Beitr Biol Pflanzen* **1968,** *45,* 113–126.

Deshmukh, N. D.; Jadhav, B. P.; Halakude, I. S.; Rajput, J. C. Identification of New Resistant Sources for Yellow Vein Mosaic Virus Disease of Okra (*Abelmoschus esculentus* L.) *Vegetable Sci.* **2011,** *38* (1), 79–81.

Dhankar, S. K. Genetic Improvement of Adopted Okra Cultivars for YVMV Disease Resistance involving Wild Relatives in Genus *Abelmoschus*. Seaveg 2012 Regional Symposium, **2012,** 24–26 January 2012.

Dhankar, S. K.; Dhankar B. S.; Yadava, R. K. Inheritance of Resistance to Yellow Vein Mosaic Virus in An Inter-specific Cross of Okra (*Abelmoschus esculentus*). *Indian J. Agric. Sci.* **2005,** *75,* 87–89.

Dhankhar, S. K.; Chohan, P. K.; Singh, S. S. Influence of Weather Parameters on Incidence of Yellow Vein Mosaic Virus in Okra. *J. Agrometeorol.* **2012,** *14* (1), 57–59.

Fajinmi, A. A.; Fajinmi, O. B. Epidemiology of Okra Mosaic Virus on Okra Under Tropical Conditions. *Int. J. Vegetable Sci.* **2010,** *16* (3), 287–296. http://dx.doi.org/10.1080/19315261003796974.

Fajinmi, A. A.; Fajinmi, O. B. Incidence of Okra Mosaic Virus at Different Growth Stages of Okra Plants (*Abelmoschus esculentus* L. Moench) Under Tropical Condition. *J. Gen. Mol. Virol.* **2010,** *2* (1), 028–031.

Fauquet, C. M.; Mayo, M. A. The 7th ICTV Report. *Arch. Virol.* 2001, *146,* 189–194.

Fondong V. N. Geminivirus Protein Structure and Function. *Mol. Plant Pathol.* **2013,** *14,* 635–649.

Gafni, Y.; Epel, B. L. The Role of Host and Viral Proteins in Intra- and Inter-cellular Trafficking of Geminiviruses. *Physiol. Mol. Plant Pathol.* **2002,** *60,* 231–241.

Gowdar, S. B.; Ramesh Babu, H. N.; Reddy, N. A. Efficacy of Insecticides on Okra Yellow Vein Mosaic Virus and Whitefly Vector, *Bemisia tabaci* (Guenn.). *Ann. Pl. Protec. Sci.* **2007,** *15* (1), 116–119.

Guddadamath, S.; Mohankumar, H. D.; Salimath, P. M. Genetic Analysis of Segregating Populations for Yield in Okra {*Abelmoschus esculentus* (L.) Moench} *Karnataka J. Agric. Sci.* **2011,** *24* (2), 114–117.

Gupta, S. K.; Thind, T. S. *Disease Problems in Vegetable Production*. Scientific Publisher: India, 2006; p 484.

Horowitz, A. R.; Podoler, H.; Gerling, D. Life Table Analysis of the Tobacco Whitefly *Bemisia tabaci* (Gennadius) in Cotton Fields in Israel. *Acta Oecologica-Oecologia Applicata* **1984,** *5,* 221–33.

Hossain, M. D, Meah, M. B.; Rahman, G. M. M. Reaction of Okra Variety to Yellow Vein Mosaic Virus and Biochemical Changes in its Infected Leaf Constituent. *Bangla. J. Plant Pathol.* **1998,** *14,* 29–32.

Jambhale, N. D.; Nerkar, Y. S. Inheritance of Resistance to Okra Yellow Vein Mosaic Disease in Interspecific Crosses of *Abelmoschus*. *Theor. Appl. Genetic* **1981,** *60,* 313–316.

Jambhulkar, P. P.; Singh, V.; Babu, S. R.; Yadav, R. K. Insecticides and Bio Products Against Whitefly Population and Incidence of Yellow Vein Mosaic Virus in Okra. *Indian J. Plant Protect.* **2013,** *41* (3), 253–256.

Jha, A.; Mishra, J. N. Yellow Vein Mosaic of Okra in Bihar. *In Proc. Bihar Acad. Agric. Sci.* **1955,** *4,* 129–130.

Jones, D. R. Plant Viruses Transmitted by Whiteflies. *Eur. J. Plant Pathol.* **2003,** *109,* 195–219.

Joseph, J. K, Nissar, V. A. M.; Latham.; Sutar, S.; Patil, P.; Malik, S. K.; Negi, K. S.; Keisham, M.; Rao, S. R.; Bhat, K. V. Genetic Resources and Crossability Relationship among Various Species of *Abelmoschus*. *Curr. Hortic.* **2013,** *1* (1), 35–46.

Joshi, A. B.; Singh, H. B.; Joshi, B. S. Is Yellow Vein Mosaic Disease a Nuisance in Your Okra, then Why Not Grow 'PusaSawani'. *Indian Frm.* **1960,** *10,* 6–7.

Karri, S. R.; Acharya, P. Incidence of Yellow Vein Mosaic Virus Disease of Okra [*Abelmoschus esculentus* (l.) Moench] under Summer and Rainy Environments. *Int. J. Curr. Res.* **2012,** *4* (05), 018–021.

Khajuria, S.; Rai, A. K.; Lata, K.; Kumar, R.; Khadda, B. S.; Jadav, J. K. Efficacy of Integrated Disease Management (IDM) Modules for Okra Yellow Vein Mosaic Virus in Semi-arid Region of Middle Gujarat. *Prog. Hortic.* **2015,** *47* (*2*), 293–295.

Kousalya, V. Introgression of Yellow Vein Mosaic Resistance from *Abelmoschus caillei* (A. cher.) Stevensinto *Abelmoschus esculentus* (L.) Moench. M.Sc. (Agri.) Thesis, Kerala Agricultural University, Thrissur, 2005.

Kulkarni, G. S. Mosaic and Other Related Diseases of Crops in the Bombay Presidency. Poona Agric. College Mag. 6 12, 1924.

Kumar, A.; Verma, R. B.; Solankey, S. S.; Adarsh, A. Evaluation of Okra (*Abelmoschus esculentus*) Genotypes for Yield and Yellow Vein Mosaic Disease. *Indian Phytol.* **2015,** *68* (2), 201–206.

Kumar, H.; Singh, R.; Gupta, V.; Zutshi, S. K. Performance of Different Germplasm, Plant Extracts and Insecticides Against Yellow Vein Mosaic of Okra (OYVMV) Under Field Conditions. *Vegetos* **2015,** *28* (1), 31–37.

Kumar, J.; Kumar, A.; Singh, S. P.; Roy, J. K.; Lalit, A. First Report of Radish Leaf Curl Virus Infecting Okra in India. *New Dis. Rep.* **2012,** *7,* 13–14.

Kumar, R.; Yadav, R. K.; Bhardwaj, R.; Baranwal, V. K.; Chaudhary, H. Studies on Genetic Variability of Nutritional Traits Among YVMV Tolerant Okra Germplasm. *Indian J. Hortic.* **2016,** *73* (2), 202–207.

Kumar, S.; Dagnoko, S.; Haougui, A.; Ratnadass, A.; Pasternak, D.; Kouame, C. Okra (*Abelmoschus* spp.) in West and Central Africa: Potential and Progress on its Improvement. *Af. J. Agric. Res.* **2010,** *5* (25), 3590–3598.

Kundu, B. C.; Biswas, C. Anatomical Characters for Distinguishing *Abelmoschus* spp. and *Hibiscus* spp. *Indian Sci. Cong.* **1973,** *60,* 295–298.

Leite, G. L. D.; Picanço, M.; Jham, G. N.; Moreira, M. D. Whitefly Population Dynamics in Okra Plantations. *Pesquisa Agropecuária Brasileira* **2005,** *40,* 10.

Meena, R. K.; Verma, A. K.; Kumar, M. Chatterjee, T.; Thakur, S. Evaluation of Okra (*Abelmoschus esculentus*) Germplasm Against Yellow Vein Mosaic Disease. *Indian Phytopathol.* **2015,** *68* (2), 226–228.

Nariani, T. K.; Seth, M. L. Reaction of *Abelmoschus* and *Hibiscus* Species to Yellow Vein Mosaic Virus. *Indian Phytopath.* **1958,** *11,* 137–140.

Nawaz-ul-Rehman, M. S.; Nahid, N.; Mansoor, S.; Briddon, R. W.; Fauquet, C. M. Post-transcriptional Gene Silencing Suppressor Activity of the Alpha-Rep of Non-pathogenical pha Satellites Associated with Begomoviruses. *Virology* **2010,** *405,* 300–308.

Nazeer, A.; Thakur, M. R.; Bajaj, K. L.; Cheema, S. S. Biochemical Basis of Resistance to Yellow Vein Mosaic Virus in Okra. *Plant Dis. Res.,* **1994,** *9* (1), 20–25.

Nerkar, Y. S.; Jambhale, N. D. Transfer of Resistance to Yellow Vein Mosaic from related Species into Okra (*Abelmoschus esculentus* (L.) Moench). *Indian J. Genetics Plant Breed.* **1985,** *45* (2), 261–70.

Pal, B. P.; Singh, H. B.; Swarup, V. Taxonomic Relationships and Breeding Possibilities of Species *Abelmoschus* Related to Okra (*Abelmoschus esculentus* (L.) Moench). *Bot. Gazzete* **1952,** *113,* 455–464.

Panda, P. K.; Singh, K. P. Resistance in Okra (*Abelmoschus esculentus* (L) Moench) Genotypes to Yellow Vein Mosaic Virus. *Vegetable Sci.* **2003**, *30* (2), 171–172.

Palanisamy, P.; Michael, P. I.; Krishnaswamy, M. Physiological Response of Yellow Vein Mosaic Virus-infected Okra (*Abelmoschus esculentus*) Leaves. *Physiol. Mol. Plant Pathol.* **2009**, *74*, 129–133.

Patil. M. N.; Jagadeesh, K. S.; Krishnaraj, P. U.; Patil, M. S.; Vastrad, A. S. Plant Growth Promoting Rhizobacteria (PGPR) Mediated Protection in Bhindi against Yellow Vein Mosaic Virus. *Indian J. Plant Protect.* **2011**, *39* (1), 48–53.

Phadvibulya, V.; Boonsirichai, K.; Adthalungrong, A.; Srithongchai, W. Selection for Resistance to Yellow Vein Mosaic Virus Disease of Okra by Induced Mutation. In *Induced Plant Mutations in the Genomics Era,* Shu, Q. Y., Ed.; Food and Agriculture Organization of the United Nations: Rome, **2009**; pp 349–351.

Prabhu, T.; Warde, S. D. Biochemical Basis of Resistance to Yellow Vein Mosaic Virus in Okra *Vegetable Sci.* **2009**, *36* (3 Suppl.): 283–287.

Prabhu, T.; Warde, S. D.; Ghante, P. H. Resistant to Okra Yellow Vein Mosaic Virus in Maharashtra. *Vegetable Sci.* **2007**, *34* (2), 119–122.

Prabu, T.; Warade, S. D.; Birade, R. M. Influence of Biochemical Factors on Incidence of Major Diseases and Pests in Okra. *J. Maharashtra Agric. Univ.* **2008**, *33* (3), 417–419.

Premnath. Breeding Vegetable Crops for Resistance to Diseases in India. SABRAO, **1975**, *7* (1), 7–11.

Priyavathi, P.; Kavitha, V.; Gopal, P. Complex Nature of Infection Associated with Yellow Vein Mosaic Disease in Okra (*Abelmoschus esculentus* (L.) Moench). *Curr. Sci.* **2016**, *111* (9), 1511–1515.

Pun, K. B.; Doraiswamy, S. Effect of Age of Okra Plants on Susceptibility to Okra Yellow Vein Mosaic Virus. *Indian J. Virol.* **1999**, *15*, 57–58.

Purseglove, J. W. Tropical Crops. In *Dicotyledons*; Longman Singapore Publishers: Singapore, 1987.

Rahid, M. H.; Yasmin, L.; Kibria, M. G.; Mollik, A. K. M. S. R.; Hossain, S. M. M. Screening of Okra Germplasm for Resistance to Yellow Vein Mosaic Virus under Field Conditions. *Pak. J. Plant Pathol.* **2002**, *1* (2–4), 61–62.

Rahoutei, J.; Garcia-Luque I.; Baron, M. Inhibition of Photosynthesis by Viral Infection: Effect on PS II Structure and Function. *Physiologia Plantarum* **2000**, *110*, 286–292.

Rajamony, L.; Chandran, M.; Rajmohan, K. In Vitro Embryo Rescue of Interspecific Crosses for Transferring Virus Resistance in Okra (*Abelmoschus esculentus* (L.) Moench). *Acta Hort.* **2006**, *725*, 235–240.

Rajmony, L.; Jessykutty, P. C.; Mohankumaran, N. Resistance to YVMV of Bhindi in Kerala. *Vegetable Sci.* **1995**, *22* (2), 116–119.

Rashida, P.; Sultan, M. K.; Khan, M. A, Noor-Ul-Islam. Screening of Cotton Germplasm Against Cotton Leaf Curl Begomovirus (CLCuV). *J. Agri. Soc. Sci.* **2005**, *3*, 35–238.

Reddy, M. T. Crossability Behaviour and Fertility Restoration Through Colchiploidy in Interspecific Hybrids of *Abelmoschus esculentus* × *Abelmoschus manihot* subsp. *tetraphyllus. Int. J. Plant Sci. Tech.* **2015**, *1* (4), 172–181.

Rishishwar, R.; Mazumdar, B.; Dasgupta, I. Diverse and Recombinant Begomoviruses and Various Satellites Are Associated with Okra Yellow Vein Mosaic Disease of Okra in India. *J. Plant Biochem. Biotechnol.* **2015**, *24* (4), 470–475.

Rojas, M. R.; Hagen, C.; Lucas, W. J.; Gilbertson, R. L. Exploiting Chinks in the Plant's Armor: Evolution and Emergence of Geminiviruses. *Ann. Rev. Phytopathol.* **2005**, *43*, 361–394.

Roychaudhary, J.; Vethannayagam, S. M.; Bhat, K.; Sinha, P. *Management of Yellow Mosaic Virus Disease in Bhindi (Abelmoschus esculentus) by Sowing Dates and with Neem Products*. In IPS Golden Jubilee International Conference. New Delhi, India, **1997**; 10–15 Nov. 1997; p 29.

Roy, B.; Chakraborty, B.; Mitra, A.; Sultana, S.; Sherpa, A. R. Natural Occurrence of Bhendi Yellow Vein Mosaicvirus on Litsea spp. in India. *New Dis. Rep.* **2015**, *31*, 7. http://dx.doi.org/10.5197/j.2044-0588.2015.031.007

Safdar Ali, Khan, M. A.; Habib, A.; Rasheed S.; Iftikhar, Y. Management of Yellow Vein Mosaic Disease of Okra Through Pesticide/Bio-pesticide and Suitable Cultivars. *Int. J. Agric. Biol.* **2005**, *7* (1), 145–147.

Samarajeewa, P. K.; Rathnayaka, R. M. U. S. K. Disease Resistance and Genetic Variation of Wild Relatives of Okra (Abelmoschus esculentus L.). *Ann. Sri Lanka Dep. Agric.* **2004**, *6*, 167–176.

Sanwal, S. K.; Singh, M.; Singh, B.; Naik, P. S. Resistance to Yellow Vein Mosaic Virus and Okra Enation Leaf Curl Virus: Challenges and Future Strategies. *Curr. Sci.* **2014**, *106* (11), 1470–1471.

Sanwal, S. K.; Venkataravanappa, V.; Singh, A. Resistance to Okra Yellow Mosaic Disease: A Review. *Indian J. Agric. Sci.* **2016**, *6* (7), 835–843.

Sarma, U. C.; Bhagabati, K. N.; Sarkar, C. R. Effect of YVMV Infection of Some Chemical Constituents of Okra [*Abelmoschus esculentus* (L.) Moench]. *Indian J. Virol.* **1995**, *11* (1), 81–83.

Sastry, K. S and Singh, S. J. Effect of Yellow Vein Mosaic Virus Infection on Growth and Yield of Okra Crop. *Indian Phytopathol.* **1974**, *27*, 295–297.

Saunders K, Norman A, Gucciardo S, Stanley J. The DNA Betasatellite Component Associated with Ageratum Yellow Vein Disease Encodes an Essential Pathogenicity Protein (betaC1). *Virology* **2004**, *324*, 37–47.

Saurabh, A.; Kudada, N.; Singh, S. K. An Approach to Management of Yellow Vein Mosaic of Okra through Chemical Insecticides and Plant Extracts. *Environ. Ecol.* **2016**, *34* (3B), 1368–1371.

Sergius, U. O.; Esther, D. U. Screening of *Abelmoschus esculentus* and *Abelmoschus callei* Cultivars for Resistance Against Okra Leaf Curl and Okra Mosaic Viral Diseases, Under Field Conditions in South Eastern Nigeria. *Afr. J. Biotechnol.* **2014**, *13* (48), 4419–4429.

Seth, T.; Chattopadhyay. A.; Dutta. S.; Hazra,P.; Singh, B. Genetic Control of Yellow Vein Mosaic Virus Disease in Okra and Its Relationship with Biochemical Parameters. *Euphytica* **2017**, *213*, 30. DOI 10.1007/s10681-016-1789-9.

Shanmugam, G.; Kumar, K.; Balasubramanian, P. Developing RNAi Strategy Against Yellow Vein Mosaic Disease (ymvd) of Okra. *Global J. Bio-sci. Biotechnol.* **2016**, *5* (1), 01–08.

Sharma, B. R.; Sharma, D. P. Breeding for Resistance to Yellow Vein Mosaic Virus in Okra. *Indian J Agric Sci.* **1984**, *54* (10), 917–920.

Sharma, B. R.; Dhillon, T. S. Genetics of Resistance to Yellow Vein Mosaic Virus in Inter Specific Crosses of Okra. *Genetica Agraria* **1983**, *37* (3/4), 267–275.

Sheikh, M. A.; Safiuddin, Khan, Z.; Mahmood, I. Effect of Bhindi Yellow Vein Mosaic Virus on Yield Components of Okra Plants. *J. Plant Pathol.* **2013**, *95* (2), 391–393.

Shetty, A. A.; Singh, J. P.; Singh, D. Resistance to Yellow Vein Mosaic Virus in Okra: A Review. *Biological Agric. Hortic.* **2013**, *29* (3), 159–164.

Sindhumole, P.; Manju, P. Genetic Architecture of Resistance to Yellow Vein Mosaic and Leaf Spot Diseases in Okra (*Abelmoschus esculentus* (L.) Moench). *Electronic J. Plant Breed.* **2015**, *6* (1), 157–160.

Singh, A. K.; Singh, K. P. Screening for Disease Incidence of Yellow Vein Mosaic Virus (YVMV) in Okra [*Abelmoschus esculentus* (L.) Moench] Treated with Gamma Rays and EMS. *Vegetable Sci.* **2000,** *27* (1), 72–75.

Singh, B.; Rai, M.; Kalloo, G.; Satpathy, S.; Pandey, K. K. Wild Taxa of Okra (*Abelmoschus* Species): Reservoir of Genes for Resistance to Biotic Stresses. *Acta Horticulturae* 2007, **752,** 323–328.

Singh, H. B.; Joshi, B. S.; Khanna, P. P.; Gupta, P. S. Breeding for Field Resistance to Yellow Vein Mosaic in Bhindi. *Indian J. Genetics* **1962,** *22* (2), 137–144.

Srinivasan, K. Surendiran, G.; Maathivanan N. Pathological and Molecular Biological Investigations on Sunflower Necrosis Virus (SNV) and ISR mediated Biological Control of SNV by PGPR Strains. Asian Conference on Emerging Trends in Plant-microbe Interaction, Chennai, India, 8–10 December, 2005.

Stanley, J.; Bisaro, D. M.; Briddon, R. W.; Brown, J. K.; Fauquet, C. M.; Harrison, B. D. (Eds.). *Virustaxonomy*, VIIIth Report of the ICTV, Elsevier/Academic Press: London, **2005;** pp 301–326.

Terell, E. E, Winters, H. F. Change in Scientific Names for Certain Crop Plants. *Hort Sci.* **1974,** *9*, 324–325.

Tiwari, S. P, Nema, S.; Khare, M. N. Whitefly A Strong Transmitter of Plant Viruses. *ESci. J. Plant Pathol.* **2013,** *02* (2), 102–120.

Thakur, M. R. Inheritance of Resistance to Yellow Vein Mosaic (YVM) in a Cross of Okra Species, *Abelmoschus esculentus* X *A. manihot ssp. manihot. SABRAO J.* **1976,** *8* (1): 69–73.

Varma, P. M. Studies on the Relationship of Bhindi Yellow Vein Mosaic Virus and its Vector, the Whitefly (*Bemisia tabaci*). *Indian J. Agric. Sci.* **1952,** *22*, 75–91.

Vashisht, V. K, Sharma, B. R, Dhillon, G. S. Genetics of Resistance to Yellow Vein Mosaic Virus in Okra. *Crop Improv.* **2001,** *28*, 18–25.

Venkataravanappa, V. *Molecular Characterization of Okra Yellow Vein Mosaic Virus*. Ph.D. Thesis, GKVK, Bengaluru, 2008.

Venkataravanappa, V.; Reddy, L. C. N.; Reddy, M. K. Begomovirus Characterization and Development of Phenotypic and DNA-based Diagnostics for Screening of Okra Genotype Resistance against Okra Yellow Vein Mosaic Virus. *Biotech* **2013,** *3*, 461–470.

Venkataravanappa, V.; Reddy, L. C. N.; Jalali, S.; Reddy, M. K. Molecular Characterization of Distinct Bipartite Begomovirus Infecting Okra (*Abelmoschus esculentus* L.) in India. *Virus Genes* **2012.** DOI 10.1007/s11262-012-0732-y.

Venkataravanappa, V.; Reddy, L. C. N.; Jalali, S.; and Reddy, M. K. Molecular Characterization of A New Species of Begomovirus associated with Yellow Vein Mosaic of Okra (Okra) in Bhubaneswar, India. *Eur. J. Plant Pathol.* **2013**. DOI: 10.1007/s10658-013-0209-4.

Disease Dynamics and Management of Vegetable Pathosystems

ABHIJEET GHATAK*

Department of Plant Pathology, Bihar Agricultural University, Sabour 813210, India

*Corresponding author. E-mail: ghatak11@gmail.com

ABSTRACT

As a succulent texture, the development and progress of a disease in the vegetable pathosystem is therefore rapid and more dreaded than the other pathosystems. In addition, the micro-environment developed in a vegetable field promotes disease and then spreads throughout the area. To counter such condition, the management package should be developed with a crop-specific module. Sometimes physical treatments are subjected to the propagating materials to reduce active propagule on or inside the material. Vegetables are infected with soil-borne, air-borne, and seed-borne pathogens. The management of air-borne pathogens is often addressed by the foliar application of chemicals. However, the diseases originated from soil and seed are not supposed to be managed with harmful chemicals. The replacement of chemicals such as fungicides could be different biocontrol agents available on the market. The biocontrol agents also induce better nutrition consumption and thus make the plant resistant to infection. Moreover, the fresh-cut vegetable products could be effectively managed with the application of plant extracts and other biological agents in order to reduce the pesticide load on the surface of a vegetable often consumed within a week of harvesting. Although it is well clear that using resistant varieties is an inexpensive way to manage diseases in the vegetable system. But the resistance breaking is a common problem in crop cultivars. Therefore, this chapter aims at a holistic approach to the management of vegetable diseases.

10.1   INTRODUCTION

The occurrence of diseases in vegetables remains a common constraint. The succulence nature of vegetables makes them a host of easy to infect for a large number of pathogens. The most drastic effect of vegetable diseases is seen under moist and warm climatic areas. However, the changing climate, which delays winter and high temperatures in winter, helps many soil-borne pathogenic problems and the recurrence of bacterial issues. This condition is responsible for many emerging diseases; these diseases were previously poorly studied because the associated pathogens did not consider causing significant loss. At this time, to counter the elevated mean global temperature, many vegetable pathosystems need to be revisited. Therefore, the previous performance of the field in respect to disease occurrence and recurrence should be understood correctly before making any definite remark and to refer a management protocol. The symptoms are sometimes ambiguous; therefore, one should be very clear about the symptoms of vegetable diseases such as root rot, leaf spots and fruit rots, and blights – the most common symptoms of vegetables.

In all of the above discussions, the primary and central issues are (1) many vegetable pathogenic problems have emerged due to climate change, (2) certain problems have repeatedly appeared, (3) understanding of the vegetable pathosystem is necessary before making a decision on management, and (4) management of vegetable pathosystems is a highly skilful job. In addition, the documented management strategies for vegetable diseases are limited. Therefore, this chapter is centralized on the management issues of the vegetable pathosystem, which focuses on various management strategies with respect to the obstruction of vegetable cultivation due to diseases.

10.2   DISEASE DYNAMICS OF VEGETABLE CROPS

Over the world, vegetables are grown in distinct agro–ecological zones. The environment plays an important role in the population build-up of soil-borne pathogens for a locality. However, the overall disease is caused by the right combination of host, pathogen and environment (Vati and Ghatak, 2015). Logically, epidemics of the soil-borne disease are often found to be severe when vegetables are cultivated under an intensive system (Gilligan et al., 1996). The intensive cropping systems of vegetable cultivation develop micro-climate to be inductive to promote disease epidemics. In such an environment, some management practices, especially biocontrol technologies, do not work potentially. The main factor of this occurrence is the environmental heterogeneity supported with numerous soil parameters, such as temperature and moisture kinetics, type and other physico-chemical properties of soil. These factors ultimately influence the soil microbiota and thus the chances of disease development in the epidemics form are always possible (White and Gilligan, 1998).

In a season, the level and multiplication of inoculum depend on the kind of selected cropping practice, for example, mono-cropping, intercropping, mixed-cropping, etc. (Gilligan et al., 1996). To this connection, it is well understood that the current season inoculum level is depending on the previous year disease dynamics in the field and the environment supporting the development and multiplication of inoculum. This condition is strongly applicable to soil-borne diseases (Ghatak et al., 2015). Various approaches toward the Management of vegetable diseases have been employed and development and manipulation in management technologies are incurred from time to time. The different principles and methods for managing the vegetable pathosystem account different phases of the pathogenic cycle; the key is to understand the type of disease (monocyclic or polycyclic) and targeting the phase of the pathogen (primary inoculum or secondary inoculum).

Various workers have been working on the development of the models dealing with the best management strategies for a pathosystem. However, to deal with this, a huge quantum of data of disease progress and environmental variability is necessarily required. As the counterpart, countless regions of the globe are untouched with such studies and lacking the data that explains the disease scenario. Many known remote areas are needed to be explored to understand the disease dynamics, particularly under changing the environment, when a pathosystem may respond differently, what it responds two decades earlier (Vati and Ghatak, 2015).

10.3 MANAGEMENT OF VEGETABLE PATHOSYSTEM

10.3.1 APPLICATION OF PHYSICAL METHOD

Physical method is not followed for vegetable crops generally, due to its nature of being succulence. But, hot water treatment is very much effective to manage *Alternaria brassicicola* and *Leptosphaeria maculans*. In cabbage, the seed is treated with water at 53°C for 10 min and at 50°C for 25–30 min reduced *A. brassicicola* infections by 92–99% and *L. maculans* infections by 87–92%, respectively (Nega et al., 2003). Likewise, infection by *A. dauci* and *A. radicina* in carrot at 50–53°C for 10 min has been effective to manage over 95% disease in a series of blotter paper and greenhouse tests (Nega et al., 2003). Exposure of warm water (44–49°C) to carrot seeds has been found to reduce *A. dauci* incidence (Hermansen et al., 1999).

10.3.2   APPLICATION OF CHEMICAL METHOD

10.3.2.1   SOIL APPLICATION

Soil fumigant methyl bromide is being used for past many decades to manage a wide range of soil-borne pathogenic fungi and nematodes (Nene and Thapliyal, 2002). Also, to reduce insect occurrence this fumigant is widely used against grub, cutworms, etc. This group of fumigants is highly effective against the oomycetes. The resting spore is known for long-duration survival, oospore is highly sensitive to methyl bromide. This fumigant is therefore used in various solanaceous vegetable pathosystems to manage seedling blight by *Pythium* spp. and late blight by *Phytophthora infestans*. The most dreaded effect of methyl bromide is its role in the depletion of the ozone layer (Spreen et al., 1995). As a replacement, metam sodium and chloropicrin are recommended, although the property to inhibit oomycetous diseases in solanaceous crops (mainly for *Phytophthora capsici*) is less than methyl bromide (Hausbek and Lamour, 2004).

The other fungicides such as captan (75% WP), thiram (75% WP) and zineb (75% WP) are used for loose-textured and fairly dry soil (Nene and Thapliyal, 2002). Apart from drenching and mixing in soil, these are also effective for furrow application. These fungicides, used as a fumigant, are effective against seed rot, damping-off and seedling blight diseases in vegetable crops. Vapam (sodium N-methyl-dithiocarbamate) is a widely used soil fumigant effective to control nematodes, soil insects and weeds (Nene and Thapliyal, 2002).

10.3.2.2   SEED TREATMENT USING CHEMICALS

Fungicides were originated from sulfur, copper, and mercury compounds previously. However, reports on the fungicidal toxicity in mammals and other vertebrates appeared repeatedly. To this consequence, a few fungicides have been banned due to its harmful effect on warm-blooded animals.

Moreover, the mercury compounds are responsible for their accumulation in the environment. These events drew the attention of scientists and policymakers for saving the ecology without disturbing agro-ecosystem. In the recent decades, some molecules have been developed that are less toxic, and therefore, contributed largely in replacement of such inorganic compounds. Many systemic fungicides are developed after the discovery of first systemic fungicide as oxathiin by von Schmelling and Kulka (1966). The most important trait of the systemic fungicides is its low quantity requirement and degrading quality, which enables such fungicides for wide adaptability.

The commonly used fungicides for seed treatment in vegetables are generally of two types: broad spectrum and narrow spectrum. As per the mechanism, there are two types of fungicides used i.e. contact and systemic (Nene and Thapliyal, 2002). The contact fungicides are highly efficient to inhibit the germination and growth of those spores found on the surface of a seed coat; however, the systemic fungicides suppress the inoculum situated at a distal part of the plant from its applied part (i.e., seed). In another way, these fungicides are much effective to manage seed-borne pathogens internally. Systemic fungicides are also effective to the inoculum that commence onset at a very later stage compared to the seed treatment stage. There are many seed-transmissible pathogens observed in the vegetable pathosystem (Table 10.1). Fungi from order Pleosporales are mainly responsible for seed transmission in vegetables. The seed transmitted fungal pathogens can be managed effectively using the systemic fungicides. In another way, the use of a combination including contact and systemic fungicide does not support the chances in building the resistance in a pathogenic population (Ghatak et al., 2015). Some new fungicides like azoxystrobin can be used against wide host range soil-borne pathogens effectively (Ghatak et al., 2017). Managing chilli fruit rot and powdery mildew, and tomato blight disease, Saxena et al. (2016) used azoxystrobin-based fungicides Onestar and Amistar, and found a reduction in disease between 70% and 78% in chilli whereas disease reduction between 69% and 71% was recorded in tomato.

10.3.2.3 FOLIAR APPLICATION

Tomato late blight can be managed effectively with the application of Mancozeb with Cymoxanil and Mancozeb with Phenamidone in combination (Ghatak et al., 2015). Apart from the leaf blight, these fungicide combinations were found effective to inhibit fruit rot incidence, which was observed in a range between 8.0% and 9.3%. This indicated the most

TABLE 10.1 List of Vegetables With Association of Seed-Transmission Fungi.

Vegetable	Pathogen	Phylum	Class	Order	Family	References
Brassica spp.	Alternaria brassicicola	Deuteromycota	Dothideomycetes	Pleosporales	Pleosporaceae	Köhl et al., 2010; Knox-Davies, 1979
	Alternaria brassicae	Deuteromycota	Dothideomycetes	Pleosporales	Pleosporaceae	Ansar and Ghatak, 2018
	Leptosphaeria maculans	Deuteromycota	Dothideomycetes	Pleosporales	Leptosphaeriaceae	Fitt et al., 2006; West et al., 2001
Lycopersicon esculentum	Fusarium oxysporum f. sp. lycopersici	Deuteromycota	Sordariomycetes	Hypocreales	Nectriaceae	Babychan and Simon, 2017
	Alternaria solani	Deuteromycota	Dothideomycetes	Pleosporales	Pleosporaceae	Mahmoud et al., 2013; Al-Askar et al., 2014
Daucus carota	Alternaria dauci	Deuteromycota	Dothideomycetes	Pleosporales	Pleosporaceae	Farrar et al. 2004; Scott and Wenham, 1973
	Alternaria radicina	Deuteromycota	Dothideomycetes	Pleosporales	Pleosporaceae	Farrar et al. 2004; Scott and Wenham, 1973
Pisum sativum	Ascochyta pinodella	Deuteromycota	Dothideomycetes	Pleosporales	Didymellaceae	Bretag et al., 1995
	Ascochyta pinodes	Deuteromycota	Dothideomycetes	Pleosporales	Didymellaceae	Bretag et al., 1995
Vicia faba	Ascochyta pisi	Deuteromycota	Dothideomycetes	Pleosporales	Didymellaceae	Bretag et al., 1995
	Ascochyta fabae	Deuteromycota	Dothideomycetes	Pleosporales	Didymellaceae	Hewett, 1973; Geard, 1962

effective fungicide combinations compared to the other chemicals for management of both leaf blight as well as fruit rot of tomato caused by *Phytophthora infestans*. The fungicides are used as the combination has the additive feature; in the pathogen, chances in development of resistance are reduced when the fungicides are used in combination.

10.3.3 APPLICATION OF BIOCONTROL METHOD

10.3.3.1 SEED TREATMENT USING BIOAGENT

Seed treatment in vegetables can be furnished by biological method, chemical method and physical method. In practicality, the chemical method is commercially performed. However, in modern agriculture, observing the environmental hazard, the biological method is also accepted. Many companies are therefore formulating biopesticides at global scale using different biocontrol agents (Anwer et al., 2017). A strain K61 (*Streptomyces griseoviridis*) was isolated from *Sphagnum* peat and used as biopesticides for various root-borne diseases caused by *Rhizoctonia, Fusarium, Pythium* and *Phytophthora* spp. This strain has rapid colonizing ability to these pathogens in the rhizoshere (Minuto et al., 2006). In bacteria, many bacilli endorse the growth promotion and protection of the plant to the pathogen. *Bacillus subtilis, B. amyloliquefaciens, B. cereus, B. pumilus, B. pasteurii, B. mycoides* and *B. sphaericus* are some species that contribute to disease severity reduction (Choudhary and Johri, 2008; Kloepper et al., 2004).

10.3.3.2 PLANT GROWTH-PROMOTING RHIZOBACTERIA

The novelty for the performance of plant growth-promoting rhizobacteria (PGPR) is to protect the plant from its numerous pathogens like oomycetes, fungi, bacteria, and viruses (Kloepper et al., 2004) and thus makes the plant more vigorous by activating induced systemic resistance (ISR). Two types of PGPR-derived plant growths and developments are observed: indirect or direct. Certain antagonistic compounds that inhibit the growth of microorganisms with pathogenic nature are produced by these bacteria. These compounds are antibiotics, certain metabolites such as siderophores, and enzymes such as glucanase or chitinase. This mechanism of pathogenic growth inhibition is categorized under indirect way (Jimenez et al., 2001). In another way, these beneficial bacteria (PGPR) facilitating the plant to absorb a nutrient by converting the nutrient from unavailable to available form (Glick et al., 1999), which can be considered as a direct way to benefit the plant. This could be exemplified with biofertilizers, which can fix nitrogen from its unavailable form in the soil. Later, the plant may uptake this available nitrogen under such condition when the soil is a deficit of this nutrient; therefore, making the plant healthy and robust. Moreover, phytohormones (such as auxin) and volatile growth stimulants (such as ethylene and 2, 3-butanediol) are the phytostimulators that can directly promote the growth of plants. In PGPR, which is sometimes called biopesticides, the biocontrol aspect is most conspicuous. The main potentiality of PGPR is to protect plants from phytopathogenic organisms (Bloemberg and Lugtenberg, 2001; Vessey, 2003; Haas and Defago, 2005). The PGPR, which mainly belongs to *Pseudomonas* and *Bacillus* spp. are antagonists of recognized root pathogens (Haas and Defago, 2005).

It is a well-known fact that PGPR induces the systemic resistance and helps the plant to achieve its normal growth. ISR is the main promoting cause generated by the PGPR. PGPR-mediated ISR is highly effective against the oomycete pathogens such as *Pythium, Phytophthora, Peronospora* etc. The use of PGPR has been well demonstrated in many vegetable pathosystems. For example, tomato late blight was saved from infection of *Phytophthora infestans* by induction of systemic resistance by a PGPR strainSE34 incorporated into the potting medium (Yan et al., 2002). Ahmed et al. (2003) used four bacterial isolates including *Bacillus subtilis* HS93 and *B. licheniformis* LS234, LS523, and LS674 that reduced *Phytophthora capsici* causing root rot of pepper by 80%. Moreover, mixtures of two isolates of *Bacillus* (Jiang et al., 2006) and *B. subtilis* ME488 (Chung et al., 2008) showed potential for the suppression of *P. capsici* on pepper under field and greenhouse conditions, respectively. An endophytic bacterium *B. megaterium* (strain IISRBP 17), isolated from black pepper stem and roots, has been found to be effective against *P. capsici* infection on black pepper (Aravind et al., 2009). Zhang et al. (2010) demonstrated the reduction of Phytophthora blight in squash using the PGPR strains GB03, SE56, 1PC-11, and 1PN-19 each applied as a seed treatment. Similarly, another strain of PGPR (1PC-11) was found to be effective for this disease management. In a different system, the PGPR strains C-9 and SE34 and T4 were used for successful management of blue mold of tobacco, caused by *Peronospora tabacina* (Zhang et al., 2002). Interestingly, reduced due to the treatment effect of the strains, the sporulation of this oomycete was significant.

10.3.3.3   PLANT EXTRACT

Two special characters can be identified for plant extracts: (1) they are of natural origin and therefore safe to the environment and (2) they are considered at low risk in the development of resistance among the target pathogens. These two typical features of plant extracts are drawing the attention of the researchers for organic agriculture. Plant extracts have natural anti-microbial properties and can be used in plant disease management (Singh et al., 2017). plant extracts can be used as an alternative to chemical fungicides for seed disinfection. These decoctions are also used in combination with physical treatments in order to eliminate the infection. Tremendous achievements for the oil obtained from clove, peppermint, laurel etc. have been countered for inhibitory activity of *Ascochyta* spp. and *Alternaria* spp. under in-vitro condition. Pretorius et al. (2003) showed the antibacterial ability of crude extracts of *Acacia karroo*, *Elephantorrhiza elephantine*, *Euclea crispa*, *Acacia erioloba*, *Senna italica* and *Buddleja saligna* that inhibited plant and non-plant pathogenic bacteria.

Such extracts contain an essential oil that has fungicidal property. Some vegetable pathogens such as *Botrytis cinerea*, *Fusarium* sp. and *Clavibacter michiganensis* subsp. *michiganensis* was controlled at relatively low concentrations (85–300 µg/mL) using oregano, thyme, dictamnus and marjoram essential oils (Daferera et al., 2003). Secondary plant metabolites are well explored for their role in the plant defense mechanism to the infection of plant pathogens (Ansar et al., 2017). Tea is also known to inhibit the food-borne pathogens, particularly the bacterial pathogens (Siddiqui et al., 2016). Different kinds of tea have been reported to be effective for different bacteria. However, tea has also the antifungal property and is well known to inhibit the growth of various food-borne fungi.

10.3.4   AGE-RELATED RESISTANCE

Less susceptibility of a crop to its pathogens, such as oomycetes, fungi, bacteria and viruses, is a key feature of that particular crop variety with age-related resistance (Develey-Riviere and Galiana, 2007; Panter and Jones, 2002; Whallen, 2005). This is just the reverse condition – resistance that is reduced by an increased age as the organs attain maturity and senescence. The age-related resistance is operated with increased age of the crop; however, the knowledge gap for the mechanism of such resistance still exists. There may be physiological manipulation in the plant system that may act to trigger the activation of the resistance mechanism in the root and stem portions. Sometimes, such varieties may have required a limited or minimum strategic chemical application for better results.

In recent years, the infection severity of *Phytophthora capsici* has notably increased particularly in cucurbit vegetable crops and caused devastating yield loss (Hausbeck and Lamour, 2004). The extent of the loss was reported to be 100% damageable. Managing this pathosystem through chemical application and cultural practices is a tedious task and often challenging to manage the disease and to reduce the loss. Moreover, resistance to some common fungicide such as mefenoxam (metalaxyl) has been detected for *P. capsici* (Lamour and Hausbeck, 2000). The strategy of crop rotation with non-host crop has no value due to the high survivability of the pathogen in the soil (Hausbeck and Lamour, 2004). The oospore of this oomycete can survive for a 10-year period. Other approaches such as architectural modification, crop coverage and wider spacing usually facilitate improved coverage of fungicide – such approaches help in avoidance of disease onset (Ando and Grumet, 2006; Ngouajio et al., 2006; Wang and Ngouajio, 2008). However, these approaches would not be applicable to rendering a profitable margin under commercial set-up. Therefore, the most useful strategy, especially for the commercial growers, is to adopt for resistance variety.

For the oomycetous pathosystems, the age-related resistance would be a better option because virulence development in oomycetes often operates through sexual reproduction. Fruit age, fruit size and exocarp color had a significant effect on the severity of infection by *P. capsici* on butternut squash, pumpkin and acorn squash (Ando et al., 2009). For all of these examined fruits, the resistance has been found to increase with increasing fruit age and size. Gevens et al. (2006) have also observed the same finding for cucumber. Age-related resistance has also been recognized in pepper plants to infection by *P. capsici* (Kim et al., 1989). The mature fruits of pepper are surrounded by cuticle with greater thickness as compared to young fruits resulting in smaller lesion size, which could be an indicator of age-related resistance (Biles et al., 1993). Pepper stems increased their age with a decrease in macro-element content that makes the pepper more resistant to *P. capsici* (Jeun et al., 1991). For potyviruses, a single gene-mediated resistance regulated by the developmental factors in cucumber seedlings was observed (Ullah and Grumet, 2002).

10.4 CONCLUSION

Disease development in a vegetable pathosystem is one of the most common and visible events. This is a recurring phenomenon because the vegetables are mostly succulent because of their physical texture. Moreover, many vegetable crops are creepers or grown in a narrow space, providing repeated contact between the healthy and diseased plants, and providing a congenial micro-environment for disease development and its progress. Therefore, combating of disease in a vegetable system is a tedious task.

Management of a disease with resistance is the cheapest way without any additive harm to the environment. Many studies showed that drastically low disease occurred under high diseases epidemics when a comparison was made with susceptible or locally grown cultivars. Therefore, disease management with resistant cultivars provides very strong support while handling vegetable pathosystem. It is well known that the sources of genetic resistance in many of the vegetable crops are limited, and also, such resistance erodes rapidly.

An integrated approach should be employed for disease management in vegetables to mitigate this problem. The modern agriculture now believes in the other cultural activities, such as soil solarization, to counteract the hazardous effect of chemicals applied in the soil to inhibit the soil-borne pathogen inoculum. This practice is effective in both pre- and post-planting conditions. This practice has been found to be compatible with biological soil amendments, which is another option to reduce the chemical load in the soil. PGPRs are also used for vegetable disease management and for making nutrients available in the deficit fields. Considering the harmful effect of chemicals on post-harvest produce (e.g., fresh fruit and vegetables), it is now well accepted that the post-harvest damages should be managed with the application of eco-friendly alternatives. The plant extracts and other biocontrol agents are the most promising alternatives for management of fresh produce under post-harvest.

It is always better to adopt precautionary principles for managing a pathosystem. Preventing the inoculum entry in the field, coupled with sanitation practices, will strongly help the other alternatives used for disease management. Once the inoculum of a pathogen enters and is amplified in the field, it is very difficult to take-care of the whole system under control. To look after this, a comprehensive management approach should be implemented. The knowledge generation and its dissemination with respect to disease development and management may be performed with various techniques and tools.

For example, on-farm experiments and workshops should be given more priority, as the growers would get to know the various pathosystems and understand the possible management strategies that could be implemented in that region. On the other hand, the crop scientists should provide the current knowledge for updating the extension personnel for more vivid information and technology sharing among the farmers in order to have a "healthy crop."

KEY WORDS

- **biocontrol**
- **disease dynamics**
- **seed-borne disease**
- **seed treatment**
- **vegetable**

REFERENCE

Al-Askar, A. A.; Ghoneem, K. M.; Rashad, Y. M.; Abdulkhair, W. M.; Hafez, E. E.; Shabana, Y. M.; Baka, Z. A. Occurrence and Distribution of Tomato Seed-borne Mycoflora in Saudi Arabia and Its Correlation with the Climatic Variables. *Microb. Biotechnol.* **2014,** *7* (6), 556–569. DOI:10.1111/1751-7915.12137.

Ando, K.; Grumet, R. Evaluation of Altered Cucumber Plant Architecture as a Means to Reduce Phytophthoracapsici Disease Incidence on Cucumber Fruit. *J. Am. Soc. Hort. Sci.* **2006,** *131,* 491–498.

Ando, K.; Hammar, S.; Grume, R. Age-related Resistance of Diverse Cucurbit Fruitto Infection by Phytophthoracapsici. *J. Am. Soc. Hort. Sci.* **2009,** *134* (2), 176–182.

Ansar, M.; Ghatak, A. Occurrence of *Alternaria brassicae* in Seed-producing Cauliflower and Its Role of Seed Infection. *Int. J. Plant Sci.* **2018**; *13* (1), 67–70.

Ansar, M.; Ghatak, A.; Ghatak, L. V.; Srinivasaraghavan, A.; Balodi, R.; Raj, C. Secondary Metabolites in Pathogen-induced Plant Defense. In *Plant Secondary Metabolites: Their Roles in Stress Ecophysiology* (Vol. 3); Apple Academic Press: USA. ISBN: 9781315207506, 2017.

Anwer, M. A. *Biopesticides and Bioagents: Novel Tools for Pest Management*; Apple Academic Press: USA. ISBN: 9781771885195, 2017.

Babychan, M.; Simon, S. Efficacy of *Trichoderma* spp. Against *Fusarium oxysporum* f. sp. *lycopersici.* (FOL) Infecting Pre and Post-seedling of Tomato. *J. Pharm. Phytochem.* **2017,** *6* (4), 616–619.

Biles, C. L.; Wall, M. M.; Waugh, M.; Palmer, H. Relationship of Phytophthora Fruit Rot to Fruit Maturation and Cuticle Thickness of New Mexican-type Peppers. *Phytopathology* **1993,** *83,* 607–611.

Bretag, T. W.; Price, T. V.; Keane, P. J. Importance of Seed-borne Inoculum in the Etiology of the Ascochyta Blight Complex of Field Peas (*Pisum sativum* L.) grown in Victoria. *Aus. J. Experiment.l Agric.* **1995,** *35* (4), 525–530.

Choudhary, D. K.; Johri, B. N. Interactions of *Bacillus* spp. and Plants: with Special Reference to Induced Systemic Resistance (ISR). *Microbiol. Res.* **2008,** *164,* 493–513.

Daferera, D. J.; Ziogas, B. N.; Polissiou, M. G. The Effectiveness of Plant Essential Oils on the Growth of *Botrytis cinerea, Fusarium* sp. and *Clavibacter michiganensis* subsp. *michiganensis. Crop Protect.* **2003,** *22,* 39–44.

Develey-Riviere, M. P.; Galiana, E. Resistance to Pathogens and Host Developmental Stage: A Multifaceted Relationship within the Plant Kingdom. *New Phytol.* **2007,** *175,* 405–416.

Farrar, J. J.; Pryor, B. M.; Davis, R. M. Alternaria Diseases of Carrot. *Plant Dis.* **2004,** *88* (8), 776–784.

Fitt, B. D. L.; Brun, H.; Barbetti, M. J.; Rimmer, S. R. World-wide Importance of Phoma Stem Canker (*Leptosphaeria maculans* and L. biglobosa) on Oilseed Rape (Brassica napus). *Eur. J. Plant Pathol.* **2006,** *114,* 3–15. DOI: 10.1007/s10658-005-2233-5.

Geard, I. D. Studies on Ascochyta, Botrytis and 'Seed Spot' of Vicia faba in Tasmania. *J. Aus. Inst. Agric. Sci.* **1962,** *28,* 218–219.

Gevens, A. J.; Ando, K.; Lamour, K.; Grumet, R.; Hausbeck, M. K. Development of a Detached Cucumber Fruit Assay to Screen for Resistance and Effect of Fruit Age on Susceptibility to Infection by Phytophthoracapsici. *Plant Dis.* **2006,** *90,* 1276–1282.

Ghatak, A.; Ansar, M.; Ghatak, L. V.; Balodi, R. Elucidation of Relationship Between Phytophthora Leaf Blight and Fruit Rot in Tomato. J. Postharvest Technol. **2015,** *3* (2), 50–57.

Ghatak, A.; Kushwaha, C.; Gupta, R. N.; Singh, K. P.; Ansar, M. Variability in Sensitivity Among Different Host Origin-*Macrophomina phaseolina* Isolates to Azoxystrobin Fungicide. *Int. J. Plant Protect.* **2017,** *10* (1), 26–33, DOI: 10.15740/HAS/IJPP/10.1/26-33.

Ghatak, A.; Shukla, N.; Ansar, M.; Balodi, R.; Kumar, J. Effect of Sowing Time, Soil Temperature and Inoculum Density on Suppression of Fusarium Wilt in Lentil (*Lens culinaris*). *Int. J. Bio-res. Stress Manage.* **2015,** *6* (4), 268–273. DOI: 10.5958/0976-4038.2015.00037.8.

Gilligan, C. A.; Simons, S. A.; Hide, G. A. Inoculum Density and Spatial Pattern of *Rhizoctonia solani* in Field Plots of *Solanum tuberosum*: Effects of Cropping Frequency. *Plant Pathol.* **1996,** *45,* 232–244.

Hausbeck, M. K.; Lamour, K. H. Phytophthora Capsici on Vegetable Crops: Research Progress and Management Challenges. *Plant Dis.* **2004,** *88,* 1292–1303.

Hausbeck, M.; Lamour, K. Phytophthoracapsici on Vegetable Crops: Research Progress and Management Challenges. *Plant Dis.* **2004,** *88,* 1292–1302.

Hermansen, A.; Brodal, G.; Balvoll, G. Hot Water Treatments of Carrot Seeds: Effects on Seed-borne Fungi, Germination, Emergence and Yield. *Seed Sci. Technol.* **1999,** *27,* 599–613.

Hewett, P. D. The Field Behaviour of Seed-borne Ascochyta fabae and Disease Control in Field Beans. *Ann. Appl. Biol.* **1973,** *74,* 287–295.

Jeun, Y. C.; Hwang, B. K. Carbohydrate, Amino Acid, Phenolicand Mineral Nutrient Contents of Pepper Plants in Relation to Age-related Resistance to *Phytophthoracapsici. J. Phytopathol.* **1991,** *131,* 40–52.

Kim, Y. J.; Hwang, B. K.; Park, K. W. Expression of Age Related Resistance in Pepper Plants Infected with Phytophthoracapsici. *Plant Dis.* **1989,** *73,* 745–747.

Kloepper, J. W.; Ryu, C. -M.; Zhang, S. Induced Systemic Resistance and Promotion of Plant Growth by *Bacillus* spp. *Phytopathology* **2004**, *94*, 1259–1266.

Knox-Davies, P. S. Relationships Between Alternaria brassicicola and Brassica Seeds. *Trans. Brit. Mycologic. Soc.* **1979**, *73* (2), 235–248.

Köhl, J.; Van Tongeren, C. A. M.; Groenenboom-de Haas, B. H.; Van Hoof, R. A.; Driessen, R.; Van Der Heijden, L. Epidemiology of Dark Leaf Spot Caused by Alternaria brassicicola and A. brassicae in Organic Seed Production of Cauliflower. *Plant Pathol.* **2010**, *59*, 358–367. DOI: 10.1111/j.1365-3059.2009.02216.x.

Lamour, K. H.; Hausbeck, M. K. Mefenoxam Insensitivity and the Sexual Stage of Phytophthoracapsici in Michigan Cucurbit Fields. *Phytopathology* **2000**, *90*, 396–400.

Mahmoud, S. Y. M.; Hosseny, M. H.; EL-Shaikh, K. A. A.; Obiadalla, A. H. A.; Mohamed, Y. A. Seed Borne Fungal Pathogens Associated with Common Bean (Phaseolus vulgaris L.) Seeds and Their Impact on Germination. *J. Environ. Stud.* **2013**, *11*, 19–26.

Minuto, A.; Spadaro, D.; Garibaldi, A.; Gullino, M. L. Control of Soil Borne Pathogens of Tomato Using a Commercial Formulation of Streptomyces Griseoviridis and Solarization. *Crop Protect.* **2006**, *25*, 468–475.

Nega, E.; Ulrich, R.; Werner, S.; Jahn, M. Hot Water Treatment of Vegetable Seed: An Alternative Seed Treatment Method to Control Seed Borne Pathogens in Organic Farming. *J. Plant Dis. Protect.* **2003**, *110*, 220–234.

Nene, Y. L.; Thapliyal, P. N. Fungicides in Plant Disease Control. Oxford & IBH Publishing Co. Pvt. Ltd.: New Delhi, 2002. ISBN 81-204-0798-9.

Ngouajio, M.; Wang, G.; Hausbeck, M. K. Changes in Pickling Cucumber Yield and Economic Value in Response to Planting Density. *Crop Sci.* **2006**, *46*, 1570–1575.

Panter, S. N.; Jones, D. A. Age-related Resistance to Plantpathogens. *Adv. Bot. Res.* **2002**, *38*, 251–280.

Pretorius, J. C.; Magama, S.; Zietsman, P. C. Growth Inhibition of Plant Pathogenic Bacteria and Fungi by Extracts from Selected South African Plant Species. *S. Afr. J. Bot.* **2003**, *69* (2), 186–192.

Saxena, A.; Sarma, B. K.; Singh, H. B. *Effect of Azoxystrobin based Fungicides in Management of Chilli and Tomato Diseases.* In Proceedings of the National Academy of Sciences, India Section B: Biological Sciences, 2016, *86* (2): 283–289.

Scott, D. J.; Wenham, H. T. Occurrence of Two Seed-borne Pathogens, Alternaria Radicina and Alternaria dauci, on Imported Carrot Seed in New Zealand. *New Zeal. J. Agr. Res.* **1973**, *16* (2), 247–250.

Siddiqui, M. W.; Sharangi, A. B.; Singh, J. P.; Thakur, P. K.; Ayala-Zavala, J. F.; Singh, A.; Dhua, R. S. Antimicrobial Properties of Teas and Their Extracts In Vitro. *Crit. Rev. Food Sci. Nutr.* **2016**, *56*, 1428–1439.

Singh, R.; Prasad, K. K.; Siddiqui, M. W.; Prasad, K. Medicinal Plants in Preventive and Curative Role for Various Ailments. In *Plant Secondary Metabolites: Biological and Therapeutic Significance* (Vol. 1). Apple Academic Press: USA, 2017; ISBN: 9781315207506.

Spreen, T. H.; Van Sickle, J. J.; Moseley, A. E.; Deepak, M. S.; Mathers, L. Use of Methyl Bromide and the Economic Impact of Its Proposed Ban on the Florida Fresh Fruit and Vegetable Industry. *Univ. Florida Tech. Bull.* **1995**, 989, pp 1-201.

Ullah, Z.; Grumet, R. Localization of Zucchini Yellow Mosaic Virus to the Veinal Regions and Role of Viral Coat Protein in Veinal Chlorosis Conditioned by the *zym* potyvirus Resistance Locus in Cucumber. *Physiol. Mol. Plant Pathol.* **2002**, *60*, 79–89.

Vati, L.; Ghatak, A. Phytopathosystem Modification in Response to Climate Change. *In Climate Dynamics in Horticultural Science: Impact, Adaptation, and Mitigation.* Choudhary, M. L., Patel, V. B. Siddiqui, M. W., Verma, R. B, Eds.; 2015; pp 161–178. ISBN: 13,978-1-77188-070-1.

Von Schmeling, B.; Kulka, M. Systemic Fungicidal Activity of 1,4-Oxathiin Derivatives. *Science* **1966,** *152* (3722): 659–660. DOI: 10.1126/science.152.3722.659.

Wang, G.; Ngouajio, M. Integration of Cover Crop, Conservation Tillage, and Low Herbicide Rate for Machine-harvested Pickling Cucumbers. *Hort Sci.* **2008,** *43*, 1770–1774.

West, J. S.; Kharbanda, P. D.; Barbetti, M. J.; Fitt, B. D. L. Epidemiology and Management of Leptoshaeria Maculans (Phoma Stem Canker) on Oilseed Rape in Australia, Canada and Europe. *Plant Pathol.* **2001,** *50*, 10–27.

Whallen, M. C. Host Defense in a Developmental Context. *Mol. Plant Pathol.* **2005,** *6*, 347–360.

White, K. A. J.; Gilligan, C. A. Spatial Heterogeneity in Three-species, Plant-parasite-hyperparasite, Systems. *Philos. Trans. R. Soc. London B – Biol. Sci.* **1998,** *353*, 543–557.

Zhang, S.; White, T. L.; Martinez, M. C.; McInroy, J. A.; Kloepper, J. W.; Klassen, W. Evaluation of Plant Growth-promoting Rhizobacteria for Control of Phytophthora Blight on Squash under Greenhouse Conditions. *Biol. Control* **2010,** *53*, 129–135.

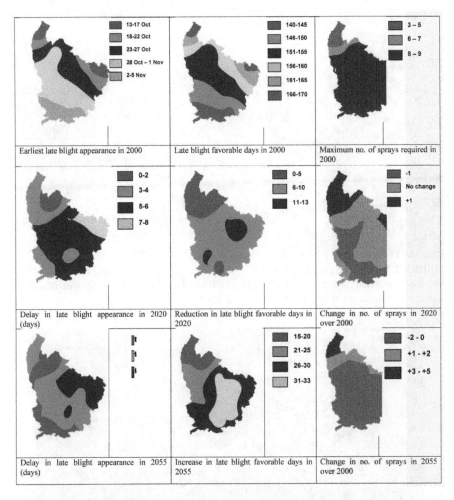

FIGURE 1.1 Distribution of late blight appearance, late blight favorable days, and delay in late blight appearance in western Uttar Pradesh in years 2000, 2020, and 2055, respectively.

FIGURE 3.1   Pointed gourd infected by vine and fruit rot disease caused by *Phytophthora melonis*. A worried farmer looks at his field which has the initial stages of the disease, which will soon overtake the whole field.

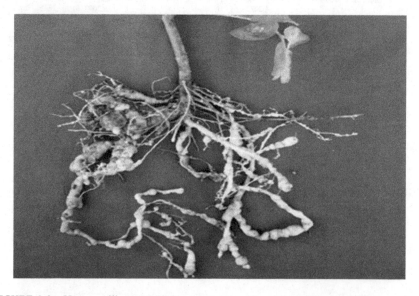

FIGURE 6.1   Heavy galling on tomato roots.

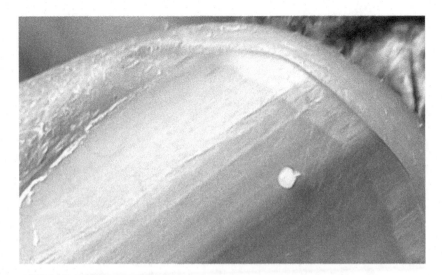

FIGURE 6.2   White mature female of root-knot nematode on thumbnail.

FIGURE 9.1   Severe symptom of YVMD of okra, covering >90% plant.

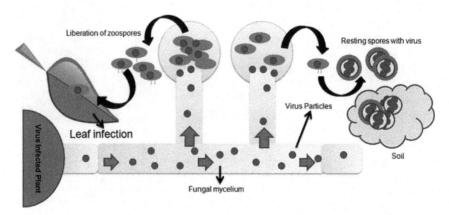

FIGURE 12.2   Schematic representation of infection process and survival in fungal transmitted virus.

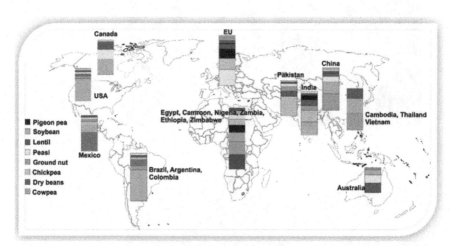

FIGURE 13.1   Major legumes and beans producing countries of the world. Production of eight important legumes and beans of country is shown in vertical bars and colors. Each bar represents different legume species and their percent production. The key for the colors associated with each legume species is given. The data is taken from FAOSTAT, USA, of year 2014 (http://faostat3.fao.org/browse/Q/QC/E).

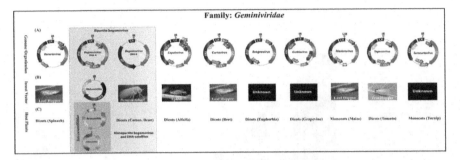

FIGURE 13.3 Genome organization of the family *Geminiviridae* **(A)**, the respective insect vectors **(B)**, and the type crop used to name each genus **(C)**. Legumoviruses are characterized by both bipartite and monopartite begomoviruses in the genus *Begomovirus* (Family *Geminiviridae*). The genus *begomovirus* (highlighted in pink background) has been named after the first begomovirus identified from common bean (*Phaseolous vulgaris*) as *Bean Golden Mosaic Virus* (Begomovirus). Some of the monopartite begomovirsues are also found associated with DNA-satellites called alphasatellite, betasatellite, and deltasatellite (Genus *Betasatellite* and *Deltasatellite*, Family *Tolecusatellitidae*), respectively. The genera *betasatellite* and *deltasatellite* are highlighted in green background.

FIGURE 14.1 Cucurbit plants exhibiting putative viral symptoms of yellowing, mosaic and leaf curling. (A) Bitter gourd leaf distortion disease on bitter gourd (*Momordica charantia*), (B) yellow mosaic disease on cucumber (*Cucumis sativus*), and (C) yellow mosaic disease on pumpkin (*Cucurbita moschata*).

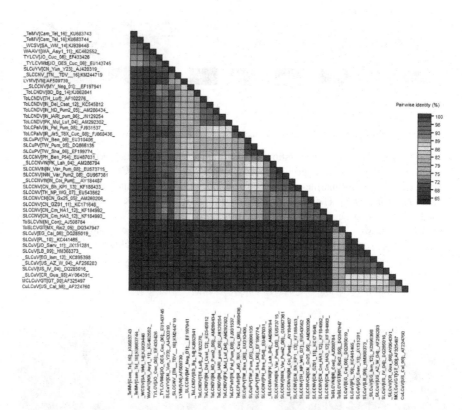

FIGURE 14.3 Color coding representation of percentage pairwise genome scores and evolutionary identity plot of full genomes of DNA A of begomoviruses infecting cucurbits. The plot was prepared using MUSCLE parameter of SDTv1.2 (Species Demarcation Tool; http://web.cbio.uct.ac.za/SDT) (Muhire et al., 2014) in two-color mode with cut-off 1, which is 94 and cut-off 2, which is is 78. The full length begomovirus sequences were downloaded from GenBank, NCBI, and tailored using VecScreen. Accession numbers of sequences are on the right of each virus name. Virus abbreviations are as recommended by Fauquet et al. 2008, Brown et al. 2015, and Zerbini et al. 2017.

FIGURE 16.1   Root rot caused by *Rhizoctonia solani* in okra.
Courtesy: Aravind T., Anand Agricultural University, Anand.

(a) Soft rot in onion   (b) Sprouting in onion

(c) Black mold in garlic   (d) Mite in garlic

FIGURE 18.1   Important Postharvest Diseases and Pests of Onion and Garlic.

CHAPTER 11

Mechanism of Microbial Infection in Vegetables Diseases

VINOD UPADHYAY[1*] and ELANGBAM PREMABATI DEVI[2]

[1]*Department of Plant Pathology, Regional Agricultural Research Station, Assam Agricultural University, Gosaaigaon 783360, Assam, India*

[2]*Department of Plant Pathology, Wheat Research Station, Sardarkrushinagar Dantiwada Agricultural University (SDAU), Vijapur 382870, Gujarat, India*

Corresponding author. E-mail: vinodupadhyay148@gmail.com

ABSTRACT

Vegetables are essential component in nutritional security and economic feasibility. Plant diseases constitute the most important problem in the production of vegetable crops as well as in its storage and processing. Farmers face serious economical yield losses both in terms of quality and quantity of these valuable vegetables crops due to several diseases which are caused by different plant pathogens such as fungi, bacteria, virus, viroid, phanerogamic parasitic plants, and nematodes. In order to overcome this pathogen attack, an appropriate management strategy should be planned against each class of pathogens. To develop proficient management ways against each particular plant pathogen, it is essential to understand their mode of mechanism of infection process in host plants. Infection is the process where pathogens derive nutrients from susceptible host plants by establishing contact with susceptible cells of the host. But, for establishing successful infection inside the host plants by plant pathogens, they must be able to make their own way into the plant systems for deriving nutrients from the host plant and also they must be able to neutralize the defense reactions of the host plant. In this chapter, an effort for simple understanding of various mechanism of infection process in each class of pathogens was made with special reference to only vegetable crops.

11.1   INTRODUCTION

Vegetables are important component in nutritional security, economic viability, and fits well in intensive cropping systems existing in different parts of India. In India, more than 60 different types of vegetables are grown in tropical, subtropical, and temperate agroclimates. During 2013–2014, production of vegetables crops in India were 162.19 million tons and procures gross worth of INR 54,629.3 million from export of vegetables. But this level of production is still insufficient for second most populous country in the world where India shares 17.5% of global population. Geographical area of India is about 2.4% of the total surface area of the world but we share only about 1% in global market of horticultural products. Plant diseases constitute the most important problem in the production of vegetables crops in the field condition itself and along with in different stages of storage and processing of horticultural crops. Farmers face serious yield losses both in quality and quantity of these valuable vegetables crops due to several diseases which are caused by different plant pathogens namely fungi, bacteria, virus, viroid, phanerogamic parasitic plants, and plant parasitic nematodes.

Infection is the process by which pathogens derive nutrients from host by establishing contact with susceptible areas of host plants. For successful establishment of infection process, a pathogen must be able to make their own way into the plant system to acquire nutrients from the host plants and must be able to neutralize the defense reactions of the host plants. For successful establishment, pathogens carry out these above-mentioned activities mainly by secretions of chemical substances like enzymes, toxin, and pathogen produced growth regulators which can affect certain host components or metabolic activities of host plants and affects structural integrity of the host plants which ultimately leads to infection. During infection, some pathogens derive nutrition from the living cells without death of the affected cells, whereas others kill cells and infect them after utilizing their contents, and still some others pathogen can destroy cells and disorganize surrounding tissues. By progressing infection process, pathogens grow, multiply, or both within the affected plant tissues and later invade then colonize the susceptible host plant into a slighter or greater amount. The ultimate result of successful infections is the symptoms development, that is, discolored, malformed, tumor, gall, leaf spot, blight, wilt, necrotic, or chlorosis areas, etc., on the host plant depending on the types of pathogens invade on them. However, some show latent infections where they do not produce acute symptoms but infection appears later on when the environmental conditions or plant get mature which turn into more favorable for disease. In most of the plant diseases, however, development of symptoms took few days to a few weeks after inoculation. De Bary (1886) while working with the Sclerotinia rot disease of carrots and other vegetables noted that host cells were killed in advance by the invading fungus hyphae and juice from rotted tissue can break down healthy tissue of host, whereas boiled juice from rotted tissue had no effect on healthy tissue. Thus, De Bary concluded that the enzymes and toxins produced by pathogen can degrade and kill plant cells for obtaining nutrients. In 1905, L. R. Jones reported association of cytolytic enzymes with several diseases of vegetables mainly in soft rot caused by bacteria. In 1915, it was reported that the pectic enzymes secreted by fungi has a key role in disease development on host plants; however, it was not until the 1940s that role of cellulases were emphasized in development of plant diseases (Agrios, 2005).

The toxins exhibit numerous distinct mechanisms of infection process, each one can affect specific sites on mitochondria, chloroplasts, plasma membranes, specific enzymes, or specific cells-like guard cells. Besides, several detailed biochemical studies were reported to portray the mechanisms of toxins which affect or kill plant cells or by which cells of resistant plants avoid or inactivate them. Initial observations in many diseases showed stunting symptoms in the affected plants, while in others they showed excessive growth, tumors, and other growth abnormalities which led many researchers to guess imbalances of levels of growth regulators in diseased plants. By the late 1950s, numerous fungi and bacteria which are pathogenic to plant were reported with the production of plant hormone indoleacetic acid (IAA) and their association with development of disease. Likewise, several studies were carried out on the different types of host–cell enzymes activated on infection, the types and amounts of metabolites accumulation and mostly, production of various kinds and amounts of phenolic compounds and phenol-oxidizing enzymes during infection. The infection process of both necrotrophs and biotrophs pathogen vary in a great extent. The necrotrophic microbial pathogens with aggressive pathogenesis strategy cause several extensive symptoms of necrosis, tissue maceration, and rotting symptoms on host plant. However, before or during colonization, some invading pathogens secrete some phytotoxins and cell wall degrading enzymes (CWDEs) into host tissues. These toxic molecules and lytic enzymes has the ability to kill the affected host cells while the process of infection and later on utilized decomposed plant tissues as their nutrients by pathogens. In contrast, biotrophic pathogens, in general, do not produce toxins and secrete very low amount of CWDEs (Oliver and Ipcho, 2004).

11.2 NATURE OF INFECTION PROCESS BY FUNGAL PATHOGENS

Infection by fungal pathogens into host plant follows a chain of sequential events, viz., contact, penetration, conidial attachment, germination, development of primary lesion, colonization, and tissue maceration afterwards sporulation. Penetration process are involved either by active or passive mechanisms. The active mechanisms comprises with formation of appressoria and enzymatic degradation. But, passive mechanisms occur through an existing wound or infection sites and sometime by stomata. Lytic enzymes employed for initial penetration in addition to toxic levels of reactive oxygen species are also involved in dismantling of cellular organization to decompose the host tissues. For ensuring the disease progression on host plant, invading fungi actively manipulate normal activity of host by suppressing defense machinery which led to successful infection (Prins et al., 2000).

After successful attachment to the plant surface under the favorable microenvironment, fungal spore start germination followed by germ tube formation and ultimately penetrates into host cell by several ways. Some pathogenic fungi form specialized structures for completion of infection which can directly breach cuticular region of plant and cell walls area mainly in epidermal cells, while others utilizes generation of cell wall degrading enzymes for digesting components of plant cell wall or combine both strategies. Appressorium put forth enormous turgor pressure and physical force on the phylloplane for enabling the fungus to breach the region of cuticular site and portion of cell wall of host plant (Choi et al., 2011; Talbot et al., 2003). In such cases, CWDE generally plays only an additional role in the penetration (Linsel et al., 2011). Branching of the single hyphae into clusters of bulbous ends results in generation of complex appressoria, so-called infection cushions, which penetrate the plant surface via penetration peg, for example, by *Sclerotinia sclerotiorum* (Erental et al., 2007) and *Rhizoctonia solani* (Pannecoucque et al., 2009). Penetration of *Botrytis cinerea* by appressoria that do not have thick melanin layer is completed by aid of hydrolytic enzymes that digest the host cell wall without relying on physical force (Choquer et al., 2007).

The adhesion process in reference to active process was reported to follow reversible attachment, also needed preformed conidial and glycoproteins released by its surface (Newey et al., 2007). In addition to the sensing of hard surfaces required for formation of appressoria, several fungi shows by acquiring a specific fibrillar and porous layer which are rich in carbohydrates and such a specialized spore coat of *Colletotrichum lindemuthianum*, the hemibiotrophic pathogen of bean, has been demonstrated to be a prerequisite for spore adhesion to hydrophobic layers (Rawlings et al., 2007).

11.2.1   ROLE OF PATHOGEN SECRETED ENZYMES IN PATHOGENESIS

Enzymes reported to be secreted by several pathogens can disintegrate the structural components of affected host cells leading to breakdown of inert food substances or affect membrane components and directly affect the protoplast, thus affecting in its system of functional process in host plants. Many pathogenic fungi depend mainly on production of cell wall degrading enzymes for their successful penetration into host plant. CWDE comprises xylanases which can degrade hemicelulose, exopolygalacturonases along with pectin methylesterases degrades pectin polymers, endoglucanases digest mainly on cellulose, and polysaccharide deacetylases which have the ability to cleave acetyl substituents in components of polysaccharide (Carapito et al., 2008). CWDE plays a significant role in the infection process of both fungal and bacterial pathogens that can break the cell wall or cause maceration of the host tissue (Keon et al., 1987; Cooper, 1984; Mishagi, 1982). The pathogens penetrations into parenchymatous tissues are being facilitated by the breakdown process of internal cell walls, which involve of pectins, cellulose, hemicelluloses, and structural proteins and for middle lamella mainly by pectins. Moreover, the result of complete disintegration of plant tissue is the breakdown of lignin (Agrios, 2005). A variety of CWDEs were noticed in the first 24 h of bean rust, *Uromyces fabae* development: cellulases (Heiler et al., 1993), proteases (Rauscher et al., 1995), pectin methylesterases (Frittrang et al., 1992), and polygacturonate lyases (Deising et al., 1995).

There are several enzymes used as weapon by various pathogenic fungi for primary infection process, some are briefly discussed below:

(a) Cutinases: Cutin comprises main component of the cuticle and waxes are admixed with the upper part of the cuticle, while its lower part merges into the outer walls of epidermal cells which is admixed with pectin and cellulose. Cutin molecules break down by cutinases and discharge monomers and oligomers component of fatty acid derivatives from the unsolvable cutin polymer, for example, *Fusarium* spp. and *Botrytis cinerea*. Cutinase being an extracellular enzyme of fungal pathogens is under consideration to be vital for the pathogenicity of various fungal species by facilitating adhesion of fungal spores to the surface of host plant (Schafer, 1993). Numerous fungi, including saprophytes, produce enzymes that attack one or more structural components of the host cell. For example, before infection, the spores of rust fungus of bean, that is, *Uromyces viciae-fabae* form structural adhesion pads on leaves where high amount of cutinase and serine esterases were found to be enclosed on its surface of these spores, and ultimately contributing the adhesion of the fungal spores to the cuticle of plant. In addition, different two cutinase isozymes and two esterases were reported on leaves of cabbage expressed during saprophytic and parasitic stages of *Alternaria brassicicola* by Koeller et al. (1995).

(b) Pectinases: The main primary components of the middle lamella are composed of pectin substances. Pectinases or pectolylic enzymes are those several enzymes which can degrade pectic substances. The first pectic enzyme is pectin methyl esterases, which can eliminate small branches of the pectin chains. The second pectic enzymes are polygalacturonases which is a chain of splitting pectinases that help in splitting the pectic chain by adding a water molecule and later on breaking the connection in between two molecules of galacturonan units. Pectin lyases can split the chain by eliminating water molecule from the linkage, followed by breaking it and liberate an unsaturated double bond as products as reported in pathogens like *Didymella bryoniae, Ralstonia solanacearum,* etc. Pectinases, particularly endopectin lyase, are virulence factors but possibly not a determinant for pathogenicity because colonization can proceed in their absence as reported in vascular wilt caused by *Verticillium albo-atrum* infection in tomato plant (Durrands and Cooper, 1998). Pectinases are determined by gene families whose size appears to differ with the type of infection. For example, *Botrytis cinerea* and *Sclerotinia sclerotiorum* produced a large set of endopolygalacturonase (endoPG) isoforms (Fraissinet-Tachet et al., 1995; Wubben et al., 1999), whereas biotrophic and hemibiotrophic pathogens like

Colletotrichum lindemuthianum have two different endoPG genes (Centis et al., 1996; Tenberge et al., 1996). An expression of a new pectate lyase gene in a pathogenic fungus, *Colletotrichum magna* in cucurbits, is sufficient to support pathogenicity (Yakoby et al., 2000).

(c) Cellulases: Cellulose is a polysaccharide comprising chains of glucose (1–4) β-D-glucan molecules and along with a large quantity of hydrogen bonds which help in holding the glucose chains. Glucose is produced by a series of enzymatic reactions, bringing out by a number of enzymes like cellulase. One cellulase (C1) attacks native cellulose by cleaving cross-linkages among chains. A second cellulose (C2) also attacks native cellulose and breaks it into shorter chains later, these are then attacked by a third group of cellulases (Cx) that degrade them to the disaccharide cellobiose. At last, with the help of enzyme β-glucosidae cellobiose is degraded into glucose. Saprophytes fungi, mainly certain groups of basidiomycetes, and saprophytic bacteria cause the breakdown of the majority of cellulose decomposed in nature. In living plant tissues, however, secretion of cellulolytic enzymes by pathogens has a very important role in the softening and breakdown of cell wall material. In *Ralstonia solanacearum*, perthotrophic pathogens relationship between production of polysaccharide hydrolases (cellulase, xylanase) and pathogenicity were reported (Kelman and Kowling, 1965; Azad, 1992).

11.2.2 ROLE OF FUNGAL TOXIN IN PATHOGENESIS

Toxins attack directly on components of host protoplast and generally interfere with the cell membrane permeability and its other functions. Toxins and phytotoxic proteins produced by necrotrophs are crucial for their virulence strategies. Toxins are metabolites which are produced by invading plant parasitic pathogens and attack directly on host protoplast, and eventually kill the affected cells of host plant. A few toxins serve as a general protoplasmic poison and influence many plant species of different families but others are toxic to only some species or varieties of plant (Table 11.1). Some non-host specific and host-specific toxins produce by parasitic plant pathogens are briefly discussed:

Non-host specific toxin or non-host selective toxin: A few phytopathogenic microorganisms produce numerous toxic substances which have been shown to produce all or part of the disease syndrome not only on host plant but also on other species of plants which are not generally attacked by the pathogen in nature. Example is Tentoxin produced by *Alternaria alternata,* which causes spots and chlorosis in many species of plants. Similarly, alternaric acid, zinniol, and alternariol are produced by *Alternaria* spp. in many diseases of mainly leaf spot in a varieties of plants; fusaric acid and lycomarasmin are being produced by *Fusarium oxysporum* in association with tomato wilt disease, etc. (Agrios, 2005).

Host-specific or host-selective toxins: Toxic substance produced by a pathogenic microorganism that shows little or no toxicity against non-susceptible plants and whose physiological concentrations influence only to the hosts of that pathogen. In the stem canker of tomato pathotype– tomato interactions, the main pathogenicity factor is the production of AAL-toxin, which induces death of affected cells but only in susceptible cultivars (Akamatsu et al., 1997; Brandwagt et al., 2000; Yamagishi, et al., 2006). It induces an apoptotic-like response in toxin sensitive tomato, as demonstrated by TUNEL positive cells, DNA laddering, and the formation of apoptotic-like bodies (Wang et al., 1996; Gilchrist, 1998). Other host-selective toxin like *Corynespora cassiicola* produces the CC toxin in tomato; destruxin B from *A. brassicae* on brassicas are vital factor for pathogenicity (Agrios, 2005).

TABLE 11.1   List of Microbial Toxins Involved in Plant Disease Development.

S.N.	Toxins	Producer organism	Host/Disease
Host-selective toxins			
1.	Victorin or HV-Toxins	*Cochliobolus victoriae*	Victoria blight of oats
2	HC-toxin race 1	*Cochliobolus carbonum*	Leaf spot of corn
3	T-toxin	*Cochliobolus heterosporus*	Southern corn leaf blight
4	HS-toxin	*Cochliobolus sacchari*	Eye spot (leaf spot) disease of sugarcane
5	AK-toxin	*Alternaria alternata*	Japanese bear
Non-host-selective toxins			
1	Tabtoxin	*Pseudomonas syringae pv.tabaci*	Wildfire disease of tobacco

TABLE 11.1   *(Continued)*

S.N.	Toxins	Producer organism	Host/Disease
2	Phaseolotoxin	*P. syringae pv. phaseolicola*	Blight of beans
3	Tentoxin	*Alternaria alternata*	Seedling chlorosis of several plants
4	Tagetoxin	*P. syringae pv. syringae*	-
5	Fumaric acid	*Rhizopus spp.*	-
6	Oxalic acid	*Sclerotinia and Sclerotium*	-
7	Fusaric acid	*Fusarium oxysporum*	Tomato, cotton wilt
8	Pyricularin	*Magnaporthe grisea*	Rice blast disease

11.2.3   ROLE OF PATHOGEN PRODUCED GROWTH REGULATORS IN PATHOGENESIS

Plant pathogens may produce the same plant growth regulators or similar inhibitors of growth regulators produced by the plant. Pathogens often cause interference in the plant's hormonal system, leading to development of abnormal responses in growth of the plant. A variety of abnormal growth responses in plant like stunting, rosetting, stem malformation, overgrowth, excessive root branching, leaf epinasty, defoliation, and suppression of bud growth are accompanied due to imbalance of hormonal system.

Growth regulators exert a hormonal effect on the cells and their increase or decrease influences their ability to divide and enlarge. Small number of groups of naturally occurring compounds that regulates plant growth and act as hormones are generally called growth regulators. The most important growth regulators are auxins, gibberellins, and cytokinins, but other compounds, such as ethylene and growth inhibitors, play significant regulatory roles in the life of the plant. Pathogens often cause an imbalance in the hormonal system of the plant and bring incompatible growth responses with the healthy development of the plant (Table 11.2). Polysaccharides play a crucial role mainly in vascular wilt diseases, where they affect passively with the translocation of water in the plants. A few pathogen produced growth regulators which play vital role in pathogenesis are discussed below:

(a) Auxins: Indole-3-acetic acid (IAA) is the natural form of auxins found in plant and related with cell elongation, differentiation, and sometime absorption of IAA to the cell membrane influences the membrane permeability. In many fungal, bacterial, viral, and nematodes infected plants, increased levels of IAA was noticed while some pathogenic pathogens may have lower amount of auxin level. *Exobasidium azalea* causing azalea and gall of flower, *Ustilago maydis* causes smut in maize, *Plasmodiophora brassicae* causes clubroot in cabbage and leafy gall of sweet pea are associated with the increased levels of IAA as wholly or partly and are described due to the decreased degradation of IAA by inhibition of enzyme, that is, IAA oxidase. *Ralstonia solanacearum* causing wilt disease of solanaceous plants mainly in bacterial wilt also induces many fold increase in level of IAA in diseased plants as compared with healthy plants by increasing cell walls plasticity due to high levels of IAA, providing the cellulose, pectin, and cell wall's proteins components more accessible to their degradation by enzymes secreted by the pathogen inhibition in lignifications of tissues by increasing levels of IAA seems to extend the exposure time of nonlignified tissues to the pathogen's cell wall degrading enzymes (Agrios, 2005).

(b) Gibberellins: These are common constituents which are mainly responsible for promoting growth of green plants. It promotes stem and root elongation of dwarf varieties to normal sizes and also helps in promote flowering and fruits growth. The bakanae disease of rice seedlings infected by pathogen *Gibberella fujikuroi* grow rapidly and become much taller than healthy plants apparently as a result of gibberellins secreted by the fungus to a considerable level is one of the important examples in unfolding role of growth regulator in disease development (Agrios, 2005).

(c) Cytokinins: There are very effective growth factors required primarily for cell growth and differentiation, and also reduce the breakdown of proteins and nucleic acids, thus causing the delay of senescence, as well as ability to direct the flow of amino acids and other nutrients from high cytokinin concentration. Its activity increases in clubroot galls, in crown galls, in smut and rust infected bean leaves and galls. It is main accountable of bacterial leafy gall disease of sweet pea caused by bacterium *Rhodococus fasciens*. Later, cytokinin activity seems to be associated to both the juvenile feature of the green islands around the infection center and the senescence outside the green island. Cytokinin plays a significant role in club root disease which is caused by pathogen, *Plasmodiophora brassicae,* with differential expression of more than 1000 genes in contrast to control plants by global gene expression analysis (Siemens et al., 2006).

(d) Ethylene: Plants naturally produced ethylene and exerts a range of effects on plants like leaf abscission, epinasty, chlorosis, stimulation of adventitious roots, and fruit ripening. In banana, the fruit infected with *Ralstonia solanacearum* showed increase in the ethylene content proportionately with the (premature) yellowing of the fruits, but no ethylene was detected in fruits which are healthy. In case of wilt of tomato caused by *Verticillium*, the presence of ethylene during infection leads to inhibit development of disease while after infection the presence of ethylene increases wilt development. Exopolysaccharides appear to be essential for several pathogens to cause normal disease symptoms either by being directly accountable for inducing symptoms or by indirectly helping pathogenesis by promoting colonization or by enhancing survival of the pathogen. The role of slimy polysaccharides in plant disease appears to be predominantly significant in vascular wilt diseases. In vascular wilts, pathogen released large polysaccharide molecules in the xylem which may be enough to cause a mechanical obstruction of vascular bundles and thus initiate wilting (Agrios, 2005).

TABLE 11.2 List of Plant Diseases Showing Abnormalities in Growth Due to Alteration in Plant Hormones.

S.N.	Plant disease	Pathogen	Growth hormones
1.	Azalea and flower gall	Exobasidium azalea	Increased level of Auxin
2.	Corn smut galls	Ustilago maydis	Increased level of Auxin
3.	Club root galls of cabbage	Plasmodiophora brassicae	Increased level of Auxin
4.	Bacterial wilt of solanaceous vegatables	Ralstonia solanacearum	Increased level of Auxin
5.	Bakanae or Foolish disease of rice	Gibberella fujikuroi	Increased level of Gibberellins
6.	Leafy gall of sweet pea	Rhodococus fasciens	Increased level of Cytokinin
7.	Cotton wilt	Verticillium dahlie	Increased level of Ethylene

11.3 NATURE OF INFECTION BY BACTERIAL PATHOGENS

In general, bacterial diseases of plants are well characterized by their morphological symptoms such as leaf and fruit spots, blights, cankers, vascular wilts, rots, tumors, and galls, etc. Phytopathogenic bacteria cause diseases in host plants after penetration into host tissues (Buonaurio, 2008). Microbial pathogenicity has often been defined as the biochemical mechanisms whereby pathogenic microorganisms cause disease in a host organism (Fuchs, 1998). Pathogenicity or virulence of Gram-negative plant pathogenic bacteria are strictly dependent on the presence of secretion apparatus in host cells, by which they secrete proteins or nucleoproteins involved in their virulence inside the apoplast or introduce these substances into host cells (Buonaurio, 2008). Bacterial pathogens contain numerous classes of genes called virulence genes which are necessary for disease development or for escalating virulence in one or more hosts. Pathogenicity factors that are encoded by pathogenicity genes (*pat*) and disease-specific genes (*dsp*) are vitally involved in diseases establishment. A few of these genes are necessary for recognition of a host by a pathogen, attachment of a pathogen to a plant's surface, development of infection structures on or within the host, penetration of the host, and colonization of host tissue. The pathogenicity genes that are involved in the synthesis and modification of the lipopolysaccharide cell wall of gram-negative bacteria may assist condition the host range of a bacterium. Virulence factors of plant pathogenic bacteria are associated with the bacterial surface or secreted into the surrounding environment. Proteins secreted by bacteria are transported through molecular systems out of bacterial cells; unrelated virulence factors usually share the similar secretion mechanism (Fuchs, 1998). Bacterial pathogenicity depends upon bacterial secretion systems (types I-IV), quorum sensing (QS), cell wall degrading enzymes of plants, toxins, hormones, polysaccharides, proteinases, siderophores, and melanin, etc. All of these systems and substances, which are required for pathogenic infection and virulence, are being produced by pathogens during bacterial pathogen–plant interactions (Agrios, 2005). They are all regulated by proteins through systems of signal transduction. A case study of the infection process of *Rhizobium radiobacter* (*Agrobacterium tumefaciens*), which caused crown gall in many fruit and forest trees, vegetables, and herbaceous dicotyledonous plants from 90 different families, showed capability of *R. radiobacter* to genetically transform healthy host cells into tumoral cells by inserting T-DNA from its tumor-inducing (Ti) plasmid into the plant genome. Binding of *R. radiobacter* to plant surfaces, a crucial stage for establishing long-term interaction with host plants is reliant upon plant factors such as bacterial polysaccharides and bacterial cellulose synthesis. In addition, infected host plants release signals that comprise definite phytochemicals in combination with acidic pH and temperatures. The decrease in the photosynthetic performance of inoculated plants was most likely resulted with decrease in the consumption of electrons released by the oxidation of water during the photochemical process (Chaves et al., 2003). A surplus of excitation energy might prefer the formation and accumulation of ROS and the generation of oxidative stress. Alteration in cell metabolism which arises from *Xanthomonas gardneri* infection in tomato might have elicited with an increase in electron flow for Mehler's reaction and benefitted the photorespiratory pathway, eventually leading to an imbalanced production of superoxide (Noctor et al., 2002; Silveira et al., 2015).

Phytopathogenic bacteria cause diseases in plants after penetrating into host tissues and process of penetration occurs by natural openings like stomata, hydathodes, lenticels, trichomes, nectarthodes, stigma, etc., or by wounds. *Pseudomonas syringae* pv. *lachrymans* on cucumber, *P. cichorii* in lettuce, *P. syringae* pv. *syringae* in bean, *P. syringae* pv. *lycopersicum* in tomato enters through stomata. *Xanthomonas campestris* pv. *campestris* in crucifers by hydathodes, *Streptomyces scabies, Erwinia carotovora* var. *atroseptica* in potato by lenticels, *Clavibacter michiganense* in tomato by trichomes, *P. syringae* pv. *glycinea* in soybean by wounds, etc. Trichomes also serve as sites of penetration for *C. michiganense, P. syringae* pv. *lyco-persici,* and *P. cichorii* (Shaykh et al., 1977; Huang, 1986). SEM of tomato leaves inoculated with *P. syringae* pv. *lycopersici* revealed that stomata and trichomes are particular sites for bacterial multiplication, predominantly substomatal cavities and trichome bases. *Pseudomonas cichorii* readily infects unwounded lettuce tissues through stomata and trichomes (Huang, 1986). Infection by trichomes often results in elongation type lesions where bacteria reproduce while infection by stomata results in either elongation or interruption type lesions. Entry of *E. carotovora* var. *atroseptica* into potato tubers through lenticels is affected by a number of environmental factors, principally by relative humidity (RH) prior to inoculation. Epidemic of bacterial blight on young soybean leaves incited by *P. syringae* pv. *glycinea* recurrently occur after wind, rain, storms (Huang, 1986). Bacteria colonize the apoplast which is intercellular spaces or xylem vessels, causing paren-chymatous and vascular or parenchymatous vascular diseases, respectively. Bacterial diseases are basically characterized by symptoms of leaf and fruit spots, cankers, blights, vascular wilts, rots, galls, and tumors. They are generally caused by gram-negative bacteria belonging to the Proteo-bacteria phylum, including *Xanthomonadaceae, Pseudomonadaceae,* and *Enterobacteriaceae* families. The severities of disease symptoms caused by bacteria in plants are determined by the production of a number of virulence factors, including phytotoxins, plant cell wall-degrading enzymes, extra-cellular polysaccharides (EPSs), and phytohormones. Phytopathogenic bacteria of the *Pseudomonas* genus, mainly *Pseudomonas syringae*, produce a wide spectrum of non-host-specific phytotoxins, that is, toxic compounds causing symptoms in many plants independently of the fact that they can or cannot be infected by the toxin-producing bacterium. *Pseudomonas* spp. has been grouped into necrosis-inducing and chlorosis-inducing phytotoxins based on their symptoms development they induce in plants, phytotoxins of. *P. syringae* pv. *syringae*, the causal agent of many diseases, and types of symptoms in herbaceous and woody plants, produces necrosis-inducing phytotoxins, lipodepsipeptides, which were based on their amino acid chain length and are usually divided in two groups: mycins (e.g., syringomycins) and peptins (e.g., syringopeptins) (Melotto et al., 2006). Chlorosis-inducing phytotoxins comprise coronatine, produced by *P. syringae* pvs. *atropurpurea, glycinea, maculicola, morsprunorum,* and *tomato,* phaseolotoxin, produced by *Pseudomonas savastanoi* pv. *phaseolicola* and *P. syringae* pv. *actinidiae*, and tabtoxin produced by the pvs. *tabaci, coronafaciens,* and *garcae* of *P. syringae* (Bender, 1990). Toxin-like coronatine, produced by *P. syringae* pv. *atropurpurea* and other forms infecting grasses and soybean; syringomycin, produced by *P. syringae* pv. *syringae* in leaf spots; thaxtomins, produced by bacterium *Streptomyces* that cause root and tuber roots, phaseolotoxin produced by bacterium *Pseudomonas syringae* pv. *Phaseolicola*, causing halo blight of bean and some other legumes, Thax-tomins cause dramatic plant cell hypertrophy and/or seedling stunting by altering the development of primary cell walls and affecting normal cell division cycles (Agrios, 2005).

Other essential virulence factors of phytopathogenic bacteria consist of EPSs, plant cell wall degrading enzymes, and phytohormones, etc. These factors are essential mainly for the pathogenesis of necrotrophic bacteria. The induction and self-catabolite repression of pectic enzymes in *Verticillium albo-atrum, Erwinia chrysanthemi* and *Fusarium oxysporum* f. sp. *lycopersici* are well established (Collmer, 1986).

Production of the phytohormones auxins (e.g., indole-3-acetic acid-IAA) and cytokinins are significant virulence factors for gall forming phytopatho-genic bacteria. The gaseous phytohormone ethylene produced by a variety of microorganisms includes plant pathogenic bacteria (Weingart and Volksch, 1997) and acts as virulence factor. *P. syringae* pv. *glycinea* and *P. savastanoi* pv. *phaseolicola* are very competent ethylene producers, generating it by using 2-oxoglutarate as the substrate and the ethylene-forming enzyme

(EFE) (Weingart et al., 1999). Weingart et al. (2001) demonstrated that when plants of bean and soybean were inoculated by the ethylene-negative (*efe*) mutants of these bacterial species, only *P. syringae* pv. *glycinea* considerably reduced its virulence. This virulence attenuation, predominantly evident in the *efe* mutant is not capable to produce the virulence factor coronatine, was characterized by weak symptoms and reduced bacterial growth *in planta*. Extracellular enzymes able to degrade plant cell walls are necessary virulence factors for necrotrophic soft rot bacteria, such as the soft rot Erwinia (e.g., *P. carotovorum* subsp. *carotovorum*, *P. atrosepticum*, *P. chrysanthemi*). A combination of extracellular enzymes like pectate lyases, pectin methylesterases, pectin lyases, polygalacturonases, cellulase, and proteases are involved in the depolymerization process of plant cell walls provoked by these bacteria (Toth et al., 2003). Proteases are released by the bacterial secretion system of type I (T1SS), whereas the rest of the above-mentioned enzymes are released by the bacterial secretion system of type II (T2SS) (Jha et al., 2005; Preston et al., 2005). Among these enzymes, pectate lyases (Pels) are mainly related in the virulence of *Pectobacterium* species which cause soft rot in plant (Toth et al., 2003). *P. chrysanthemi* has five major Pel isoenzymes, encoded by the *pelA*, *pelB*, *pelC*, *pelD*, and *pelE* genes, which are organized in two clusters, *pelADE* and *pelBC* (Huguvieux et al., 1996).

EPSs may be related with the bacterial cell as a capsule, be released as fluidal slime, or be present in both forms (Denny, 1999). EPSs are significant pathogenicity or virulence factors, primarily for bacterial vascular wilt. *Ralstonia solanacearum* causing wilting of several hundred plant species (e.g., potato, tomato, tobacco, peanut, and banana) is another phytopathogenic bacterium whose virulence largely depends on EPS production. Among the EPSs it produces, EPS1 which is heteropolysaccharide having an acidic high molecular mass and is also responsible primary virulence factor of bacterium because *eps* mutants showed severe reduction in systemic colonization when they were inoculated by unwounded roots of tomato plants and they were not able to cause wilt symptoms even after inoculated directly into stem wounds (Denny, 1999). Yun et al. 2006 have demonstrated that xanthan which is a major exopolysaccharide of *Xanthomonas* spp. also plays a vital role in *X. campestris* pv. *campestris* pathogenesis. They observed that a mutant of xanthan minus and a mutant that can produce truncated xanthan fails to develop disease in both tested plants of *Nicotiana benthamiana* and *Arabidopsis*. They also confirmed that xanthan suppresses deposition of callose in plant cell wall which acts as primary form of resistance against colonization of bacterial pathogens.

11.3.1   ROLE OF BACTERIAL SECRETION SYSTEM IN BACTERIAL PATHOGENESIS

Pathogenicity and/or virulence of Gram-negative plant pathogenic bacteria are strictly dependent on the presence of secretion apparatus in their cells, by which they secrete proteins or other nucleoproteins involved in their virulence in apoplast or inject into host cell. Till date, five secretion systems, numbered from I to V have been illustrated in both animal and plant gram-negative pathogenic bacteria (Desvaux et al., 2004). Moreover, it is now emerging that the expression of virulence factors is under the control of quorum sensing, a communication mechanism by which bacteria control the expression of certain genes in response to their population density. A number of biotrophic gram-negative bacteria are able to cause diseases in plants through a type III secretion system, composed of a protruding surface appendage, known as the hrp pilus, through which the bacterium injects effector proteins into the host cells and influence plant cells particularly to suppress plant defenses. The pathogenicity numerous biotrophic gram-negative bacteria including genera like *Pseudomonas, Xanthomonas, Ralstonia, Erwinia,* and *Pantoea* are mainly due to their ability to produce a type III secretion system (T3SS), also called injectisome (Desvaux et al., 2004), by which the bacterium injects proteins involved in its virulence into host plant cells (T3SS effectors). The T3SS is a complex structure composed of numerous subunits made up of approximately 20 bacterial proteins in sequences and these proteins which make up the T3SS apparatus termed structural proteins. Extra proteins called translocators serve up the function of translocating an additional set of proteins into the host cell cytoplasm. The translocated proteins are known as effectors since they are the virulence factors that influence the alteration in the host cells, allowing the invading pathogen to inhabit, multiply, and in some cases regularly persist in the host. The structural components of the T3SS and the process of translocation are skilfully reassessd by Ghosh (2004).

11.4   NATURE OF INFECTION BY VIRAL PATHOGENS

Near about 1000 number of viruses is presently known to be potentially capable of infecting plants. Even after large number of possible combinations, the development of disease is rare rather than a common outcome. Therefore, in the majority cases, plants are capable to defend against the harmful effects of viruses. This resistance in plants is due to the absence of necessary host susceptibility factors or due to the presence of several defense layers that virus has to conquer. First, the virus needs to defeat a sequence of pre-existing physical and chemical barriers in plants. If pathogenic viruses succeed in overcoming this first line of defense, it would have to countenance the nonspecific defensive reactions with which the plant responds to some molecular patterns that are common to different pathogens (Jones and Dangl, 2006). If a virus has evolved to attain virulence factors to work against this basal defense, it is in a position to be able to prompt infection. In many cases, however, plants are able to distinguish these virulence factors and create a new, highly specific resistance layer that is only induced when counteract with viruses expressing this virulence factor (Jones and Dangl, 2006). A virus can cause fruitful infection only in those plants that have not developed specific self-protective responses to its virulence factors. Viruses are generally established directly through the plant cells by insects, thus they do not apply mechanical forces. Viruses must seize the cellular machinery and prevent or counteract the plant defenses in order to establish a successful infection. To understand the process of viral infection in susceptible plants, it is prerequisite to identify and determine the role of the host genes selectively expressed during infection (Czosnek et al., 2013). The most essential functions of viral genes control are specific host infectivity, cell to cell virus movement, replication of the virus, long-distance transport of the virus in the plant, plant to plant virus transmission, and coat protein production of the virus. Each of these functions is essential for the pathogenicity and survival of the virus, although the variations in the level of each of these functions are carried out, influencing the virulence of the virus (Agrios, 2015). There are many viral diseases in which infections progress efficiently with no symptom development, induction of plant defense mechanisms, their inhibition by counteracting viral strategies, and the host factors required for replication and movement of virus can confer a pathological character upon the viral infection. Viral RNA such as ssRNA, RNases, Dicer-type dsRNA, RNases bring together in RNA-induced silencing complexes (RISC), RNA polymerases, and RNA helicases induces definite plant defense responses (Dunoyer and Voinnet, 2005). For the virus to be successful, it has to get away with this antiviral response, which is one of the manifestations of a complex set of cellular processes known as RNA silencing. Although viruses may take up various strategies to achieve this, it is believed that the most general approach viruses adopt is that of generating silencing suppressors (Li and Ding, 2006; Valli et al., 2009). These suppressors influence antiviral defense and also interfere with physiological processes of plant that depend on RNA silencing which in turn may include significantly to the different viral pathogenicity. Virus not only wants to get away the defenses that plants exerts but also should attempt various processes to complete its productive cycle (Maule et al., 2002). The RNA silencing machinery cut viral dsRNA structures, results in formation of small interfering RNAs (siRNA) that leads to RISC complexes which degrade viral ssRNA, and/or to reduce its translation. RISC loaded with viral siRNAs can also be the target of mRNAs of a few host genes, and it has been assumed that these downregulated genes can contribute to suppressing antiviral defenses and/or bringing forth disease symptoms (Moissiard and Voinnet, 2006). Certainly, it has been confirmed newly that disease symptoms caused by CMV satellite RNA are the outcome of siRNA-directed RNA silencing of the chlorophyll biosynthetic gene CHLI (Shimura et al., 2011; Smith et al., 2011). As different silencing suppressors act with distinct mechanisms and affect different stages of the silencing process, it is not amazing that more severe symptoms are produced in infections where viruses from different families are mixed, and that they reach high viral titres than in individual infections. The involvement of the potyvirus HCPro protein particularly in mixed infections with PVX (Gonzalez-Jara et al., 2005; Pruss et al., 1997; Saenz et al., 2001; Yang and Ravelonandro, 2002), and the collective action of several silencing suppressors could cause severe pathological symptoms in many viral diseases, some of which are of significant economic and social importance (Cuellar et al., 2008; Mukasa et al., 2006; Scheets, 1998). Associations between the interactions of specific virus factors with cell components and alterations in hormone synthesis and signalling have been well known (Culver and Padmanabhan, 2007), for example, interactions among the helicase domain of the TMV replicase and several members of the auxin/indole acetic acid (Aux/IAA) protein family (Padmanabhan et al., 2006). The subcellular localization of Aux/IAA proteins accumulation levels is lesser in the presence of the TMV replicase. On the other hand, their partial downregulation through virus-induced gene silencing gives rise to symptoms similar to those of TMV infection, and a mutation that reduces the ability of the replicase to interact with the Aux/IAA proteins considerably lowered accumulation of virus in older plant leaves (Padmanabhan et al., 2008). The viral coat protein plays a vital role in host recognition, uncoating and releasing nucleic acid, helps in nucleic acid replication, virus movement between cells and organs, virus movement via a vector between plants, and modification of symptoms. Besides, movement protein facilitates viruses to move between cells and/or throughout the phloem system by changing the plasmodesmata properties. On the other hand, a few movement proteins open movement channels for the virus; they also obstruct a defense molecule, the suppressor of virus silencing by the plant cell activated by the viral infection (Agrios, 2005). The biochemical changes of cellular constituents are reported to be directly related to morphological divergence of virus infected plants and the degree of crop loss is mainly determined by visible symptoms. The specificity of symptoms and its severity are concerned with the changes of specific cellular components due to virus infection. Various reports proposed that virus multiplication inside the plant cell alters different biochemical constituents of plants and interrupt the physiological process like photosynthesis, transpiration, and respiration of the infected plants which influence the growth and yield. They also accounted that in order to understand the activities of the host cell and the nature and level of damage caused by the virus, the determination of cellular constituents in virus infected plant is very essential. The metabolic processes are distorted due to variation in organic acids, ascorbic acids, and nucleic acid contents, concentration and position of each organic acids and endogenous auxin content in fruits of Tomato yellow leaf curl virus (TYLCV) infected plants, which may be the reason of decline in yield as well as quality deterioration of tomato (Tajul et al., 2011). The P6 protein of cauliflower mosaic virus (CaMV) is responsible for the formation of inclusion bodies (IBs), which are the sites for viral gene expression, replication, and virion assembly (Rodiguez et al., 2014).

11.5   NATURE OF INFECTION BY VIRIOD PATHOGENS

Among vegetables of economic interest, potato, tomato, pepper, and cucumber are vulnerable to numerous natural viriod diseases where infections are induced by various members of the genus Pospiviroid as cucumber is susceptible to a disease elicited by Hop stunt viriod (HSVd), within the same family Pospiviroidae. Viriod may also naturally infect vegetables without eliciting visible symptoms, as illustrated by citrus exocortis viriod (CEVd) and turnip (Fagoaga and Duran-Vila, 1996.) and eggplant latent viriod (ELVd) in its natural host (Fadda et al., 2003). They are single-stranded, covalently closed, circular, highly structured noncoding RNAs replicating autonomously and move systemically in host plants with the aid of host machinery. After entering into host cell, the RNA of viroid must be preferably gone to its site of replication, either the nucleus or the chloroplast, for the generation of progeny and releases to the cytoplasm to invade neighboring cells. The primary steps in the colonization of a host plant by viroids are their ability to move within the cell to the site of replication (Takeda and Ding, 2009; Zhu et al., 2001) than cell to cell through plasmodesmata (Qi et al., 2004), and, ultimately, systemically through vascular tissues, particularly through phloem cells that occurs in relation to the photosynthates from source to sink organs (Palukaitis, 1987; Flores et al., 2005). Proteins and nucleic acids, either endogenous or of viral origin, have been stated to move cell to cell via the plasmodesmata, the plant organelle providing cytoplasmic connections between cell (Flores et al., 2005). Translocations of viroids through the phloem are reported to be most likely facilitated by host proteins. Similarly, in the systemic spread of PSTVd, involvement of VirP1 and PP2 from tomato and cucumber, respectively, has been reported (Flores et al., 2005). RNA binding host proteins which may assist movement of viroids have been isolated using several approaches (Zhu et al., 2002; Gomez and Pallas, 2001; Maniataki et al., 2003). Characterization of host factors interacting with the viroid RNA may also contribute to the explanation of RNA related movement pathways of host plants. Moreover, RNA signatures on viroid molecules regulate cell to cell and long distance movement of viroids (Qi et al., 2004; Zhong et al., 2007) and the nuclear targeting of PSTVd (Harders et al., 1989; Zhao et al., 2001; Qi and Ding, 2003). Viroid pathogenicity is a complex phenomenon which is influenced by both the viroid and host genomes, for example, infection of various strains of viroid on the same host plant can lead to the development of latent or asymptomatic infections (Flores et al., 2006) or mild to severe symptoms (Ding 2009; Flores et al., 2005; O'Brien and Raymer, 1964; Owens and Hammond, 2009). In absence of capacity of protein-coding, viroids ought to arouse disease by direct interaction of their genomic RNA or derivatives thereof with host factors (proteins or nucleic acids) (Flores et al., 2005). Macroscopic symptoms as a result of infection by viriod are somewhat similar to those associated with many plant viral infections, and consist of stunting, sterility, vein discoloration and clearing, epinasty, chlorotic or necrotic spots, leaf distortion and mottling cankers, scaling and cracking of bark, malformation of tubers, flowers, and fruits, and rarely, death of the plant, etc. Since after replication of viroids, they set forward effects of pathogenic in exclusive of encoding proteins and its implications followed by mediating functions of viriod by the signals of both structural and sequences. Sequence analysis of viroid variants and reverse genetics on infectious cDNA copies of viroids with the introduction of mutations has exposed that there are complex associations between viroid sequence/structure and function (Flores et al.,

2012; Owens and Hammond, 2009). Although single nucleotide changes in the P domain of PSTVd can lead to pathogenicity, and naturally occurring sequence variants, or introduced mutations in that region, lead from mild to severe symptoms (Schnolzer et al., 1985; Hammond, 1985). Sano et al. (1992) confirmed that multiple structural regions might be accountable for pathogenicity of pospiviroids. A comprehensive analysis of the differential gene expression patterns of tomato plants at various stages of infection by mild and severe strains of PSTVd revealed that both of these strains changed expression of genes encoding products involved in defense/stress response, cell wall structure, chloroplast function, protein metabolism, signalling of hormone, and some other functions (Itaya et al., 2002; Owens et al., 2012). Analysis of proteomic studies of viriod–host interactions of citrus exocortis viriod (CEVd) in tomato, carried out using two-dimensional PAGE and mass spectroscopy, resulted in the identification of differentially expressed proteins, such as defense-related proteins, translation elongation factors, and translation initiation factors, and provided confirmation that pathogenicity may involve gentisic acid (GA) signalling and interaction of CEVd with eukaryotic elongation factor 1a (eEF1a) (Lison et al., 2013; Dube et al., 2009). In tomato plant, infection of PSTVd has been reported to modify the state of phosphorylation of its host-encoded protein p68, which resulted in activation of its dsRNA-dependent protein kinase activity (Hiddinga et al., 1988). Diener et al. (1993) confirmed the differential activation of the mammalian p68 by intermediate and mild strains of PSTVd, and suggested that activation of a plant enzyme homologous to mammalian p68 protein kinase might represent the triggering event in viroid pathogenesis. Hammond and Zhao (2000) identified a specific protein kinase gene (pkv) that is transcriptionally activated in plants infected with PSTVd. This encoded PKV protein is a new member of signal-transducing protein kinases under AGC VIIIa group. Expression of antisense RNA of pkv proteins by utilizing a viral-based vector has led to marked suppression of symptoms of viroid infection in tomato plants (Hammond and Zhao, 2000).

11.6 NATURE OF INFECTION BY PHANAROGEMIC PLANTS

Above 2500 species of different higher plants are well known to survive parasitically on some other plants. Their most important and common characteristic is that these parasites are vascular plants with well-developed specific organs which penetrate the tissues of other (host) vascular plants, establish connection to the vascular elements of host plants, and suck up nutrients from them. These parasitic plants produce flowers and seeds and belong to numerous widely separated botanical families. Relatively, only few of the well-known parasitic higher plants resulted with causing the important diseases on some agricultural crops or forest trees. Among these, species of the genus *Orobanche* (broomrapes) are holoparasitic weeds that cause devastating losses in many economically important vegetables crops (Parker and Riches, 1993). *Orobanche aegyptiaca* and *O. ramosa* are important scourges of vegetable like tomato, potato, and other field crops in large parts of Asia and the Middle-East (Kreutz, 1995; Goldwasser et al., 2000). Normal development of the parasite begins with germination in response to the reception of a chemical stimulus exudates from host roots (Yokota et al., 1998; Sato et al., 2003; Yoneyama et al., 2001). The particular characteristics of this pathogenic weed, that is, underground development and their attachment to the host roots help in hindering in development of the most effective control strategies. Moreover, 500,000 seeds can be released by a single plant of broomrape which also remains viable for decades in the soil without host plant. This provides an immense genetic adaptability of parasite to environmental changes, including agronomical practices, host resistance, and herbicide treatments, thus generating a great challenge in management (Joel et al., 2007). But the seed responds in the presence of conditioned stimuli, that is, exposed to water and presence of suitable temperatures for several days (Joel et al., 1995). Then, germinated seedling start infecting host roots with the development of haustorium which helps in penetrating the root system of host plant and derives nutrients from host plant and eventually act as a physiological bridge in between the host–parasite interactions. The young parasites are associated with development of a tubercle with certain adventitious roots and shoot. Many important developmental processes of parasitic plants like seed germination and differentiation of infection organs are being controlled by the signals exerted from the host plant (Bouwmeester et al., 2003). Seed germination was shown before to be stimulated by root exudate of host plant or by presence of synthetic strigol analogues such as GR24 (Joel and Losner-Goshen, 1994; van Woerden et al., 1994). Once the seeds get germinated, chemical or physical signals are required for the development of their attachment organ (Chang and Lynn, 1986; Albrecht et al., 1999; Keyes et al., 2000). The first acts during germination are mediated by strigol analogues such as GR24, whereas secondly need a distinct class of signalling compounds, usually benzoquinones, which serves as initiators of attachment organ (Press et al., 1990). It has been suggested that the benzoquinone signal may be generated during oxidation of phenols of host origin by peroxidases present in parasitic plants (Frick et al., 1996; Kim et al., 1998; Albrecht et al., 1999). The parasite has potentials of recognizing hosts which are crucial in the host differentiation process. The seeds of broomrapes are able to respond to specific germination stimulants released by their hosts (Perez-de-Luque et al., 2000; Galindo et al., 2002), thus chemical communication is significant for survival of parasite since it allows the seeds to recognize the presence of nearby suitable host prior to germination, avoiding a suicidal germination. At the molecular level, research has been mainly focused on identifying and characterizing the signals of host plant which can induce germination and induction of haustoria development (Yoder, 1999; Galindo et al., 1992). A number of reports had also revealed that several typical defense responses of host plant against pathogenic microorganisms are also being induced in response to parasitic plant infection which include phytoalexin induction (Serghini et al., 2001), increase peroxidase activity, lignifications, cell-wall phenolic deposition (Goldwasser et al., 1999), pathogenesis-related protein (Joel and Portnoy, 1998), and 3-hydroxy-3- methylglutaryl CoA reductase encoding gene induction (Westwood et al., 1998).

11.7 NATURE OF INFECTION BY PLANT PARASITIC NEMATODE

The agricultural production in India is very much hampered by the parasitism and predation of different pathogen and pests. Among them, plant–parasitic nematodes (PPN) present an alarming problem for different crops including vegetables, fruits, field crops, ornamentals, and common weeds (Saxena and Singh, 1997; Anes and Gupta, 2014; Ngele and Kalu, 2015). PPN cause approximate annual crop losses of $8 billion in the United States and $78 billion worldwide (Barker et al., 1998). Damage caused by PPN on 24 vegetable crops in the USA was estimated to be 11% (Feldmesser et al., 1971).

PPN are vital limiting factor in vegetable production. Most species of PPN attack and feed on plant roots and underground plant parts. Out of various PPN species, *Meloidogyne spp.* (root-knot nematodes) and *Globodera* spp. (potato cyst nematodes) are particularly destructive, contributing greatly to an estimated annual yield loss of $100 billion. Coevolution of host plant and parasitic nematodes from years has developed the ability of nematode to recognize and respond to chemical signals of host origin. Invasion of roots and the migration of nematodes to their feeding sites result in changed root architecture and significant reductions in nutrient and water uptake and consequent crop yields. The plant–nematode relationship is governed by a complex network of interactions resulting in the formation of feeding sites. The infective second-stage juveniles (J2) of *Meloidogyne* spp. after penetrating the root epidermis migrate intercellularly between cortical cells until they find a suitable root cell to form their feeding site. In contast, J2 of *Globodera* spp. penetrate roots and migrate intracellularly toward the vascular cylinder and establish an intimate nutritional relationship with their host through the development of syncytial feeding sites. Root cells around the nematode's head are stimulated to go through repeated rounds of mitosis uncoupled from cytokinesis, leading to multinucleated giant cells (Gheysen and Fenoll, 2002). Finally, the nematode feeding site is formed as the result of nematode-induced changes in plant gene expression (Gheysen and Fenoll, 2002). Feeding cell formation is presumably initiated in response to signal molecules released by the J2, but the nature of the primary stimulus is unknown. The most widely held hypothesis is that the necessary metabolic reprogramming of root cells is triggered by specific nematode secretions, which presumably interact with membrane or cytoplasmic receptors in the plant to switch on cascades of gene expression that alter cell development (Williamson and Hussey, 1996). Syncytia and giant cells require repeated stimulation from the nematode to maintain their function. Cyst and root-knot nematodes depend entirely on functional feeding cells to complete their life cycles. Many parasitic nematodes enter their hosts by active invasion and their transmission success is often based on a mass production of infective stages which show a highly specific host-finding behavior. These infective stages recognize their hosts via complex sequences of behavioral patterns with which they successfully respond to various environmental and host cues (Haas et al., 1997; Haas, 2003). Endoparasitic plant nematodes do not feed during their migrations in soil and in roots and their survival depends on finding a food source or sexual partner without unnecessary energy expenditure. They rely on stored lipid reserves to provide the energy for their movement and their infective stages recognize their hosts via complex sequences of behavioral patterns in response to various environmental cues (Haas et al., 1997; Haas, 2003). Plant signals are essential for nematodes to locate hosts and feeding sites, nematodes with > 60 % of their lipid reserves depleted are no longer capable of directed movement (Robinson et al., 1987).

Plant roots exude various compounds into the rhizosphere which intervene belowground interactions with pathogenic and beneficial soil organisms. These compounds include secretion of ions, free oxygen and water, enzymes, mucilage, and a diverse array of primary and secondary metabolites (Bertin et al., 2003). Even though the functions of most root exudates have not been determined, several compounds play important roles in biological processes (Bais et al., 2006). Root compounds are secreted into the surrounding rhizosphere or released from root border cells which separate from the roots as they grow. Chemical components of root exudates may alter nematode behavior and can either attract nematode to the roots or result in repellence, motility inhibition, or even death (Robinson, 2002; Wuyts et al., 2006; Zhao et al., 2000). It seems that a combination of signals in a given set of exudates determines the nematode behavior in a given plant–nematode interaction.

The role of plant signals in synchronizing host and parasite lifecycle can be best explained with the two species of potato cyst nematodes (*Globodera* spp.), as these nematodes almost completely depend on root exudates for hatching. Hatching is an important step in the nematode lifecycle, leading to parasitism and this dependence of the *Globodera* spp. on hatching factors have been exploited to reduce the number of nematodes in infested fields (Timmermans et al., 2007). *Solanum sisymbriifolium* is used successfully in Europe as a trap crop for potato cyst nematodes, as this plant stimulates hatch of second stage juveniles but does not support their development and is completely resistant to *G. Pallid* and *G. rostochiensis* (Timmermans et al., 2007). The ability to orientate toward plant roots increases survival of nematode. Long distance attractants' can enable nematodes to locate roots, whereas attractants that cause the nematode to move to individual host roots may be termed as "short distance attractants" and the orientation to the preferred site of invasion in the root tip might be mediated by "local attractants" (Perry, 2001). To date, including *M. incognita,* only CO_2 has been identified as a prime long distance attractant for plant parasitic nematodes, (Robinson, 2002). Additional attractants are aminoacids, sugars, and metabolites (Bird, 1959; Robinson, 2002; Perry, 2001; Prot, 1980). The chemotactic response of *Bursaphelenchus xylophilus*, the pine wood nematode, to a variety of chemicals varied according to the nematodes lipid content. Nematodes with the lowest lipid content were attracted to the pine wood volatile α-myrcene, while nematodes with the highest lipid content were attracted to a toluene hydrocarbon in the cuticle of the beetle vector for the nematode (Stamps and Linit, 2001). Thus, nematode lipid contents seem to be an important factor in the nematode's response to host signals. Plant roots also trigger *C. elegans* behavioral response by emitting volatile signals attracting nematodes to the root proximity (Horiuchi et al., 2005).

Plant roots produce various allelochemicals to defend themselves against potential soil-borne pathogens and a number of plant secondary metabolites have been shown to function as nematode antagonists (Guerena, 2006). For example, the triterpenoid compound, cucurbitacin A, from cucumber plants repels nematodes and a β-terthienyl compound from *Tagetes erecta* is repellent and nematotoxic (Castro et al., 1988). Allelochemicals present in the root exudates of maize such as cyclic hydroxamic acid have been shown to affect the behavior of *Pratylenchus zea, M. incognita,* and *Xiphinema americanum.* The methoxylated hydroxamic acid (DIMBOA, dihydroxy-7-methoxy-1,4-benzoxaxin-3(4H) increased attraction and its degradation product (MBOA, 6-methoxy-2-benzoxazolinone) induced repellence to roots (Friebe et al., 1997; Zasada et al., 2005). Root exudates not only contain compounds that induce nematode hatching, attraction, and repellence to roots but also compounds that induce characteristic nematode exploratory behavior, including stylet thrusting, release of secretions in preparation for root penetration, aggregation and increase in nematode mobility.(Clarke and Hennessy, 1984; Curtis, 2007; Grundler et. al., 1991; Robinson, 2002).

The outermost cuticular layer of nematodes, that is, epicuticle is covered in many species by a fuzzy coating material termed the "surface coat" (SC). The SC is composed mainly of proteins, carbohydrates, and lipids (Spiegel and McClure, 1995). Nematode secretions and surface coat antigens are possibly to be the first signals perceived by the plant and they probably have important roles in the host–parasite relationship (Jones and Robertson, 1997). They may help with the initial penetration, migration of the second stage juveniles in the plant tissues, and digestion of host cell contents (Williamson and Hussey, 1996) and are also possibly involved in the protection of the nematode from plant defense responses (Prior et al., 2001; Robertson et al., 2000). Surface composition can also change within a single stage during entry of parasitic nematodes into a new host or host tissue (Modha et al., 1995; Proudfoot et al., 1993). One of the most interesting features of the nematode SC is its dynamic nature, there is a continuous turnover of the surface associated antigens, which are shed and replaced (Blaxter and Robertson, 1998). Plant signals present in root exudates trigger a rapid alteration of the surface cuticle of *Meloidogyne incognita* and *Globodera rostochiensis* and the same changes were also induced by phytohormones in particular IAA to *M. incognita* but not *G. rostochiensis* (Akhkha et al., 2002, 2004; Curtis, 2007; Lopez de Mendoza et al., 2000). This surface change induced by plant signals might allow *M. incognita* to adapt and survive plant defense processes. However, more specific host cues from root exudates of solanaceous plants were responsible for increasing the lipophilicity of the surface cuticle of infective juveniles of *Globodera* species (Akhkha et al., 2002). In vitro, IAA has also been shown to induce the production of nematode secretions (Duncan et al., 1995) and an increase in nematode mobility (Curtis, 2007). High concentrations of IAA have also been detected around the root cells surrounding the nematode head (Karczmark et al., 2004). It is possible that IAA acts as a signal that orientates the nematode on the root surface in the rhizosphere and inside the root tissue and thereby promotes nematode infection.

The nervous system of nematode is the conduit between stimulus, reception, and behavioral output. The main nematode chemosensory organs involved in host-recognition processes are two bilaterally symmetrical amphids in the nematode head and two paired pore-like phasmids located in the lateral field of the nematode tail. Nematodes have the ability to chemo-orientate using a combination of head to tail chemosensory sensors to simultaneously compare the intensities of the stimulus across their body length (Hilliard et al., 2002). The amphids contain a number of dendritic processes which are surrounded by secretions produced by the glandular sheath cell. Disruptions of chemoresponses render nematodes disorientated in the soil and unable to find a host (Fioretti et al., 2002; Perry 2005; Zuckerman, 1983). Chemo-orientation in a concentration gradient is vital for nematode survival and is essential for detection of host plant exudates, food stimulants, food deterrents, and sex pheromones. The free living nematode *Caenorhabditis elegans* changes its surface in response to environmental chemical signals detected by the nematodes's chemosensory organs (Olsen et al., 2007). Plant nematodes also rely on chemoreception, for example, to find a host in the soil, and when a root is encountered, its surface is explored for a suitable penetration site. IAA binds to the chemosensory organs of *M. incognita* (Curtis, 2007).

The effector proteins of nematodes originate in the secretory gland cells and their synthesis as well as secretion is regulated throughout the parasitic cycle of the nematode (Wyss and Zunke, 1986; Hussey, 1989; Davis et al., 2008). It is hypothesized that these secretions modify the plant cells into complex feeding structures (Burgess and Kelly, 1987; Davis et al., 2000; Gheysen and Fenoll, 2002; Hewezi and Baum, 2012). Gao et al. (2003) studied the transcriptome of nematode secretions and reported 51 gland-expressed candidate parasitism genes from *H. glycines*. In addition, the secretome of *M. incognita is* resulted in the identification of 486 proteins (Bellafiore et al., 2008). Some nematode proteins might have functions of regulating the plant cell cycle or growth while others could reprogram the genetic events taking place in the host cells to facilitate compatible interactions (Bellafiore et al., 2008). Secretome contains products of diverse parasitism genes. The genes encode for plant cell wall degrading/modifying enzymes including cellulases or β-1,4-endoglucanases (Smant et al., 1998; Goellner et al., 2000; Gao et al., 2002; Gao et al., 2004), chitinases (Gao et al., 2002), xylanases (Opperman et al., 2008), pectate lyases and pectinases (Abad et al., 2008; Opperman et al., 2008), and expansins (Qin et al., 2004; Kudla et al., 2005). These cell wall degrading enzymes help J2 the infective nematodes to migrate through the plant roots by softening the cell wall. The effector proteins include CLAVATA3 (CLV3)/ESR (CLE)–like proteins (Wang et al., 2005; Patel et al., 2008; Lu et al., 2009; Wang et al., 2010a; Wang et al., 2010b), SPRY domain-containing (SPRYSEC) effector proteins (Huang et al., 2006; Rehman et al., 2009; Sacco et al., 2009; Postma et al., 2012), Mi-EFF1 (Jaouannet et al., 2012), calreticulin Mi-CRT (Jaubert et al., 2002; Jaubert et al., 2005; Jaouannet et al., 2013), and Hs19C07 (Lee et al., 2011). These proteins are involved in plant–nematode compatible interactions by suppressing the defense mechanisms of host plants in some cases. A *M. incognita* effector protein 7H08 imported into the nucleus of plant cells has the ability to activate transcription of plant genes (Zhang et al., 2015). This is the first report on nematode effector protein with transcriptional activation activity. However, some nematode effectors proteins are also involved in incompatible interactions, such as *M. incognita* secreted map-1.2 protein shows specifically increased expression in *Mi-1* resistant tomato lines. It implies that the *map-1.2* might be a putative avirulent (Avr gene) response to Mi gene (Castagnone-Sereno et al., 2009). However, another *M. incognita* gene, *Cg-1*, was confirmed as an *Avr* gene corresponding to the *Mi-1* resistance gene in tomato (Gleason et al., 2008). Likewise, *SPRYSEC-19* (Rehman et al., 2009; Postma et al., 2012), Gp-Rbp-1 (Blanchard et al., 2005; Sacco et al., 2009), and *Gr-VAP1* (Lozano-Torres et al., 2012) are some other examples of nematode effector proteins involved in incompatible plant–nematode interactions.

11.8   CONCLUSION

Plant diseases constitute a major problem in the production of vegetables crops and as well as storage and processing of agricultural products. However, the farmers face heavy economic yield losses both in terms of quality and quantity of these valuable vegetables crops due to the various plant diseases which are caused by several parasitic plant pathogens. Different pathogens such as fungi, bacteria, virus, viroid, phanerogamic parasitic plants, and nematodes were found to be associated with development of several diseases in valuable vegetables crops. Therefore, it is very necessary to develop an appropriate management strategy against each class of pathogens in order to conquer the pathogen attack. To develop proficient ways of management against several attacks of plant pathogens, it is necessary to understand the several mechanism of infection of each class of pathogens. For a successful infection by pathogens, it must be able to make its way into and through the plant, obtain nutrients from the host plant, and must be able to counteract the defense reactions of the host. Failure of these pathogens to invade suitable host cells will prevent them from disease development in host plant. Therefore, a complete perceptive of mechanism of infection process is necessary to artificially intervene the plant parasitic pathogen at different stages of infection. At the present time, with the increase accessibility of modern molecular methods to facilitate high-throughput study will certainly improve our knowledge of mechanism of infection which in turn unlocks the various ways for promising disease management strategies.

KEYWORDS

- **cell wall degrading enzymes (CWDEs)**
- **infection**
- **pathogenesis**
- **toxins**
- **phytohormones**

REFERENCES

Abad, P., J.; Gouzy, J. M.;Aury, P.; Castagnone-Sereno, E. G. J.; Danchin, E.; Deleury, L.; Perfus-Barbeoch, V.; Anthouard, F.; Artiguenave, V. C.; Blok, M.C.; Caillaud, P.M.; Coutinho, C.; Dasilva, F.; De Luca, F.; Deau, M.; Esquibet, T.; Flutre, J.V.; Goldstone, N.; Hamamouch, T.; Hewezi, O.; Jaillon, C.; Jubin, P.; Leonetti, M.; Magliano, T. R.; Maier, G. V.; Markov, P.; McVeigh, G.; Pesole, J.; Poulain, M.; Robinson-Rechavi, E.; Sallet, B.; Segurens, D.; Steinbach, T.; Tytgat, E.; Ugarte, C.; van Ghelder, P.; Veronico, T. J.; Baum, M.; Blaxter, T.; Bleve-Zacheo, E. L.; Davis, J. J.; Ewbank, B.; Favery, E.; Grenier, B.; Henrissat, J. T.; Jones, V.; Laudet, A. G.; Maule, H.; Quesneville, M. N.; Rosso, T.;

Schiex, G.; Smant, J.; Weissenbach and Wincker P. Genome Sequence of the Metazoan Plant–Parasitic Nematode *Meloidogyne incognita*. *Nat. Biotechnol*. **2008**, *26*, 909–915.

Agrios, G. N. *Plant Pathology*. Academic Press: San Diego, 2005, pp 948.

Akamatsu, H.; Itoh, Y.; Kodama, M.; Otani, H.; Kohmoto, K. AAL-toxin Deficient Mutants of *Alternaria Alternata* Tomato Pathotype by Restriction Enzyme-Mediated Integration. *Phytopathology* **1997**, *87*, pp 967–972.

Akhkha, A.; Curtis, R.; Kennedy, M.; Kusel, J. The Potential Signaling Pathways Which Regulate Surface Changes Induced By Phytohormones in the Potato Cyst Nematode (*Globodera rostochiensis*). *Parasitology* **2004**, *128*, 533–539.

Akhkha, A.; Kusel, J.; Kennedy, M.; Curtis R. H. C. Effects of Phytohormones on the Surfaces of Plant Parasitic Nematodes. *Parasitology* **2002**, *125*, 165–175.

Albrecht, H.; Yoder, J. I.; Phillips, D. A. Flavonoids Promote Haustoria Formation in the Root Parasite Triphysaria Versicolor. *Plant Physiology* **1999**, *119*, 585–591.

Anes, K. M.; Gupta, G. K. Distribution of Plant Parasitic Nematodes in the Soybean (*Glycine Max*) Growing Areas in India. *Ind. J. Nematol*. **2014**, *40* (2), 227–231.

Bais, H. P.' Weir, T. L.; Perry, L. G.; Gilroy, S.; Vivanco, J. M. The Role of Root Exudates in Rhizosphere Interactions with Plants and Other Organisms. *Ann. Rev. Pl. Biol*. **2006**, *57*, 233–266.

Barker, K. R.; Pederson, G. A.; Windham, G. L. *Plant and Nematode Interactions*. Agronomy Monograph 36. American Society of Agronomy: Madison, WI, 1998.

Bellafiore, S.; Shen, Z. X.; Rosso, M. N.; Abad, P.; Shih, P.; Briggs, S. P. Direct Identification of the *Meloidogyne incognita* Secretome Reveals Proteins with Host Cell Reprogramming Potential. *Plos Pathol*. **2008**, *4*, e1000192.

Bender, C. L. Chlorosis-inducing Phytotoxins Produced by *Pseudomonas syringae*. *Eur. J. Plant Pathol*. **1999**, *105*, 1–12.

Bertin, C.; Yang, X. H.; Weston, L. A. The Role of Root Exudates and Allelochemicals in the Rhizosphere. *Plant Soil* **2003**, *256*, 67–83.

Bird, F. The Attractiveness of Roots to the Plant Parasitic Nematodes *Meloidogyne javanica* and *M. hapla*. *Nematologica* **1959**, *42*, 322–335.

Blanchard, A.; Esquibet, M.; Fouville, D.; Grenier, E. Ranbpm Homologue Genes Characterised in the Cyst Nematodes Globodera Pallida and Globodera 'Mexicana'. *Physiol. Mol. Plant Pathol*. **2005**, *67*, 15–22.

Blaxter, M.; Robertson, W. The cuticle, In: *Free-living and Plant Parasitic Nematodes;* Perry, R.N, Ed; CAB International: Wallingford, UK, 1999, pp 25–48.

Bouwmeester, H. J.; Matusova, R.; Zhongkui, S.; Beale, M. H. Secondary Metabolite Signalling in Host–Parasitic Plant Interactions. *Curr. Opin. Plant Biol*. **2003**, *6*, 358–364.

Brandwagt, B. F.; Mesbah, L. A.; Takken, F. L. W.; Laurent, P. L.; Kneppers, T. J. A.; Hille, J.; Nijkamp, H. J. J. A Longevity Assurance Gene Homolog of Tomato Mediates Resistance to *Alternaria Alternate*. *F. Sp. Lycopersici* Toxins and Fumonisin B1. *P. Natl. Acad. Sci. USA* **2000**, *97*, 4961–4966.

Buonaurio, R. Infection and Plant Defense Responses During Plant-Bacterial Interaction. *Plant Microbe Interactions* **2008**, *8*, 169–197.

Burgess, T. L.; Kelly, R. B. Constitutive and Regulated Secretion of Proteins. *Ann. Rev. Cell Biol*. **1987**, *3*, 243–293.

Carapito, R.; Hatsch, D.; Vorwerk, S.; Petkovski, E.; Jeltsch, J. M.; Phalip, V. Gene Expression in *Fusarium Graminearum* Grown on Plant Cell Wall. *Fungal Genet. Biol*. **2008**, *45*, 738–774.

Castro, C. E.; Belser, N. O.; Mckinney, H. E.; Thomason, I. J. Quantitative bioassay for chemotaxis with plant parasitic nematodes. Attractant and Repellent Fractions for *Meloidogyne Incognita* from Cucumber Roots. *J. Chem. Ecol.* **1988,** *15,* 1297–1309.

Centis, S.; Dumas, B.; Fournier, J.; Marolda, M.; Esquerre-Tugaye, M. T. Isolation and Sequence Analysis of Clpg1, A Gene Coding for an Endopolygalacturonase of the Phytopathogenic Fungus *Colletotrichum lindemuthianum. Gene.* **1996,** *170,* 125–129.

Chang, M.; Lynn, D. G. The Haustorium and the Chemistry of Host Recognition in Parasitic Angiosperms. *J. Chem. Ecol.* **1986,** *12,* 561–579.

Chaves, M. M.; Maroco, J. P.; Pereira, J. S. Understanding Plant Responses to Drought from Genes to the Whole Plant. *Funct. Plant Biol.* **2003,** *30,* 239–264.

Choi, J.; Kim, K. S.; Rho, H. -S.; Lee, Y. -H. Differential Roles of the Phospholipase C Genes in Fungal Development and Pathogenicity of *Magnaporthe oryzae. Fungal Genet. Biol.* **2011,** *48,* 445–455.

Choquer, M.; Fournier, E.; Kunz, C.; Levis, C.; Pradier, J. M. *Botrytis cinerea* Virulence Factors: New Insights Into A Necrotrophic and Polyphageous Pathogen. *FEMS Microbiol. Lett.* **2007,** *277,* 1–10.

Clarke, A. J.; Hennessy, J. Movement of *Globodera rostochiensis* Juveniles Stimulated by Potato Root Exudate. *Nematologica,* **1984,** *30,* 206–212.

Collmer, A. The Role of Pectic Enzymes in Plant Pathogenesis. *Ann. Rev. Phytopathol.* **1986,** *24,* 383–409.

Cooper, R. M. The Role of Cell Wall Degrading Enzymes in Infection and Damage. In: *Plant Diseases: Infection, Damage and Loss*; Wood, R. K. S.; Jellis, G. J., Eds; Oxford, Blackwell Scientific Publications: Oxford, 1984, pp 13–27.

Crum, C. J.; Hu, J.; Hiddinga, H. J; Roth, D. A. Tobacco Mosaic Virus Infection Stimulates the Phosphorylation of a Plant Protein Associated with Double-stranded RNA-dependent Protein Kinase Activity. *J. Biol. Chem.* **1988,** *263,* 13440–13443.

Cuellar, W. J.; Tairo, F.; Kreuze, J. F.; Valkonen, J. P. T. Analysis of Gene Content in Sweet Potato Chlorotic Stunt Virus RNA1 Reveals the Presence of the P22 RNA Silencing Suppressor in Only A Few Isolates: Implications for Viral Evolution and Synergism. *J. Gen. Virol.* **2008,** *89,* 573–582.

Culver, J. N.; Padmanabhan, M. S. Virus-induced Disease: Altering Host Physiology One Interaction at a Time. *Ann. Rev. Phytopathol.* **2007,** *45,* 221–243.

Curtis, R. H. C. Do Phytohormones Influence Nematode Invasion and Feeding Site Establishment? *Nematology* **2007,** *9,* 155–160.

Czosnek , H.; Eybishtz , A.; Sade, D.; Gorovits, R.; Sobol , I.; Bejarano, E.; Rosas-Diaz, T.; Lozano-Duran, R. Discovering Host Genes Involved in the Infection by the Tomato Yellow Leaf Curl Virus Complex and in the Establishment of Resistance to the Virus Using Tobacco Rattle Virus-Based Post Transcriptional Gene Silencing. *Viruses* **2013,** *5,* 998–1022.

Davis, E. L.; Hussey, R. S.; Baum, T. J.; Bakker, J.; Schots A. Nematode Parasitism Genes. *Ann. Rev. Phytopathol.* **2000,** *38,* 365–396.

Davis, E. L.; Hussey, R. S.; Mitchum, M. G.; Baum, T. J. Parasitism Proteins in Nematode-Plant Interactions. *Curr. Opin. Plant Biol.* **2008,** *11,* 360–366.

De Bary, A. Ubereinige Sclerotinien und Sclerotienkrankheiten. *Botanique Zeitung* **1886,** *44,* 377–474.

Denny, T. P. Autoregulator-dependent Control of Extracellular Polysaccharide Production in Phytopathogenic Bacteria. *Eur. J. Plant Pathol.* **1999,** *105,* 417–430.

Deising, H.; Frittrang, A. K.; Kunz, S.; Mendgen, K. Regulation of Pectin Methylesterase and Polygalacturonase Activity During Differentiation of Infection Structures in *Uromyces vicae-fabae*. *Microbiology* **1995**, *141*, 561–571.

Desvaux, M. N. J.; Parham, A.; Scott-Tucker; Henderson, I. R. The General Secretory Pathway: A General Misnomer? *Trends Microbiol*. **2004**, *12*, 306–309.

Diener, T. O.; Hammond, R. W.; Black, T.; Katze, M. G. Mechanism of Viriod Pathogenesis: Differential Activation of the Interferon-Induced, Double-Stranded RNA-Activated, Mr 68,000 Protein Kinase by Viroid Strains of Varying Pathogenicity. *Biochimie* **1930**, *75*, 533–538.

Ding, B. The Biology of Viroid–Host Interactions. Ann. Rev. Phytopathol. *47*,105–131.

Dube, A.; Bisaillon, M.; Perreault, J. P. Identification of Proteins from *Prunus Persica* that Interact with Peach Latent Mosaic Viroid. *J. Virol*. **2009**, *83*, 12057–12067.

Dunoyer, P.; Voinnet, O. The Complex Interplay Between Plant Viruses and Host RNA-Silencing Pathways. *Curr. Opin. Plant Biol*. **2005**, *8*, 415–423.

Durrands, P. K.; Cooper, R. The Role of Pectinases in Vascular Wilt Disease as Determined by Defined Mutants of *Verticillium albo-atrum* **1998**, *32* (3), 363–371.

Erental, A.; Harel, A.; Yarden, O. Type 2A Phosphoprotein Phosphatase is Required for Asexual Development and Pathogenesis of *Sclerotinia sclerotiorum*. *Mol. Plant Microbe Interaction* **2007**, *20*, 944–954.

Fadda, Z.; Dar Os, J. A.; Fagoaga, C.; Flores, R.; Duran-Vila, N. Eggplant Latent Viroid, the Candidate Type Species for a New Genus Within the Family Avsunviroidae (hammerhead viroids). *J. Virol*. **2003**, *77*, 6528–6532.

Fagoaga, C.; Duran-Vila, N. Naturally Occurring Variants of Citrus Exocortis Viroid in Vegetable Crops. *Plant Pathol*. **1996**, *45*, 45–53.

Feldmesser, J.; Edwards, D. I.; Epps, J. M.; Heald, C. M.; Jenkins, W. R.; Johnson, H. J. B.; Lear, C. W.; McBeth, C. W.; Nigh, E. L.; Perry, V. G. *Estimated Crop Losses from Plant-Parasitic Nematodes in the United States. Comm. Crop losses*. Soc. Nematologists: Hyattsville, Maryland, 1971.

Fioretti, L.; Porter, A.; Haydock, P. J.; Curtis, R. H. C. Monoclonal Antibodies Reactive With Secreted-excreted Products from the Amphids and the Cuticle Surface of *Globodera Pallid* Affect Nematode Movement and Delay Invasion of Potato Roots. *Int. J. Parasitol*. **2002**, *32*, 1709–1718.

Flores, R.; Delgado, S.; Rodio, M. E.; Ambros, S.; Hernandez, C.; Serio, F. D. Peach Latent Mosaic Viroid: Not So Latent. *Mol. Plant Pathol*. **2006**, *7*, 209–221.

Flores, R.; Hernandez, C.; Martinez de Alba, E.; Daros, J. A.; Di Serio, F. Viroids and Viroid–Host Interactions. *Ann. Rev. Phytopathol*. **2005**, *43*, 117–139.

Flores, R.; Serra, P.; Minoia, S.; Di Serio, F.; Navarro, B. Viroids: From Genotype Tophenotype Just Relying on RNA Sequence and Structure Motifs. *Front. Microbiol*. **2012**, *3*, 217.

Frick, E.; Frahne, D.; Wegmann, K. Biochemical Synthesis of 2, 6-Dimethoxy-Para-Benzoquinone, A Haustorial Stimulant of Striga Asiatica (L.) *Kuntze. Natural Products Lett*. **1996**, *9*, 153–159.

Fraissinet-tachet, L.; Reymond-cotton, P.; Fevre, M. Characterization of a Multigene Family Encoding an Endopolygalacturonase in *Sclerotinia sclerotiorum*. *Curr. Genet*. **1995**, *29*, 96–99.

Friebe, A.; Lever, W.; Sikora, R.; Schnabl, H. Allelochemical in Root Exudates of Maize. Effects on Root Lesion Nematode *Pratylenchus Zea*. *In*: Phytochemical Signals and Plantmicrobe Interactions. *Recent Adv. Phytochem*. **1997**, *32*, 71–93.

Frittrang, A. K.; Deising, H.; Mendgen, K. Characterization and Partial Purification of Pectinesterase, A Differentiation-Specific Enzyme of *Uromyces viciae-fabae*. *J. Gen. Microbiol*. **1992**, 138, 2213-2218.

Fuchs, T. M. Molecular Mechanisms of Bacterial Pathogenicity. *Naturwissenschaften* **1998**, 85, 99–108.

Gadea, J.; Mayda, M. E.; Conejero, V.; Vera, P. Characterization of Defense-Related Genes Ectopically Expressed in Viroid-Infected Tomato Plants. *Mol. Plant Microbe Interaction* **1996**, 9, 409–415.

Galindo, J. C. G; Perez de Luque, A.; Jorrın, J.; Macıas, F. A. SAR Studies Of Sesquiterpene Lactones as *Orobanche Cumana* Seed Germination Stimulants. *J. Agric. Food Chem*. **1992**, 50, 1911–1917.

Gamacho, A. H.; Sanger, H. L. Purification and Partial Characterization of the Major Pathogenesis-Related Tomato Leaf Protein P14 from Potato Spindle Tuber Viroid (Pstvd)-Infected Tomato Leaves, *Arch. Virol*. **1984**, 81, 263–284.

Gao, B.; Allen, R.; Davis, E. L.; Baum, T. J.; Hussey, R. S. Molecular Characterisation and Developmental Expression of a Cellulose-Binding Protein Gene in the Soybean Cyst Nematode Heterodera Glycines. *Int. J. Parasitol*. **2004**, 34, 1377–1383.

Gao, B.; Allen, R.; Maier, T.; Davis, E. L.; Baum, T. J.; Hussey, R. S. Identification of a New Beta-1,4-Endoglucanase Gene Expressed in the Esophageal Subventral Gland Cells of *Heterodera glycines*. *J. Nematol*. **2002**, 34, 12–15.

Gao, B. L.; Allen, R.; Maier, T.; Davis, E. L.; Baum, T. J.; Hussey, R. S. The Parasitome Of the Phytonematode Heterodera Glycines. *Mol. Plant Microbe Interaction* **2003**, 16, 720–726.

Gheysen, G.; Fenoll, C. Gene Expression in Nematode Feeding Sites. *Annu. Rev. Phytopathol*. **2003**, 40, 191–219.

Ghosh, P. Process of Protein Transport by the Type III Secretion System. *Microbiol. Mol. Biol. Rev*. **2004**, 68, 771–795.

Gilchrist, D. G. Programmed Cell Death In Plant Disease: The Purpose And Promise Of Cellular Suicide. *Ann. Rev. Phytopathol*. **1998**, 36, 393–414.

Gleason, C. A.; Liu, Q. L.; Williamson, V. M. Silencing a Candidate Nematode Effector Gene Corresponding to the Tomato Resistance Gene Mi-1 Leads to Acquisition of Virulence. *Mol. Plant Microbe Interaction* **2008**, 21, 576–585.

Goellner, M.; Smant, G.; De Boer, J. M.; Baum, T. J.; Davis, E. L. Isolation of Beta-1,4-Endoglucanase Genes from *Globodera Tabacum* and Their Expression During Parasitism. *J. Nematol*. **2000**, 32, 154–165.

Goldwasser, Y.; Plakhine, D.; Yoder, J. I. Arabidopsis Thaliana Susceptibility to *Orobanche spp*. *Weed Sci*. **2000**, 48, 342–346.

Goldwasser, Y.; Hershenhorn, J.; Plakhine, D.; Kleifeld, Y.; Rubin, B. Biochemical Factors Involved in Vetch Resistance to *Orobanche aegyptiaca*. *Physiol. Mol. Plant Pathol*. **1999**, 54, 87–96.

Gomez, G.; Pallas, V. Identification of An In Vitro Ribonucleoprotein Complex Between a Viroid RNA and a Phloem Protein from Cucumber Plants. *Mol. Plant Microbe Interaction* **2001**, 14, 910–913.

Gonzalez-Jara, P.; Atencio, F. A.; Martı´nez-Garcia, B.; Barajas, D.; Tenllado, F.; Diaz-Ruiz, J. R. A Single Amino Acid Mutation in the Plum Pox Virus Helper Component-Proteinase Gene Abolishes both Synergistic and RNA Silencing Suppression Activities. *Phytopathology* **2005**, 95, 894–901.

Grundler, F.; Schnibbe, L.; Wyss, U. *In Vitro* Studies on the Behaviour of Second Stage Juveniles of *Heterodera Schachtii* in Response to Host Plant Root Exudates. *Parasitology* **1991**, *103*, 149–155.

Guerena, M. Nematodes: Alternative Control. National Sustainable Agriculture Information Service. ATTRA publication #IP287, 2006, pp 1–20.

Haas, W. Parasitic Worms: Strategies of Host Finding, Recognition and Invasion. *Zoology* **2003**, *106*, 349–364.

Haas, W.; Diekhoff, D.; Koch, K.; Schmalfuss, G.; Loy, C. *Schistosoma mansoni* cercariae: Stimulation of Acetabular Gland Secretion is Adapted to the Chemical Composition of Mammalian Skin. *J. Parasitol.* **1997**, *83*, 1079–1085.

Hammond, R. W.; Zhao, Y. Characterization of a Tomato Protein Kinase Gene Induced by Infection by Potato Spindle Tuber Viroid. *Mol. Plant Microbe Interaction* **2000**, *13*, 903–910.

Hammond, R. W. Analysis of the Virulence-Modulating Region of Potato Spindle Tuber Viroid (Pstvd) by Site-Directed Mutagenesis. *Virology* **1992**, *187*, 654–662.

Harders, J.; Lukacs, N.; Robert-Nicoud, M.; Jovin, J. M.; Riesner, D. Imaging of Viroids in Nuclei from Tomato Leaf Tissue by In Situ Hybridization and Confocallaser Scanning Microscopy. *EMBO J.* **1989**, *8*, 3941–3949.

Heiler, S.; Mendgen, K.; Deising, H. Cellulolytic Enzymes of the Obligately Biotrophic Rust Fungus *Uromyces Viciae-Fabae* are Regulated Differentiation-Specifically. *Mycol. Res.* **1993**, *97*, 77–85.

Hewezi, T.; Baum, T. Manipulation of Plant Cells by Cyst and Root-Knot Nematode Effectors. *Mol. Plant Microbe Interaction* **2012**, *26*, 9–16.

Hiddinga, H. J.; Crum, C. J.; Hu, J.; Roth, D. A. Viroid-induced Phosphorylation of a Host Protein Related to a Dsrna-dependent Protein Kinase. *Science* **1988**, *241*, 451–453.

Hilliard, M. A.; Bargmann, C. I.; Bazzicalupo, P. *C. Elegans* Responds to Chemical Repellents by Integrating Sensory Inputs from the Head and Tail. *Curr. Biol.* **2002**, *12*, 730–734.

Horiuchi, J.; Prithiviraj, B.; Bais, H.; Kimball, B. A.; Vivanco, J. M. Soil Nematodes Mediate Positive Interactions Between Legume Plants and Rhizobium Bacteria. *Planta* **2005**, *222*, 8480857.

Huang, G. Z.; Dong, R. H.; Allen, R.; Davis, E. L.; Baum, T. J.; Hussey, R. S. A Root-knot Nematode Secretory Peptide Functions as a Ligand for a Plant Transcription Factor. *Mol. Plant Microbe Interaction* **2006**, *19*, 463–470.

Huang, J. Ultrastructure of Bacterial Penetration in Plants. *Ann. Rev. Phytopathol.* **1986**, *24*,141–157.

Hugouvieux-Cotte-Pattat, N.; Condemine, G.; Nasser, W.; Reverchon, S. Regulation of Pectinolysis in *Erwinia chrysanthemi*. *Ann. Rev. Microbiol.* **1996**, *50*, 213–257.

Hussey, R. S. Disease-inducing Secretions of Plant-Parasitic Nematodes. *Ann. Rev. Phytopathol.* **1989**, *27*, 123–141.

Itaya, A.; Matsuda, Y.; Gonzales, R. A.; Nelson, R. S.; Ding, B. Potato Spindle Tuber Viroid Strains of Different Pathogenicity Induces and Suppresses Expression of Common and Unique Genes in Infected Tomato. *Mol. Plant Microbe Interaction* **2002**, *15*, 990–999.

Jaouannet, M.; Magliano, M.; Arguel, M. J.; Gourgues, M.; Evangelisti, E.; Abad, P.; Rosso, M. N. The Root-knot Nematode Calreticulin Mi-CRT is a Key Effector in Plant Defense Suppression. *Mol. Plant Microbe Interaction* **2013**, *26*, 97–105.

Jaouannet, M.; Perfus-Barbeoch, L.; Deleury, E.; Magliano, M.; Engler, G.; Vieira, P.; Danchin, E. G.; Da Rocha, M.; Coquillard, P.; Abad, P.; Rosso, M. N. A Root-knot

Nematode-Secreted Protein is Injected into Giant Cells and Targeted to the Nuclei. *New Phytol.* **2012,** *194,* 924–931.

Jaubert, S.; Ledger, T. N.; Laffaire, J. B.; Piotte, C.; Abad, P.; Rosso, M. N. Direct Identification of Stylet Secreted Proteins from Root-Knot Nematodes by A Proteomic Approach. *Mol. Biochem. Parasit.* **2002,** *121,* 205–211.

Jaubert, S.; Milac, A. L.; Petrescu, A. J.; de Almeida-Engler, J.; Abad, P.; Rosso, M. N. In Planta Secretion of a Calreticulin by Migratory and Sedentary Stages of Root-Knot Nematode. *Mol. Plant Microbe Interaction* **2005,** *18,* 1277–1284.

Jha, G.; Rajeshwari, R.; Sonti, R. V. Bacterial Type Two Secretion System Secreted Proteins: Double-Edged Swords for Plant Pathogens. *Mol. Plant Microbe Interaction* **2005,** *18,* 891–898.

Joel, D. M.; Losner-Goshen, D. The Attachment Organ of the Parasitic Angiosperms *Orobanche cumana* and *O. aegyptiaca* and its Development. *Canadian J. Bot.* **1994,** *72,* 564–574.

Joel, D. M.; Portnoy, V. H. The Angiospermous Root Parasite *Orobanche Ramosa* L. (Orobanchaceae) Induces Expression of A Pathogenesis Related (PR) Gene in Susceptible Tobacco Roots. *Ann. Bot.* **1998,** *81,* 779–781.

Joel, D. M.; Hershenhorn, J.; Eizenberg, H.; Aly, R.; Ejeta, G.; Rich, P. J.; Ransom, J. K.; Sauerborn, J.; Rubiales, D. Biol. Managem. Weedy Root Paras. *Hort. Rev.* **2007,** *33,* 267–349.

Jones, J.; Robertson, W. Nematodes Secretions. In: *Aspects of Plant Parasitic Nematode Interactions*; Fenoll C., Grundler, F. M., Ohl, S. A., Eds; Kluwer Academic Publishers: The Netherlands, 1997, pp 98–106.

Jones, J. D. G.; Dangl, J. L. The Plant Immune System. *Nature* **2006,** *444,* 323–329.

Karczmark, A.; Overmars, H.; Helder, J.; Goverse, A. Feeding Cell Development by Cyst and Root-Knot Nematodes Involves a Similar Early, Local and Transient Activation of A Specific Auxin-Inducible Promoter Element. *Mol. Plant Pathol.* **2004,** *5,* 343–346.

Keon, J. P,. R.; Byrde, R. J. W.; Cooper, R. M. Some Aspects of Fungal Enzymes That Degrade Cell Walls. In: *Fungal Infection of Plants*; Pegg, G. F., Ayres, P. G., Eds, Symposium of the British Mycological Society Cambridge University Press: Cambridge, 1987, pp 133–157.

Keyes, W. Y; O'Malley, R. C.; Kim, D.; Lynn, D. G. Signaling Organogenesis in Parasitic Angiosperms: Xenognosin Generation, Perception, and Response. *J. Plant Growth Reg.* **2000,** *19,* 217–231.

Kim, D.; Kocz, R.; Boone, L.; Keyes, W. J.; Lynn, D. G. On Becoming a Parasite: Evaluating the Role of Wall Oxidases in Parasitic Plant Development. *Chem. Biol.* **1998,** *5,*103–117.

Koeller, W.; Yao, C.; Trial, F.; Parker, D. M. Role of Cutinase in the Invasion of Plants. *Canadian J. Bot.* **1995,** *73,* 1109–1118.

Kreutz, C. A. J. *Orobanche: The European Broomrape Species.* Maastricht: Stichting Natuurpublicaties, Limburg, 1995.

Kudla, U.; Qin, L.; Milac, A.; Kielak, A.; Maissen, C.; Overmars, H.; Popeijus, H.; Roze E.; Petrescu, A.; Smant, G.; Bakker, J.; Helder, J. Origin, Distribution and 3D-Modeling of Gr-EXPB1, an Expansion from the Potato Cyst Nematode Globodera rostochiensis. *Febs Lett.* **2005,** *579,* 2451–2457.

Lee, C.; Chronis, D.; Kenning, C.; Peret, B.; Hewezi, T.; Davis, E. L.; Baum, T. J.; Hussey, R.; Bennett, M.; Mitchum, M. G. The Novel Cyst Nematode Effector Protein 19C07 Interacts with the Arabidopsis Auxin Influx Transporter LAX3 to Control Feeding Site Development. *Plant Physiol.* **2011,** *155,* 866–880.

Li, F.; Ding, S. W. Virus Counterdefense: Diverse Strategies for Evading the RNA-Silencing Immunity. *Ann. Rev. Microbiol.* **2006**, *60*, 503–531.

Linsel, K. J; Keiper, F. J; Forgan, A.; Oldach, K. H. New Insights into the Infection Process of *Rhynchosporium Secalis* in Barley Using GFP. *Fungal Genet. Biol.* **2014**, *48*, 124–131.

Lison, P.; Tarraga, S.; Lopez-Gresa, P.; Sauri, A.; Torres, C.; Campos, L.; Belles, J. M.; Conejero, V.; Rodrigo, I. A Non Coding Plant Pathogen Provokes Both Transcriptional and Posttranscriptional Alterations in Tomato. *Proteomics* **2013**, *13*, 833–844.

Lopez De Mendoza, M. E.; Modha, J.; Roberts, C.; Curtis, R. H. C.; Kusel, J. Observations of the Changes of the Surface Cuticle of Parasitic Nematodes Using Fluorescent Probes. *Parasitology* **2000**, *120*, 203–209.

Lozano-Torres, J. L.; Wilbers, R. H. P.; Gawronski, P.; Boshoven, J. C.; Finkers-Tomczak, A.; Cordewener, J. H. G.; America, A. H. P.; Overmars, H. A.; Van't Klooster, J. W.; Baranowski, L.; Sobczak, M.; Ilyas, M.; Van der Hoorn, R. A. L.; Schots, A.; De Wit, P. J. G. M.; Bakker, J.; Goverse, A.; Smant, G. Dual Disease Resistance Mediated by the Immune Receptor Cf-2 in Tomato Requires a Common Virulence Target of a Fungus and a Nematode. *Proc. Natl. Acad. Sci. USA* **2012**, *109*, 10119–10124.

Lu, S. W.; Chen, S. Y.; Wang, J. Y.; Yu, H.; Chronis, D.; Mitchum, M. G.; Wang, X. H. Structural and Functional Diversity of CLAVATA3/ESR (CLE)-Like Genes from the Potato Cyst Nematode *Globodera rostochiensis*. *Mol. Plant Microbe Interaction* **2009**, *22*, 1128–1142.

Maniataki, E.; Tabler, M; Tsagris. M. Viroid RNA Systemic Spread May Depend on the Interaction of a 71-Nucleotide Bulged Hairpin with the Host Proteinvirp1. *RNA* **2003**, *9*, 346–354.

Maule, A.; Leh, V.; Lederer, C. The Dialogue Between Viruses and Hosts in Compatible Interactions. *Curr. Opin. Plant Biol.* **2002**, *5*, 279–284.

Melotto, M.; Underwood, W.; Koczan, J.; Nomura, K.; He, S. Y. Plant Stomata Function in Innate Immunity Against Bacterial Invasion. *Cell* **2006**, *126*, 969–980.

Mishagi, I. J. The Role of Pathogen Produced Cell-Wall Degrading Enzymes in Infection and Damage. In: *Physiology and Biochemistry of Plant Pathogen Interactions*; Mishagi, I. J., Ed; Plenum Press: New York and London, 1982.

Modha, J., Kusel J.R. and Kennedy, M. (1995). A role for second messengers in the control of activation of the surface of *Trichinella spiralis* infective larvae. Molecular Biochemical Parasitology, 72, 141.

Moissiard, G.; Voinnet, O. RNA Silencing of Host Transcripts by Cauliflower Mosaic Virus Requires Coordinated Action of the Four Arabidopsis Dicer-like Proteins. *Proc Natl Acad Sci U S A* **2006**, *103*, 19593–19598.

Mukasa, S. B.; Rubaihayo, P. R.; Valkonen, J. P. T. Interactions Between a Crinivirus, an Ipomovirus and a Potyvirus in Coinfected Sweetpotato Plants. *Plant Pathol.* **2006**, *55*, 458–467.

Newey, L. J.; Caten, C. E.; Green, J. R. Rapid Adhesion of *Stagonospora Nodorum* Spores to a Hydrophobic Surface Requires Pre-formed Cell Surface Glycoproteins. *Mycol. Res.* **2007**, *111*, 1255–1267.

Ngele, K. K.; Kalu, U. N. Studies on Different Species of Plant Parasitic Nematodes Attacking Vegetable Crops Grown in Afikpo North L.G.A, Nigeria. *Dir. Res. J. Agric. Food Sci.* **2015**, *3* (4), 88–92.

Noctor, G.; Veljovic-Jovanovic, S.; Driscoll, S.; Novitskaya, L.; Foyer, C. H. Drought and Oxidative Load in Wheat Leaves: A Predominant Role for Photorespiration. *Ann. Bot.* **2002**, *89*, 841–850.

O'Brien, M. J.; Raymer, W. B. Symptomless Hosts of the Potato Spindle Tuber Virus. *Phytopathology* **1964**, *54*, 1045–1047.

Oliver, R. P.; Ipcho, S. V. Arabidopsis Pathology Breathes New Life Into the Necrotrophs Vs Biotrophs Classification of Fungal Pathogens. *Mol. Plant Pathol.* **2004**, *5*, 347–352.

Olsen, D. P.; Phu, D.; Libby, L. J. M.; Cormier, J. A.; Montez, K. M.; Ryder, E. F.; Politz, S. M. Chemosensory Control of Surface Antigens Switching in the Nematode *Caenorhabditis Elegans*. *Genes, Brain Behaviour* **2007**, *6*, 240–252.

Opperman, C. H.; Bird, D. M.; Williamson, V. M.; Rokhsar, D. S.; Burke, M.; Cohn, J.; Cromer, J.; Diener, S.; Gajan, J.; Graham, S.; Houfek, T. D.; Liu, Q.; Mitros, T.; Schaff, J.; Schaffer, R.; Scholl, E.; Sosinski, B. R.; Thomas, V. P.; Windham, E. Sequence and Genetic Map of *Meloidogyne Hapla*: A Compact Nematode Genome For Plant Parasitism. *Proc. Natl. Acad. Sci. USA* **2008**, *105*, 14802–14807.

Owens, R. A.; Hammond, R. Viroid Pathogenicity: One Process, Many Faces. *Viruses* **2009**, *1*, 298–316.

Owens, R. A.; Tech, K. B.; Shao, J. Y.; Sano, T.; Baker, C. J. Global Analysis of Tomato Gene Expression During Potato Spindle Tuber Viroid Infections Reveals A Complex Array of Changes Affecting Hormone Signaling. *Mol. Plant Microbe Interaction* **2012**, 25, 582–598.

Padmanabhan, M. S.; Kramer, S. R.; Wang, X.; Culver, J. N. Tobacco Mosaic Virus Replicase-Auxin/Indole Acetic Acid Protein Interactions: Reprogramming the Auxin Response Pathway to Enhance Virus Infection. *J. Virol.* **2008**, *82*, 2477–2485.

Padmanabhan, M. S.; Shiferaw, H.; Culver, J. N. The Tobacco Mosaic Virus Replicase Protein Disrupts the Localization and Function of Interacting Aux/IAA Proteins. *Mol. Plant Microbe Interaction* **2006**, *19*, 864–873.

Palukaitis, P. Potato Spindle Tuber Viroid: Investigation of the Long-Distance, Intra-Plant Transport Route. *Virology* **1987**, 158, 239–241.

Pannecoucque, J.; Hofte, M. Interactions Between Cauliflower and Rhizoctonia Anastomosis Groups with Different Levels of Aggressiveness. *BMC Plant Biol.* **2009**, *9*, 95.

Parker, C.; Riches, C. R. *Parasitic Weeds of the World: Biology and Control.* CABI: Wallingford, UK, 1993, pp 114–116.

Patel, N.; Hamamouch, N.; Li, C. Y.; Hussey, R.; Mitchum, M.; Baum, T.; Wang, X. H.; Davis, E. L. Similarity and Functional Analyses of Expressed Parasitism Genes in *Heterodera schachtii* and *Heterodera glycines*. *J. Nematol.* **2008**, *40*, 299–310.

Perez-de-Luque, A.; Galindo, J. C. G.; Macıas, F. A.; Jorrın, J. Sunflower Sesquiterpene Lactone Models Induce *Orobanche cumana* Seed Germination. *Phytochemistry* **2000**, *53*, 45–50.

Perry, R. N. An Evaluation of Types of Attractants Enabling Plant-Parasitic Nematodes to Locate Plant Roots. *Russian J. Nematol.* **2001**, *13*, 83–88.

Postma, W. J.; Slootweg, E. J.; Rehman, S.; Finkers-Tomczak, A.; Tytgat, T. O.; van Gelderen, K.; Lozano-Torres, J. L.; Roosien, J.; Pomp, R.; van Schaik, C.; Bakker, J.; Goverse, A.; Smant, G. The Effector SPRYSEC-19 Of *Globodera Rostochiensis* Suppresses CC-NB-LRR-Mediated Disease Resistance in Plants. *Plant Physiol.* **2012**, *160*, 944–954.

Press, M. C.; Graves, J. D.; Stewart, G. R. Physiology of the Interaction of Angiosperm Parasites and Their Higher Plant Hosts. *Plant Cell Environ.* **1990**, *13*, 91–104.

Preston, G. M.; Studholme, D. J; Caldelari, I. Profiling the Secretomes of Plant Pathogenic Proteobacteria. *Fems Microbiol. Rev.* **2005**, *29*, 331–360.

Prins, T. W.; Tudzynski, P.; Tiedemann, A. V. Tudzynski, B.; Ten, H. A.; Hansem, M. E.; Tenberge, K.; Van Kan, J. A. L. Infection Strategies of *Botrytis Cinerea* and Related Necrotrophic Pathogens. In: Fungal Pathology, Kronstad, J. W., Ed; Kluwer: Dordrecht, pp 33–64.

Prior, A.; Jones, J.; Block, V.; Beauchamp, J.; Mcdermott, L.; Cooper, A.; Kennedy. M. A. Surface-associated Retinol- and Fatty Acid Protein (Gp-FAR-1) From the Potato Cyst Nematode *Globodera Pallida*: Lipid Binding Activities, Structural Analysis and Expression Pattern. *Biochem. J.* **2001**, *356*, 387–394.

Prot, J. C. Migration of Plant-Parasitic Nematodes Towards Roots. *Revue de Nematologie* **1980**, *3*, 305–318.

Proudfoot, L.; Kusel, J. R.; Smith, H. V.; Kennedy, M. External Stimuli and Intracellular Signalling in the Modification of the Nematode Surface During Transition to the Mammalian Host Environment. *Parasitology* **1993**, *107*, 559–566.

Pruss, G.; Ge, X.; Shi, X. M.; Carrington, J. C.; Bowman Vance, V. Plant Viral Synergism: The Potyviral Genome Encodes a Broadrange Pathogenicity Enhancer that Transactivates Replication of Heterologous Viruses. *Plant Cell* **1997**, *9*, 859–868.

Qi, Y.; Ding, B. Differential Sub Nuclear Localization of RNA Strands of Oppositepolarity Derived From an Autonomously Replicating Viriod. *Plant Cell* **2003**, *15*, 2566–2577.

Qi, Y.; Pelisser, T.; Itaya, A.; Hunt, E.; Wassengger, M.; Ding, B. Direct Role of a Viroid RNA Motif in Mediating Directional RNA Trafficking Across a Cellular Boundary. *Plant Cell* **2004**, *16*, 1741–1752.

Qin, L.; Kudla, U.; Roze, E. H. A.; Goverse, A.; Popeijus, H.; Nieuwland, J.; Overmars, H.; Jones, J. T.; Schots, A.; Smant, G.; Bakker, J.; Helder J. Plant Degradation: A Nematode Expansion Acting on Plants. *Nature* **2004**, *427*, 30–30.

Rauscher, M.; Mendgen, K.; Deising, H. Extracellular Proteases of the Rust Fungus *Uromyces viciae-fabae*. *Exp. Mycol.* **1995**, *19*, 26–36.

Rawlings, S. L.; O'connell, R. J.; Green, J. R. The Spore Coat of the Bean Anthracnose Fungus *Colletotrichum Lindemuthianum* is Required for Adhesion, Appressorium Development and Pathogenicity. Physiol. *Mol. Plant Pathol.* **2007**, *70*, 110–119.

Rehman, S.; Postma, W.; Tytgat, T.,; Prins, P.; Qin, L.; Overmars, H.; Vossen, J.; Spiridon, L. N.; Petrescu, A. J.; Goverse, A.; Bakker, J.; Smant, G. A Secreted SPRY Domain-Containing Protein (SPRYSEC) from the Plant-Parasitic Nematode *Globodera rostochiensis* Interacts with a CC-NB-LRR Protein from a Susceptible Tomato. *Mol. Plant Microbe Interaction* **2009**, *22*, 330–340.

Robertson, L.; Robertson, W. M.; Sobczak, M.; Helder, J.; Tetaud, E.; Ariyanayagam, M. R.; Fergusson, M. A. J.; Fairlamb, A.; Jones J. Cloning, Expression and Functional Characterisation of a Peroxiredoxin from the Potato Cyst Nematode *Globodera Rostochiensis*. *Mol. Biochem. Parasitol.* **2000**, *111*, 41–49.

Robinson, F. Nematodes Behaviour and Migration Through Soil and Host Tissue. In: *Basis of Behaviour*; Chen, C., Chen, S., Dickson, D. W., Eds; Recent Nematology Topics, 2002, pp 331–401.

Robinson, M. P.; Atkinson, H. J.; Perry, R. N. The Influence of Soil Moisture and Storage Time on the Motility, Infectivity and Lipid Utilization of Second Stage Juveniles of the Potato Cyst Nematodes *Globodera rostochiensis* and *G. Pallida*. *Revue de Nematologie* **1987**, *10*, 343–348.

Rodriguez, A.; Carlos, A.; Lutz, A. L.; Leisner, S. M.; Nelson, R. S.; Schoelz, J. E. Association of the P6 Protein of Cauliflower Mosaic Virus with Plasmodesmata and Plasmodesmal Proteins. *Plant Physiol.* **2014**, *166*, 1345–1358.

Saenz, P.; Quiot, L.; Quiot, J. -B.; Candresse, T.; Garcia, J. A. Pathogenicity Determinants in the Complex Virus Population of a Plum Pox Virus Isolate. *Mol. Plant Microbe Interaction* **2001**, *14*, 278–287.

Sacco, M. A.; Koropacka, K.; Grenier, E.; Jaubert, M. J.; Blanchard, A.; Goverse, A.; Smant, G.; Moffett, P. The Cyst Nematode SPRYSEC Protein RBP-1 Elicits Gpa2-and RanGAP2-Dependent Plant Cell Death. *Plos Pathol.* **2009**, *5*, e1000564.

Sano, T.; Candresse, T.; Hammond, R. W.; Diener, T. O.; Owens, R. A. Identification of Multiple Structural Domains Regulating Viroid Pathogenicity. *Proc. Natl. Acad. Sci. USA* **1992**, *89*, 10104–10108.

Sato, D.; Awad, A. A.; Chae, S. H.; Yokota, T.; Sugimoto, Y.; Takeuchi, Y.; Yoneyama, K. Analysis of Strigolactones, Germination Stimulants for Striga and Orobanche, by High Performance Liquid Chromatography/Tandem Mass Spectrometry. *J. Agric. Food Chem.* **2003**, *51*, 1162–1168.

Saxena, R.; Singh, R. Survey of Nematode Fauna of Groundnut, *Arachis Hypogea* In and Around Bareilly Region, U.P., India. *Curr. Nematol.* **1997**, *8* (1&2), 93–97.

Schafer, W. The Role of Cutinase in Fungal Pathogenicity. *Trends in Microbiol.* **1993**, *1* (2), 69–17.

Scheets, K. Maize Chlorotic Mottle Machlomovirus and Wheat Streak Mosaic Rymovirus Concentrations Increase In the Synergistic Disease Corn Lethal Necrosis. *Virology* **1998**, *242*, 28–38.

Schnolzer, M.; Haas, B.; Ramm, K.; Hofman, H.; Sanger, H. L. Correlation Between Structure and Pathogenicity of Potato Spindle Tuber Viroid (PSTV). *EMBO J.* **1985**, *4*, 2181–2190.

Serghini, K.; Perez de Luque, A.; Castejon Munoz, M.; Garcia-Torres, L.; Jorrín, J. V. Sunflower (*Helianthus annuus* L.) Response to Broomrape (*Orobanche cernua* Loefl.) Parasitism: Induced Synthesis and Excretion of 7-hydroxylated Simple Coumarins. J. Exp. Bot. **2001**, *364*, 2227–2234.

Shaykh, M.; Soliday, C.; Kolattukudy, P. E. Proof for the Production of Cutinase by *Fusarium Solani F. Pisi* During Penetration into its Host, *Pisum sativum*. Plant Physiol. **1977**, *60* (1), 70–72.

Shimura, H.; Pantaleo, V.; Ishihara, T.; Myojo, N.; Inaba, J. I.; Sueda, K.; Burgyan, J.; Masuta, C. A Viral Satellite RNA Induces Yellow Symptoms on Tobacco by Targeting a Gene Involved in Chlorophyll Biosynthesis Using the RNA Silencing Machinery. *PLoS Pathol.* **2011**, *7*, e1002021.

Siemens, J.; Keller, I.; Sarx, J.; Kunz, S.; Schuller, A.; Naggel, W.; Schmulling, T.; Parniske, M.; Ludwig-Muller, J. Erratum: Transcriptome Analyses of Arabidopsis Clubroots Indicate a Key Role for Cytokinins in Disease Development. *Mol. Plant -Microbe Interactions* **2006**, *19*, (5), 480–494.

Silveira, P. R.; Nascimento, K. J. T.; Andrade, C. C. L.; Bispo, W. M. S.; Oliveira, J. R.; Rodrigues, F. A. Physiological Changes in Tomato Leave Arising from *Xanthomonas gardneri* infection. *Physiol. Mol. Plant Pathol.* **2015**, i,130–138.

Smant, G.; Stokkermans, J. P. W. G.; Yan, Y. T.; De Boer, J. M.; Baum, T. J.; Wang, X. H.; Hussey, R. S.; Gommers, F. J.; Henrissat, B.; Davis, E. L.; Helder, J.; Schots, A.; Bakker, J. Endogenous Cellulases in Animals: Isolation of Beta-1,4-Endoglucanase Genes from

Two Species of Plant-Parasitic Cyst Nematodes. *Proc. Natl. Acad Sci. USA* **1998**, *95*, 4906–4911.

Smith, N. A.; Eamens, A. L.; Wang, M. -B. Viral Small Interfering Rnas Target Host Genes to Mediate Disease Symptoms in Plants. *PLoS Pathol.* **2011**, *7*, e1002022.

Stamps, W. T.; Lint, M. Interaction of Intrinsic and Extrinsic Chemical Cues in the Behaviour of *Bursaphelenchus xylophilus* (Aphelenchida: Aphelenchoididae) in Relation to Its Beetle Vectors. *Nematology* **2001**, *3*, 295–301.

Tajul, M. I.; Naher, K.; Hossain, T.; Siddiqui, Y.; Sariah, M. Tomato Yellow Leaf Curl Virus (TYLCV) Alters the Phytochemical Constituents in Tomato Fruits. *Austr. J. Crop Sci.* **2011**, *5* (5), 575–581.

Takeda, P.; Ding, B. Viroid Intercellular Trafficking: RNA Motifs, Cellular Factors and Broad Impacts. Viruses **2009**,*1*, 210–221.

Talbot, N. J. On the Trail of a Cereal Killer: Exploring the Biology of *Magnaporthe grisea*. *Ann. Rev. Microbiol.* **2003**, *57*, 177–202.

Tenberge, K. B.; Homann, V.; Oeser, B.; Tudzynski, P. Structure and Expression of Two Polygalacturonase Genes of Claviceps Purpurea Oriented in Tandem and Cytological Evidence for Pectinolytic Enzyme Activity During Infection of Rye. *Phytopathology* **1996**, *86*, 1084–1097.

Timmermans, B. C. G.; Vos, J.; Stomph, T. J.; Van Nieuwburg, J.; Van Der Putten, P. E. L. Field Performance of *Solanum Sisymbriifolium*, a Trap Crop for Potato Cyst Nematodes. II. Root Characteristics. *Ann. Appl. Biol.* **2007**, *150*, 99–106.

Toth, I. K.; Bell, K. S.; Holeva, M. C.; Birch, P. R. J. Soft Rot Erwiniae: From Genes to Genomes. *Mol. Plant Pathol.* **2003**, *4*, 17–30.

Valli, A.; Lopez-Moya, J. J.; Garcıa, J. A. *RNA Silencing and Its Suppressors in the Plant-Virus Interplay*. John Wiley & Sons Ltd: Chichester, UK, 2009.

Van Woerden I. C.; van Ast, A.; Zaitoun, F. M. F.; Ter Borg, S. J. Root Exudates of Resistant Faba Bean Cultivars are Strong Stimulants of Broomrape Germination. In: *Third International Workshop on Orobanche and Related Striga Research*; Pieterse, A. H.; Verkleij, J. A. C.; Ter Borg, S. J. Eds; The Royal Tropical Institute: Amsterdam, The Netherlands, 1994.

Wang, H.; Li, J.; Bostock, R. M.; Gilchrist, D. G. Apoptosis: A Functional Paradigm for Programmed Plant Cell Death Induced by a Host-Selective Phytotoxin and Invoked During Development. *Plant Cell* **1996**, *8*, 375–391.

Wang, J.; Hewezi, T.; Baum, T. J.; Davis, E. L.; Wang, X.; Mitchum, M. G. Trafficking of Soybean Cyst Nematode Secreted CLE Proteins in Plant Cells. *Phytopathology* **2010a**, *100*, S132–S132.

Wang, J. Y.; Lee, C.; Replogle, A.; Joshi, S.; Korkin, D.; Hussey, R.; Baum, T. J.; Davis, E. L.; Wang, X. H.; Mitchum, M. G. Dual Roles for the Variable Domain in Protein Trafficking and Host-Specific Recognition of *Heterodera glycines* CLE Effector Proteins. *New Phytol.* **2010b**, *187*, 1003–1017.

Wang, X. H.; Mitchum, M. G.; Gao, B. L.; Li, C. Y.; Diab, H.; Baum, T. J.; Hussey, R. S.; Davis, E. L. A Parasitism Gene from a Plant-Parasitic Nematode with Function Similar to CLAVATA3/ESR (CLE) of *Arabidopsis thaliana*. *Mol. Plant Pathol.* **2005**, *6*, 187–191.

Weingart, H.; Volksch, B. Ethylene Production by *Pseudomonas syringae* pathovars In Vitro and In Planta. *Appl. Environ. Microbiol.* **1997**, *63*,156–161.

Weingart, H.; Ullrich, H.; Geider, K.; Volksch, B. The Role of Ethylene Production in Virulence of *Pseudomonas syringae* pvs. *glycinea* and *phaseolicola*. *Phytopathology* **2001**, *91*, 511–518.

Weingart, H.; Volksch, B.; Ullrich, M. S. Comparison of Ethylene Production by *Pseudomonas syringae* and *Ralstonia solanacearum*. *Phytopathology* **1999**, *89*, 360–365.

Westwood, J. H.; Yu, X.; Foy, C. L.; Cramer, C. L. Expression of a Defense-related 3-hydroxy-3 Methylglutaryl CoA Reductase Gene in Response to Parasitization by *Orobanche spp*. *Molecular Plant-Microbe Interactions* **1998**, *11*, 530–536.

Williamson, V. M.; Hussey, R. S. Nematode Pathogenesis and Resistance in Plants. *Plant Cell* **1996**, *8*, 1735–1745.

Wubben, J. P.; Mulder, W.; ten Have, A.; van Kan, J. A.; Visser, J. Cloning and Partial Characterization of Endopolygalacturonase Genes from *Botrytis cinerea*. *Appl. Environ. Microbiol.* **1999**, *65*, 1596–1602.

Wuyts, N.; Swennen, R.; Waele, D. Effects of Plant Phenylpropanoid Pathway Products and Selected Terpenoids and Alkaloids on the Behaviour of the Plant Parasitic Nematodes *Radopholis similis*, *Pratylenchus penetrans* and *Meloidogyne incognita*. *Nematology* **2006**, *8*, 89–101.

Wyss, U.; Zunke, U. Observations on the Behaviour of Second Stage Juveniles of *Heterodera Schachtii* Inside Host Roots. *Revue de Nematologie* **1986**, *9*, 153–165.

Yamagishi, D.; Akamatsu, H.; Otani, H.; Kodama, M. Pathological Evaluation of Host-Specific AAL-Toxins and Fumonisin Mycotoxins Produced by *Alternaria* and *Fusarium* Species. *J. Gen. Plant Pathol.* **2006**, *72*, 323–326.

Yang, S.; Ravelonandro, M. Molecular Studies of the Synergistic Interactions Between Plum Pox Virus HC-Pro Protein and Potato Virus X. Arch. Virol. **2002**, *147*, 2301–2312.

Yakoby, N.; Freeman, S.; Dinoor, A.; Keen, N. T.; Prusky, D. Expression of Pectate Lyase from *Colletotrichum gloeosporioides* in *Collectotrichum magna* Promotes Pathogenicity. *Mol. Plant Microbe Interaction* **2000**, *13*, 887–891.

Yoder, J. I. Parasitic Plant Responses to Host Plant Signals: A Model for Subterranean Plant-Plant Interactions. *Curr. Opin. Plant Biol.* **1999**, *2*, 65–70.

Yokota, T.; Sakai, H.; Okuno, K.; Yoneyama, K.; Takeuchi, Y. Alectrol and Orobanchol, Germination Stimulants for Orobanche Minor, from Its Host Red Clover. *Phytochemistry* **1998**, *49*, 1967–1973.

Yoneyama, K.; Takeuchi, Y.; Yokota, T. Production of Clover Broomrape Seed Germination Stimulants by Red Clover Root Requires Nitrate, But is Inhibited By Phosphate and Ammonium. *Physiologia Plantarum* **2001**,*112*, 25–30.

Yun, M. H.; Torres, P. S.; El Oirdi, M.; Rigano, L. A.; Gonzalez-Lamothe, R.; Marano, M. R.; Castagnaro, M. A.; Dankert, K.; Bouarab; Vojnov, A. A. Xanthan Induces Plant Susceptibility by Suppressing Callose Deposition. *Plant Physiol.* **2006**, *141*,178–187.

Zasada, I. A.; Meyer, S. L. F.; Halbrendt, J. M.; Rice, C. Activity of Hydroxamic Acids from *Secale Cereale* Against the Plantparasitic Nematodes *Meloidogyne incognita* and *Xiphinema americanum*. *Phytopathology* **2005**, *95*, 1116–1121.

Zhang, L.; Davies, L. J.; Elling, A. A. A *Meloidogyne incognita* Effector is Imported Into the Nucleus snd Exhibits Transcriptional Activation Activity in Planta. Mol. Plant Pathol. **2015**, *16*, 48–60.

Zhao, X.; Schmitt, M.; Hawes, M. Species-dependent Effects of Border Cell and Root Tip Exudates on Nematode Behaviour. *Nematology* **2000**, *90*, 1239–1245.

Zhao, Y.; Owens, R. A.; Hammond, R. W. Use of a Vector Based on Potato Virus X in a Whole Plant Assay to Demonstrate Nuclear Targeting of Potato Spindle Tuber Viroid. *J. Gen. Virol.* **2001,** *82,* 1491–1497.

Zhong, Y.; Tao, X.; Stombaugh, J.; Leontis, N.; Ding, B. Tertiary Structures and Function of an RNA Motif Required for Plant Vascular Entry to Initiate Systemic Trafficking. *EMBO J* **2007,** *26,* 3836–3846.

Zhu, Y.; Qi, Y.; Yun, Y.,; Owens, R.; Ding, B. Movement of Potato Spindle Tuber Viroid Reveals Regulatory Points of Phloem-Mediated RNA Traffic. *Plant Physiol.* **2002,** *130,*138–146.

Zhu, Y.; Green, L.; Woo, Y. M.; Owens, R.; Ding, B. Cellular Basis of Potato Spindle Tuber Viroid Systemic Movement. *Virol.* **2001,** *279,* 66–77.

Zuckerman, B. M. Hypotheses and Possibilities of Intervention in Nematode Chemoresponses. *J. Nematol.* **1983,** *15,* 173–182.

CHAPTER 12

Virus Diseases: An Inimitable Pathosystem of Vegetable Crop

MOHAMMAD ANSAR*, ANIRUDDHA KUMAR AGNIHOTRI, and
SRINIVASARAGHAVAN A.

*Department of Plant Pathology, Bihar Agricultural University, Sabour
813210, Bihar, India*

Corresponding author. E-mail: ansar.pantversity@gmail.com

ABSTRACT

Viruses are most important cause of loss in many vegetable crops. Usually, complex relationship between the virus, host plant, and the vector creates trouble in formulation of efficient management system. Characterization of causal viruses and understanding their epidemiology is the key to assess the severity and economic impact of the diseases and efficient management. The upshot of interactions among virus, host, vector, and environmental dynamics decides the nature of epidemics. Each epidemic can be considered a distinct pathosystem in which each of the components contributes too. Characterization of virus associated with the different pathosystems and their vectors is the basis for developing management strategies. In the perspective of virus infection in vegetable crop, we categorized 11 pathosystem which has different modes of infection. Each pathosystem has been discussed based on modes of transmission and dissemination *viz.* seed or vegetative propagules carrying virus, insect vector, nematodes, and fungal transmission. On the basis of spread mode in various pathosystem, different management strategies applicable are discussed.

12.1 INTRODUCTION

Virtually, a pathosystem is subsystem of an ecosystem and it is explicit by the event of parasitism. Under phytopathosystem, one end host species is a plant and in another end, the parasite spends a significant part of its lifespan inhabiting on the host individual. The parasite may be fungi, bacteria, virus, insects, and nematodes. The concept of pathosystem includes its analysis and management. On the basis of system control, pathosystem is broadly divided in two distinct classes, natural and crop pathosystem. Natural pathosystem is considered as wild pathosystem which is independent. It is primarily due to communication between basic components of the pathosystem, for example, host, pathogen, environment, and linked vector. In artificial or crop pathosystem, human role has an additional component influencing the other three components intensely. Under crop pathosystem, cultivars deferred from wild host, cultivation also deferred from wild ecosystem and pathogen population can be unnaturally controlled. Viral diseases are serious limiting factor of productivity and profitability of a wide range of vegetable crops. Characterization of causal viruses and understanding their epidemiology is the key to ballpark figure of incidence and economic impact, and formulate control strategies. The interactions between virus, host plant, vector, and environmental factors influence epidemics, and each epidemic can be well thought out of a unique pathosystem in which each of the components leads to rise in and none are limiting. Diagnosis of pathogen is the key to managing the disease, and we listed vegetable crops, along with associated virus with pathosystem is depicted. Here, we described 11 pathosystem on the basis of the mode of transmission or dissemination, and the different line of action as applicable to each virus discussed.

Plant viruses are clustered into 73 genera and 49 families. However, these figures relate only to cultivated plants that represent only a small portion of the total number of plant species. Viruses in wild plants have been studied poorly, but existing studies almost overwhelmingly shows such interactions between wild plants and their viruses do not appear to cause disease in the host plants. Transmission of virus from one plant to another and from one plant cell to another, plant viruses must use strategies that are usually different from animal viruses. Plants do not move, therefore, plant to plant transmission usually involves vectors. The virus genus belonging to various families are interacting by various means like insect vector which play a significant role in dissemination and dispersal. Among virus groups, whitefly-transmitted geminiviruses, aphid-transmitted potyviruses, and thrips-transmitted tospoviruses covered a wide range of solanaceous, cucurbitaceous, and leguminous hosts. Viral diseases are detected by a varying range of symptoms which is generally distinct from other microbes (Agrios, 2005). It can be detected through a variety of techniques including mechanical inoculation or vector transmission to indicator hosts, electron microscopy, or serological assays (Davis and Rua-bete, 2010). At present, preference has been given to various molecular tests such as the polymerase chain reaction (PCR), rolling circle amplification (RCA) for circular genome, and next generation sequencing (NGS). PCR assays are robustly used in many virus genera within a short period of time. Moreover, it is possible to obtain sequence information in order to allow classification of virus at genus or species level. Detection of RNA viruses is taken into consideration by applying cDNA protocol through reverse transcription step in PCR and downstream applications. Demarcation of species, genus, and family is important because it allows relating with similar viruses and helps to envisage economic and biological properties relevant to pathosystem components.

12.2 EPIDEMICS AND PATHOSYSTEM

The viruses have unique biological properties which determine its chain of events in disease development. Every virus species followed specific mode of dissemination from one plant to another. Interaction of factors like virus, host plant, associated vector, environment, time, and human involvement regulate the degree of epidemics. An epidemic well thought out an ecosystem in which the essential components coincide to allow the virus dissemination. The vegetable pathosystem is the subsystems of ecosystem in which viral pathogens are the prime element of interest, and it can be considered an epidemic as a plant virus pathosystem. When occurrence of epidemic is considered a pathosystem, control approaches can be followed at any of its components in order to subvert and downregulating the progress. To understand clearly, we explain 11 classes of virus pathosystem, and make use of descriptors for vegetable crops (Table 12.1). Each pathosystem is distinguished by the means of spread of the virus. All are having certain biological implications influencing the progress of an epidemic of particular virus. In the illustration of pathosystem B1, infected vegetatively propagated plants will produce infected offspring up to 100%. Moreover, if a virus passes in seed propagules and is also vector transmissible, the pathosystem will be linked with other pathosystem and incidence aggravated by vector ecology.

TABLE 12.1 Description of Pathosystem (A1–F1) Showing Their Modes of Spread Between Infector and Infectee.

Viral-pathosystem	Description
Pathosystem-A1	Seed: Horizontal cell–cell movement of virus within somatically identical clone
Pathosystem-A2	Pollen: Vertical cell–cell spread of virus via gametes between sexually compatible host plants
Pathosystem-B1	Tuber/vegetative plant part: Horizontal cell–cell movement of virus
Pathosystem-C1	Mechanical: Horizontal spread of free virus particle without a vector
Pathosystem-C2	Dodder: Horizontal movement of virus through bridge
Pathosystem-D1	Insect vector (non persistent): Horizontal spread by vector without a latent period
Pathosystem-D2	Insect vector (semi persistent): Horizontal spread by vector without a latent period
Pathosystem-D3	Insect vector (persistent): Horizontal spread by vector with a latent period
Pathosystem-D4	Insect vector (propagative): Horizontal spread by a vector in which the virus replicates or is also vertically transmitted to progeny
Pathosystem-E1	Nematode vector: Horizontal spread under soil
Pathosystem-F1	Fungi: Horizontal spread under soil by adsorptive mode

12.3   ASSOCIATED VIRUS GROUPS IN VEGETABLE PATHOSYSTEM

12.3.1   PATHOSYSTEM A 1 AND 2

The three-way interplay of genetic mechanism of the virus, infected host maternal, and its progeny results in seed transmission of plant viruses. The passage of plant virus from generation to generation occurs in about 20% of viruses. As viruses are transmitted by seeds causes infection in generative cells and prolong in the germ cells and occasionally but less often, in seed coat. Under unfavorable condition when growth and development of plants are delayed, there is a greater chance of virus infections in seeds. The concept of mechanisms involved in the transmission of plant viruses via seeds is not well known, even though it is recognized as environmentally inclined and seed transmission takes place as direct incursion of the embryo via the ovule or by an indirect means as embryo mediated by infected gametes.

12.3.1.1 CAULIFLOWER MOSAIC VIRUS

Several vegetables can be infected through seeds but it is not limited to the families like *Leguminaceae, Solanaceae, Compositae, Rosaceae, Cucurbitaceae, Gramineae.* Under Caulimoviridae seven genera including *Caulimovirus, Soymovirus, Petuvirus, Cavemovirus, Solendovirus, Badnavirus,* and *Tungrovirus* are considered for seed transmission. All have a circular, noncovalently closed dsDNA genome of 7–9 kb and either isometric (50–52 nm) or bacilliform (30 × 130–150 nm) virions (King et al., 2012). In tropics, the *Caulimoviridae* reaches maximum diversity even though its best-known representative *Caulimovirus* is found in temperate regions. The latter four genera are almost exclusively distributed in tropical regions. Members of the *Caulimoviridae* are also known to infect vegetatively propagated crops. Therefore, most effective management strategy must be use of clean and healthy planting material. The use of disease resistant or tolerant cultivars is the best option for sustainable control.

12.3.1.2 CUCUMBER MOSAIC VIRUS

Cucumber mosaic virus (CMV) belonging to genus *Cucumovirus* (*family Bromoviridae*) is an important virus of vegetable crops. Owing to its extensive host range (>1000 plant species) and its economic impact, CMV has been considered as one of the most important virus group. The genome is organized into three single-stranded messenger sense genomic RNAs (RNAs 1, 2, and 3). RNAs 1 and 2 codes for a component of the replicase complex whereas RNA2 also codes for the 2b protein involved in the suppression of gene silencing. The 3a protein encoded by RNA3 is essential for the virus movement and the coat protein (CP) is expressed from subgenomic RNA4. The CP has an important role not only in the shape of the viral particles but also in virus movement, vectors transmission (aphid) and symptom expression (Crescenzy, 1993; Palukaitis et al., 1992).

Some CMV strains also link with satellite RNA molecules which are small, linear, noncoding and single-stranded RNA molecules that depends fully on the helper virus (CMV) for replication, encapsidation, and transmission but has almost no sequence similarity with the helper CMV genome. CMV strains can be divided in two subgroups, I and II, the former being further divided into subgroups IA and IB based on their biological, serological, and molecular properties. Asian strain suggested to subgroup IB, whereas other members of subgroup I have been kept under subgroup IA. Sequence identity between CMV subgroup II and I strains in nucleotides ranges from 69 to 77%, whereas > 90% identity within subgroup (Palukaitis et al., 2003).

12.3.1.3   CUCUMBER MOSAIC VIRUS INFECTING VEGETABLE CROPS

Cucurbits: Most of cucurbits are vulnerable to CMV with varying in symptoms severity. Severe epinasty, downward twisting of the petiole and leaf surface along with leaf reduction frequently appears in early season of infection in summer squash. Plants infected at premature stage are severely stunted and leaves are malformed and fruits are unmarketable because of distinct roughness on the fruit surface. Infection of vining crops like musk-melon, shows severely stunted growing tips even though fruit may not show symptoms of poor quality.

Pepper: Foliar symptoms of pepper plants may vary according to stage of infection. The initial flush of symptoms naturally includes a chlorosis of newly emerged leaves that may occur on basal portion and whole leaf. Subsequently, leaves developed chlorosis and chlorotic mosaic may have varied degrees of deformation including sunken interveinal lamina with protruding primary veins. These leaves also have a dull light green look as contrasting to the dark green, rather shiny leaves of healthy pepper plants. Symptomatic patterns vary in severity depending on the age of the plant. Seed coat infection of CMV ranged from 53 to 83% while the embryo ranged from 10 to 46% (Ali and Kobayashi, 2009). In seed growth tests transmission recorded approximately 10–14%.

Spinach: Expression of blight as a result of CMV infection in spinach. The symptoms can vary depending upon the variety, plant age when infection started, temperature, and virus strain. Characteristic symptom includes leaf chlorosis, which can increase to cause severe blighting of the growing point and ultimate plant death. In addition to chlorotic mottle, leaves can show narrowing, wrinkling with vein distortion, and inward leaf roll appearance.

Lettuce: CMV infection in lettuce having leaf mottling, severe bumpiness of the leaf and occasional necrosis within the leaf tissue. Plants are usually stunted if infected at an early stage of growth.

Tomato: Tomato plants at early stage become yellow, bushy appearance, and considerably stunted. The leaves may show a mottle which resembles to *Tomato mosaic virus* (*ToMV*). The most characteristic symptom of CMV is filiform or shoestring-like leaf blades. The symptoms can be temporary, with the bottom leaves or newly developed top leaves showing severe symptoms, while the middle leaves shows almost normal. Severely affected plant exhibits fewer numbers of fruits, which are usually small, often mottled or necrotic, with delayed maturity.

Bean: Leaf curl, green mottle and blistering, roughness along the main veins involving only a few leaves. Foliar symptoms are most clear. Moreover, pod formation stage loss is higher when plants are infected before bloom. Early infected plants may yield fewer pods because of flower abortion and abnormal development. The pods are mostly twisted, mottled, and reduced in size.

The transmission of CMV is also vectored by aphids having shortest acquisition period (5–10 s) and an inoculation period of about a minute. However, inoculation largely decreases after 2 min and within 2 h no longer ability to transmit by that particular vector. This virus is nonpersistent and is stylet borne. CMV is carried by 60–80 different species of aphids besides seed transmission. Therefore, CMV is linked with two pathosystem in order to spread from one host to another.

12.3.2 PATHOSYSTEM-B1

In order to carry forward the virus, vegetative propagules play a significant role. Tubers, rhizomes, stolons, corms, bulbs, and buds of economically important crops like cassava, potato, sugarcane, banana, sweet potato, beet root, onion, and majority of fruits and ornamental plants act as a carrier of virus.

12.3.2.1 POTATO VIRUS X (MILD MOSAIC)

In potato cultivation, propagation is undertaken through seed tubers which harbor number of viruses without expressing symptom. Yield losses due to the viruses are not confined to the year of infection but is sustained to increase gradually upto saturation with one or more viruses. Moreover, *Potato Virus A* is linked with PVX which play a key role in reducing the yield and affects seed production (Halterman et al. 2012; Yardmic et al., 2015). A mixed infection of PVX and A was detected in terai region of Uttarakhand state of India which affected severely seed production plots (Ansar and Singh, 2016). To distinguish the associated virus on the basis of symptomatology is a very difficult task. PVX belongs to the family of *Alphaflexiviridae* and the order *Tymovirales*. It is the type species of the genus *Potexvirus*. PVX is infecting potatoes and is only transmitted mechanically. There are no insect or fungal vectors documented for this virus. The virus induces mild symptoms sometimes without apparent expression in most potato varieties. Moreover, mixed infection with *Potato virus Y* synergistically causes severe symptoms. Virus indexing and restricted generation production of potato, with starting of disease-free tissue culture plantlets almost eliminated this virus from many countries.

12.3.2.2   GINGER CHLOROTIC FLECK VIRUS

The disease is marked by appearance of chlorotic flecks on the leaves of infected ginger (*Zingiber officinale:* Zingiberaceae). However, in rhizomes, no obvious symptoms are found. The virus is transmitted through the rhizome of infected plants. Vector transmissibility is still unknown. First documented report of this virus was from New Zealand Queensland, Australia (Thomas, 1986). At present, the disease is found in several parts of world in imported lines of ginger. *Ginger chlorotic fleck virus* has many properties in similar with viruses of sobemovirus group (Matthews, 1982), a predominant ssRNA species and a limited host range. The virus is serologically unrelated to possible members of the Sobemovirus group, including *Cocksfoot Mottle virus*, *Lucerne transient streak virus*, *Solanum nodiflorum mottle virus*, *Southern bean mosaic virus*, *Sowbane mosaic virus*, *Turnip rosette virus,* and *Velvet tobacco mottle virus*. The best approach to manage the disease is use of virus-free planting material (rhizomes), especially in areas where disease is previously not found. The use of virus-free planting rhizomes seems to be the only applied option as the vector is not known.

12.3.3   PATHOSYSTEM-C1

The plant viruses which are highly contagious are grouped under this pathosystem. Transmission by this means occurs by taking away of the sap from a diseased plant to healthy ones through artificial or natural means of rubbing. Different means involved in mechanical transmission like contact of healthy and infected leaves brought about by wind, abrasion of sap from diseased leaves over the surface of the healthy ones, cultivation tools, and grafting infected buds on the healthy plants. The pathosystem involves several viruses along with viroids.

12.3.3.1   TOBACCO MOSAIC VIRUS

TMV has a very wide host range. It is known to infect members of nine plant families, and at least 125 individual species, including tobacco, tomat o, pepper (majority of Solanaceae family), cucumbers, and a number of orna- mental flowers (Invasive Species Compendium TMV, CABI 2017). *Tobacco mosaic virus* is a genus in the family Virgaviridae. The name *Tobamovirus* is an acronym, denoting host and symptoms of the first virus discovered. There are four informal subgroups within this genus, tobamoviruses infecting bras- sicas, cucurbits, malvaceous, and solanaceous hosts. Principally, the differences between groups are genomic sequences and particular range of host plants. The first symptom of this virus is light green coloration between the veins of young leaves, followed by quick development of a "mosaic" or mottled pattern. Besides, mosaic rugosity may also be seen where the infected plant leaves exhibited localized random wrinkles. Under the Tobamovirus, *Cucumber green mottle mosaic virus* (CGMMV), *Cucumber virus* 4 (CV4), and *Tomato mosaic virus* (ToMV) are definitive member. However, *Beet necrotic yellow vein virus* (BNYVV), *Nicotiana velutina mosaic virus* (NVMV), and *Potato mop-top virus* (PMTV) are tentative members linked with vegetable pathosystem.

12.3.3.2   SOUTHERN BEAN MOSAIC VIRUS

Southern bean mosaic virus is still in unassigned family. Virions are stable icosahedra with a single-positive sense RNA encompassing four open reading frames. Small circular satellite RNAs (single-stranded) are associ- ated with the virus and encapsidated by five species (King et al., 2012). Around 16 sobemoviruses have been described which is infecting both mono and dicotyledonous plant species. The major transmission cause of sobemoviruses is mechanical wounding of host plants. Several species of insects may act as vectors of sobemoviruses. The first identified vectors responsible for transmission of sobemoviruses like *Southern bean mosaic virus, Turnip rosette virus, Solanum nodiflorum mottle virus,* and *Southern cowpea mosaic virus* were different leaf feeding beetles (Walters and Henry, 1970; Gerber, 1981; Holing and Stone, 1973). The insect vectors, a garden flea-hopper (*Halticus citri*), pea leaf miner (*Liriomyza langei*), beet leaf-hopper (*Circulifer tenellus*) and a green peach aphid (*Myzus persicae*), but not beetles, had been investigated as vectors of SoMV (Bennet and Costa, 2009).

12.3.3.3   TOMATO BUSHY STUNT VIRUS

Tombusvirus is a genus in the family Tombusviridae. There are currently 17 species in this genus including the type species *Tomato bushy stunt virus*. Symptoms are associated with this genus appearance mosaic (ICTV Report 2014). The name of the genus is derived by the type species *Tomato bushy stunt virus*. Most of the species have a narrow host range, all are mechanically transmissible, many are vectored by beetles, and there are examples of natural transmission by aphid and mirid vectors. Few members are seed transmissible. The virus is associated with three different pathosystem infecting vegetable crop like Pathosystem-A1, Pathosystem-C1, and Pathosystem-D1.

Following members are associated with vegetable pathosystem:

(a) *Tomato bushy stunt virus*
(b) *Artichoke mottled crinkle virus*
(c) *Cucumber Bulgarian virus*
(d) *Cucumber necrosis virus*
(e) *Eggplant mottled crinkle virus*
(f) *Moroccan pepper virus*
(g) *Pelargonium leaf curl virus*
(h) *Pelargonium necrotic spot virus*

12.3.3.4   VIROIDS

One of the most intriguing groups of plant pathogens is viroids that have been infecting several plant species. It may infect both herbaceous and woody species. Several diseases considered to economic importance are incited by viroid. In case of potato crop yield, losses can be high due to infection of *Potato spindle tuber viroid* (PSTVd). Moreover, other viroids like *Chrysanthemum stunt viroid* (CSVd) in chrysanthemum, *Citrus exocortis viroid* (CEVd) in citrus, *Coconut cadang-cadang viroid* in coconut, *Avocado sunblotch viroid*

(ASBVd) in avocado causes severe loss. The causal agent of potato spindle tuber disease was first example of a novel class of pathogens. It has small single-stranded covalently closed circular RNA molecules without protective protein capsid. Viroids are easily transmissible by mechanical inoculation and effectively disperse by contact with contaminated tools, farm implements, and human hands. It may also spread by graft transmission and foliar contact among neighboring plants. For potatoes, PSTVd set up in to a field by planting infected tubers or true potato seed. Vertical transmissibility has been recognized for ASBVd in avocado (Whitesell, 1952) and PSTVd in tomato and pepino (Benson and Singh, 1964; Hollings and Stone, 1973). However, vertical transmissibility is not found in tomato of *Tomato planta macho viroid* (Galindo, 1987). There are eight viroids that are demonstrated which act as principal component of vegetable pathosystem (Table 12.2). Principally, exclusion and eradication of infected planting materials is the most efficient means of sustainable control.

TABLE 12.2   Recognized Viroid Species Under Vegetable Pathosystem.

Viroid genus	Host	Emerging host
Citrus exocortis (CEVd)	Citrus, tomato	Tomato
Columnea latent (CLVd)	*Columnea, Brunfelsia, Nemathanthus*	Tomato
Potato spindle tuber (PSTVd)	Potato	Tomato, Avocado
Tomato apical stunt (TASVd)	Tomato	Tomato
Tomato chlorotic dwarf (TCDVd)	Uncertain (tomato?)	Tomato
Tomato planta macho (TPMVd)	-	Tomato
Hop stunt (HSVd)	Citrus, Grapevine, *Prunus* spp.	Hop, Cucumber
Eggplant latent (ELVd)	Eggplant	Eggplant

12.3.4   PATHOSYSTEM-C2

Under this pathosystem, *Cuscuta spp.* act as green bridge which plays a vital role in dissemination of virus from one host to another. Major transmissibility has been found with the dodder (*Cuscuta*). In the genus, 100–170 s pecies are known to be parasitic on flowering plants. The genus is found all over the temperate, subtropical, and tropical regions of the world. Fewer genus are reported in cool temperate climates, only four species belongs to northern Europe. It can be identified by thin stems, leafless appearance, and minute scales. The discovery of virus transmissibility has been noticed

75 years back from infected to a healthy plant which was parasitizing both. It is well known that dodder haustoria establish cellular link with the host plant. Different viruses can be transmitted through bridge formed between two plants by coiling the stem. The *Cuscuta* spp. transmitting green strain of *Cucumber mosaic virus* on *Nicotiana glutinosa, Lucern (alfa alfa) mosaic virus* to tobacco, and *Potato stem mottle virus* to tobacco. Moreover, dodder transmissibility has also been found in *Sugar beet curly top virus* (Bennett, 1944) and *Cucumber mosaic virus* (Francki et al., 1979). The results of Powell et al. (1984) indicated that dodder possibly has a role in epidemiology of Prunus stem pitting, making possible *Tomato ring spot virus* transmission in weed species.

12.3.5   PATHOSYSTEM-D1

The genus Potyvirus (named as type member, *Potato virus Y*) is the largest genus of the family Potyviridae. It contains at least 200 described and tentative species (Berger et al., 2005) which cause significant losses in agricultural, horticultural plants (Ward and Shukla, 1991).They affect a wide range of mono and dicotyledonous plant species and have been found in all parts of the world (Gibbs and Ohshima, 2010). They are transmitted broadly by aphids in a nonpersistent manner (Gibbs et al., 2003; Poutaraud et al., 2004; Fauquet et al., 2005). Potyviruses are aphid borne, ipomoviruses are whitefly borne, both are transmitted in a nonpersistent manner. Fewer in number they may also be transmitted through the seeds of their hosts (Johansen et al., 1994). Furthermore, they are transmitted through infected planting materials such as cuttings and tubers (Shukla et al., 1994). Even though potyvirus distribution is worldwide, they are most common in tropical and subtropical countries (Shukla et al., 1998). The viruses are flexuous filaments 650–900 × 11–15 nm with a single-stranded, positive sense RNA genome of approximately 9.7 kb. The genome encodes a polyprotein that is self-cleaved into around 10 functional proteins. A unique feature shared by all potyviruses is the induction of characteristic pinwheel or scroll-shaped inclusion bodies in the cytoplasm of the infected cells (Edwardson, 1974). Inclusion bodies are cylindrical, formed by a virus-encoded protein and can be considered as the most important phenotypic decisive factor for assigning viruses to the potyvirus group (Milne, 1988; Shukla et al., 1994; Ford, et al., 1989). Many potyviruses also induce cytoplasmic amorphous inclusion bodies and some form nuclear inclusions. The virion RNA is infectious and provide as both the genome and viral messenger RNA. The genomic RNA is encoded into polyproteins which are subsequently processed by the action of three viral-encoded proteinases into functional products. Approximately, more than 4000 potyvirus sequences are available in database of Genbank.

As published database on NCBI, 55 potyviruses are recorded in India, in which 40 are most important. These potyviruses infect a range of economically important crops, including potato, brinjal, capsicum, papaya, cowpea, common bean, tuber crops, cucurbits (Table 12.3) and caused significant economic losses (Mali and Kulthe, 1980; Bhat et al., 1999; Sreenivasulu and Gopal, 2010; Babu et al., 2012). The virus caused dark green mottling and distortion symptoms on leaves in chilli growing areas of eastern Uttar Pradesh. The disease was distributed throughout area and the incidence ranged from 5 to75%. A virus inducing severe mosaic disease of gherkin (*Cucumis anguria* L.) in south India was identified (Srinivasulu et al., 2010). It is an important cucurbitaceous vegetable crop grown in the southern states of India like Andhra Pradesh, Karnataka, and Tamil Nadu for slicing and pickling. *Pepper veinal mottle virus* affects *Capsicum annuum* L. crops in India (Nagaraju and Reddy, 1980) which causes considerable economical loss.

12.3.5.1 ZUCCHINI YELLOW MOSAIC VIRUS

Zucchini yellow mosaic virus (ZYMV) under genus of Potyvirus has flexuous filaments with a modal length of 750–800 nm, containing a single-stranded positive-sense RNA (Lisa and Lecoq, 1984). The virus was first described in zucchini squash in Italy (Lisa et al., 1981). *ZYMV* is also known as muskmelon yellow stunt virus (Lecoq et al., 1981). Twenty-two isolates have been grouped into three pathotypes according to the reaction on muskmelon line PI 414723 (Pitrat and Lecoq, 1984). ZYMV is geographically distributed in many areas of the world where cucurbits crops are grown. In India, it was reported in Pune, Maharashtra (Verma et al., 2004) However, in Africa, ZYMV has been recorded in Algeria, Egypt, Mauritius, Morocco (Brunt et al., 1990).

ZYMV causes severe damage in zucchini squash (*Cucurbita maxima*), muskmelon (*Cucumis melo*), cucumber (*Cucumis sativus*), and watermelon (*Citrullus lanatus*). The leaf symptoms include mosaic, yellowing, shoestring, and stunting. Fruits are deformed, twisted, and covered with protuberances.

Host range: The host range of ZYMV is narrow which includes mainly cucurbit species.

ZYMV is transmitted in a nonpersistent manner by different aphids species. *Myzus persicae*, *Aphis gossypii*, and *Macrosiphum euphorbia* are common vectors.

12.3.5.2   PEPPER VEINAL MOTTLE VIRUS

Pepper veinal mottle virus (PVMV), genus Potyvirus, consists of particles 770 nm long and 12 nm wide (Brunt and Kenten, 1971). PVMV was first recognized as a distinctive member of a group of viruses which was originally designated as the Potato virus Y group but was later renamed the Potyvirus group (Harrison et al., 1971). PVMV occurs mainly in Africa, although it affects *Capsicum annuum* L. crops in Afghanistan (Singh et al., 1988) and India (Nagaraju et al., 1980). PVMV also occurs in Capsicum spp. in Sierra Leone and Zaire, (Huguenot et al., 1996). Strains of the virus are also experimentally transmissible to atleast 35 species under the family Solanaceae and nine species of five others belonging to aizoaceae, amaranthaceae, apocynaceae, chenopodiaceae, and rutaceae.

Symptoms: The nature and severity of symptoms is dependent on type of host, virus strain, and environmental conditions. PVMV causes mottle, leaf distortion, veinal chlorosis and vein banding, and mosaic spots in tomato. Severe strains may cause leaf and stem necrosis in tomato.

Host range: The primary hosts of PVMV are sweet pepper, tomato, and eggplant (*Solanum melongena*).

TABLE 12.3   Species of the Potyviruses Associated with Vegetable Pathosystem in India.

Sl. no.	Virus species	Host
1	*Amaranthus mosaic virus*	Red amaranth
2	*Bean common mosaic virus*	Bean, tomato
3	*Blackeye cowpea mosaic virus*	Cowpea
4	*Chilli veinal mottle virus*	Pepper
5	*Dasheen streak mosaic virus*	Taro, elephant food yam
6	*Lettuce mosaic virus*	Lettuce
7	*Onion yellow dwarf virus*	Allium, garlic
8	*Papaya ring spot virus*	Watermelon
9	*Potato virus A*	Potato
10	*Potato virus Y*	Potato, tomato
11	*Sweet potato feathery mottle virus*	Sweet potato
12	*Turnip mosaic virus*	Radish
13	*Yam mild mosaic virus*	Yam
14	*Zucchini yellow mosaic virus*	Gherkin

12.3.6   PATHOSYSTEM-D3

Across the globe, geminiviruses are infecting temperate, tropical, and subtropical vegetable crops. Virions are paired icosahedra measuring 18×30 nm and comprising a single structural protein which encapsidates a circular single-stranded DNA genome of 2.5–3 kb. There are currently over 360 species in this family, divided among nine genera. The genera Becurtovirus, Begomovirus Capulavirus, Curtovirus, Eragrovirus, Grablovirus Mastrevirus, Topocuvirus, and Turncurtovirus (Zerbini et al., 2017, ICTV report 10th) are distinguished by vector, host range, and genome properties (King et al., 2012). These are transmitted persistently and circulated in insect vectors body. They are identified by PCR and RCA using different restriction enzymes subjected to cloning and sequence of the DNA components.

Begomoviruses are the leading and most important genus in the family with single-stranded DNA, having characteristic geminate incomplete icosahedral particles. Begomoviruses (type species: *Bean golden mosaic virus*) are transmitted by whiteflies and encompass either a monopartite (a single DNA) or a bipartite (with two DNA components: DNA-A and DNA-B) genome organization, infecting dicotyledonous plants (Gutierrez 1999, 2000; Mansoor et al., 2003; Jeske, 2009). The DNA-A of bipartite and the single component of monopartite begomoviruses contain five open reading frames (ORFs), but occasionally it may be six, one (AV1) or two (AV1 and AV2) in the viral sense (V-sense) strand, and four (AC1 to AC4) in the complementary sense (C-sense) strand. Both the DNA-A and DNA-B are approximately 2.7 kb in size. The DNA-B contains two ORFs (BV1 and BC1) in V-sense and C-sense strand, respectively. Extensive work has been carried out on these viruses on their sequence analysis, phylogeny, infectivity, functions of viral proteins, virus–host interactions, virus-derived transgenic resistance coupled with satellites DNA. Important begomoviruses infecting vegetable crop producing varying range of symptom coupled with betasatellite are presented in Table 12.4.

Whitefly transmitted severe curling and mosaic has been reported form different parts of world, which was suspected to be caused by begomoviruses. Diverse symptom has been reported from mild to severe mosaic in different crops. Emergence of several begomoviruses in major cucurbits like sponge gourd, bitter gourd, pumpkin, and ridge gourd are severely affected by producing shortening of internodes, mottling, stunting, puckering of leaf lamina, and fruit deformities. In summer and kharif season, cucurbits covered a major part among vegetables. Protected cultivation provides

TABLE 12.4   Molecular Description of Important Begomoviruses Associated with Vegetable Pathosystem.

Name of the virus	Crop infected	Mono-/bipartite	Association of satellite(s)	Symptoms produced	Reference(s)
Bhendi yellow vein mosaic virus	Bhendi/Okra	Monopartite	Betasatellite	Vein clearing, yellowing Reduced size of leaves and fruits	Jose and Usha, 2003; Kulkarni, 1924
Bittergourd yellow mosaic virus	Bittergourd	-	-	Leaf yellowing and mosaic	Raj et al., 2005
Chili leaf curl virus	Chilli	Monopartite	Betasatellite	Yellowing, leaf curling,, stunting and blistering, shortening of internodes	Shih et al. 2006; Chattopadhyay et al., 2008
Cucumis yellow mosaic disease associated virus	Cucumis	Bipartite (suspected)	—	Leaf yellowing and mosaic	Raj and Singh, 1996
Dolichos yellow mosaic virus	Frenchbean, cowpea, dolichos bean	Bipartite	-	Leaf yellowing	Varma and Malathi, 2003; Balaji et al., 2004; Girish et al., 2005
French Leaf curl virus	Frenchbean,	Monopartite	Betasatellite	Severe leaf curling	Naimuddin et al., 2014
Capsicum Leaf curl virus	Capsicum	Monopartite	-	Curling and chlorosis	Unpublished
Tomato leaf curl New Delhi virus, Tomato leaf curl Bangalore virus, Tomato leaf curl Gujarat virus, Tomato leaf curl Karnataka virus	Tomato	North Indian isolates bipartite, south Indian isolates monopartite	Betasatellite	Vein clearing, stunting; infection at the seedling stage can make plants sterile	Saikia and Muniyappa, 1989; Srivastava et al., 1995; Chatchawankanphanich et al., 1993; Chakraborty, 2003; Pratap et al., 2011

shelter to whiflies which harbors the Begomovirus by infecting tomato, capsicum, and cucumber crop. A large number of begomoviruses associated with vegetable is probably due to its warm tropical condition supporting the year round survival of the whiteflies and rigorous crop cultivation.

12.3.7   PATHOSYSTEM-D4

The genus Tospovirus of the family Bunyaviridae is the emerging problem for the vegetable cultivation across the world (Pappu et al., 2009; Kunkalikar et al., 2011). The genus name is derived from the name of its first member, *Tomato spotted wilt virus* (TSWV), firstly observed in Australia in 1915. The spotted wilt disease of tomato was later shown to viral origin (Brittlebank, 1919; Samuel et al., 1930) that was reported to be transmitted by thrips (Pittman, 1927). However, the reports describing the virus under different names, viz., *Tomato bronzing virus*, Kromnek virus, *Pineapple yellow wilt virus*, *Makhorka tip chlorosis virus,* and *Viracabeca virus* (Best, 1968; Sakimura, 1963 and Smith, 1972). Difference in nomenclature showed wide variation in disease symptoms, host species, and geographical regions where disease has been reported. TSWV is known to infect 1090 different plant species, belonging to more than 85 distinct botanical families (Sherwood et al., 2000; Parrella et al., 2003). At present, the genus is known to have more than 20 different viruses from all over the world (Pappu et al., 2009).

Biology and transmission: In plant viruses, tospoviruses, have unique particle structure, genome organization, and expression strategies. The pleomorphic virus particles are 80–120 nm in diameter and have surface projections composed of two viral glycoproteins. The genome includes three RNAs referred to as large (L), medium (M), and small (S). The L-RNA is in negative sense while the M- and S-RNAs are ambisense. L-RNA codes for the RNA-dependent RNA polymerase (RdRp) and M-RNA for the precursor of two glycoproteins (GN and GC) and a nonstructural protein (NSm). The S-RNA codes for the N protein and another nonstructural protein (NSs). NSm and NSs were shown to function as movement protein and silencing suppressor, respectively (Tsompana and Moyer, 2008). The NSm of TSWV was recently shown to act as an avirulence determinant during the interaction between TSWV and resistant pepper containing the *Tsw* gene. The three genomic RNAs are tightly linked with the N protein to form ribonucleo proteins (RNPs). These RNPs are encased within a lipid envelope consisting of two virus-coded glycoproteins and a host-derived membrane.

Tospoviruses are spread from plant to plant by several species of thrips in circulative and propagative manner (Whitfield et al., 2005). Thrips are minute insect found in multiple habitats around the world. Thrips species *Frankliniell aoccidentalis* (western flower thrips), *F. fusca* (tobacco thrips), *Thrips tabaci* (onion thrips), and *Thrips* palmi (melon thrips) are some of the major vectors of the tospoviruses. Few other thrips species (not described here) may be more or less important as vectors on a regional basis in different parts of the world. As many insect vector–virus associations, the thrips–tospovirus relationship is very specific. Only few members among known thrips species are able to acquire and transmit tospoviruses in vegetable. Till now there are at least 10 species of thrips that have been reported to transmit the virus (Pappu et al., 2009) and the mode of transmission is very unique, that is, only the larval stages can acquire the virus and only adults can transmit the virus. Invasion of new species in a continent or major growing areas, crop intensification, and development of resistance against systemic insecticides may be the major cause. The transmission ability of several thrips species to tospoviruses or the tomato spotted wilt group has appeared obvious increase in world agriculture scenario (Fig. 12.1).

Tospoviruses are emerged as persistent limiting factor for the vegetable production in India (Kunkalikar et al., 2011). Four out of five tospoviruses reported in India are known to be seriously affecting vegetable cultivation (Table 12.5). The oldest tospovirus reported in India *Peanut/Groundnut Bud Necrosis virus* (PBNV/GBNV) is known to have very huge host range infecting most of the solanaceous and other vegetables, viz., tomato (*Lycopersicum eculentum*), chilli (*Capsicum annuum*), potato (*Solanum tuberosum*), brinjal (*Solanum melongena*), pea (*Pisum sativum*), Amaranthus (*Amarnthus sp.*) (Mondal et al., 2012; Sharma and Kulshrestha, 2014). The recent documentation in India are *Iris yellow spot virus* (IYSV) infecting on onion (*Allium cepa*) (Ravi et al., 2006) and garlic (*Allium sativum*) (Gawande et al., 2010), and *Capsicum chlorosis virus* (CaCV) on tomato (Kunkalikar et al., 2007) and chili (Krishnareddy et al., 2008). Cucurbitaceous crops especially watermelon is being seriously affected by *Watermelon bud necrosis virus* (WBNV) (Jain et al., 1998). Early grown potato crop is severely challenged by stem necrosis disease caused by GBNV in terai ecology of Uttarakhand and plateau region of Rajasthan (Ansar et al., 2015). All four tospoviruses infecting vegetables GBNV and WBNV is distributed throughout the subcontinent. The increasing reports of natural infection of GBNV on new crops and its geographical expansion is alarming.

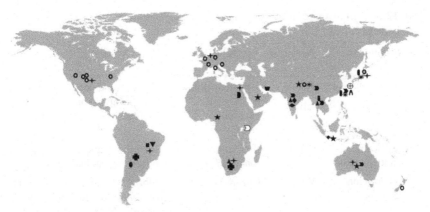

FIGURE 12.1   Geographical distribution of tospoviruses.

◆   *Groundnut (peanut) bud necrosis virus (GBNV)*
✚   *Groundnut ringspot virus (GRSV)*
○   *Impatiens necrotic spot virus (INSV)*
▲   *Groundnut Yellow spot virus (GYSV)*
●   *Tomato chlorotic spot virus (TCSV)*
★   *Tomato spotted wilt virus (TSWV)*
☀   *Watermelon silver mottle virus (WSMoV)*
■   *Zucchini lethal chlorosis virus (ZLCV)*
▶   *Capsicum chlorosis virus (CaCV)*
▼   *Chrysanthemum stem necrosis virus (CNSV)*
+   *Iris yellow spot virus (IYSV)*
❚   *Melon yellow spot virus (MYSV)*
∧   *Groundnut chlorotic fan-spot virus*
◆   *Watermelon bud necrosis virus (WBNV)*
◤   *Tomato yellow fruit ring virus*
⊕   *Calla lily chlorotic spot virus (CCSV)*
◗   *Soybean necrosis virus (SNV)*
◆   *Peanut bud necrosis virus*
∑   *Tomato yellow ring virus*

12.3.8   PATHOSYSTEM-E1

The viral pathosystem linked with plant parasitic nematodes commonly inhabit the region around the roots in the soil. Few members also feed on the aerial parts of the plants like the stem, leaves, grains, etc. Nematode population increase slowly and the diseases caused by them are usually not of abrupt epidemic type. Several species of nematode act as vector of viruses which causes loss in economically important crops. The relations

TABLE 12.5   List of Tospoviruses Infecting Vegetable Crops in India and Their Symptoms and Host Range.

S. no.	Tospovirus	Acronym	Vegetable host	Major symptoms on vegetable crops
1.	Groundnut bud necrosis virus	PBNV/ GBNV	Tomato, chilli, pea, brinjal, cucurbits, potato, etc.	Chlorotic and necrotic ring spot chlorotic apical bud necrosis and stem necrosis
2.	Capsicum chlorosis virus	CaCV	Capsicum, *Amaranthus* sp.	Chlorotic ring spot, necrotic ring spot and apical bud necrosis
3.	Watermelon bud necrosis virus	WBNV	Cucurbits, tomato	chlorotic mottling, yellow spots/ patches, mild crinkling, Bud necrosis, dieback vines
4.	Irish yellow spot virus	IYSV	Onion, garlic	Yellow stripes, spindle-shaped chlorotic lesions/rings

between plant parasitic nematodes and viruses may be divided into two set: a specific interrelationships between definite ectoparasitic nematode species and few viruses which they transmit. Under second relationship, general effects of plant viruses on different parasitic nematodes through host plant. Under this pathosystem, currently 30 nematode species are accepted that are known to transmit 15 viruses. The viruses are of two types that are earlier referred as either Nepoviruses, with spherical particles, or Tobraviruses, with rod-shaped particles. Nepovirus (nematode-transmitted polyhedral viruses-NEPO) and Tobravirus (tobacco rattle viruses-TOBRA), have nematode vectors. Nepoviruses belongs to Comoviridae family, whereas Tobraviruses are still in unassigned family. Nepoviruses are isometric (polyhedral) particles around 30 nm in diameter. This is only known vectors in the genera *Xiphinema* and *Longidorus*. Tobraviruses are straight tubular particles with two different size range, 180–210 nm and 45–115 nm. Known species *Trichodorus* and *Paratrichodorus* are the virus vectors. In vector specificity, 11 species of *Xiphinema* transmit 13 NEPO viruses, 11 species of *Longidorus* transmit 10 NEPO viruses. Fourteen species of *Trichodorus* transmit various strains of two TOBRA viruses like tobacco rattle and pea early browning. Since several *Trichodorid* spp. transmit the same virus, and that both viruses are transmitted by the same nematode, vector specificity is less developed in Trichodorids than in longidorids as suggested by Lamberti (1987). Nematodes act a vector of virus in vegetable pathosystem described in Table 12.6.

TABLE 12.6   Plant Virus Transmitting Nematode Groups.

Virus transmitting group	Genus	Virus retention site	Nematode-transmitted viruses associated in vegetable pathosystem
Nepovirus	Longidorus	Odontostyle	*L. attenuates- Tomato black ring virus* (Italian isolate) *L. apulus –Artichoke Italian latent virus* *L. elongates- Tomato black ring virus* (Greek isolate) *L. fasciatus - Artichoke Italian latent virus* (Greek isolate)
	Xiphinema	Odontostyle	*X. americanum sensu lato - Tomato ring spot virus* *X. bricolensis* -do- *X. californicum* -do- *X. intermedium* -do- *X. rivesi* -do- *X. tarjanense* -do-
Tobravirus	Paralongidorus	Anterior food canal to oesophagous tract	-
	Paratrichodorus	-do-	*P. anemones- Pea early browning virus*
	Trichodorus	-do-	*Trichodorus spp- Pea early browning virus*

12.3.9   PATHOSYSTEM-F1

Role of fungi as plant virus vector was first reported in 1960s (Teakle, 1964). The fungal group is related to Chytrids (*Olpidaceae*) or Plasmodiophorids (*Plasmodiophoraceae*). Both fungi are habitually obligated parasites infecting root system of cool and wet soil ecology. Most of the viruses tranmistted by these fungi are more rampant in temperate climate. Under Chytrids, three species like *Olipidium brassicae, O. barnovanus,* and *Synchytrium endobioticum* are recognized as vectors of viruses (Campbell, 1996). These fungi are producing zoospores which are posteriorly uniflagellate and set apart by single celled resting spores. In case of Plasmodiophorids, three species, viz., *Polymyxa graminis, P. betae,* and *Spongospora subterranean* are key transmitters of viruses. Such fungi having biflagellate zoospores and have single celled resting spores.

Transmission of virus through zoospores is recognized by two ways. Nonpersistent transmission: A virus particle appears to enter zoospores by infection tube but it does not enter in resting spores during its formation. In case of Persistent transmission, viruses are acquired by fungi from virus infected plant cells and persist in resting spores (Fig. 12.2). The virus remains present in the soil and viable for the year even if soil loses the moisture. The virus becomes released when zoospores get germinated. In isometric viruses, *Cucumber necrosis virus* (*O. redicales*) and *Lettuce big vein virus* (*O. brassicae*) are known to be transmitted by respective species. Moreover, *Cucumber necrosis virus* is found to be transmitted by *O. cucurbitacaerum* (Dias, 1970); furthermore, *Polymyxa betae* by *Beet necrotic yellow vein virus* (Lemaire et al., 1988) and *Spongospora subterranean* transmits *Potato mop top virus* (Harrison and Jones, 1970; Jones and Harrison, 1972).

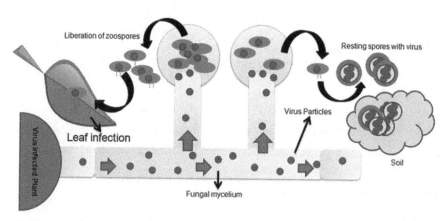

FIGURE 12.2   (See color insert.) Schematic representation of infection process and survival of fungal transmitted virus.

12.4   SIGNIFICANCE OF PATHOSYSTEM FOR INTEGRATED MANAGEMENT

Depending upon strategies followed by viruses for infection and spread, efficient management approaches may be designed. It may achieved by using virus-free planting material, delaying, or reducing spread by direct control of vector. Different virus genera are placed in separate pathosystem. However, several viruses share their role in two pathosystem or more in such condition management practice can be modified well. If virus spread through infected seeds like *Pea seed borne mosaic virus* and *Cucumber mosaic virus*, basic method for management is use of virus-free seeds obtained from certified agencies. In case of vegetative propagation like potato tuber infected with PVX and PVA, use of virus-free tubers procured from specialized organizations may be planted.

The viruses categorized under highly infectious group (RNA viruses) are mechanically disseminated from infected to healthy plant. Sanitary and preventive measures become very effective in case of Tobamoviruses infecting tomato, cucumber, and beet. The viruses transmitted by dodder, close mowing is an effective management practice. Burning of infested residues, invaded tissue of host plants along with dodder prevents regeneration from embedded haustoria.

Natural spread of virus via vectors influenced both the behavior of the vector and the period for which the vector maintain infectivity. Viruses under the pathosystem (nonpersistent) are typically carried by actively migrating noncolonizing aphids which probe for a short period on source plants and inoculate within minutes on target. Insecticidal application of plants is not much effective because the killing time is shorter than the inoculation threshold, consequently cultural control measures are applied (Hull, 2002).

Viruses in pathosystems D3 and D4 (circulative and propagative) can be effectively managed by striking the vector. For the management of sucking insect vectors which are transmitting viruses, common practices generally followed are:

Insecticide application: Several systemic insecticides are available for the effective control of insect vectors. An integrated approach of seed treatment with imidacloprid followed by spraying of systemic insecticides for the control of thrips vector is the most effective method for management of *Watermelon bud necrosis virus* disease in Watermelon (Kamanna et al., 2010). Effect of six insecticides (cyazypyr, flupyradifurone, pyrafluquinazon, and sulfoxaflor) on transmission of *ToYLCV* by the *Bemisia tabaci* biotype-B to compare them with two established insecticides, pymetrozine and a zeta-cypermethrin/bifenthrincombination. Percentage of tomato seedlings expressing virus symptoms tended to be lowest in seedlings treated with flupyradifurone (Smith and Giurcanu, 2014).

Cultural practices: Cultural management practices are the actions undertaken by human beings to prevent and control disease by manipulating cropping system. Practices that lower down the initial levels of inoculum include selecting appropriate planting materials, destruction of crop residues, elimination of viral affected plants that act as source of inoculum, and crop rotation. Several agricultural systems are characterized by intense plant populations with genetic homogeneity (monocultures). Once a virus infects plant community, it can rapidly spread to epidemic extent. In order to control the epidemic, diverse range of plant community may be incorporated in cropping system.

Effect of planting modification on viral incidence: Altering in planting date is an effective method to avoid/escape the viral incidence. In tomato crop, middle October transplanting severely infested by whiteflies, resulting in high pressure of *ToLCV*. However, middle November planted crop exhibited lower incidence (unpublished finding). However, in chilli, the incidence was found low in early planting (last week of March), but disease was much increased in delayed planted crop with 43.7% in 2013–2014 cropping season (unpublished finding).

Application of insecticides with cultural practices: A field experiment for management of okra yellow vein mosaic disease, integrated application of insecticide with border cropping with maize showed good result in minimizing the disease. A treatment consisting seed treatment and spray with imidacloprid and neem oil was found effective to reduce the vector population (1.73/plant). However, sequential sprays of different insecticides minimize the whitefly population and reducing the disease in comparison to check plots. (Ansar et al., 2014).

12.5 SUMMARY

Vegetable crops are severely hampered by various biotic factors, among them viruses play a significant role. Within the last decade or so, a large number of viruses have been investigated, encompassing legumes, root crops, and vegetables. A complex relationship has been found in virus, host-plant, and the vector which creates unique pathosystem. In the current perspective of virus infection in vegetable crop, we sort out 11 pathosystems which are linking host by different modes. On the basis of dissemination of virus from one host to another, every pathosystem discussed like seed carrying virus, vector-mediated transmission greatly focused particularly whitefly-transmitted geminivirus in different vegetable crops. Moreover, thrips transmissibility of tospoviruses has been discussed with respect to different host across the globe. Role of nematode as vector of Nepovirus and Netuvirus group coupled with vegetable pathosystem major emphasis is given. Two major groups of fungi Chytrids and Plasmodiophorids having *Olipedium spp.* and *Polymyxa spp.* as important members respectively are infecting economically important vegetable crop as well as acting as major vectors. Integrated approaches are essential to justify the management execution programme. Preventive application of insecticides is routine approach to controlling insect vectors. However, prolonged application single insecticide leads to development of resistance. Alternative application of different molecules will be helpful in managing vectors in an efficient way. Avoidance of vector population in order to manipulate planting time coupled with insecticidal application explains sustainable management of vector-borne disease.

KEYWORDS

- **epidemiology**
- **pathosystem**
- **transmission**
- **vegetable**
- **virus**

REFERENCES

Agrios, G. N. *Plant Pathology*. Elsevier Academic Press: Burlington, MA, USA, 5th ed, 2005.

Ali, A.; Kobayashi, M. Seed Transmission of Cucumber Mosaic Virus in Pepper. *J. Virol. Methods* **2009**, *163* (2), 234–237. DOI: 10.1016/j.jviromet.2009.09.026.

Ansar, M.; Singh, R. P. Field Diagnosis and Temporal Progress of PVX and A in Tarai Region of Uttarakhand. *J. Mycol. Pl. Pathol.* **2016**, *46* (4), 368–373.

Ansar, M.; Akram, M..; Singh, R. B.; Pundhir, V. S. Epidemiological Studies of Stem Necrosis Disease in Potato Caused by Groundnut Bud Necrosis Virus. *Indian Phytopath.* **2015**, 68 (3), 321–325.

Ansar, M.; Saha, T.; Sarkhel, S.; Bhagat, A. P. Epidemiology of Okra Yellow Vein Mosaic Disease and its Interaction with Insecticide Modules. *Trend Biosci.* **2014**, *7* (24), 4157–4160.

Balaji, V.; Vanitharani, R.; Karthikeyan, A. S.; Anbalagan, S.; Veluthambi, K. Infectivity Analysis of Two Variable DNA B Components of Mungbean Yellow Mosaic Virus-Vigna in *Vigna mungo* and *Vigna radiata*. *J. Biosci.* 29 297–308.

Bennet, C. W. Studies of Dodder Transmission of Plant Viruses. *Phytopathology* **1944**, *34*, 905–932.

Bennet, C. W.; Costa, A. S. Sowbane Mosaic Caused by a Seed-Transmitted Virus. *Phytopathology* **1961**, 51, 546–550.

Berger, P. H.; Adams, M. J.; Barnett, O. W.; Brunt, A. A.; Hammond, J.; Hill, J. H.; Jordan, R. L.; Kashiwazaki, S.; Rybicki, E.; Spence, N. et al. Potyviridae. In: *Virus Taxonomy. VIIIth Report of the International Committee on Taxonomy of Viruses*; Fauquet, C. M.; Mayo,

M. A.; Maniloff, J.; Desselberger, U.; Ball, L. A., Eds; Elsevier/Academic Press: London, 2005, pp 819–841.

Best, R. J. Tomato Spotted Wilt Virus. In: *Advances in Virus Research*, 1968, Vol. 13, pp 65–146.

Brittlebank, C. C. A New Tomato Disease, Spotted Wilt. *J. Agric., Victoria* **1919**, *27*, 231–235.

Brunt, A. K. C.; Gibbs, A. *Viruses of Tropical Plants*. CAB International: Wallingford, UK, 1990, pp 707.

Campbell, R. N. Fungal Transmission of Plant Viruses. *Ann. Rev. Phytopathol.* **1996**, 34, 87–108.

Chakraborty, S.; Pandey, P. K.; Banerjee, M. K.; Kalloo, G.; Fauquet, C. M. Tomato Leaf Curl Gujarat Virus, s New Begomovirus Species Causing a Severe Leaf Curl Disease of Tomato in Varanasi, India. *Phytopathology* **2003**, *93*, 1485–1495

Chatchawankanphanich, O.; Chiang, B. T.; Green, S. K.; Singh, S. J.; Maxwell, D. P. Nucleotide Sequence of a Geminivirus Associated with Tomato Leaf Curl from India. *Plant Dis*. **1993**, *77*, 1168.

Crescenzy, A. Cucumber Mosaic Cucumovirus Populations in Italy under Natural Epidemic Conditions and After a Satellite Mediated Protection Test. *Plant Dis.* **1993**, *77*, 28–33.

Dias, H. F. Transmission of Cucumber Necrosis Virus by *Olpidium cucurbitacaerum* Barr & Dias. *Virology* **1970**, 40, 828–839.

Fauquet, C. M.; Mayo, M. A.; Maniloff, J.; Desselberger, U.; Ball, L. A. Virus Taxonomy. In: *Eighth Report of the International Committee on Taxonomy of Viruses*. Elsevier Academic Press: San Diego, CA, 2005.

Francki, R. I. B.; Mossop, D. W.; Hatta, T. Cucumber Mosaic Virus. CMI/AAB Descriptions of Plant Viruses, 1979, Vol. 213.

Galindo-Alonso, J. In Temas en Virologia II; Alvizo Villasana, H. F.; Lozoya-Saldana, H., Eds; Sociedad Mexicana de Fitopatologia, CONACyT: Mexico City, Mexico, 1987.

Gawande, S. J.; Khar, A.; Lawande, K. E. First Report of Irish Yellow Spot Virus on Garlic in India. *Plant Dis.* **2010**, 94,1066.

Gibbs, A.; Ohshima, K. Potyviruses and the Digital Revolution. *Ann. Rev. Phytopathol.* **2010**, *48*, 205–223.

Gibbs, A. J.; Mackenzie, A. M.; Gibbs, M. J. The "Potyvirid Primers" Will Probably Provide Phylogenetically Informative DNA Fragments from All Species of Potyviridae. *J. Virol. Methods* **2003**, *12*, 41–44.

Girish, K. R.; Usha, R. Molecular Characterization of Two Soybean-Infecting Begomoviruses from India and Evidence for Recombination among Legume-Infecting Begomoviruses from South-East Asia. *Virus Res.* **2005**, 108, 167–176.

Gopal, K.; Muniyappa, V.; Jagadeeshwar, R. Management of Peanutbud Necrosis Disease in Groundnut by Botanical Pesticides. *Arch.Phytopathol. Pl. Prot.* **2010**, *44* (13), 1233–1237.

Harrison, B. D.; Finch, J. T.; Gibbs, A. J.; Hollings M.; Shepherd, R. J.; Valenta, V.; Wetter, C. Sixteen Groups of Plant Viruses. *Virology* **1971**, *45*, 356–363.

Harrison, B. D.; Jones, R. A. C. Host Range and Some Properties of Potato Mop-Top Virus. *Ann. Appl. Biol.* **1970**, *65* (3), 393–402.

Hollings, M.; Stone, O. Turnip Rosette Virus. In: *CMI/AAB Description of Plant Viruses*. Commonwealth Mycological Institute: Kew, Surrey and Association of Applied Biologists: Wellesbourne, Warwick, UK, 1973.

Huguenot, C.; Furneaux, M. T.; Clare, J.; Hamilton, R. Sero-diagnosis of Pepper Veinal Mottle Viru in West Africa Using Specific Monoclonal Antibodies in DAS-ELISA. *J Phytopathol.* **1996,** *144*, 29–32.

Hull, R. *Matthews' Plant Virology.* Academic Press: San Diego, 2002, 4th ed.

ICTV. "Virus Taxonomy: 2014 Release." Retrieved 15 June, 2015.

Jain, R K, Pappu, H R, Pappu, S S, Krishanareddy, M, and Vani, A (1998). Watermelon bud necrosis tospovirus is a distinct virus species belongingto serogroup IV. *Arch. Virol.* 143:1637-1644.

Jones, R. A. C.; Harrison, B. D. Ecological Studies on Potato Mop-Top Virus in Scotland. *Ann. Appl. Biol* **1972,** *71* (1), 47–57.

Jose, J.; Usha, R. Bhendi Yellow Vein Mosaic Disease in India Caused by Association of a DNA B Satellite with a Begomovirus. *Virology* **2003,** *305*, 310–317.

Krishnareddy, M.; Usha Rani, R.; Kumar, A.; Madhavireddy, K.; Pappu, H. R. First report of Capsicum Chlorosis Virus (Genus Tospovirus) Infecting Chili Pepper (*Capsicum annuum*) in India. *Plant Dis.* **2008,** *92*,1469.

Kulkarni, G. S. Mosaic and Other Related Diseases of Crops in the Bombay Presidency. *Poona Agric. College Mag. 1924, 6*, 12.

Kunkalikar, S. R.; Poojari, S.; Arun, B. M.; Rajagopalan, P. A.; Chen, T. C.; Yeh, S. D.; Naidu, R. A.; Zehr, U. B.; Ravi, K. S. Importance and Genetic Diversity of Vegetable-Infecting Tospoviruses in India. *Phytopathology* **2011,** *101*, 367–376.

Kunkalikar, S. R.; Sudarsana, P.; Rajagopalan, P.; Zehr, U. B.; Naidu, R. A.; Ravi, K. S. First Report of Capsicum Chlorosis Virus in Tomato in India. *Plant Health Prog.* **2007.** doi:10.1094/PHP-2007-1204-01.

Lecoq, H.; Pitrat, M.; Clement, M. Identification et Caractérisation D'un Potyvirus Provoquant La Maladie du Rabougrissement Jaune du Melon. *Agronomie* **1981,** *1*, 827–834.

Lemaire, O.; Merdinoglu, D.; Valentin, P.; Putz, C.; Ziegler-Graff, V.; Guilley, H.; Jonard, G.; Richards, K. Effect of Beet Necrotic Yellow Vein Virus RNA Composition on Transmission by *Polymyxa betae. Virology* **1988,** *162* (1), 232–235.

Lisa, V.; Lecoq, H. *Zucchini Yellow Mosaic Virus.* CMI/AAB Descriptions of Plant Viruses No. 282. Association of Applied Biologists: Wellesbourne, UK, 1984.

Lisa, V. G.; Boccardo, G. D.; Agostino, G.; Dellavalle and Aquilio, M. D. Characterization of a Potyvirus that Causes Zucchini Yellow Mosaic. *Phytopathology* **1981,** 71, 667–672.

Milne, R .G. Taxonomy of the Rod-shaped Filamentous Viruses. In: *The Plant Viruses. The Filamentous Plant Viruses*; Milne, R. G., Ed, Plenum Press: New York, Vol.4, pp 3–50.

Nagaraju, R.; Reddy, H. R. Occurrence and Bell pepper Viruses Distribution of Around Bangalore. *Curr Res.* **1980,** *10* (155), 6.

Naimuddin, K.; Akram, M.; Pratap, A.; Yadav, P. Characterization of a New Begomovirus and A Beta Satellite Associated With The Leaf Curl Disease Of French Bean In Northern India. *Virus Genes* **2014,** *46*, 120–127. DOI 10.1007/s11262-012-0832-8

Palukaitis, P.; Garcia-Arenal, F. Cucumoviruses. *Adv Virus Res.* **2003,** *62*,241–323. doi: 10.1016/S0065-3527(03)62005-1.

Palukaitis, P.; Roossinck, M. J.; Dietzgen, R. G.; Francki, R. I. B. Cucumber Mosaic Virus. *Adv Virus Res.* **1992,** *41*, 281–349.

Pappu, H. R.; Jones, R. A. C.; Jain, R. K. Global Status of Tospovirus Epidemics in Diverse Cropping Systems: Successes Achieved and Challenges Ahead. *Virus Res.* **2009,** *141*, 219–236.

Parrella, G.; Gonalons, P.; Gebre-Selassie, K.; Vovlas, C.; Marchoux, G.An Update of the Host Range Of Tomato Spotted Wilt Virus. *J. Pl. Pathol.* **2003**, *85*, 227–264.

Pitrat, M.; Lecoq, H. Inheritance of Zucchini Yellow Mosaic Virus Resistance in *Cucumis melo* L. *Euphytica* **1984**, *33*, 57–61.

Pittman, H. A. Spotted Wilt of Tomatoes. *J. Aust. Counc. Sci. Ind. Res.* **1927**, *1*, 74–77.

Poutaraud, A.; Desbiez, C.; Lemaire, O.; Lecoq, H.; Herrbach, E. Characterisation of a New Potyvirus Species Infecting Meadow Saffron (*Colchicum autumnale*). *Arch Virol.* **2004**, *149*, 1267–1277.

Powell, C. A.; Forer, L. B.; Stouffer, R. F.; Cummins, J. N.; Gonsalves, D.; Rosenberger, D. A.; Hoffman, J.; Lister, R. N. Orchard Weeds As Hosts Of Tomato Ring Spot and Tobacco Ring Spot Viruses. *Plant Dis.* **1984**, *68*, 242–244.

Pratap, D.; Kashikar, A. R.; Mukherjee, S. K. Molecular Characterization and Infectivity of A Tomato Leaf Curl New Delhi Virus Variant Associated with Newly Emerging Yellow Mosaic Disease Of Eggplant in India. *Virol. J.* **2011**, *8*, 305.

Raj, S. K.; Singh, B. P. Association of Geminivirus Infection with Yellow Green Mosaic Disease of Cucumis Sativus: Diagnosis by Nucleic Acid Probes. *Indian J. Exp. Biol.* **1996**, *34*, 603–605.

Raj, S. K.; Khan, M. S.; Singh, R.; Kumari, N.; Praksh, D. Occurrence of Yellow Mosaic Geminiviral Disease on Bitter Gourd (*Momordica Charantia*) and Its Impact on Phytochemical Contents. *Intl. J. Food Sci. Nutr.* **2005**, *56*, 185–192.

Ravi, K. S.; Kitkaru, A. S.; Winter, S. Iris Yellow Spot Virus Inonions: A New Tospovirus Record Form India. *Plant Pathol.* **2006**, *55*, 288.

Saikia, A. K.; Muniyappa, V. Epidemiology and Control of Tomato Leaf Curl Virus in Southern India. *Trop. Agric.* **1989**, *66*, 350–354.

Sakimura, K. Frankliniella Fusca An Additional Vector for the Tomato Spotted Wilt Virus, With Notes on Thrips Tabaci, Another Vector. *Phytopathology* **1963**, *53*, 412.

Samuel, G.; Bald, J. G.; Pittman, H. A. Investigations of Spotted Wilt of Tomatoes. *CSIR, Bull.* **1930**, *44*, 1–64.

Sharma, A.; Kulshrestha, S. First Report of Amaranthus Sp. As A Natural Host Of Capsicum Chlorosis Virus In India. *Virus Dis.* **2014**, 25(3):412–413.

Sherwood, J. L.; German, T. L.; Moyer, J. W.; Ullman, D. E.; Whitefield, A. E. Tospoviruses. In: *Encycloperdia of Plant Pathology*; Maloy, O. C.; Murray, T. D., Eds; John Wiley & Sons: New York, 2000, pp 1034–1040.

Shih, S. L.; Tsai, W. S.; Green, S. K.; Singh, D. First report of Tomato Leaf Curl Joydebpur Virus Infecting Chilli in India. *New Dis. Rep.* **2006**, *14*, 17.

Shukla, D. D.; Ward, C. W.; Brunt, A. A. *The Potyviridae*. CAB International University Press: Cambridge, 1994.

Shukla, D. D.; Ford, R. E.; Tosic, M.; Jilka, J.; Ward, C. W. Possible Members of the Potyvirus Group Transmitted or Whiteflies Share Epitopes with Aphid-Transmitted Definitive Members of the Group. *Arch Virol.* **1989**, *105*, 143–151.

Singh, S. J.; Krishnareddy, M. Watermelon Bud Necrosis: A New Tospovirus Disease. *Acta Hortic.* **1996**, *431*, 68–77.

Smith, H. A.; Giurcanu, M. C. New Insecticides for Management of Tomato Yellow Leaf Curl, A Virus Vectored By the Silver Leaf Whitefly (*Bemisia tabaci*). *J. Insect Sci.* **2014**, *14* (1), 183.

Smith, K. M. *A Textbook of Plant Virus Diseases* Academic Press: New York, 3rd ed, pp 897.

Srinivasulu, M.; Sarovar, B.; Anthony Johnson, A. M.; Sai Gopal, D. V. R. Association of a Potyvirus with Mosaic Disease of Gherkin (Cucumis anguria L.) in India. *Indian J. Microbiol.* **2010**, *50*, 221–224.

Srivastava, K. M.; Hallan, V.; Raizada, R. K.; Chandra, G.; Singh, B. P.; Sane, P. V. Molecular Cloning of Indian Tomato Leaf Curl Virus Genome Following A Simple Method of Concentrating the Supercoiled Replicative Form of DNA. *J. Virol. Methods* **1995**, *51*, 297–304.

Teakle, D. S.; Hiruki, C. Vector Specificity in *Olpidium*. *Virology* **1964**, *24*, 539–544.

Tsompana, M.; Moyer, J. W. Tospoviruses In: *Encyclopedia of Virology*; Mahy, B. W. J.; Van Regenmortel, M. H. V., Eds; Elsevier Ltd.: Oxford, UK, 2008, 3ʳᵈ ed., Vol. 5, pp 157–162.

Varma, A.; Malathi, V. G. Emerging Geminivirus Problems: A Serious Threat to Crop Production. *Ann. Appl. Biol.* **2003**, *142*, 145–164.

Virus Taxonomy: The Classification and Nomenclature of Viruses. Geminiviridae ICTV Online (10th) Report, 2017 (www.talk.ictvonline.org/ictv-reports/ictv_online_report).

Walters, H. J.; Henry, D. G. Bean Leaf Beetle As A Vector of the Cowpea Strain of Southern Bean Mosaic Virus. *Phytopathology* **1970**, *60*, 177–178.

Ward, C. W.; Shukla, D. D. Taxonomy of Potyviruses: Current Problems and Some Solutions. *Intervirology* **1991**, *32*, 269–296.

Whitfield, A. E.; Ullman, D. E.; German, T. L. Tospovirus-thrips Interactions. *Ann. Rev. Phytopathol.* **2005**, *43*, 459–489.

Zerbini, F. M.; Briddon, R. W.; Idris, A.; Martin, D. P.; Moriones, E.; Navas-Castillo. J.; Rivera-Bustamante, R.; Roumagnac, P.; Varsani, A. ICTV Report Consortium, ICTV Virus Taxonomy Profile: Geminiviridae. *J. Gen. Virol.* **2017**, *98* (2), 131–133. PMID 28284245. doi:10.1099/jgv.0.000738.

CHAPTER 13

Begomovirus Diversity and Management in Leguminous Vegetables and Other Hosts

MUHAMMAD NAEEM SATTAR[1*] and ZAFAR IQBAL[2,3]

[1]*Department of Environment and Natural Resources, College of Agriculture and Food Science, King Faisal University, Alhafuf, Kingdom of Saudi Arabia*

[2]*Central Laboratories, Box 400, King Faisal University, Alhafuf, Kingdom of Saudi Arabia*

[3]*Department of Plant Pathology, University of Florida, Florida, United States of America*

Corresponding author. E-mail: naeem.sattar1177@gmail.com

ABSTRACT

Geminiviruses (family *Geminiviridae*) are plant viruses with circular single-stranded DNA genome encapsidated in twinned semi-icosahedrons. Family *Geminiviridae* has been classified into nine genera viz., *Becurtovirus, Begomovirus, Capulavirus, Curtovirus, Eragrovirus, Grablovirus, Mastrevirus, Turncurtovirus,* and *Topocuvirus*. Out of these nine genera, *Begomovirus* is the largest genus and poses a major threat to worldwide agricultural crops, particularly to legumes in tropical and subtropical regions of the world. The members of the genus *Begomovirus* have either bipartite (DNA-A and DNA-B) or monopartite (DNA-A) genome. Some of the monopartite begomoviruses are also associated with subviral DNA satellites known as alphasatellite, betasatellite, and deltasatellite (members of genus *Betasatellite* and *Deltasatellite* family *Tolecusatellitidae*), respectively. All begomoviruses are transmitted by ubiquitous whitefly *Bemisia tabaci* in a circulative persistent manner. The high diversity and variability among begomoviruses are because of several factors including mutation, recombination, pseudo-recombination, and satellite capture. Legume-infecting begomoviruses are collectively called legumoviruses, which severely affect worldwide legumes production. Generally, legumoviruses induce leaf yellowing, curling, yellow mosaic, growth stunting, blistering, and reduced pod production. Legumo-viruses can be subdivided into the new world legumoviruses and the old world legumoviruses based on their genome organization and phylogenetic analysis. Although legumoviruses are considered as "host-specific" but because of recombination and component capture strategy, they can spread to the new non-leguminous host plants. A stable and efficient host plant resis-tance, using either conventional approaches or modern molecular biology approaches such as CRISPR/Cas9, may offer a good controlling strategy to counter legumovirus spread in the legumes.

13.1   INTRODUCTION

Taxonomically, all the legumes (>18,000 species) belong to the family *Legu-minosae* or *Fabaceae* and subdivided into four subfamilies: *Papilionoideae*, *Mimosoideae*, *Swartzioideae*, and *Caesalpinoideae*. The major food legumes belong to *Papilionoideae* and include soybean, common bean, chickpea, cowpea, pigeonpea, pea, fababean, mungbean, urdbean, and lentils. The majority of these legumes are cultivated in tropical to temperate regions of the world, the areas that are most suitable for the geminiviruses proliferation. Legumes have been cultivated for millennia, and their use as a staple food started in Asia, the Americas (common bean), and Europe (broad beans) by 6000 BC. Carbon dating shows that domestication of lentils, chickpeas, and peas were started in the Eastern arc in the 7th millennium BC (Smartt, 1990). Legumes vary greatly in their diversity, nutritional value, and importance as a food in various parts of the world. In Latin America, common beans and legumes including soybean, cowpea, and chickpea are chief source of food, whereas mungbean, lentils, urdbean, cowpea, and pigeonpea are widely used in South Asia. Cowpea and common beans are consumed widely in Africa. In North Africa and Middle East, lentil, chickpea, and fababeans are the most important legumes. According to Food and Agriculture Organization (FAO), USA production statistics 2014, total area covered in the world by groundnut, beans, soybean, lentil, peas, pigeonpea, chickpea, and cowpea is ~219 million ha (FAOSTAT, 2014; Fig. 13.1). The worldwide production data shows that production of grain legumes rank third after cereals and oilseeds. According to FAO, about 180 million ha (or 12–15%) of earth's arable land is covered by grain and forage legumes (Table 13.1) (source: FAO Database—http://apps.fao.org/page/collections). Collectively, both these crops share 27% of the world's crop production, whereas legumes alone provides upto 33% of the dietary protein demands of humans. Besides meeting the human and livestock protein demands, legumes are helpful in maintaining the soil nitrogen balance through their virtue of symbiotic association with *Rhizobium* bacteria, which makes them an indispensable component for the sustainability of the ecosystem. The legumes are of prime importance in developing and underdeveloping countries because of their good proteinaceous value. The demand of legumes is rising with the rise in world's population; the scenario has increased the pressure on agriculture sector to meet increasing demands. Both biotic and abiotic factors are major constraints to legume production in tropical and subtropical areas. Among the biotic constraints, the disease caused by viruses (especially begomoviruses) is a top ranked menace to legumes production.

Soybean is the chief grain legume, comprising three quarters of the world production of grain legumes (308 million tons) grown primarily in China, USA, Brazil, Argentina, India, Paraguay, and Uruguay (Fig. 13.1). The enormous increase in worldwide soybean production has been evident in last few decades compared with just 50% for other grain legumes. Lentils has been very important dietary constituent since old civilizations; it is being grown in Australia, Canada, Pakistan, India, EU, USA, some African countries, and Mexico (FAOSTAT, 2014). China, USA, Nigeria, India, Cambodia, Thailand, Vietnam, Sudan, Argentina, Tanzania, and Indonesia are among major groundnut producers. Mungbean is cultivated in Pakistan, Thailand, India, China, Bangladesh, Vietnam, Indonesia, Myanmar, and in the hot and dry regions of Southern USA and Southern Europe. Beans (dry) are an important production of Mexico, Canada, China, Africa, Thailand, Vietnam, Cambodia, and USA. Cowpea is an important leguminous crop and over 80% of cowpea production is confined to European Union, China, Niger, Nigeria, and Burkina Faso. Pigeon pea is predominately cultivated in the Indian subcontinent, EU, and in African countries. Chickpea is the third most important pulse crop and grown in almost all parts of the word; however, its main producers are Pakistan, India, Australia, Canada, USA, EU, and Tanzania. The world production of all other legumes amounts to 108 million tons, growing primarily in southern Asia and China (Fig. 13.2).

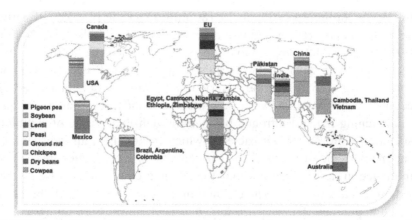

FIGURE 13.1   (See color insert.) Major legumes and beans producing countries of the world. Production of eight important legumes and beans of country is shown in vertical bars and colors. Each bar represents different legume species and their percent production. The key for the colors associated with each legume species is given. The data is taken from FAOSTAT, USA, of year 2014 (http://faostat3.fao.org/browse/Q/QC/E).

Although cultivated areas for major legumes and beans crops have been increased from 200 million ha to 219 million ha (in just few years), but the overall production has been hampered by certain abiotic and biotic stresses including the viruses. Almost all the economically important legumes are susceptible to viral infections and 150 types of viruses, belonging to different genera, have been reported to infect legumes. Among these viruses, begomoviruses are important phytopathogens infecting huge numbers of economical important crops, vegetables including the legumes grown in tropical to temperate regions of the world. Whitefly is an invasive pest, which has risen to international prominence because of its disease transmittance ability. Whitefly (*B. tabaci*) is a highly destructive pest, has high diversity, broad host range, and can infect more than 600 species in 74 plants families (Wang et al., 2010) and these include herbaceous plants, shrubs, and trees. Legume, bean, beet, brinjal, cassava, cotton, chillies, tomato, and tobacco are the major food crops affected by whitefly (Monsalve-Fonnegra et al., 2001). *B. tabaci* is a species complex comprising of ~12 genetic groups that can be identified easily morphologically and biologically by using DNA markers. The main habitants of whiteflies are tropical and subtropical regions of the world but their population is increasingly being present in more temperate regions also. Several virus families including *Comoviridae, Geminiviridae Potyviridae,* and few genera *Closterovirus* and *Carlaviruses* are vectored by whiteflies in persistent (circulative) and non-persistent manner. In 1889, for the first time, it was categorized as agricultural pest in Greece and had significantly damaged various crops including tobacco. In America, the presence of whitefly was noted on a crop of sweet potatoes. In Asia, whitefly was declared as pest in 1919, it has proliferated much from last few decades and massive losses to different agroeconomical crops (including legumes, cotton, tomato, radish, brinjal, papaya, cucumber, other vegetables, and ornamental plants) have been accredited to whitefly-transmitted diseases. Whitefly damaged the host plants in two ways (Stansley, 2010), first, by piercing the phloem followed by sucking plant sap. During this feeding activity, affected area of the host plant gets damaged; moreover, whitefly produces honeydew which provides suitable environment for sooty mold growth. Second problem caused by whitefly is its ability to transmit a huge number of viruses. Whitefly has been principally involved, directly or indirectly, in the spread of several plant diseases including mung bean yellow mosaic disease (MYMD), cotton leaf curl disease (CLCuD), tomato leaf curl disease (ToLCuD), cassava mosaic disease (CMD), cassava brown streak virus disease (CBSD), bean yellow mosaic disease (BYMD), and many such other diseases by transmitting the viruses responsible for these diseases.

Legumoviruses (and all begomoviruses) are vectored by a single species of whitefly (*B. tabaci*) in circulative and semipersistent fashion. During whitefly feeding (from phloem), these viruses are acquired through sap, egested with saliva, and vectored in highly organized way. During phloem feeding, whitefly inserts its stylet (which is extended into cibarium and oesophagus) into the phloem and ingests sap. Virions are acquired during this activity and enter the esophagus and get encapsulated by GroEL protein produced by endosymbionts, then move to digestive tract. GroEL protein protects the virions from the digestive enzymes in digestive canal. From the digestive tract, virions get penetrated into gut membranes through clathrin-mediated endocytosis and reach into the hemolymph. At last, virions move to salivary glands and transmitted or egested with saliva during feeding (Ghanim and Czosneck, 2016).

Whitefly is a vital agriculture pest in the world and helps in transmission of 10% of total plant viruses in the world. For the sustainability of worldwide agriproductivity, several whitefly management strategies are being employed to minimize the losses caused by whitefly invasion. Several cultural, biological, and insecticidal control strategies are being used. In cultural control, use of insect control net, mulching, crop placement, and intercropping are used. While in biological control, genetically modified crops, use of natural whitefly enemies, and cytoplasmic incompatibility (CI) are used.

13.2   FAMILY GEMINIVIRIDAE

Geminiviridae is the largest plant infecting family of small and circular single-stranded DNA (cssDNA) viruses whose genome is encapsidated in a geminate particle, from which family derived its name and this has been the unique feature of this family. This geminate particle is a twinned quasi-icosahedra and comprises 22 pentameric capsomeres (22 × 38 nm) formed by 110 coat protein subunits. The geminate shape morphology was worked out in 1979 through electron microscope when *Chloris striate mosaic virus* (CSMV) particles were revealed to have a 1.7 nm electron-dense cleft clearly dividing the particle into two halves. The genomes (ssDNA) of these viruses are present in both halves and form 19% of particle. The members of the family *Geminiviridae* are transmitted exclusively through their respective insect vectors and infect important agroeconomical species of the plants including herbs, shrubs, mono-cots, and dicots globally, including the legumes, thus posing a major threat to worldwide agricultural productivity especially in the tropical to increasing temperate areas of the world. The history of these viruses have been known from the anthology of poem written in 752 AD by Empress Koken of Japan, describing the pleasing vein yellowing and mosaic symptoms in *Eupatorium makinoi* plant. Although history of these viruses is quite old but they came into limelight in 20th century when many diseases were reported, which are now known to have geminivirus disease etiology. Several geminiviral disease outbreaks have been recorded earlier, a mosaic disease of cassava in Africa in 1894, spread of beat curly top disease in California in 1899, streak disease of maize in South Africa in 1901, and CLCuD in Asia and Africa in 1990. The geminivirus group was first established as a plant virus group in 1979 and later curing 1995, it was revisited as a new plant virus family *Geminiviridae*. These viruses are smallest known viruses that encode only few proteins; therefore, they are dependent heavily on the host-encoded factors for their replication, movement, proliferation, and spread. Based upon genome organization, host plant and type of insect vector, members of this family have been classified into nine genera including *Becurtovirus, Begomovirus, Capulavirus, Curtovirus, Grablovirus, Eragrovirus, Mastrevirus, Turncurtovirus,* and *Topocuvirus* (Zerbini et al., 2017; Brown et al., 2015).

13.2.1   GENUS BEGOMOVIRUS

Out of nine genera, *Begomovirus* is principally the leading genus whose members are exceeding 500 species. The members of this genus are highly diversified, economically important, and geographically the most widespread. These have been found infecting only dicotyledonous plants in the world. This genus derived its name from its oldest known member *Bean golden mosaic virus* (BGMV), now referred as *Bean golden yellow mosaic virus* (BGYMV). Generally, on the basis of genome components, begomoviruses are classified into either bipartite begomoviruses, having two genomic components (DNA-A and DNA-B), of same size (~2800 nt). Monopartite begomovirus genome is homologue of DNA-A of bipartite begomoviruses. Furthermore, some of the monopartite begomoviruses are associated with DNA-satellite molecules, that is, alphasatellite, betasatellite, and deltasatellite, former two are present frequently although (Fig. 13.3). Geographically, begomoviruses are distributed into two groups, those originating from western hemisphere [new world (NW)] and the other one from eastern hemisphere [old world (OW)]. Considerable differences occur among the viruses of both hemispheres such as most of the begomoviruses in NW are bipartite, whereas in OW most of them are either monopartite or found associated with DNA-satellite molecules. Another striking difference is absence of AV2 gene in all NW begomoviruses. The NW begomoviruses CP has a conserved amino acid sequence motif (PWRLMAGT), which is absent in OW begomoviruses. Begomoviruses are transmitted from one plant to the next through the whitefly (*B. tabaci*) in both the hemispheres of the world.

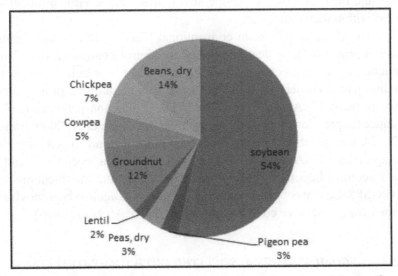

FIGURE 13.2 Worldwide production of eight selected legumes and beans. Graph is based on the data available for the year 2014 at FAO, USA website (http://faostat3.fao.org/browse/Q/QC/E).

13.2.2   MONOPARTITE AND BIPARTITE BEGOMOVIRUSES

The majority of legume-infecting begomoviruses are bipartite in nature and have genome of ~5200–5700 nt and comprises two almost equally sized cssDNA components, denoted as DNA-A and DNA-B. Typically six genes, separated by small intergenic region (IR), are encoded on DNA-A in both virion and complementary sense (Fig. 13.3). Four genes are transcribed on the complementary strand whereas two genes on the virion sense strand. The DNA-B of bipartite begomoviruses encodes two genes (movement protein [MP] and nuclear shuttle protein [NSP]) in the opposite directions.

Both DNA-A and DNA-B do not share any sequence homology except for a small ~200 nt common region (CR). This region is conserved among whole members of the family *Geminiviridae* containing origin of replication, the nonanucleotide (TAATATT/AC) sequence. In bipartite legumoviruses, DNA-A encodes factors involved in viral genome replication [the replication associated protein (Rep)], replication enhancer (REn), transcriptional activation (TrAP), encapsidation/insect transmission [the coat protein (CP)], precoat protein (AV2), and suppressor of host–defense mediated by post-transcriptional gene silencing (PTGS) (AC4). For *mungbean yellow mosaic India virus* (MYMIV), the AC5 protein encodes a protein that have role in viral DNA replication. Although only two proteins are encoded on DNA-B, the MP and the NSP, both, act cooperatively to move the virus from cell-to-cell and within the whole plant.

A range of techniques from conventional breeding to advance genome editing approaches have been employed to control begomoviruses. These techniques include conventional breeding approaches, molecular approaches including pathogen-derived resistance (both protein and non-protein mediated resistance), DNA interference (Tahir, 2014), and non-pathogen derived resistance (expression of *GroEL* and *cry1Ac*) (Edelbaum et al., 2009; Rana et al., 2012). Besides this, whitefly management through advanced biological techniques like RNAi and *Wobachia*-mediated CI has become a leading focus to control begomoviruses. Very recently, genome modifications tool like CRISPR/Cas9 has been proposed and employed against begomoviruses to gain a comprehensive control (Iqbal et al., 2016; Ali et al., 2016).

13.2.3   BEGOMOVIRUSES ASSOCIATED WITH DNA-SATELLITES

True satellites are nucleic acids (either DNA or RNA) molecules dependent on their cognate virus for replication, movement, encapsidation, and transmission by sharing little or no sequence homology to the helper virus. The first ever-identified DNA satellite associated with a begomovirus was a small (~680 nt) cssDNA molecule called tomato leaf curl virus satellite-DNA (ToLCV sat-DNA) that was sharing no sequence homology to the helper ToLCV except nonanucleotide sequence and a putative Rep-binding motif.

Till now, DNA-satellite molecules have been categorized as alphasatellite and two members, betasatellite, and deltasatellite of family *Tolecusatellitidae* (Adams et al., 2017).

13.2.4   ALPHASATELLITE

Many monopartite begomoviruses are associated with nanovirus-like components called "alphasatellites" (previously known as DNA-1). Following the definition of satellites, alphasatellites are not true satellites but can be referred as "satellite-like" because of their self-replicating nature. Alphasatellites have maximum structural resemblance to Rep-encoding components of nanoviruses (Family, *Nanoviridae*) and the hairpin loop structure of alphasatellites including nonanucleotide sequence (TAGTATTAC) also resembles those of the nanoviruses. It is, thus, presumed that during a mix infection alphasatellites were captured by begomoviruses.

Alphasatellites are usually found associated with monopartite begomovirus–betasatellite complexes but occasionally also present with monopartite begomoviruses. Nevertheless, alphasatellites have been shown empirically to be supported by a bipartite begomovirus and a curtovirus. Additionally, alphasatellite can be transmitted between plants by a leafhopper together with a leafhopper-transmitted curtovirus. Recently, the NW alphasatellites have also been found associated with bipartite begomoviruses under field condition.

It has been assumed that alphasatellites may attenuate the virus infection, thus ensuring the host plant survival and further transmission of the whole virus–satellite complex. The Rep of few alphasatellites has PTGS suppressor activity, suggesting that alphasatellites may contribute to overcome host–plant defense.

13.2.5   BETASATELLITE

Betasatellites are very recently classified into genus *Betasatellite* (family *Tolecusatellitidae*) and first ever reported betasatellite was from *Ageratum conyzoides* plant in association with *Ageratum yellow vein virus* (AYVV). This satellite was about half the size of helper virus (AYVV) and was named DNA-β (which was later renamed as betasatellite). Shortly, another betasatellite was found associated with CLCuD from the infected cotton plants in Pakistan. Since then, several hundreds of such complexes have been characterized from different cultivated and noncultivated plant species in the OW. Betasatellites are true satellites that exacerbate phenotypic symptoms and are half the size (~1350 nt) of their helper virus genome. These perform diverse functions like alter the symptoms induction in host plants, upregulate the replication of associated helper begomoviruses, interacts with host-encoded factors, essential for pathogenesis, increase helper virus DNA titer, regulate the micro-RNAs level, complement the missing functions of DNA-B of bipartite begomovirus *Tomato leaf curl New Delhi virus* (TLCNDV), suppress host–plant defense, extend the host range, and may impart role in movement (Amin et al., 2011; Kon et al., 2007).

Experimental analyses have revealed the structural conservation of these satellites including adenine rich (A-rich) region, a 100 nt satellite conserved region, and a single gene (βC1) in complementary sense, responsible for all the functions. The nonanucleotide sequence in betasatellite is TAA/GTATTAC. The major center of betasatellite diversity is China and Indian subcontinent, the possible center of origin of betasatellites. Phylogenetic analysis revealed the two major groups of betasatellites, the first isolated from plants of *Malvaceae* family (such as cotton, hibiscus, okra, and hollyhock), whereas the second isolated from non-malvaceous plants (such as *Ageratum*, chillies, honeysuckle, tomato, zinnia, etc.) (Briddon et al., 2003).

13.2.6   *DELTASATELLITES*

These newly characterized non-coding molecules are approximately quarter the size of their helper begomovirus genome and belong to genus *Deltasatellite* (family *Tolecusatellitidae*). These newly identified molecules have been named "deltasatellites" (Lozano et al., 2016). Deltasatellites have been found associated with ToLCV in Australia (Dry et al., 1997), Sida golden yellow vein virus (SiGYVV-Ma) in the Caribbean (Fiallo-Olivé et al., 2012), and swepoviruses in Venezuela and Spain (Lozano et al., 2016). Deltasatellites share structural and sequence similarities with betasatellites but contain some additional features as well. Vector-enabled metagenomics approach has discovered some of the deltasatellites in *B. tabaci* from Florida (Ng et al., 2011). The nonanucleotide (TAATATTAC) in their stem loop form a part of the loop. Moreover, they contain an A-rich region but do not code any protein. Additionally, they encompass a second predicted stem loop, which is present next to the iteron sequences and a TATA motif. A short region of deltasatellites is also identical to the SCR of betasatellites. These deltasatellites are supposed to be evolved from betasatellites because of their common features. Nevertheless, their parental betasatellites are yet to be identified or those might be extinct now. Although different subclasses of deltasatellites have been identified but phylogenetically these are not identical to each other (Lozano et al., 2016).

13.3 DIVERSITY OF LEGUME-INFECTING BEGOMOVIRUSES IN DIFFERENT COUNTRIES

The geographical distribution of legumoviruses is very diverse across the legume cultivating countries (Fig. 13.4). Based on the phylogenetic analysis and their genome arrangement, legumoviruses can be classified into two major groups OW (including Europe, Africa and Asia) and the NW (the Americas) legumoviruses (Fig.13.5, Table 13.2). Country-wise diversity of legume-infecting begomoviruses was assessed from the submitted sequences in the GenBank (http://www.ncbi.nlm.nih.gov/) and presented in the Figures 13.5a and b.

TABLE 13.1 Worldwide Cultivation of Legume Crops in Year 2014 (FAOSTAT, 2014).

Crop	Area (mha)
Soybean	117.72
Pigeon pea	6.68
Lentil	4.52
Cowpea	12.52
Chickpea	14.80

13.3.1 INDIA

Legumes and pulses are main dietary constituent in India and it falls among major legume producing countries. However, yield losses because of begomoviruses were projected to surpass $300 m in India (Varma and Malathi, 2003). Among the major constraints to legume production in India, MYMD is principally highly ranked. This disease has been caused by four species of begomoviruses, also referred as legume yellow mosaic viruses (LYMVs), including MYMIV, *Mungbean yellow mosaic virus* (MYMV), *Horsegram yellow mosaic virus* (HYMV), and *Dolichos yellow mosaic virus* (DYMV). This disease was reported for the first time from western India in 1940s from Lima bean, whereas in 1960 it was reported from mungbean in northern India. This disease causes immense losses to different legume crops, upto 75.8% reductions in yield of the soybean and 83.9% maximum yield loss for mungbean and cowpea have been reported. Out of four LYMVs, MYMIV is most prevalent in southern and western Indian states, whereas MYMV is predominant in central, northern, and eastern parts of India. Both these viruses infect many legumes including blackgram (*Vigna mungo*), mungbean (*Vigna radiata*), soybean (*Glycine max*), pigeonpea (*Cajanus cajan*), common bean (*Phaseolus vulgaris*), moth bean (*Vigna aconitifolia*), yardlong bean (*Vigna sesquipedalis*), and hyacinth bean (*Lablab purpureus*).

Beside these LYMVs, other begomoviruses have also been reported infecting different legume crops in India. For example, *Cowpea golden mosaic virus* (CPGMV—a bipartitie begomovirs) from cowpea, *Velvet bean severe mosaic virus* (VBSMV—bipartite begomovirus) from velvet bean, *Rhynchosia yellow mosaic India virus* (RhYMIV—a bipartite begomivirus) from *Rhynchosia minima, French bean leaf curl virus* (FbLCV) from French bean, and *Tomato leaf curl Karnataka virus* (ToLCKnV) from soybean.

13.3.2   PAKISTAN

In the Indian subcontinent, Pakistan is second largest legume producing country. Among the legume and bean crops, chickpea is cultivated on an area about 1.1 mha and its annual production has been exceeded from 0.75 million tons per annum. The yield of lentils is ~500 kg/ha with annual production of 8600 tons. Soybean is least cultivated crops (FAOSTAT, 20014). Overall, lower yield of all leguminous crops is because of adoption of poor agricultural practices and begomoviruses infection. The intensity of begomoviral disease is high in beans. Yellow mosaic disease (YMD) was reported from Pakistan in 1942 from the vicinity of Faisalabad. The diversity of legumoviruses in Pakistan have been studied thoroughly by Ilyas et al. (2010) and results showed that MYMV, MYMIV, *Rhynchosia yellow mosaic virus* (RhYMV), *Pedilanthus leaf curl virus* (PeLCV), and *Papaya leaf curl virus* (PaLCuV) were found to be infecting different legumes. Contrasting to India where MYMV is chiefly responsible for infecting legumes, MYMIV is the most prevalent begomovirus and is responsible for YMD of legumes, and has been isolated from mungbean, soybean, blackgram, cowpea, and leguminous weeds across Pakistan, whereas MYMV is found to be associated with just few weeds (*Rhynchosia* spp). Reports from India depict that almost all the cultivated soybean suffers with LYMVs, suggesting soybean cultivation exacerbating LYMV problems in other legume crops. Although MYMV has been potentially infecting soybean in Pakistan, however its occurrence cannot signify the soybean cultivation in Pakistan.

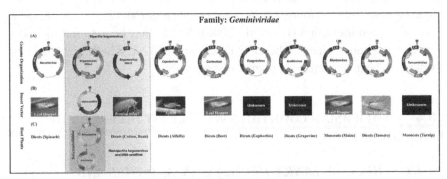

FIGURE 13.3   (See color insert.) Genome organization of the family *Geminiviridae* **(A)**, the respective insect vectors **(B)**, and the type crop used to name each genus **(C)**. Legumoviruses are characterized by both bipartite and monopartite begomoviruses in the genus *Begomovirus* (Family *Geminiviridae*). The genus *begomovirus* (highlighted in pink background) has been named after the first begomovirus identified from common bean (*Phaseolous vulgaris*) as *Bean Golden Mosaic Virus* (Begomovirus). Some of the monopartite begomovirsues are also found associated with DNA-satellites called alphasatellite, betasatellite, and deltasatellite (Genus *Betasatellite* and *Deltasatellite*, Family *Tolecusatellitidae*), respectively. The genera *betasatellite* and *deltasatellite* are highlighted in green background.

13.3.3   BANGLADESH

In Bangladesh, almost all major legumes and beans are cultivated across the country and collectively they occupy second largest crop area after rice. Lentils, soybean, beans, and groundnuts are major crops of Bangladesh and their annual production is 0.98, 0.70, 0.51, and 0.50 million tons, respectively (FAOSTAT, 2014). The major cause for low productivity of legumes and beans is the susceptibility of the available germplasm to numerous viral diseases, of which YMD ranked high. The first disease incidence of YMD was reported in 1960s characterized by MYMV and MYMIV. Recently, another legumovirus, DoYMV, has been isolated from legumes in Bangladesh (Maruthi et al., 2006).

13.3.4   VIETNAM

The major legumes cultivated in Vietnam are mungbean and soybean. Strenuous efforts have resulted in an increase of groundnut yield up to 0.45 million tons per annum from a total of 0.21 mha arable land, while per annum yield of soybean is 0.16 million tons (FAOSTAT, 2014). MYMD was first reported in 1970s in Vietnam and Thailand where it caused severe yield losses. The chief causal agents for MYMD is MYMV; however, two other legumoviruses *Mimosa yellow leaf curl virus* (MiYLCV) and KuMV have also been identified (Ha et al., 2008). KuMV has been found to infect soybean and kudzu plants, whereas MYMV and MiLYCV have been isolated from mungbean and mimosa plants, respectively.

13.4   OTHER ASIAN COUNTRIES WITH LEAST DIVERSITY OF LEGUMOVIRUSES

13.4.1   CHINA, SRI LANKA, NEPAL, JAPAN, AND THAILAND

China produces 16 million tons of groundnut, 12.2 million tons of soybean, and is the largest producer of fababean and prominent producer of lentils in the world (FAOSTAT, 2014). However, legumoviruses diversity is either very low in this region or has not been explored yet, only KuMV and *Tomato yellow leaf curl china virus* (TYLCCNV) have been found to infect legumes in China.

In Sri Lanka, soybeans, cowpeas, and fababeans are being cultivated but only one type of legumovirus *Horsegram yellow mosaic virus* (HgYMV) has been found infecting fababeans.

In Japan, Thailand, and Cambodia, a common virus, that is, MYMV is the causative agent that infects mungbean, whereas in Nepal, MYMIV is the only virus found infecting fababeans and lama beans.

13.4.2   ARGENTINA, BRAZIL, COLOMBIA

Soybean, chickpea, lentil, and peas are the major cultivated leguminous crops in Brazil, Argentina, and Colombia. Brazil dominates in soybean with arable area, soybean cultivation is 30.2 mha and annual production is 86.8 million tons. Argentina secured second position followed by Colombia with 53.4 and 0.09 million tons per annum, respectively (FAOSTAT, 2014). In

Brazil, the variability of yield is 26%, which is the highest among the top bean-producing countries. Collectively, these countries produce more than 120 million tons soybean of the world.

The sequence analysis of a begomovirus infecting soybean in Northwestern Argentina showed that it was closely related to *Sida mottle virus* (SiMoV). In addition, soybean crop has also been shown to host BGMV, *Soybean blistering mosaic virus* (SbBMV), and *Tomato yellow spot virus* (ToYSV) in Argentina (Alemandri et al., 2012). In Argentina, the rate of yield losses and incidence of BGMV occurrence is highest, followed by SbBMV and ToYSV. Since 1960s, BGMV, a predominant NW bipartite begomovirus in Argentina and Brazil, has been an important phytopathogen causing massive yield losses ranging from 40% to 100% (Morales, 2006). However, the annual average yield losses because of BGMV were estimated to be 20%. BGMV has very narrow host range and has been isolated from common beans, soybean, and *Macroptillium lathyroides*. Besides BGMV, other begomoviruses infecting legumes in this region are *Macroptilium yellow spot virus* (MaYSV), *Sida micrantha mosaic virus* (SiMMV), *Macroptilium yellow vein virus* (MaYVV), SiMoV, *Soybean chlorotic spot virus* (SbCSV), and SbBMV. MaYSV induces yellow spotting, leaf crumpling, and distortion and occurs in the Northeast region of Brazil. SiMMV was isolated from common bean where it was associated with golden mosaic, chlorotic spots, and leaf distortion. Furthermore, in the field conditions, MaYSV and SiMMV have also been found infecting common bean, *M. lathyroides*, *Calopogonium mucunoides*, and *Canavalia sp.* and legumenous weed hosts in northeastern Brazil. Moreover, SbCSV was found in the field samples of *M. lathyroides*, whereas MaYVV was only reported to infect *M. lathyroides* plants besides a lima bean field.

13.4.3   MEXICO

Mexico is the largest country in Latin America, with an area of 1,958,201 sq/km. Soybean is the second largest crop and is cultivated on 0.20 mha, and 0.39 million tons is produced annually. In Mexico, all the legume crops are infested by different begomoviurses including BGYMV, *Bean calico mosaic virus* (BCaMV), *Bean yellow mosaic Mexico virus* (BYMMxV), *Rhynchosia golden mosaic virus* (RhGMV), *Vigna yellow mosaic virus* (ViYMV), *Rhynchosia Yellow mosaic Yucatan virus* (RhYMYuV), and *Sida yellow mosaic Yucatan virus* (SiYMYuV) (Fig. 13.5). In 1974, bean golden mosaic disease was first time evident in the northwestern state of Sinaloa. In

1977–1980, BGYMV emerged on the Gulf Coast of Veracruz, Las Huastecas, and Chiapas, where it affected common beans and other legumes. Finally, in 1980, this disease spread to the Yucatán Peninsula and state of Sonora, and was found to be associated with a distinct virus, BCaMV, that causes severe golden mosaic symptoms that eventually lead to bleaching effect.

13.4.5 PUERTO RICO

Puerto Rico is situated in Latin America and the major dietary constituent of Puerto Rico people are legumes and beans. Local production of legumes and beans does not meet the demand; therefore, a major proportion of beans is imported from the United States. The annual production of beans is 0.04 million tons and this is being cultivated on 0.04 mha, whereas 0.24 million tons of pigeon peas are cultivated from 0.29 mha (FAOSTAT, 2014). Although this is very small area but biodiversity of legumoviruses have been widely demonstrated in this country (Fig. 13.5).

Golden mosaic disease (GMD) in Puerto Rico is caused by BGYMV, which is the major constraint to bean production not only in Puerto Rico but also found to impair bean production in other Caribbean countries and in the USA. Besides BGYMV infection, the legume crops in Puerto Rico has been found infected with *Kudzu Mosaic virus* (KuMV), RhGMV, *Macroptilium mosaic Puerto Rico virus* (MacMPRV), and *Rhynchosia mild mosaic virus* (RhMMV). The YMD of pigeon peas, although with very low disease incidence, has also been reported from Puerto Rico, where it causes maximum yield losses up to 40%.

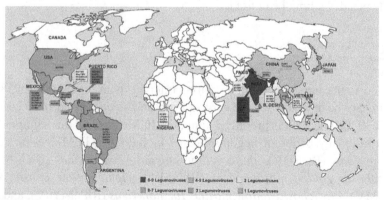

FIGURE 13.4 Geographical diversity and distribution of legumoviruses in the world. The map shows the distribution of legumoviruses and the number of species reported in each country. The key for the colors associated with each legumovirus species is given.

13.4.6 CUBA

Cuba is a small country whose land comprises mountains, plains, and basins. Legumes and beans are two major cultivated crops. Although legumes and beans have limited cultivation yet few begomoviruses including BGYMV, *Rhynchosia golden mosaic Havana virus* (RhGMHaV), *Macroptilium yellow mosaic virus* (MacYMV), and *Rhynchosia rugose golden mosaic virus* (RhRGMV) has been reported from these crops.

TABLE 13.2 List of Begomoviruses Infecting Legumes and Their Host Names and Country of Origin.

	Country	Viruses	Crop		Country	Viruses	Crop
	Bangladesh	MYMIV	-		Argentina	SbBMV	Soybean
		DoYMV	-		Brazil	CeYSV	Centrosema brasilianum
	Cambodia	MYMV				MaYNV	Macroptilium
		MYMV	Mothbean			MaYSV	French bean
	China	KuMV	Kudzu				Macroptilium
	India	MYMV	Soybean			SoCSV	Soybean
			Blackgram				Macroptilium
			Mothbean			MaYVV	Macroptilium
		HgYMV	Lima bean			BGMV	French bean
			Horsegram				Soybean
			Cowpea				Lima bean
			French bean				Macroptilium
New World		CPGMV	Cowpea	New World	Columbia	BDMV	French bean
		MYMIV	French bean		Cuba		Rhynchosia
			Mungbean				Rhynchosia
			Soybean				French bean
			Blackgram				French bean
		VBSMV	Velvet bean		Dominican Republic	BGYMV	French bean
		RhYMV	French bean		Guatemala	BGYMV	French bean
		DoYMV	Dolichos		Honduras	RhGMV	Rhynchosia
		FbLCV	French bean		Jamaica	MacGMV	Wissadula amplissima
		SHLDV	Crotalaria			MacYMV	Macroptilium
					Mexico	RhGMV	Sinaloa
	Indonesia	MYMIV	Soybean				Rhynchosia
			Cowpea				Soybean

TABLE 13.2   *(Continued)*

Country	Viruses	Crop	Country	Viruses	Crop
Nepal	MYMIV	French bean			Tobacco
		Lima bean		BYMMxV	French bean
Nigeria	SbMMV	Soybean		BGYMV	French bean
	SbCBV	Soybean		ViYMV	Velvet bean
		Lima bean	Puerto Rico	RhMMV	Rhynchosia
	CPGMV			MacMPRV	Macroptilium
Pakistan	MYMV	Soybean			French bean
					Pumpkin
	MYMIV	Mungbean		BGYMV	French bean
		Cowpea	USA	BGYMV	French bean
		Soybean		MacYMFV	Macroptilium
			Venezuela	BYCV	French bean
Sri Lanka	HgYMV	French bean		BWCMV	French bean
Thailand	MYMV	Mothbean			
Vietnam	MYMV	Mothbean			
	KuMV	Soybean			

In Cuba, intensive use of agricultural practices just after World War II welcomed the whitefly and so do begomoviruses. In 1970s, plant viruses were reported from common beans and it became an epidemic in late 1980s when production of common bean was severely affected. Bean golden mosaic disease (caused by BGYMV) has been a threat to common bean cultivation in Cuba since early 1970s. The epicenter of the problem was Velasco town in the western Holguín, which has spread to the neighboring provinces including Guantanamo, Havana, and Las-Tunas. Ciego de Avila, Holguín, Las-Tunas, and Camaguey are the recent centers of BGYMV infestation in Cuba.

13.4.7   NIGERIA

In Nigeria, cowpeas and soybean are major legume crops encompassing 3.70 and 0.72 mha with annual production of 2.10 and 0.68 million tons, respectively (FAOSTAT, 2014). In Africa, Nigeria is the leading soybean producer and it accounts for 51% of soybean production in Africa. Increasing soybean cultivation has faced a major constraint to soybean production, that is, legumoviruses. In Nigeria, three begomoviruses causing YMD have been isolated from soybean: *African Cassava mosaic virus* (ACMV), *Soybean mild mottle virus* (SbMMV), and *Soybean chlorotic blotch virus* (SbCBV) (Alabi et al., 2010).

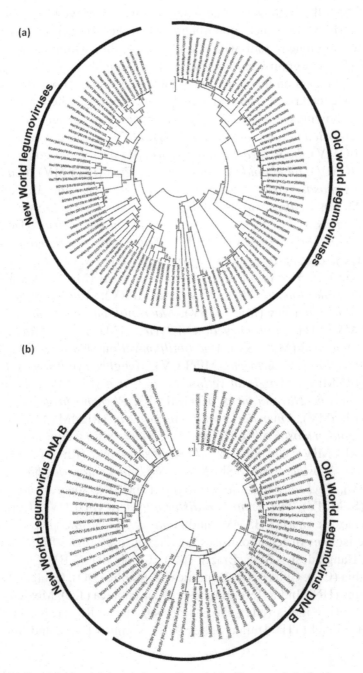

FIGURE 13.5 Neighbor-joining phylogenetic analysis of legumes infecting begomoviruses DNA-A **(a)** and DNA-B **(b)**.

All the full-length nucleotide sequences were retrieved from the NCBI database (http://www.ncbi.nlm.nih.gov/) and then aligned using pairwise sequence alignment tool with ClustW algorithm available in Mega7. The phylodendogram was constructed using neighbor-joining method in Mega7. Numbers on the nodes are showing percentage bootstrap support values for each branch. Vertical branches are arbitrary, whereas horizontal lines are showing calculated mutation distances. The begomovirus isolates used were: *Bean calico mosaic virus* (BCaMV), *Bean chlorotic mosaic virus* (BCMV), *Bean dwarf mosaic virus* (BDMV), *Bean golden mosaic virus* (BGMV), *Bean golden yellow mosaic virus* (BGYMV), *Bean white chlorosis mosaic virus* (BWCMV), *Bean yellow mosaic Mexico virus* (BYMMxV), *Centrosema yellow spot virus* (CeYSV), *Cowpea golden mosaic virus* (CPGMV), *Dolichos yellow mosaic virus* (DoYMV), *French bean leaf curl virus* (FbLCV), *French bean severe leaf curl virus* (FbSLCV), *Horsegram yellow mosaic virus* (HgYMV), *Kudzu mosaic virus* (KuMV), *Macroptilium golden mosaic virus* (MacGMV), *Macrop-tilium mosaic Puerto Rico virus* (MacMPRV), *Macroptilium yellow mosaic Florida virus* (MacYMFV), *Macroptilium yellow mosaic virus* (MacYMV), *Macroptilium yellow net virus* (MacYNV), *Macroptilium yellow spot virus* (MacYSV), *Macroptilium yellow vein virus* (MacYVV), *Mimosa yellow leaf curl virus* (MiYLCV), *Mungbean yellow mosaic India virus* (MYMIV), *Mungbean yellow mosaic virus* (MYMV), *Rhynchosia golden mosaic Havana virus* (RhGMHaV), *Rhynchosia golden mosaic virus* (RhGMV), *Rhynchosia mild mosaic virus* (RhMMV), *Rhynchosia rugose golden mosaic virus* (RhRGMV), *Rhynchosia yellow mosaic India virus* (RhYMIV), *Rhynchosia yellow mosaic virus* (RhYMV), *Soybean blistering mosaic virus* (SbBMV), *Soybean chlorotic blotch virus* (SbCBV), *Soybean chlorotic spot virus* (SbCSV), *Soybean mild mottle virus* (SoMMV), *Sun hemp leaf distortion virus* (SHLDV), *Velvet bean severe mosaic virus* (VBSMV), and *Vigna yellow mosaic virus* (ViYMV). All the isolates of legumoviruses are shown to originate from Argentina (AR), Bangladesh (BD), Brazil (BZ), Cambodia (CB), China (CH), Colombia (CO), Cuba (CU), Dominican Republic (DO), Guatemala (GT), Honduras (HN), India (IN), Indonesia (ID), Jamaica (JM), Mexico, (MX), Nepal (NP), Nigeria (NG), Pakistan (PK), Puerto Rico (PR), Sri Lanka (SL), Thailand (TH), United States (US), Vietnam (VN), and Venezuela (VZ).

13.5   FACTORS AFFECTING CROSS-CONTINENT GENETIC DIVERSITY OF BEGOMOVIRUSES

Diversity of viruses is an ever-growing phenomenon, which is directly affected by social activities, global trade of plant hosts and insect vectors, adoption of new agricultural methods, and climate changes. The bego-moviruses have become more diversified worldwide for their origin and distribution particularly during previous couple of decades (Navas-Castillo et al., 2011). Their diversification categorized them into the OW and the NW but apparently both niches share some common origin. For example, their genomic organization is very similar except additional *V2* gene in the OW begomoviruses, which is missing in the NW begomovirus genomes. The erroneous replication of plant viruses usually cause genotypic varia-tion, which ultimately are imprinted in their predecessors. Similar to RNA viruses, geminiviruses have high nucleotide substitution rates because geminivirus replication is based upon two processes, that is, rolling circle replication (RCR) and recombination dependent replication (RDR) for their effective multiplication. However, the ultimate factor in diversification is recombination among these viruses.

Geographically, begomoviruses can be subdivided into seven major groups. In the OW, they consists of African or Mediterranean, Indian, Asian, and legume-infecting begomoviruses whereas, in NW classification, they are based upon two groups Latin American and Meso American. The 7th group called sweet potato-infecting begomoviruses (Sweepoviruses) is present in OW and NW. These aforementioned seven groups have 34 sub-populations. The major cause of these groupings is the occurrence of geographical barriers. Although the Asian and African begomoviruses faces frequent recombination and spatial associations between them, genetically these populations are distinctively isolated (Prasanna et al., 2010).

13.5.1   DIVERSIFICATION

According to modern concept of diversification, geminivruses have been assigned to eight centers of diversification including Australia, Central America, Indian sub-continent, Japan, Mediterranean–European region, South America, South China, and Sub-Saharan Africa (Nawaz-ul-Rehman and Fauquet, 2009). Most of the genetic diversity of geminiviruses and DNA-satellites are confined to Chinese and Indian diversification centers. The centers of plant origin and the centers of the geminivurses diversification are not always the same. For example, it differs for Australian, African, and South American centers of diversification. The center of origin for cassava and cotton is in the NW but these plant species do not host an equal number of begomovirus species comparing to their current geographical areas after their introduction (Faquet et al., 2008). For example, cotton (*Gossypium hirsutum*) was introduced from the United States in to the agroecological zones of Indian subcontinent during 19th century. The crop has been facing a major challenge of CLCuD including at least five different species of monopartite and bipartite begomoviruses in Indian subcontinent (Sattar et al., 2013; Iqbal et al., 2016; Sattar et al., 2017). In USA, it is only infected by *Cotton leaf crumple virus* (CLCrV), which has a bipartite genome. Similarly, cassava was introduced into Africa during 16th century from USA; however, it has never been reported to host any geminivirus in USA. Nevertheless, in Africa, CMD has been a major constraint to cassava crop, which is a complex of 11 cassava mosaic geminiviruses (CMGs) to date (Fondong, 2017). Thus, introduction of new crops in an area does not necessarily introduce new begomovirus species but rather the introduced crops are invaded by the local begomovirus species in that particular area.

The climatic conditions of OW (including Indo–Pak subcontinent and Africa) are tropical to subtropical, which support high genetic diversity of begomoviruses; However, comparatively, the Indo–Pak subcontinent have more genetic diversity of the monopartite begomoviruses than bipartite begomoviruses. On the contrary, in Africa, both types of begomoviurses are equally identified. The diversity of DNA-satellites is much more in the OW as compared with NW, where a few DNA-satellites have been recently reported.

13.5.2   MUTATION

Several biological processes such as replication, exposure to UV light, use of chemical mutagens, etc., results into mutation. Usually, point mutations arise during duplication of genetic material when a noncomplementary nucleotide is incorporated into the developing DNA or RNA strand, which modifies the genetic information. The replication of RNA viruses depends upon RNA-dependent RNA polymerase (RdRp) and thus their mutation rate is believed to be more than DNA viruses because of error-prone RdRp. Contrarily, DNA viruses manipulate the replication mechanism of the host plants but evidently, DNA viruses are usually deprived of DNA-repair machinery from the host plant. Ultimately, any mismatches or error during replication get imprinted in the new progenitors of the virus evident. The mutation rate of *Tomato yellow leaf curl virus* (TYLCV), TYLCCNV, and *East African cassava mosaic virus* (EACMV) (i.e., $\sim 10^{-4}$ substitution/site/year) is almost similar to the mutation rate of RNA viruses, which supports this hypothesis. Among DNA viruses, the mutation rate of geminiviruses is very high both in cultivated crops and noncultivated wild host plants. For example, the mutation frequency of *Cotton leaf curl Multan virus* (CLCuMuV) isolates at three ORFs (i.e., CP, Pre-CP, and C4) and exceeds many plant and animal infecting viruses. This suggests that besides the polymerase fidelity, the chances of mutation in plant viruses depend upon several other factors. These factors include virus genome architecture, rapidity of replication, type of host plant, and virus inoculum. Although recombination is considered as main driving force in deciding virus diversity, however, accumulation of point mutations in a plant virus population also affects the viral diversity.

13.5.3 RECOMBINATION

It is well documented that the exchange of genetic material or genetic recombination occurs among the viral diversity pool in a particular disease epidemic infecting diverse host plants (Silva et al., 2014). Genetic recombination has a pivotal role in the evolution of diversified virus families, particularly in the emergence of new geminivirus species (Nagy, 2008). Geminiviruses evolve with nucleotide substitution rate almost similar to RNA viruses, recombination evidently occurs in their diversity spectrum (Lefeuvre et al., 2009). Interspecific recombination serves as an active evolutionary force to widen virus diversity in a particular niche. The interspecific recombination, even between distantly related geminiviruses, ensures that the genome architecture of the newly evolved progeny does not change to ensure that the adaptive capacity of the recombinant is not compromised. The genus *Topocuvirus* was originated as a result of interspecific recombination of a mastrevirus and a begomovirus (Rojas et al., 2005). The recombinant begomovirus species are the primary source of many disease epidemics, for example, CMD, CLCuD, and tomato yellow leaf curl disease (TYLCD). Another epidemic of *Tomato yellow leaf curl Malaga virus* (TYLCMaV) has also been reported from Spain in beans (Monci et al., 2002). Because of recombination, the recombinant virus species have a selective advantage over its parental viruses to be established in its particular niche or to explore new habitats by infecting new host plants. CLCuD appeared in Pakistan during 1988–1989 and then spread into the whole cotton growing areas afterwards. But the disease was fully controlled by establishment of new resistant cotton cultivars. However, after few years, a recombinant begomovirus species called cotton leaf curl Burewala virus [(now called Burewala strain of *Cotton leaf curl Kokhran virus* (CLCuKoV)] was evolved by recombination between CLCuMuV and CLCuKoV together with a recombinant betasatellite molecule. As a result of these recombinants in the field, the areas remained disease free during this course of time and became disease infected (Sattar et al., 2013). Many other crops in Indo–Pak subcontinent are thought to be infected with emerging begomovirus disease complexes and likely have a link with recombination (Venkataravanappa et al., 2011). Apart from interspecific recombination, begomoviruses do recombine with the DNA-B and/or DNA-satellites, that is, alphasatellites and betasatellites. Such a recombination event, if exists, further complicates the situation (see the next section).

The origin of recombinant viral species depends upon the genetic similarity of the begomoviurses coinfecting a certain host together. Naturally, the CRs of begomoviruses are most prone to recombination. Some of the best known examples where CRs were exchanged are *Potato yellow mosaic virus* (PYMV) and *Potato yellow mosaic Panama virus* (PYMPV), DNA-A of *Sri Lankan cassava mosaic virus* (SLCMV) and DNA-B of Indian cassava mosaic virus (ICMV). It has been speculated that the hairpin loop in the IR of geminiviruses may serve as a recombination signal that may lead to replication recess and regulate recombination (George et al., 2015).

Usually the parental strains of a recombinant virus have high mutational load as compared with the recombinant progeny. Mostly, mixed infections and lack of cross-protection in begomoviruses give rise to recombination. During a recombination event, several different recombinant viruses may arise but not all of them are responsible for the disease epidemics (Saleem et al., 2016). The ability of the recombinant viruses to become pandemic in a specific region depends upon the perfect combination and a positive selection pressure. The majority of the recombinant viruses likely have poor fitness but those in favorable combinations outcompete their parental viruses and overdominate the population.

Occurrence of recombination in geminivirus is possibly linked to the presence of compound single sequence repeats (cSSR) or microsatellites in their genomes. George et al. (2015) found that out of 28% of geminivirus species that lack cSSR motifs in their genomes, the recombination was absent in 15%. Moreover, the distribution pattern of cSSRs within geminiviral genomes are in harmonically similar to the recombination host spots in the majority of the geminiviruses. Thus, it is proposed that the type, length, and polymorphism of cSSR motifs in a geimivirus genome may directly contribute to the recombination process. Usually, DNA-A of monopartite begomoviruses predominantly support this association as compared with the DNA-A of bipartite begomoviruses. Possibly, the host recombination machinery comparatively supports the replication and recombination of monopartite begomoviruses. However, such inferences need experimental validation to get approved. Nonhomologous recombination can apparently be noticeable at genus level, whereas interspecies or interstrain homologous recombination is a major phenomenon through which almost every gemini-virus was evolved (Lefeuvre et al., 2010). Although recombination hot-spots and cold-spots are very much conserved, yet recombination breakpoints are distributed non-randomly in geminivirus genomes. The plant viruses are known to pursue homologous recombination in host plants and thus gemi-niviruses may disrupt the recombination machinery of the host to promote viral recombination (Richter et al., 2014).

The recombinants under natural selection may exhibit a host-shift as compared with their parental viruses and thus may broaden their host range. For example, *Tomato yellow leaf curl Sardinia virus* (TYLCSaV) and TYLCV are member of TYLCD infecting tomato crop in Spain. However, the emergence of their recombinant TYLCMaV caused a serious epidemic to *P. vulgaris* in Italy and Spain (Garcia-Andres et al., 2007).

For legumoviruses, the information about recombination is only limited to MYMIV. However, the component capture or pseudorecombination is a common attribute in the viral diversification of legumoviruses.

13.5.4 PSEUDORECOMBINATION

The begomoviruses usually have two genomic components, that is, DNA-A and DNA-B but they are frequently found associated (mostly in the OW) with DNA satellites. The packaging of these genomic components or the associated DNA satellites and later cotransmission into the host plant cells may provide opportunities for a genomic reassortment or pseudorecombination. There are several known examples of such component reassortment between the begomoviruses and DNA-B or DNA satellites (Briddon et al., 2010; Lefeuvre et al., 2007; Chen et al., 2009). The DNA-B of bipartite begomoviruses and betasatellites require a compatibility with the Rep expressed by DNA-A of the helper begomoviurs. However, a variation exists in terms of the specificity with which different Reps interact with these components. For instance, betasatellites may interact with distantly related cognate begomovirues as compared with the DNA-B components and are therefore more inclined to reassortment than DNA-B molecules. In doing so, the begomovirus may exchange its CR with the newly adopted molecule through heterologous recombination. A mixed infection by bipartite begomoviruses coinfecting the same host provides a good opportunity for heterologous recombination or pseudorecombination. Similarly, new virulent betasatellites with a functional βC1 gene may arise because of the recombination between SCR of a betasatellite and CR of a begomovirus (Nawaz-ul-Rehman and Fauquet, 2009). The begomoviruses may also exchange their CRs with the alphasatellite molecules as well. Leke et al. (2012) identified two types of defective begomoviruses with complete IR and partial CP, Pre-CP, and Rep from *Ageratum leaf curl Cameroon virus* (ALCCMV) and an incomplete Rep of Ageratum leaf curl Cameroon alphasatellite (ALCCMA). The appearance of such defective forms of recombinant begomoviruses may broaden the ability of begomoviruses in exploring new geographical niches and infecting new host plants under particular environments.

BCaMV is known a "host-restricted" begomovirus with a narrow host range infecting only bean crops. An interspecific reassortment between a "non-host restricted" Melon chlorotic leaf curl virus (MCLCuV) DNA-A and "host-restricted" BCaMV DNA-B resulted into a viable reassortant, which can infect bean successfully (Idris et al., 2008). Thus, bean crops can serve as a diversification driver for the adapted begomoviruses by providing a melting pot for reassortment.

Both interspecies and intraspecies pseudorecombination is a major factor of host adaptation by LYMVs. DoYMV, MYMV, and MYMIV are predicted to contain similar iterons (GGTGT), whereas the iteron motif of HGYMV is different (GGTAT) and thus HGYMV may not exchange its components with LYMVs. In such a case, the one of both the DNA-A and DNA-B may get exchange between distinct species and may give rise to a viable reassortant. Most of the begomoviruses promote extensive interspecific recombination, thus giving rise to new species and/or broaden their host ranges. However, the evidence for the interaction between LYMVs and non-legume-infecting begomoviruses is not well available. This is because most of the LYMVs are genetically isolated from other begomoviruses and possess very narrow host range encompassing only the legume hosts. Hence, the opportunities of mixing with other begomoviruses are only the whitefly vectors.

Pseudorecombination has also been proved experimentally for legume-infecting and non-legume-infecting begomoviruses successfully.

For example, the pseudorecombinants between *Bean dwarf mosaic virus* (BDMV) and *Tomato mottle virus* (ToMoV) gave rise to viable reassortants, which after three successive generations became highly virulent in laboratory experiments with *P. vulgaris*.

Quite recently, betasatellites has been found associated with legumoviruses. Association of *Papaya leaf curl betasatellite* (PaLCB) with MYMV (Satya et al., 2013), *Tobacco curly shoot betasatellite* (TbCSB) with MYMIV (Ilyas et al., 2010), and PaLCB with FbLCV (Kamaal et al., 2013) has been found recently infecting legume crops in India and Pakistan, respectively. Although betasatellites are not found in legumes quite often yet, the association of betasatellites with legume-infecting begomoviruses in legumes may indicate more complicated disease pattern in legumes. The mechanism by which bipartite legumoviruses have started acquiring betasatellites during their natural infection in legumes is difficult to explain. Usually, legumoviruses are mostly host restricted and rarely found from other cultivated crops. But association of legumoviruses with the betasatellites may open up new avenues for their future by broadening their host range beyond legumes.

Pseudorecombination of begomoviruses with DNA satellites may predominantly drive a host shift and adaptation (Ranjan et al., 2014). Such pseudorecombinants are also involved in modulating the expression pattern of a disease in host-dependent way.

13.6 ROLE OF VIRUS–VECTOR–HOST INTERACTIONS IN BEGOMOVIRUS DIVERSITY

Begomoviruses interact with each other either synergistically or antagonistically. Apart from these interactions, begomoviruses do interact with their host plants as well as with their insect vector during transmission.

13.6.1 ROLE OF WHITEFLY AS A VECTOR

During course of feeding, whiteflies acquire a subtle virus population from the infected host plant. Thus, multiple feedings on infected host plants add many different types of virus populations to the viral pool inside the insect. From here on, these multiple types of virus species are transmitted to other hosts. It has been estimated that a single insect vector may dock more than 600 million virions alone (Czosnek et al., 2002). Whitefly can retain the virus in its gut over its complete life span; however, the virus may lose its transmission efficiency during this time. TYLCSV has been found in the subsequent whitefly progenies but the new infants lost their infection ability (Bosco et al., 2004).

The manifestation of extremely polyphagous and invasive type of white-flies *B. tabaci* in the fields contributed to a broader host range of begomovi-ruses by providing opportunities for diversification through mixed infection and recombination. The ever-spreading diversification of begomoviruses is interlinked to the high fecundity and population density of the vector *B. tabaci*. Consequently, the begomoviruses not only spread within crops but also move from non-cultivated hosts and weeds into crop plants. Several begomovirus epidemics are reportedly spread because of the introduction of new whitefly species in a particular region. The global invasion of CLCuD complex, CMD pandemic, and TYLCD was directly because of a discrete species of whitefly.

The virus–vector association specifically depends upon interaction between virus-encoded CP and whitefly-encoded unknown proteins in the insect gut and salivary gland membranes, which is required for efficient virus transmission. After acquisition from plant, the begomoviruses require GroEL protein (produced by endosymbiotic bacteria) for their durable subsis-tence inside the insect midgut. According to Zhang et al. (2012), the βC1 of *Tomato yellow leaf curl China betasatellite* (TYLCCNB) helps in enhancing whitefly's performance on host plants by suppressing the jasmonic acid (JA) pathway of the infected host plants.

After gradual continent drift, the whitefly–begomovirus interaction was more refined and the geographically separated regions coadapted insect–virus combinations (Bradeen et al., 1997).

Usually, a broader host range of geminiviruses decides the opportuni-ties for recombination to multiply the viral genetic diversity. Prior to 1990, a single species of begomovirus was known to infect tomato in Brazil but now 15 different species have been isolated from Brazilian tomato crop (Albuquerque et al., 2010). This is because of the introduc-tion of an invasive whitefly species that have an extended host range than the indigenous species. Thus, it overcomes the transmission barrier between those geminiviruses and their potential hosts in this region. Nevertheless, most of the newly isolated tomato-infecting begomovi-ruses were the recombinants.

13.6.1 ROLE OF HOST PLANT

The begomovirus genomes are highly adaptable to the dynamic cropping schemes by rapidly evolving to counter variable environmental conditions. It seems the begomovirus, insect vector, and the host plant together decides the virus population in a distinct geographical area. For example, the cultivation of smooth leaved cotton varieties may promote different whitefly population in comparison to the hairy leaved varieties and, consequently, creates a selection pressure for a discrete whitefly species (Luqman, 2010). It ultimately affects the adaptability of host-specific begomovirus species/ strains. The begomoviruses differ in their choice of host and thus cultivation of a particular crop may directly or indirectly influence the prevalence of a particular virus species or strain in that region. For example, the recombinant strain of TYLCV and *Tomato yellow leaf curl Sardinia virus* (TYLCSV) in Spain was highly pathogenic with a wide host range comparing to both parents (Monci et al., 2002). Moreover, the synergistic or antagonistic interaction between two begomoviruses during a coinfection also depends upon the type of host they are infesting. For example, *Pepper huasteco yellow vein virus* (PHYVV) and *Pepper golden mosaic virus* (PepGMV) interact synergistically in tobacco and *Nicotiana benthamiana*, whereas the same viruses showed antagonistic interaction in pepper (Mendez-Lozano et al., 2003). Certain begomoviral complexes are well known to become pandemic on newly introduced crops in an area. This does not mean that the virus was already associated with that particular crop but instead the virus may have been existing the surroundings by the time but only get quickly adapted to the new host (Nawaz-ul-Rehman and Fauqet, 2009). Cassava in not a native crop in Africa but upon its introduction from the NW into Africa, it experienced the biggest epidemic of CMD in Africa although it has never been existed in the NW,the native origin of cassava.

The non-cultivated hosts (weeds and wild plants) also play a crucial role in begomovirus diversity by offering a "melting pot" to govern recombination and genetic assortment between begomoviruses. Many new begomovirus species have been emerged from the weed hosts, which subsequently adapted to the crop plants cultivated in that vicinity. The widespread cultivation of resistant cultivars against begomoviruses may also play its role in reshaping the viral diversity of a region. High level of resistance in the crop plants may shift the selection pressure toward the virus, which may lead toward the evolution of more aggressive strains. After the onset of CMD in Africa, the use of resistant cultivars gave rise to more virulent Uganda strain of EACMV-EU in resistant cultivars. The cultivation of virus-tolerant cultivars may promote less virulent virus strains with high accumulation rate in the tolerant cultivars.

Evidently, transfer of genetic material occurs between geminiviruses and the host plant genome or vice versa. Begomoviruses and their associated subgenomic DNA satellites are well established to contain nonviral DNA fragments, which are presumed to be derived from the host plant genome they reside in. Such a recombinational DNA transfer has also been demonstrated experimentally by transferring Cp transgenes into the geminivirus from the host plant genome (Frischmuth and Stanley, 1998). Similarly, the host plants are also reported to have geminivirus-like sequences in their genome (Martin et al., 2011). Such nonhomologous recombinations between the viral and the host plant genomes may explain the evolutionary dilemma of distinct geminivirus lineages present in the NW and OW (Lefeuvre et al., 2011).

Noncultivated wild host plants (such as weeds and ornamental plants) have a diverse role in the begomovirus epidemiology by providing a "mixing reservoir" for recombination and/pseudorecombination to occur. They also serve as "sheltered hosts" for the begomoviruses, which otherwise will extinct in the absence of their natural cultivated hosts. Noncultivated host plants are genetically much more variable than the cultivated crop plants; however, this variability does not affect the genetic variability of the begomoviruses. The host plants do not affect the genetic variability in begomoviruses rather the variability is prone to recombination, which helps shaping the genetic structure of a given population.

13.7   MANAGEMENT STRATEGIES TO LIMIT LEGUMOVIRUSES

Diseases caused by legumoviruses have significant impact on the production of legumes in tropical to temperate regions of the world. Among the causal viruses, MYMV, MYMIV, BGMV, BGYMV, RhGMV, RhYMV, and few other legumoviruses are economically very important. They not only infect more than one crop but also are present in many parts of the world. Insect-transmitted viruses of legumes have quarantine importance. In the past few decades, these viruses have proliferated much because of intensive cropping, climate changes, insect resistance, overuse of pesticides, genetic recombination, mutation in viruses, pseudorecombination, component capture, and some of the factors aiding insect vector multiplication and spread. For the efficient management of legumoviruses, several controlling strategies/

options have been used previously and few are being proposed here. These may include cultivation of resistant varieties, better agronomical practices, wise use of insecticidal chemicals, biological controls of insect vectors, genetically improved new virus-resistant crops through conventional and/ or nonconventional breeding methods, and use of modern biotechnological techniques such as RNA interference (RNAi), pathogen-derived resistance (PDR), non-pathogen-derived resistance (NPDR), CRISPR/Cas9 and inducing CI in whitefly vector could boost the worldwide legume production. Finally, sensible integration of more than one method, to circumscribe the legumoviruses, would be a best choice. It should, however, be noted that every strategy be affordable to the farmer, environmental friendly, and socially acceptable.

13.7.1 INTEGRATED PEST MANAGEMENT

The whitefly (*B. tabaci*) is a dominant pest of the major food and fiber crops (including legumes) and responsible for transmission of 10% of total plant viruses in the world. Whitefly infestation causes massive losses to plants in three ways: direct damage imposed by phloem feeding, which causes leaf chlorosis and leaf drop and ultimately a prolonged feeding may cause considerable yield reduction. While feeding, whiteflies also secrete honey-dews, which provides an enriched medium to promote the growth of black sooty mold, which may affect the lint and fiber quality . Most importantly, *B. tabaci* help in the transmission of begomoviruses, thus constituting a major threat to the legume production. Several techniques can be used to control the whitefly spread and multiplication of legumoviruses.

13.7.1.1 WHITEFLY MANAGEMENT THROUGH INSECTICIDE

Primarily, whiteflies are managed by applying different sorts of insecticides. This is a better choice of controlling the whitefly populations and so do the spread of viruses. To curtail the legumoviruses, several Integrated pest management programs have been applied successfully around the globe. A wide and diverse range of insecticides are available to control whiteflies. On the basis of their mode of action, these insecticides are divided into 11 different groups. All these groups vary in their efficiency and targets, for example, some target the whitefly eggs, or immatures, and/or adults; therefore, a very careful and wise use of insecticide is required to obtain the optimal control. Furthermore, timing of application and use of insecticide in rotation with different modes of action can yield the optimum control. Of all the insecticides available, neonicotinoids is a group that has had offer the best control on the management of whitefly-transmitted viruses. The use of this group has expanded greatly from last couple of decades because of their rapid effect on whitefly that can lead to quick control over virus transmission through whiteflies. Neonicotinoides are generally applied as soil drench and/or sprayed over plants. The major drawback associated with use of neonicotinoids is its adverse effects on pollinators such as the bumble bees, European honeybee, birds, and they have been reported to contribute to colony collapse disorder (CCD) (Blacquie et al., 2012; Mineau and Palmer, 2013). In the EU, three neonicotinoids (clothianidin, imidacloprid, and thiamethoxam) have been banned from use for 2 years on flowering crops where bees actively forage but allowed on crops where bees are less active. In the United States, the response has been more pragmatic and piecemeal. Although the US Environmental Protection Agency has been sued and petitioned to limit the use of neonicotinoids, a ban on use of three insecticides (clothianidin, imidacloprid, and thiamethoxam) belonging to the neonicotinoid family has been imposed. All the pesticides belonging to neonicotinoids are being applied either by spraying, granular form into soil and by seed coatings.

13.7.1.2   WHITEFLY MANAGEMENT THROUGH CULTURAL PRACTICES

The secondary way of controlling whitefly is use of cultural practices that play a role in control regimes, specifically, use of cultural practices can reduce dependence on the use of insecticides. Diverse array of activities falls under this regime. Some of the most important and widely applied cultural practices for the whiteflies managements are as under.

13.7.1.2.1   Polyethylene Soil Mulches

The use of polyethylene (plastic) soil mulches is relatively cheap, popular, and inexpensive strategy for protection of crops against whiteflies, aphids, and thrips (Weintraub and Berlinger, 2004). Generally, two main mulching approaches are being observed; using yellow colored plastic to attract the whiteflies to the mulch instead of the host, and second is use of coated plastic mulches that strongly reflect light, which restricts the invading whiteflies.

Both types of soil mulch interfere with the insect's ability to find the crop. The yellow plastic mulches are used extensively in Israel to protect tomato plants from the whitefly-borne TYLCV (Polston and Lapidot, 2007). The successful use of reflective plastic mulch to delay the onset of whitefly infestations and infection of whitefly-transmitted viruses in open-field production has been well established (Simmons et al., 2010).

13.7.2   CROP PLACEMENT—IN SPACE AND IN TIME

Two abilities of whiteflies, to fly over several kilometers distance and transmission of viruses in persistent or semi-persistent manner, have put the nearby crops vulnerable to be infected by viruses present in the neighboring older fields. In order to minimize the risk of contamination of a new crop, it must be grown either in an isolated field or the field must be located at the upwind direction of the inoculum source. The incidence of BGMV and BGYMV has successfully been reduced by cultivation at a distant locality from the preferred hosts of the whitefly (Morales, 2003).

Generally, whitefly population increases in warm weather and when host plants are in young and rapidly growing stage. Keeping this in view, the sowing dates can be changed to reduce whitefly populations, thus incidence of legumovirus can be reduced (Abdallah, 2012).

13.7.3   INTERCROPPING

Although cultivating more than one crop in a single field is not a very common practice in commercial agricultural environment; however, mixing crops can render the whiteflies to infest the crops. Experimentally, intercropping has been considered as a potential approach to control whitefly infestation. For cowpea in Cameroon, intercropping of maize with cowpea has shown modest reductions in YMD incidence (Fondong et al., 2002). Intercropping of cotton in wheat field has been started in Pakistan, although not extensively, but a clear reduction of CLCuD incidence has been observed.

13.7.4   HOST PLANT RESISTANCE TO LEGUMOVIRUSES

Legumoviruses cause major losses in legumes and other leguminous crops. The one way of controlling these viruses is to breed elite cultivars with specific gene(s) for genetic resistance to legumoviruses. Several molecular breeding approaches, for example, expressed sequence tags (ESTs), molecular markers (RFLP, RAPD, AFLP, SSRs), molecular genetic maps, functional molecular markers, functional and comparative genomics, linkage-disequilibrium (LD)-based association mapping have been initiated to develop genomic resources.

In the beginning, morphological and isozyme loci were exploited to develop chickpea linkage map. Currently, several hundred SSR markers are being used in chickpea (Lichtenzveig et al., 2005) to construct the genetic map after interspecific cross. Currently, about 500 SSR markers are available that could potentially be employed to draw a genetic linkage map to identify the resistance related genes in chickpea. In 1994, genetic diversity in wild species of pigeonpea was assessed through RFLPs using nuclear DNA probes (Nadimpalli et al., 1994). Recently, AFLP analysis revealed high level of polymorphism in different cultivars. About 10 SSR and 39 microsatellite markers are accessible that could be used to construct the genetic map of pigeonpea. The International Institute of Tropical Agriculture, Nigeria, have found the varieties Bossier, Pelican, and the germplasm accessions TGm 119 and TGm 662 resistant to African soybean dwarf virus (ASDV) (Hartman et al., 1999). The YMD resistance of soybean cultivars was assessed in the Indian subcontinent under field conditions; however, the promising lines must be further employed in the current breeding programs to develop YMD-resistant soybean cultivars.

The BGMV resistance in black-seeded common bean cultivars in South America has been successfully used in breeding programs in Guatemala and Mexico, respectively. In this study, 188 resistance sources, different from previously known, were identified from 1660 accessions of bean germplasm. Resistance sources against BGYMV have also been identified in black-seeded Mesoamerican genotypes and have shown tolerance against BGYMV infection under field conditions. Few other BGYMV resistance genes, "bgm-1" and "bgm-2," have been identified from two different bean genotypes belonging to Mexican and Andean origin (Morales, 2003). Additionally, both these genes have been used successfully in the development of BGYMV-resistant cultivars.

From 1977 to 2003, in Punjab, India, more than 10,000 mungbean germplasm lines were screened against MYMV and 31 germplasm lines were found to have resistance against MYMV (Singh et al., 2003). Maiti et al. used two marker loci, YR4 and CYR1, in multiplex PCR to screen urdbean germplasm against MYMV and found that CYR1 has strong linked with

MYMIV resistance (Maiti et al., 2011). QTL Mapping for MYMIV resistance in a mungbean cultivar generated from a cross between NM10-12-1 (MYMIV resistance) and KPS2 (MYMIV susceptible) have been performed using F8 RIL marker, and field trials of this cultivar have been done in India and Pakistan (Kitsanachandee et al., 2013).

13.7.5   USE OF DISEASE RESISTANCE (NBS-LRR) GENE HOMOLOGS

Every plant genome contains several resistance genes and among them, the most abundant class contains a centrally located nucleotide-binding site (NBS) and a carboxy terminal leucine-rich repeat (LRR) domain. More than 3000 such resistance genes from different legumes (such as cowpea, mungbean, common bean, pigeonpea, chickpea, and lentil) have been identified (Zhang et al., 2016). These genes could be used purportedly in each of the target species, thus providing tools for molecular breeding as well as more fundamental studies of resistance gene identification and manipulation. These genes may later be incorporated into the current breeding programs for crop plant resistance to legumoviruses.

13.7.6   MOLECULAR BIOLOGY APPROACHES

Molecular biology and biotechnology approaches have the potential to resolve and address the problems caused by legumoviruses to legumes. Such modern techniques have been used successfully in some cereal and other plant species, thus legumes improvement can be benefited by exploring such approaches. Soybean and common bean expressing Rep gene of BGMV (Tollefson, 2011), and mungbean expressing CP, Rep-sense, Rep-antisense, truncated Rep, NSP, and MP of MYMV (Haq et al., 2010) have been genetically engineered successfully. Of these, only genetically engineered common bean expressing Rep gene of BGMV and resistant to BGMV has been commercialized in Brazil (Tollefson, 2011). Transient gene expression of MYMV AC4 gene in *N. benthamiana* plants conferred resistance against MYMV by reducing the virus accumulation in the transgenic tobacco plants (Sunitha et al., 2013). RNAi-based approaches have been employed in chickpea to develop virus resistance (Nahid et al., 2011). Besides all the advancements in genomics of legumes and plant transformations techniques, incorporation of broad spectrum legumovirus resistance to elite legumes cultivars is yet to be accomplished. Furthermore, a latest technology CRISPR/Cas9 is a versatile approach that can also be employed as described by Iqbal et al. (2016) to circumvent the legumoviruses. Quite recently, CRISPR/Cas9 system has been employed quite successfully to confer resistance against *Bean yellow dwarf virus* (BeYDV) in beans (Baltes et al., 2015).

13.8   WHITEFLY MANAGEMENT THROUGH CI

The whitefly *B. tabaci* was found to be widely infected with an endosymbiont *Wolbachia*. This endosymbiont can cause different reproductive abnormalities especially CI in a wide range of arthropod hosts. CI is a phenomenon which arises because of changes in gamete cells caused by *Wolbachia* (living in the cytoplasm of the host cell) that leads to inability to form viable offspring (Stouthamer et al., 1999). Thus, CI leads to embryonic mortality in mattings between insects having different *Wolbachia* infection. CI can be either unidirectional or bidirectional; in unidirectional CI, an infected male mate with an uninfected female, whereas reciprocal cross is compatible and no CI is observed. Bidirectional CI usually occurs in mattings between infected individuals harboring different strains of *Wolbachia*. In some haplodiploid species (e.g., *B. tabaci*), CI could lead to production of male offspring; thus, a shift in sex ratio would be helpful in controlling the whitefly population and inevitably lead to extinction of whitefly populations over generations (Zhou and Li, 2016).

13.9   SUMMARY

Legumes provide a vital source of protein, consumed widely around the globe and play pivotal role in the sustainability of global economy. Geminiviruses are a major threat to the crop productivity in a number of legume crops worldwide. The genetic variability of begomoviruses is their intrinsic property, which helps to shape the viral populations in an area. Mutations, recombinations, pseudorecombinations, and DNA satellite capture (during mixed infections) are major driving forces for genetic variability in begomoviruses. Like other begomoviruses, most of the legume-infecting begomovirus populations are also strongly structured by their geographic distribution and the host plants. The role of wild and/or non-cultivated plants hosts in the genetic variability framework of begomoviruses cannot be ruled out. Such botanical families may enhance the species diversity of legumoviruses, thus playing crucial epidemiological role by acting as reservoir hosts for these viruses. Therefore, to get a broader overview about the diversity of legume-infecting begomoviruses in a particular niche, the role of other crops and weeds should also be focused. Furthermore, we entail comprehensive futuristic studies to explore the geographical distribution of legumes infecting begomoviruses, their genetic structure, and the potential to spread into other commercial crops. The resistance against geminiviruses through classical breeding approaches requires long time and is not durable because of the emergence of new virulent viral species. The deployment of newly emerged technologies for introgression of resistance against multiple viral pathogens is a prerequisite to enhance the yield and quality of legume crops. The molecular mechanisms of host resistance against viral invasions has been well progressed during last couple of decades. The advent of RNAi- and CRISPR-based approaches further strengthen our toolkits to better understand and explore the host plant resistance against geminiviruses. Such tools would provide hands-on and unprecedented platform to engineer our crop plants for durable and effective virus resistance.

KEYWORDS

- **beans**
- **diversification**
- *Geminiviridae*
- **legumes**
- **legumoviruses**
- **mutation**
- **resistance**

REFERENCES

Abdallah, Y. E. Y. Effect of Plant Traps and Sowing Dates on Population density of Major Soybean Pests. *J. Basic Appl. Zool.* **2012,** *65*, 37–46.

Adams, M. J.; Lefkowitz, E. J.; King, A. M. Q.; Harrach, B., et al. Changes to Taxonomy and the International Code of Virus Classification and Nomenclature Ratified by the International Committee on Taxonomy of Viruses. *Arch. Virol.* **2017,** *162*, 2505–2538.

Alabi, O.; Lava Kumar, P.; Mgbechi-Ezeri, J.; Naidu, R. Two New 'Legumoviruses' (Genus *Begomovirus*) Naturally Infecting Soybean in Nigeria. *Arch. Virol.* **2010**, *155*, 643–656.

Albuquerque, L. C.; Martin, D. P.; Avila, A.C; Inoue-Nagata, A. K. Characterization of *Tomato Yellow Vein Streak Virus*, a Begomovirus from Brazil. *Virus Genes* **2010**, *40*, 140–147.

Alemandri, V.; Rodriguez Pardina, P.; Izaurralde, J.; Medina, S. G.; Caro, E. A.; Mattio, M. F. Incidence of Begomoviruses and Climatic Characterization of *Bemisia Tabaci*-Geminivirus Complex in Soybean and Bean in Argentina. *AgriScientia* **2012**, *26*, 31–39.

Ali, Z.; Ali, S.; Tashkandi, M.; Zaidi, S. S. A.; Mahfouz, M. M. *CRISPR/Cas9-Mediated Immunity to Geminiviruses: Differential Interference and Evasion*; Scientific Reports, 2016.

Amin, I.; Patil, B.; Briddon, R.; Mansoor, S.; Fauquet, C. Comparison of Phenotypes Produced in Response to Transient Expression of Genes Encoded by Four Distinct Begomoviruses in *Nicotiana Benthamiana* and their Correlation with the Levels of Developmental miRNAs. *Virol. J.* **2011**, *8* (1), 238.

Baltes, N. J.; Hummel, A. W.; Konecna, E.; Cegan, R.; Bruns, A. N.; Bisaro, D. M. Conferring Resistance to Geminiviruses with the CRISPR–Cas Prokaryotic Immune System. *Nat. Plants* **2015**, *1*, 15145.

Blackmer, J.; Byrne, D. N. Flight Behaviour of Bemisia Tabaci in a Vertical Flight Chamber: Effect of Time of Day, Sex, Age and Host Quality. *Physiological Entomol.* **1993**, *18*, 223–232.

Blacquière, T.; Smagghe, G.; Cornelis A. M. G.; Mommaerts, V. Neonicotinoids in Bees: A Review on Concentrations, Side Effects and Risk Assessment. *Ecotoxicol.* **2012**, *21* (4), 973–992.

Bosco, D.; Mason, G.; Accotto, G. P. TYLCSV DNA, but Not Infectivity, Can be Transovarially Inherited by the Progeny of the Whitefly Vector *Bemisia Tabaci* (Gennadius). *Virol.* **2004**, *323*, 276–283.

Bradeen, J. M.; Timmermans, M. C.; Messing, J. Dynamic Genome Organization and Gene Evolution by Positive Selection in Geminivirus (*Geminiviridae*). *Mol. Biol. Evol.* **1997**, *14* (11), 1114–1124.

Briddon, R. W.; Bull, S. E.; Amin, I.; Idris, A. M.; Mansoor, S.; Bedford, I. D.; Dhawan, P.; Rishi, N.; Siwatch, S. S.; Abdel-Salam, A. M.; Brown, J. K.; Zafar, Y.; Markham, P. G. Diversity of DNA β: A Satellite Molecule Associated with Some Monopartite Begomoviruses. *Virol.* **2003**, *312*, 106–121.

Briddon, R. W.; Patil, B. L.; Bagewadi, B.; Nawaz-ul-Rehman, M. S.; Fauquet, C. M. Distinct Evolutionary Histories of the DNA-A and DNA-B Components of Bipartite Begomoviruses. *BMC Evolutionary Biol.* **2010**, *10*, 97.

Brown, J.; Zerbini, F. M.; Navas-Castillo, J.; Moriones, E.; Ramos-Sobrinho, R.; Silva, J. F.; Fiallo-Olivé, E.; Briddon, R.; Hernández-Zepeda, C.; Idris, A.; Malathi, V. G.; Martin, D.; Rivera-Bustamante, R.; Ueda, S.; Varsani, A. Revision of Begomovirus Taxonomy Based on Pairwise Sequence Comparisons. *Arch. Virol.* **2015**, *160*, 1593–1619.

Chen, L. F.; Rojas, M.; Kon, T.; Gamby, K.; Xoconostle-Cazares, B.; Gilbertson, R. L. A Severe Symptom Phenotype in Tomato in Mali is Caused by a Reassortant Between a Novel Recombinant Begomovirus (Tomato Yellow Leaf Curl Mali Virus) and a Betasatellite. *Mol. Plant Pathol.* **2009**, *10*, 415–430.

Czosnek, H.; Ghanim, M; Ghanim, M. The Circulative Pathway of Begomoviruses in the Whitefly Vector *Bemisia Tabaci*-Insights from Studies with *Tomato Yellow Leaf Curl Virus*. *Annals Appl. Biol.* **2002**, *140*, 215–231.

Dry, I. B.; Krake, L. R.; Rigden, J. E.; Rezaian, M. A. A Novel Subviral Agent Associated with a Geminivirus: The first Report of a DNA Satellite. *Proc. Nat. Acad. Sci.* **1997**, *94*, 7088–7093.

FAOSTAT. Statistics. Rome: FAO: Food and Agriculture Organization of the United Nations, 2014. http://faostat3.fao.org/browse/Q/QC/E

Fauquet, C. M.; Briddon, R. W.; Brown, J. K.; Moriones, E.; Stanley, J.; Zerbini, M.; Zhou, X. Geminivirus Strain Demarcation and Nomenclature. *Arch. Virol.* **2008**, *153* (4), 783–821.

Fiallo-Olive, E.; Martinez-Zubiaur, Y.; Moriones, E.; Navas-Castillo, J. A Novel Class of DNA Satellites Associated with New World Begomoviruses. *Virol.* **2012**, *426*, 1–6.

Fondong, V. N.; Thresh, J. M.; Zok, S. Spatial and Temporal Spread of Cassava Mosaic Virus Disease in Cassava Grown Alone and When Intercropped with Maize and/or Cowpea. *J. Phytopathol.* **2002**, *150*, 365–374.

Fondong V.N. The Search for Resistance to Cassava Mosaic Geminiviruses: How Much We Have Accomplished, and What Lies Ahead. *Frontiers in Plant Sci.* 2017, doi.org/10.3389/fpls.2017.00408.

Frischmuth, T.; Stanley, J. Recombination Between Viral DNA and the Transgenic Coat Protein Gene of African Cassava Mosaic Geminivirus. *J. Gen. Virol.* **1998**, *79*, 1265–1271.

Garcia-Andre, S.; Accotto, G. P.; Navas-Castillo, J.; Moriones, E. Founder Effect, Plant Host, and Recombination Shape the Emergent Population of Begomoviruses that Cause the Tomato Yellow Leaf Curl Disease in the Mediterranean Basin. *Virol.* **2007**, *359*, 302–312.

George, B.; Alam, C. M.; Kumar, R. V.; Granasekaran, P;Chakraborty, S. Potential Linkage Between Compound Microsatellites and Recombination in Geminiviruses: Evidence from Comparative Analysis. *Virol.* 2015, *482*, 41–50.

Ghanim, M.; Czosnek, H. Interactions Between the Whitefly Bemisia tabaci and Begomoviruses: Biological and Genomic Perspectives. In: *Management of Insect Pests to Agriculture: Lessons Learned from Deciphering their Genome, Transcriptome and Proteome*; Henryk, C.; Murad, G., Eds. Springer International Publishing: Cham, Switzerland, 2016, pp 181–200.

Ha, C.; Coombs, S.; Revill, P.; Harding, R.; Vu, M.,; Dale, J. Molecular Characterization of Begomoviruses and DNA Satellites from Vietnam: Additional Evidence that the New World Geminiviruses were Present in the Old World Prior to Continental Separation. *J. Gen. Virol.* **2008**, *89*, 312–326.

Haq, Q. M. I.; Ali, A.; Malathi, V. G. Engineering Resistance Against Mungbean Yellow Mosaic India Virus Using Antisense RNA. *Ind. J. Virol.* **2010**, *21*, 82–85.

Hartman, G. L.; Sinclair, J. B.; Rupe, J. C. *Compendium of Soybean Diseases,* 4th ed; American Phytopathological Society Press: St Paul, MN, USA, 1999.

Idris, A. M.; Mills-Lujan, K.; Martin, K.; Brown, J. K. Melon Chlorotic Leaf Curl Virus: Characterization and Differential Reassortment with Closest Relatives Reveal Adaptive virulence in the Squash Leaf Curl Virus Clade and Host Shifting by the Host-Restricted Bean Calico Mosaic Virus. *J. Virol.* **2008**, *82* (4), 1959–1967.

Ilyas, M.; Qazi, J.; Mansoor, S; Briddon, R. W. Genetic Diversity and Phylogeography of Begomoviruses Infecting Legumes in Pakistan. *J. Gen. Virol.* **2010**, *91* (8), 2091–2101.

Iqbal, Z.; Sattar, M. N.; Shafiq, M. CRISPR/Cas9: A tool to Circumscribe Cotton Leaf Curl Disease. *Frontiers in Plant Sci.* **2016**, *7*, 475.

Kamaal, N.; Akram, M.; Pratap, A.; Yadav, P. Characterization of a New Begomovirus and a Beta Satellite Associated with the Leaf Curl Disease of French Bean In Northern India. *Virus Genes* **2013**, *46*, 120–127.

Kitsanachandee, R.; Somta, P.; Chatchawankanphanich, O.; Akhtar, K. P.; Shah, T. M.; and Nair, R. M. Detection of Quantitative Trait Loci for *Mungbean Yellow Mosaic India Virus* (MYMIV) Resistance in Mungbean (*Vigna radiata* (L) Wilczek) in India and Pakistan. *Breeding Sci.* **2013,** *63,* 367–373.

Kon, T.; Sharma, P.; Ikegami, M. Suppressor of RNA Silencing Encoded by the Monopartite Tomato Leaf Curl Java Begomovirus. *Arch. Virol.* 2007, *152* (7)**,** 1273–1282.

Lefeuvre, P.; Harkins, G. W.; Lett, J. -M.; Briddon, R. W.; Leitch, A. R.; Chase, M. W.; Moury, B.; Martin, D. P. Evolutionary Time-scale of Begomoviruses: Evidence from Integrated Sequences in *Nicotiana* Genome. *PLoS ONE* **2011,** *6,* e19193

Lefeuvre, P.; Lett, J. M.; Varsani, A;Martin, D. P. Widely Conserved Recombination Patterns Among Single-Stranded DNA Viruses. *J. Virol.* **2009,** *83,* 2697–2707.

Lefeuvre, P.; Martin, D. P.; Harkins, G.; Lemey, P.; Gray, A. J. A.; Meredith, S.; Lakay, F.; Monjane, A.; Lett, J. M.; Varsani, A. The Spread of Tomato Yellow Leaf Curl Virus from the Middle East to the World. *PLoS Pathogen* **2010,** *6,* e1001164.

Lefeuvre, P.; Martin, D. P.; Hoareau, M.; Naze, F.; Delatte, H.; Thierry, M.; Varsani, A.; Becker, N.; Reynaud, B.; Lett, J. M. Begomovirus 'melting Pot' in the South-West Indian Ocean Islands: Molecular Diversity and Evolution Through Recombination. *J. Gen. Virol.* **2007,** *88,* 3458–3468.

Leke, W. N.; Brown, J. K.; Ligthart, M. E.; Sattar, N.; Njualem, D. K.; Kvarnheden, A. (2012). *Ageratum Conyzoides*: A Host to a Unique Begomovirus Disease Complex in Cameroon. *Virus Res.* **2012,** *163* (1), 229–237.

Lichtenzveig, J.; Scheuring, C.; Dodge, J.; Abbo, S.; Zhang, H. B. Construction of BAC and BIBAC Libraries and their Applications for Generation of SSR Markers for Genome Analysis of Chickpea (*Cicer arietinum* (L.)). *Theor. Appl. Genetics* 2005, *110,* 492–510.

Lozano, G.; Trenado, H. P.; Fiallo-Olive, E.; Chirinos, D.; Geraud-Pouey, F.; Briddon, R. W.; Navas-Castillo, J. Characterization of Non-coding DNA Satellites Associated with Sweepoviruses (Genus *Begomovirus, Geminiviridae*)—Definition of a Distinct Class of Begomovirus-Associated Satellites. *Frontiers in Microbiol.* **2016,** *7,* 162.

Luqman, A. *Characterization and Infectivity Analysis of Begomovirus Components Found in Cotton-Based Agro-Ecosystem.* Quaid-i-Azam University: Islamabad, 2010.

Maiti, S.; Basak, J.; Kundagrami, S.; Kundu, A.; Pal, A. Molecular Marker-Assisted Genotyping of Mungben Yellow Mosaic India Virus Resistant Germplasm of Mungbean and Urdbean. *Mol. Biotechnol.* **2011,** *47,* 95–104.

Martin, D. P.; Biagini, P.; Lefeuvre, P.; Golden, M.; Roumagnac, P.; Varsani, A. Recombination in Eukaryotic Single Stranded DNA Viruses. *Viruses* **2011,** *3,* 1699–1738.

Maruthi, M. N.; Manjunatha, B.; Rekha, A. R.; Govindappa, M. R.; Colvin, J.; Muniyappa, V. *Dolichos yellow mosaic virus* Belongs to a Distinct Lineage of Old World Begomoviruses: Its Biological and Molecular Properties. *Ann. Appl. Biol.* **2006,** *149,* 187–195.

Mendez-Lozano, J.' Torres-Pacheco, I.; Fauquet, C. M.; Rivera-Bustamante, R. Interactions Between Geminiviruses in a Naturally Occurringmixture: Pepper Huasteco Virus and Pepper Golden Mosaic Virus. *Phytopathol.* **2003,** *93* (3), 270–277.

Mineau, P.; Palmer C. *the Impact of the Nation's Most Widely Used Insecticides on Birds.* American Bird Conservancy: USA, 2013.

Monci, F.; Sanchez-Campos, S.; Navas-Castillo, J.; Moriones, E. A Natural Recombinant Between the Geminiviruses Tomato Yellow Leaf Curlsardinia Virus and Tomato Yellow Leaf Curl Virus Exhibits a Novel Pathogenic Phenotype and is Becoming Prevalent in Spanish Populations. *Virology* **2002,** *303,* 317–326.

Monsalve-Fonnegra Z; Argüello-Astorga G.; Rivera-Bustamante R. Geminivirus replication and gene expression. In: *Plant Viruses as Molecular Pathogens*; Khan, J. A.; Dijstra, J., Eds. The Haworth Press: New York, 2001; pp 257–277.

Morales, F. J. Common Bean. In: *Virus and Virus-like Diseases of Major Crops in Developing Countries*; Loebenstein, G.; Thottappilly, G., Eds.; Kluwer Academic Publishers: The Netherlands, 2003, pp 425–445.

Nadimpalli, R. G.; Jarret, R. L.; Phatak, S. C; Kochert, G. Phylogenetic Relationships of the Pigeon Pea (*Cajanus Cajan*) Based on Nuclear Restriction Fragment Length Polymorphism. *Genome* **1994**, *36*, 216–223.

Nagy, P. D. Recombination in Plant RNA Viruses. In: *Plant Virus Evolution*; Roossinck, M. J., ed.; Springer-Verlag: Berlin, Heidelberg, 2008, pp 133–156.

Nahid, N.; Amin, I.; Briddon, R. W.; Mansoor, S. RNA Interference Based Resistance Against a Legume Mastrevirus. *Virol. J.* **2011**, *8*, 499.

Nariani, T. K. Yellow Mosaic of Mung (*Phaseolus aureus* L.). *Ind. Phytopathol.* **1960**, *13*, 24–29.

Navas-Castillo, J.; Fiallo-Olive, E.; Sanchez-Campos, S. Emerging Virus Diseases Transmitted by Whiteflies. *Annual Rev. Phytopathol.* **2011**, *49*, 219–248.

Nawaz-ul-Rehman, M. S.; Fauquet, C. M. Evolution of Geminiviruses and their Satellites. *FEBS Lett.* **2009**, *583* (12), 1825–1832

Ng, T. F.; Duffy, S.; Polston, J. E.; Bixby, E.; Vallad, G. E.; Breitbart, M. Exploring the Diversity of Plant DNA viruses and their Satellites Using Vector-Enabled Metagenomics on Whiteflies. *PLoS ONE* **2011**, *6*, e19050.

Polston J. E.; Lapidot, M. Management of Tomato Yellow Leaf Curl Virus—US and Israel Perspectives. In: *Tomato Yellow Leaf Curl Disease: Management, Molecular Biology, Breeding for Resistance*; Czosnek H, ed.; Kluwer Academic Publishers Group: Dordrecht, The Netherlands, 2007, pp 251–62.

Prasanna, H. C.; Sinha, D. P.; Verma, A.; Singh, M.; Singh, B.; Rai, M.; Martin, D. P. The Population Genomics of Begomoviruses: Global Scale Population Structure and Gene Flow. *Virol. J.* **2010**, *7*, 220.

Ranjan, P.; Singh, A. K.; Kumar, R. V.; Basu, S.; Chakraborty, S. Host Specific Adaptation of Diverse Betasatellites Associated with Distinct Indian Tomato-Infecting Begomoviruses. *Virus Genes* **2014**, *48*, 334–342.

Richter, K. S.; Kleinow, T.; Jeske, H. Somatic Homologous Recombination in Plants is Promoted by a Geminivirus in a Tissue-Selective Manner. *Virology* **2014**, *452*, 287–296.

Rojas, M. R.; Hagen, C.; Lucas, W. J; Gilbertson, R. L. Exploiting Chinks in the Plant's Armor: Evolution and Emergence of Geminiviruses. *Annual Rev. Phytopathol.* **2005**, *43*, 361–394.

Saleem, H.; Nahid N.; Shakir, S.; Ijaz, S.; Murtaza, G.; Khan, A. A.; Mubin, M.; Nawaz-ul-Rehman, M. S. Diversity, Mutation and Recombination Analysis of Cotton Leaf Curl Geminiviruses. *PLOS One* **2016**, *11* (3), e015116.

Sattar, M. N.; Kvarnheden, A.; Saeed, M.; Briddon, R. W. Cotton Leaf Curl Disease—An Emerging Threat to Cotton Production Worldwide. *J. General Virol.* **2013**, *94*, 695–710.

Sattar, M. N.,; Iqbal, Z.; Tahir, M. N.; Ullah, S. The Prediction of a New Clcud Epidemic in the Old World. *Frontiers in Microbiol.* **2017**, *8*, 631.

Satya, V. K.; Malathi, V. G.; Velazhahan, R.; Rabindran, R.; Jayamani, P.; Alice, D. Characterization of Betasatellite Associated with the Yellow Mosaic Disease of Grain Legumes In Southern India. *Acta Virologica* **2013**, *57*, 405–414.

Silva, F.; Lima A.; Rocha, C.; Castillo-Urquiza, G.; Alves-Junior, M.; Zerbini, F. Recombination and Pseudorecombination Driving the Evolution of the Begomoviruses *Tomato Severe Rugose Virus* (Tosrv) and Tomato Rugose Mosaic Virus (Tormv): Two Recombinant DNA-A Components Sharing the Same DNA-B. *Virol. J.* **2014**, *11*, 66.

Simmons, A. L.; Chandrasekar, S.; Amnon K. L. Combining Reflective Mulch and Host Plant Resistance for Sweetpotato Whitefly (*Hemiptera: Aleyrodidae*) Management in Watermelon. *Crop Protection,* **2010**, *29*, 898–902.

Singh, G.; Sharma, Y. R.; Shanmugasundaram, S.; Shih, S. L.; Green, S. K. Status of *Mungbean Yellow Mosaic Virus* Resistant Breeding. In: *Final Workshop and Planning Meeting on Mungbean*, 2003, pp 204–212).

Siwatch, S. S.; Abdel-Salam, A. M.; Brown, J. K.; Zafar, Y.; Markham, P.G. Diversity of DNA β: A Satellite Molecule Associated with Some Monopartite Begomoviruses. *Virology* **2003**, *312*, 106–121

Smartt, J. *Grain Legumes: Evolution and Genetic Resources.* Cambridge University Press: Cambridge Cambridge, UK, 1990.

Stansley, P. A.; Naranjo, S. E. *Bemisia: Bionomics and Management of a Global Pest.* Springer: Dordrecht, The Netherlands, 2010.

Stouthamer, R.; Breeuwer, J. A. J.; Hurst, G. D. D. *Annu. Rev. Microbiol.* **53**, 71–102.

Sunitha, S.; Shanmugapriya, G.; Balamani, V.; Veluthambi, K. *Mungbean Yellow Mosaic Virus* (MYMV) AC4 Suppresses Post-Transcriptional Gene Silencing and an AC4 Hairpin RNA Gene Reduces MYMV DNA Accumulation in Transgenic Tobacco. *Virus Genes* **2013**, *46*, 496–504.

Tollefson, J. Brazil Cooks up Transgenic Bean. *Nature* **2011**, *478*, 168.

Varma, A.; Malathi, V.G. Emerging Geminivirus Problems: A Serious Threat to Crop Production. *Ann. Appl. Biol.* **2003**, *142*, 145–164.

Venkataravanappa, V.; Lakshminarayana Reddy, C.; Swaranalatha, P.; Jalali, S.; Briddon, R. W.; Reddy, M. K. Diversity and Phylogeography of Begomovirus-associated Betasatellites of Okra in India. *Virol. J.* **2011**, *8*, 555.

Wang X. W; Luan J. B; Li J. M; Bao Y. Y; Zhang C. X; Liu S. S. De Novo Characterization of a Whitefly Transcriptome and Analysis of its Gene Expression During Development. *BMC Genomics* **2010**, *11* (1), 400.

Weintraub, P. G.; Berlinger, M. J. Physical Control in Greenhouses and Field crops. In: *Novel Approaches to Insect Pest Management*; Horowitz, A. R.; Ishaaya, I., Eds.; Springer Science: Berlin, 2004, pp 301–318.

Zerbini, F. M.; Briddon, R. W.; Idris, A.; Martin, D. P.; Moriones, E.; Navas-Castillo, J.; Rivera-Bustamante, R.; Roumagnac, P.; Varsani, A. ICTV Virus Taxonomy Profile: Geminiviridae. *J. Gen. Virol.* **2017**, *98* (2), 131–133.

Zhang, T.; Luan, J. B.; Qi, J. F.; Huang, C. J.; Li, M.; Zhou, X. P.; Liu, S. S. Begomovirus-whitefly Mutualism is Achieved Through Repression of Plant Defenses by a Virus Pathogenicity Factor. *Mol. Ecol.* **2012**, *21* (5), 1294–1304.

Zheng, F.; Wu, H.; Zhang, R.; Li, S.; He, W.; Wong, F. L.; Li, G.; Zhao, S.; Lam, H. -M. Molecular Phylogeny and Dynamic Evolution of Disease Resistance Genes in the Legume Family, *BMC Genetics* **2016**, *17*, 402.

Zhou, X. F.; Li, Z. X. *Establishment of the Cytoplasmic Incompatibility—Inducing Wolbachia Strain Wmel in an Important Agricultural Pest Insect*; Scientific Reports, 2016.

CHAPTER 14

Biology and Molecular Epidemiology of Begomovirus Infection on Cucurbit Crops

BHAVIN S. BHATT[1], FENISHA D. CHAHWALA[2], SANGEETA RATHORE[2], BIJENDRA SINGH[3], and ACHUIT K. SINGH[3*]

[1]*Department of Environmental Science, Shree Ramkrishna Institute of Computer Education and Applied Sciences, Surat 395001, Gujarat, India*

[2]*School of Life Sciences, Central University of Gujarat, Gandhinagar 382030, Gujarat, India*

[3]*Division of Crop Improvement, ICAR—Indian Institute of Vegetable Research, Varanasi 231304, Uttar Pradesh, India*

*Corresponding author. E-mail: achuits@gmail.com

ABSTRACT

Various economically important plants are infected by single-stranded DNA viruses in subtropical and tropical region. Emergence of new species covered with wide range of cropping system among cucurbit is severely affected by producing mosaic and curling symptom. Under the family *Cucurbitaceae* melons, squash and gourds is significantly hampered by the begomovirus infection. Few reports available mention that Begomovirus also infect cucurbits in new world and old world. *Cucurbits leaf crumple virus, Squash leaf curl virus, Melon chlorotic leaf curl virus, Squash yellow mild mottle virus,* and *Squash mild leaf curl virus* are known to cause a huge economic deprivation in the family. Epidemiology of begomovirus infecting cucurbit plants is also distinct. In the western part of world, *Squash mild leaf curl virus* and *Squash leaf curl virus* are more prominent, whereas *Watermelon chlorotic stunt virus* is aggressive in the eastern part of world. Philippines, China, and Vietnam were severely affected by *squash leaf curl china virus,* whereas *Squash leaf curl Yunnan virus* has been present localized only in southern China. *Loofa yellow mosaic virus* is geographically restricted and is endemic species in southern Vietnam. Recombinant viruses are present in nature, which play an important role in virus evolution. Sometimes, recurrent or mixed infection provides a vital platform for viral DNA molecules for recombination. These molecular events are basis of enormous molecular diversity of plant viruses. Because of diverse host range and diversity in symptoms produced by same group of viruses onto different plants, classification of plant viruses is very thoughtful process. Here, we illustrate molecular diversity and epidemiology of begomoviruses and their satellites molecules in cucurbits crops which is seriously threatened.

14.1   INTRODUCTION

Plant viruses are most devastating and related to economic loss in the fields worldwide. It affects the food crops, non-food crops, as well as weeds plants. Weeds often act as a reservoir of viruses and stay in dormant condition when their host plants are not available. Because of the lack of awareness for the infection of plant viruses, it is a little hard to figure out the total financial loss in agriculture. More than \$30 million estimated loss are reported annually because of plant virus infection. During past two decades, losses occured because of geminivirus infection were US\$ 5 billion for cotton in Pakistan by cotton leaf curl virus (Briddon and Markham, 2001) in 1992-97, US\$ 2.3 billion for cassava by African cassava mosaic virus (Thresh and Cooter, 2005), US\$ 300 million for grain legumes in India (Varma and Malathi, 2003), and US\$ 140 million in Florida, USA, for Tomato (Moffat, 1999). Yield constraint, tremendous losses, and economical outbreaks leads researchers and policy makers gained interest in geminiviruses. Sugar beet infection by *Beet curly top virus*, cassava infection by *African cassava mosaic virus*, cotton infection by *Cotton leaf curl virus*, *Bean golden mosaic virus* of common bean, maize infection by *Maize streak virus*, and finally tomato infection by *Tomato leaf curl virus* are past pandemics that causes huge loss in production of respective crops in various parts of the world (Legg and Fauquet, 2004; Morales and Anderson, 2001).

The major viruses that infect cucurbits are *Squash mosaic virus* (SqMV), *Watermelon mosaic virus* (WMV), *Cucumber mosaic virus* (CMV), *Zucchini yellow mosaic virus* (ZYMV), and *Papaya ring spot virus* (PRSV). The viruses differ in the range of host plants they infect, the ways in which they are transmitted, and how they survive between crops. Since control programs are based on this information, identification of viruses is very important. This virus complex has caused growers in certain areas to stop cultivation. In this chapter, we will discuss molecular biology and epidemiology of begomoviruses on cucurbits.

14.1.1 PATHOSYSTEM IN CUCURBIT CROPS

Because of their sessile nature, plants are more prone to attack with various factors, biotic factor and abiotic factor. Among abiotic factors, high temperature, low rainfall, high heavy metal concentration, etc., contribute majorly affecting yield. In biotic factors, crop yield decreases up to 60% (Subramanya, 2014). Under some drastic situation, a huge yield loss had been observed in past. Apart from these factors, parasites, predators, and weeds also contribute heavily for loss of production. As other economic important food crops, cucurbits also suffer from various diseases. Several phytopathogens such as bacterial, fungal, and nematode infect worldwide and reduces the yield drastically.

Bacterial diseases result in development of small, angular, and water-soaked leaf spots and straw-like structure with hole that is known as angular leaf spot diseases. It infects almost all cucurbits where high humid condition is available. Bacterial fruit blotch/seedling blight mostly infects watermelon and causes significant loss in the field. Fungal infections mainly distort leaf and fruit structure which ultimately result in their saleability. Several fungal pathogen viz, Altenaria, Fusarium, Colletotrichum are interacting with the crop and causes significant yield loss. They produced leaf blight, spots on leaf and stem, black root rot, blue mold rot, fruit rot, crown and foot rot, damping-off, downy mildew, graymold, Phomopsis black stem, Pythium fruit rot, etc. Cucurbits are extremely susceptible with root nematode infection which leads to root dysfunction, resulting in interruption of water uptake. Moreover, some of the diseases related are lesion, ring, root-knot, stubby-root, stunting, etc. (Dorman and Nelson, 2012). Among the phyto-pathogens, majority of crop loss is accountable to viruses. Losses because of virus infection on plants is estimated 47%, which is a greater than the loss encountered by fungi, bacteria, phytoplasma, or nematodes put together (Anderson et al., 2004).

14.1.2   THE BEGOMOVIRUS: A POTENTIAL ROLE IN VEGETABLE PATHOSYSTEM

Geminivirideae characterized as single-stranded DNA virus is the largest among all plant viruses, where begomovirus is the largest genara in the family with more than 300 species. Geminiviruses are characterized genomes that are encapsidated in twinned quasi-isometric particles of about 18 × 30 nm (Goodman, 1977; Howarth and Vandemark, 1989). The *Geminiviridae* family has been divided on basis of genome organization, host range, and insect vector into seven genera: Mastrevirus, Curtovirus, Topocuvirus, Begomovirus, Becurtovirus, Eragrovirus, and Turncurtovirus (Fauquet et al., 2008, Brown et al., 2012). Recently, family *Geminiviridae* is further classified and two new genera has been added: *Grablovirus* and *Capulavirus* (Adams et al., 2016, Varsani et al., 2017, Al Rwahnih et al., 2017). All the genera of family *Geminiviridae* use distinct vector system for their transmission and dispersal. Begomovirus are transmitted by whiteflies whereas Becurtovirus, Curtovirus, Grablovirus, Mastrevirus, and Turncurtovirus are transmitted by specific leafhoppers. Topocuvirus is transmitted by a treehopper and one member of the genus Capulavirus is transmitted by an aphid.

Subgroup III is the largest genera of geminiviruses, includes whitefly transmitted viruses with dicotyledons as their host range (Mansoor et al., 2003). Currently, this group comprises 288 distinct virus species. The members of this subgroup contain monopartite (~2.7 kb), monopartite with some subviral satellite molecule (~1.3 kb), or bipartite genome component. Begomoviruses are credited with most complex genome organization among the Geminiviruses. The genome majorly is bipartite with two single-stranded circular DNA, of 2600 nucleotides each, is encapsided independently, and designated as DNA A and DNA B. However, in few Begomoviruses, the genome is monopartite and satellite DNA, named betasatellite, is associated with it (Briddon et al., 2003, 2008). In case of bipartite genome, both genome components are essential for efficient disease transmission and systemic infection. DNA A component encodes for major proteins for virus replication and multiplication inside the host cell, whereas DNA B cares for intracellular- and intercellular movement of virus particles. The movement protein genes of viruses with monopartite genomes belonging to subgroups I, II, and III are encoded on the single genome component.

<H1>14.2 BEGOMOVIRUS INFECTION ON CUCURBITS

Evolutionary begomovirus infection of cucurbits is kind of new world infection where there is negligible role of satellite molecules. Here, we will discuss about epidemological study of begomovirus infection on cucurbits.

Begomoviruses infecting cucurbit crops are mostly under the new world. There are only few reports of old world begomovirus infection on cucurbits. Very few reports show the presence of alpha and/or beta satellites during begomovirus infection on cucurbits (Shahid et al., 2015). Some of Begomoviruses which are reported to infect cucurbit crops are *Squash leaf curl virus, Melon chlorotic leaf curl virus, Squash yellow mild mottle virus, Cucurbit leaf crumple virus, Euphorbia mosaic virus, Squash leaf curl China virus, Chayote yellow mosaic virus, Coccinia mosaic virus, Bitter gourd yellow vein virus, Squash mild leaf curl virus, Loofah yellow mosaic virus,* and *Melon chlorotic leaf curl virus.* Non-host viruses that infect cucurbits are *Tomato leaf curl virus, Rhynchosia mosaic virus, Pepper golden mosaic virus, Tomato leaf curl Palampur virus, Tomato leaf curl New Delhi virus,* and *Croton yellow vein mosaic virus.* Common symptoms include yellow mosaic and mottling of leaves, yellow vein, vein clearing, leaf curling, leaf crumpling, small fruits, etc. (Fig. 14.1).

FIGURE 14.1 **(See color insert.)** Cucurbit plants exhibiting putative viral symptoms of yellowing, mosaic and leaf curling. (A) Bitter gourd leaf distortion disease on bitter gourd (*Momordica charantia*), (B) yellow mosaic disease on cucumber (*Cucumis sativus*), and (C) yellow mosaic disease on pumpkin (*Cucurbita moschata*).

14.2.1   CUCURBIT LEAF CRUMPLE VIRUS

In 1998, the production of watermelon was affected in imperial valley of south California. Later investigations confirmed that causal agent for lower production of watermelon is due to infection by *Cucurbit leaf crumple virus*

(CuLCrV). It was an evolving and potentially economically important bipartite begomovirus (Guzman et al., 2000). Field surveys showed that CuLCrV has become established in the Imperial Valley. It was most devastating and infects cucurbits, musk melon, squash, and watermelon. This virus expresses the symptoms of dwarfism and crinkling of leaves, curling, and chlorosis. CuLCrV can transmit by agroinoculation method, which showed the susceptibility to other cucurbits plants such as squash, watermelon, cantaloupe, and honeydew melon. There are number of resistant honeydew melon and muskmelon cultivar available against this virus. Germ plasm screening studies play potential role to find out resistant variety. *Cucurbit leaf crumple virus* isolate (CuLCrV-CA) has bipartite genome of ~2.7 kbp each for DNA A and DNA B with characteristic begomoviral signature nonanucleotide sequence at origin of replication site. Molecular phylogenetic analysis showed CuLCrV-CA has close relationship with *Squash leaf curl virus* (SLCV), which is also bipartite new world begomovirus.

Cucurbit leaf crumple virus was also found in Mexico during 2002. It was also found in squash as well as zuchini in Florida that infects cucurbits as well as other vegetable crops. It was also found to infect *P. vulgarisin* Florida (Adkins et al., 2009).

Tomato severe leaf curl virus (ToSLCV) associated with *Pepper golden mosaic virus* (PepGMV) was responsible for new yellow leaf crumple disease of cucumber in Maharashtra, India. Mixed infection with presence of more than one distinct begomovirus indicates emergence of severe new strain of begomovirus (Suresh et al., 2013).

14.2.2 MELON CHLOROTIC LEAF CURL VIRUS

Melon chlorotic leaf curl virus (MCLCuV) is another new world bipartite begomovirus infecting cucurbits. The MCLCuV genome was cloned in the year 2000 in Guatemala on the basis of symptoms appeared on leaves (Brown et al., 2001). Sequence analysis and phylogenetic relationship showed that MCLCuV DNA A and DNA B sequence analysis showed closest identity with a papaya isolate of MCLCuV from Costa Rica at ~90 and 81%, respectively. The DNA A components of MCLCuV shared 88.8% nucleotide identity. Furthermore, both strain of MCLCuV-GT and MCLCuV-CR are phylogenetically placed with other western hemisphere cucurbit-infecting begomoviruses cluster. (Idris et al., 2008).

Mutation, recombination, pseudorecombination, synergism, reassortment, and transcomplementation are common features of Geminiviruses

(García-Arenal et al., 2001). These viruses often adopt such "molecular arrangements" and are major source of genetic variation in plant viruses. Viruses can exchange or rearrange their genetic material during multiple infection in a host through this mechanism/s, which results in diverse host range (Rajeshwari et al., 2005). Sequence annotation of DNA A and DNA B were identical with MCLCuV-GT, which shares less than 86% nucleotide identity with the respective DNA A and DNA B CR of MCLCuV-CR. This is a striking example of evolution of new strain by molecular rearrangements as the common region (which is less prone to such molecular rearrangements and comparatively considered as "conserved") shares less nucleotide identity and other genomic components shares high nucleotide identity with MCLCuV-CR. Conclusively, recombination and substantial accumulation of mutation resulted into evolutionary diverse strains of MCLCuV. As other Begomoviruses infecting cucurbits, MCLCuV was also grouped in SLCV clade at the species level, which is localized in the southwestern United States and Mexico. MCLCuV has comparatively higher host range and can also infect crops of *Cucurbitaceae*, *Fabaceae*, and *Solanaceae*. Idris et al. (2008) also studied the origin of MCLCuV. It showed MCLCuV is the *Bean calico mosaic virus* (BCaMV), other cucurbits-infecting species SLCV and SMLCV. The epidemiological data suggests precise yet overlapping host ranges for cucurbit viruses with diverse symptomology.

14.2.3 CUCURBIT LEAF CURL VIRUS

CuLCV) a geminivirus that is transmitted via whitefly was previously partially characterized from the southwestern part of United States and northern part in Mexico. It was identified as bipartite begomovirus species. This distinct virus showed sequence identity with partially characterized *Cucurbit leaf crumple virus* from California (Brown et al., 2002). On the basis of experiments and its natural host range, CuLCV has a comparatively diverse host range in *Cucurbitaceae* family and it also infects tobacco and bean. *Cucurbit leaf curl virus* was isolated from the Arizona, named as CuLCV-AZ. Bipartite genome of CuLCV-AZ DNA A and DNA B were introduced by biolistic method to pumpkin. Progeny virus was able to transmit via the whitefly vector, *Bemisia tabaci*, which completed Koch's postulates. *Squash leaf curl Virus* (SLCV-R and SLCV-E) and *Bean calico mosaic virus* (BCaMV) shares highest nucleotide sequence similarity with CuLCV-AZ DNA A component with 84% and 80%, respectively (Brown et al., 2002).

14.2.4   PUMPKIN YELLOW VEIN MOSAIC VIRUS

Pumpkin is commercially an important vegetable grown abundantly for its big and fleshy fruit. In the year 1940, northern India was severely affected by yellow vein mosaic disease of pumpkin (Vasudeva and Lal, 1943) and later it was reported all over in India (Varma, 1955, Capoor and Ahmad, 1975). The disease symptoms included vein yellowing with chlorotic patches, stunting, and premature drop of flowers and thus greatly reducing yields. The virus has been identified as PYVMV which was spread readily in an incessant manner by the whitefly, *Bemisia tabaci* (Maruthi et al., 2007).

Begomovirus infection produces mosaic and yellow vein symptoms on pumpkin (*Cucurbita moschata*), which is a serious problem for the production around India (Jaiswal et al., 2012). Symptomatic leaf samples were collected around Varanasi region of Northern India. Biodiversity study of different begomoviruses indicate mixed infection of the three begomoviruses which could be three species named as *Tomato leaf curl Palampur virus* (ToLCPV), *Squash leaf curl China virus* (SLCCV), and *Tomato leaf curl New Delhi virus* (ToLCNDV). Based on complete nucleotide sequence similarity results, PYVMV was most nearest with bipartite *Tomato leaf curl New Delhi virus* and designated as a new strain of ToLCNDV (Muniyappa et al., 2003).

PYVMD was endemic to south India in the year 2004 with 100% viral spread and significant yield loss. The occurrence of Biotype B of whiteflies was prevalent in the field. The transmission of begomovirus insect vectors plays an important role. *Bemicia tabaci* are of three biotypes, viz. biotype A, biotype B, and biotype C. Biotye B is dominant and becomes more aggressive to cause infection.

14.2.5   CHAYOTE YELLOW MOSAIC VIRUS

Chayote (*Sechium edule*) is an abundantly consumed vegetable with high nutritional value. Disease symptoms include yellow mosaic spots on leaves and upward curling with occasional enations. *Chayote yellow mosaic virus* shares maximum nucleotide identity (95%) with previously characterized ToLCNDV from Pakistan. ChaYMV also infects other members of cucurbits viz. bitter gourd, cucumber, and squash (Mandal et al., 2004).

14.2.6   TELFAIRIA MOSAIC VIRUS

Fluted pumpkin (*Telfairia occidentalis*) is the foremost and most popular leafy vegetable in the diet of Nigerians, where it is cultivated by marginal farmers for household use. Disease symptoms include severe mosaic, leaf distortion, and malformation. The whole genome sequence of virus infecting *Telfairia occidentalis* was determined and characterized. The genome is bipartite in nature with DNA A and DNA B is of 2700 bp and 2600 bp in size, respectively. Computational analysis confirmed that coding regions are arranged and oriented as other bipartite begomoviruses. Phylogenetic analysis of whole genome shared maximum nucleotide similarity with other annotated bipartite begomoviruses. DNA A component matches maximum nucleotide identity with *Chayote yellow mosaic virus* (ChaYMV) of 78%. The DNA B component has more close evolutionary relationship to that of *Soybean chlorotic blotch virus* (SbCBV) at 64%. As per species demarcation criteria by ICTV for begomovirus, this newly discovered bipartite virus was named as *Telfairia mosaic virus* (TelMV) (Leke et al., 2016).

14.2.7   COCCINIA MOSAIC VIRUS

Ivy gourd (*Coccinia grandis* L.), a vegetable crop, is also host for the begomovirus and associated with mosaic diseases in Tamil Nadu, India. The genome components were cloned by using rolling circle amplification. Presence of DNA A and DNA B component and sequence analysis study was found to be distinct of old world bipartite begomovirus. There was no presence of betasatellite during infection. The DNA A and DNA B components were cloned from infectious ivy gourd plant. Phylogenetic analysis of complete nucleotide sequence of DNA A component showed maximum nucleotide similarity with *Loofah yellow mosaic virus* (LYMV-[VN]-AF509739) at 73.4%, whereas DNA B component was closely matched with *Tomato leaf curl New Delhi virus* (ToLCNDV) at 55%. Phylogenetic analysis clearly showed an event of transcomplementation. This novel virus species was named as *Coccinia mosaic virus* (CoMoV Ivy gourd [TN-TDV-Coc1]). Recombination studies evaluate that there was no recombination event in the genome. CoMoV ivy gourd [TN-TDV-Coc1] is novel and different begomovirus that infect *Coccinia* plant (Nagendran et al., 2016).

14.2.8   WATERMELON CHLOROTIC STUNT VIRUS

Watermelon chlorotic stunt virus (WmCSV) is a bipartite begomovirus (genus *Begomovirus*, family *Geminiviridae*) infecting cucurbits, extensively watermelon, inhabiting the Middle east and North Africa. Disease symptoms include foliate curling, yellowing, and dwarfism which are typical of a begomovirus infection (Bananej et al., 2002; Ali-Shtayeh et al., 2014).

14.2.9   BITTER GOURD YELLOW VEIN VIRUS

Momordica charantia (bitter gourd), a vegetable crop, is commonly cultivated for vegetable and medicinal purpose. Symptoms include yellow vein and mosaic patches on leaves (Raj et al., 2005). The virus genome is bipartite and transmitted by whitefly. DNA A component showed the nearest sequence identity (86.9%) with *Tomato leaf curl New Delhi virus* (ToLCNDV). As per ICTV species demarcation criteria, it is new virus which was named as *Bittergourd yellow vein virus* (BGYVV). DNA B showed highest sequence similarity (97.2%) with Indian strain of *Squash leaf curl China virus* (Tahir et al., 2010).

14.3   NON-HOST VIRUSES INFECTING CUCURBITS

Begomovirus is responsible to cause huge impact on tomato production worldwide since it is known as tomato leaf curl disease (TLCD) and tomato yellow leaf curl disease (TYLCD). One of *Tomato leaf curl virus*, known as *Tomato leaf curl Palampur virus* (ToLCPMV), is responsible for the diseases on cucurbits and melon   (Chigurupati et al., 2012). *Tomato yellow leaf curl virus* (TYLCV) and *Tomato leaf curl New Delhi virus* (ToLCNDV) infection was found on watermelon and zuchini in Tunisia, respectively (Mnari-Hattab et al., 2014, 2015).

In northern India, mosaic and yellow veining symptoms were found on ridged gourd (*Luffa acutangula*) and sponge gourd (*Luffa cylindrica*). Sequence similarity and phylogenetic identities confirmed symptomatology was due to presence of *Tomato leaf curl New Delhi virus* (Tiwari et al., 2011).

Warm and humid atmosphere is ideal for cultivation of bottle gourd (*Luffa siceraria*). Summer-rainy season is best for its cultivation in India. Chlorotic curly stunt disease (CCSD) was observed on bottle gourd during 2003–2006. Whitefly act as vector for spread so molecular studies and phylogenetic relationship was studied. By using comparison of complete nucleotide sequencing and amino acid sequencing methods, genome was associated with coat protein and replication initiator protein (Rep). CCSD comes under the cluster of ToLCNDV of begomovirus (Sohrab et al., 2010). Kundru (*Cucumis indica*) is a creeper plant which has medicinal property and used for treatment of diabetes mellitus. Begomovirus infection was distrusted on *Cucumis indica* which expresses the symptoms of severe yellow mosaic, leaf curling, and blistering. Shahid et al. in 2015 reported that *Tomato leaf curl New Delhi virus* cause yellow mosaic disease on *Cucumis indica*. They also reported association of alphasatellite with ToLCV. *Cucurbita maxima* plants also express characteristic begomoviral symptoms like curling and yellowing of leaves indicated presence of a begomovirus and DNA B component or a betasatellite was not associated with an infection. Molecular and sequence analysis studies revealed that the Israel strain of tomato yellow leaf curl virus associated with Sida yellow vein China alpha satellite are causal agents of disease. This is the only report of association of satellite molecule with begomoviral disease on cucurbits.

Mosaic disease of cucumber (*Cucumis sativus*) was observed during 1993–1994. The begomovirus association with yellow mosaic disease of *Cucumis sativus* was suspected as (1) characteristic mosaic symptoms of begomovirus infection and (2) prevalence of whitefly in the field. The begomovirus association was diagnosed by nucleic acid probes derived from DNA B of *Tomato leaf curl virus* (ToLCV) and *Tobacco leaf curl virus* (TobLCV). Based on positive detection signals of the DNA isolated from infected *Cucumis sativus* leaves, with both the DNA B probes of ToLCV and TobLCV, both viruses are present infecting *Cucumis sativus* causing dark green mosaic and leaf curl symptoms (Raj and Singh, 1996).

Tomato leaf curl Palampur virus (ToLCPV) is causal agent of yellow mosaic disease on wild melon (*Cucumis melo*). The infected plants show yellow mosaic on leaves, stunting and reduced fruit yield BLAST analysis phylogenetic relationships of the sequence data indicate closest 99% sequence identities with ToLCPV.

Disease symptoms of yellow mosaic disease were observed on bitter gourd associated with leaf curling and severe yellow mosaic pattern with smaller fruit size in eastern UP, India. Based on complete sequence similarities and phylogenetic association, an isolate of *Pepper leaf curl Bangladesh virus* (PepLCBV) was identified to cause diseases (Raj et al., 2010). *Pepper golden mosaic virus* (PepGMV) extends the host range to include *C. moshata*, *C. pepo* and chayote in western hemisphere.

Bitter gourd plants are also shown to be infected by begomovirus. Symptoms include yellow leaves, leaf curling, and stunting and vein enation. The complete nucleotide sequencing of DNA A and DNA B components showed closest nucleotide similarity with *Squash leaf curl china virus* [Pumpkin-Lucknow]. Bitter melon leaf curl disease (BMLCD) was observed in Lucknow, India. Symptoms were similar to begomovirus symptoms like upward leaf curling and distortion, deformed and small fruits. PCR and hybridization analysis proved the presence of *Tomato leaf curl virus* and *Cotton leaf curl virus* for BMLCD (Khan et al., 2002).

Bitter gourd is also prone to Indian cassava mosaic virus (ICMV). In Tamil Nadu, India, symptoms were expressed as yellow mosaic pattern (Rajinimala and Rabindran, 2007). The symptoms are alike of yellow mosaic disease of bitter gourd. It mainly includes mosaic patches on young leaves and mottling only. The edges of leaf started with mottling patches and as disease progress, the foliate became chlorotic with few patches of intermittent green tissue. This was the first reports of ICMV infection on bitter gourd in India (Rajinimala and Rabindran, 2007).

Pointed gourd (*Trichosanthes dioica* Roxb.), commonly known as "Parwal," is a tropical perennial vegetable cash crop for marginal farmers. Mosaic symptoms were observed on pointed gourd in 2008. Leaf mosaic and curling symptoms were prevalent with small and malformed fruits which are nonmarketable. Begomovirus was detected from symptomatic leaf samples by specific coat protein primers. The complete genome sequence was cloned and compared with previously characterized begomoviruses. Analysis showed close phylogenetic relationships with *Ageratum enation virus* (AgEV) (Raj et al., 2011). On the basis of genome organization, sequence analysis of cucurbit infecting begomoviruses were compared with other begomoviruses. Sequence alignments were performed with ClustalW program of Megalign in Lasergene 10 coresuite. Phylogenies of total genome of DNA A (Fig. 14.2 (A)) and DNA B (Fig. 14.2 (B)) were constructed with the best-fit model of nucleotide substitution. Bootstrap trials of 1000 with 100 seed value was used to test consistency and robustness of trees created. Color coded matrix of pairwise sequence identity based on full-length sequences of genome components DNA A of representative begomoviruses was constructed for species demarcation (Fig. 14.3).

14.4   MANAGEMENT OF BEGOMOVIRUS INFECTION

Begomoviruses are continued to cause measurable loss since four decades of cucurbits. Begomovirus are not only responsible for significant loss in cucurbits but also for other vegetable crops as well, which ultimately affects

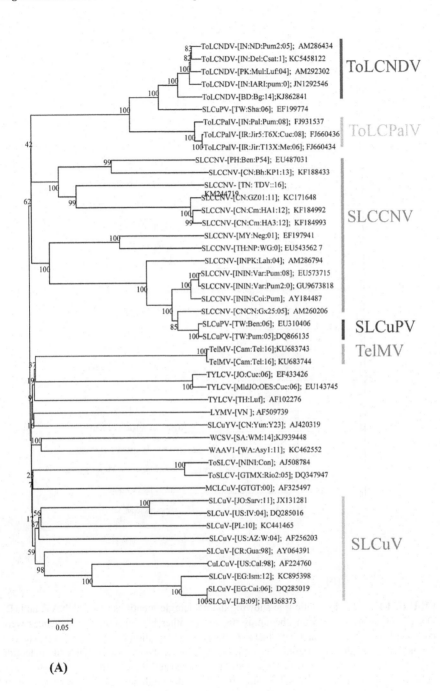

FIGURE 14.2 *(Continued)*

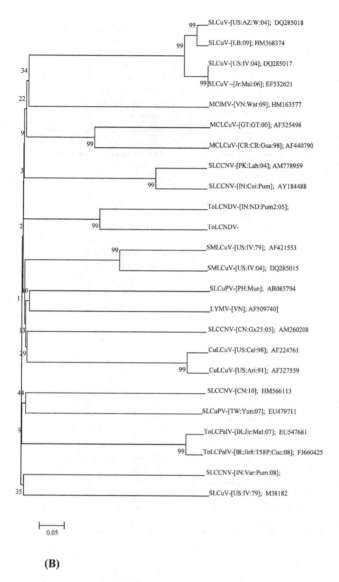

(B)

FIGURE 14.2 Phylogenetic tree showing the nucleotide identities of (A) DNA A and (B) DNA B of cucurbits infecting begomoviruses with other begomoviruses. Sequences were computed and matched using the Clustal W algorithm (MegAlign DNAstar), the vertical axis is random and length of horizontal axis represents in percentage the occurrence of nucleotide substitution × 100. The numbers at the nodes indicate posterior probability values (Bootstrap values) in which the given branch is supported. The full length begomovirus sequences were downloaded from GenBank, NCBI, and tailored using VecScreen. Accession numbers of sequences are on the right of each virus name. Virus abbreviations are as recommended by Fauquet et al., 2008, Brown et al., 2015, and Zerbini et al., 2017.

economy of the country which is agriculture oriented. So, there is strong need for the management for begomovirus infection worldwide. Following are the possible ways for the management of begomovirus infecting cucurbits.

Strategies for the management of any viruses include two types of methods, conventional method and nonconventional method. Conventional methods include good agricultural practice and powerful breeding strategies to avoid the infection. This is very strong and natural methods to avoid the contamination. But success of this method is based on the severe outburst or in conditions where the crop has a genetic liable to be infected. In this situation, one needs to think about the nonconventional methods.

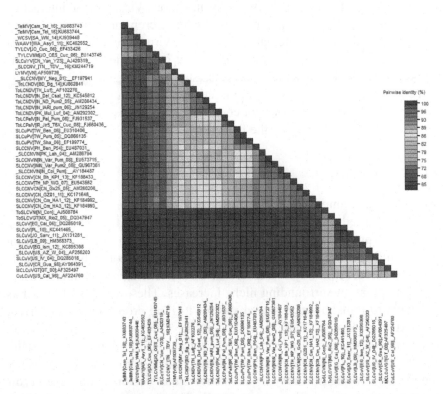

FIGURE 14.3   (See color insert.) Color coding representation of percentage pairwise genome scores and evolutionary identity plot of full genomes of DNA A of begomoviruses infecting cucurbits. The plot was prepared using MUSCLE parameter of SDTv1.2 (Species Demarcation Tool; http://web.cbio.uct.ac.za/SDT) (Muhire et al., 2014) in two-color mode with cut-off 1, which is 94 and cut-off 2, which is is 78. The full length begomovirus sequences were downloaded from GenBank, NCBI, and tailored using VecScreen. Accession numbers of sequences are on the right of each virus name. Virus abbreviations are as recommended by Fauquet et al. 2008, Brown et al. 2015, and Zerbini et al. 2017.

14.4.1 CONVENTIONAL METHOD

14.4.1.1 VECTOR

Begomovirus requires specific vector, whitefly, *Bemisia tabaci*, for efficient spread in the field. Mostly these viruses are not sap or seed transmitted. Three biotypes are available, where biotype B causes significant loss in the field. These vectors are also responsible for the spread of the virus from one field to other region (Mayer et al., 2002). Also, it is responsible for mixed infection of different viruses in one. So, very first and prior step to avoid the infection is to control the whitefly infection. Virus vector interaction is changed according to epidemiology, habitat, crop, and viruses. Hence, control of viral vectors automatically controls virus spread in the field. For the management of vectors, there are many insecticides available in the market since 1930. But some reports also showed the development of resistance in biotype B against specific insecticide. Hence, it becomes crucial to think on other strategies for vector control. In some cases, farmers use net to prevent the whitefly attack in the field.

14.4.1.2 WEEDS

Weeds are unwanted growth of different plant in the field. It has been shown that weeds are prospective storehouse for dormant stage of viruses. Weeds also provide favorable platform for mobilization and recombination during multiple infections of begomovirus and their cognate components (Mubin et al., 2010; Jyothsna et al., 2011). During favorable season, vector population surrounding niche increases. These vectors often act as a vehicle to transfer weed stored viruses to field crop. Once transferred, the viruses establish themselves and initiate the infection cycle. Soon, due to availability of large amount of vectors, viruses spread in the field, resulting in significant disease incidence and loss of crop yield. Many reports are coming nowadays, which shows that vectors act as a reservoir for the begomovirus (Khan et al., 2003). Timely removal of weeds from cultivated fields has been found to be successful technique to reduce the viral incidences in cucumber and celery.

Rhynchosia minima, Jatropha gossypifolia, Sorghum vulgare, Croton bonplandianum, Coccinia grandis, Malvesrtum coromandalianum, Ageratum conyzoides, Nicotiana plumbaginifolia, etc., are examples of some common weeds, which grow near the crop field and reported to transmit begomoviruses. Therefore, elimination of such weed plants from the culti-vated fields significantly reduce begomovirus transmission and infection to crop plants.

The best ways to get rid of virus are the use of virus-free plants and seeds and regular inspection for disease symptoms at early stages of plants in and around the field. Regular cleaning and removal of commonly growing weeds are the good approach as they also compete for the nutritional prospect.

14.4.1.3   REMOVAL OF INFECTED PLANTS

One possible way to harboring the disease is regular checking of all plant in the field and bright watch for the development of symptoms. At the same time, one should be vigilant enough to demark symptoms of biotic stress with abiotic stress. If found symptomatic, one should diagnose the virus infection and removal of plant as early as possible. The best way to give heat treatment (34–50°C) and burring of infected plant (Raychaudhuri and Verma, 1977). But it is very cost-effective and in developing country it becomes difficult to diagnose plant infection in early stages.

14.4.1.4   DEVELOPMENT OF RESISTANT/TOLERANT VARIETIES

This is also one possible approach to develop resistance or tolerant variety of plant. Resistance/tolerant variety can be developed by breeding strate-gies (Pelham et al., 1970). Less susceptible can be used for breeding so that possible tolerant variety can be developed. But, as it is very time-consuming method and there is always possibility of genetic reversal, it not widely applicable method for diverse group of plants. The next one is development of genetically engineered transgenic plant. Both strategies are useful to some extent, however, recombination of virus breakdown resistance many times.

14.4.1.5   WHITEFLY CONTROL STRATEGIES BY USING INSECTICIDES APPLICATION

This strategy comprises controlling carrier vector (whitefly) population in a field by concurrent insecticide or pesticide applications. Begomoviruses are vector transmitted viruses, hence controlling their vector will control the virus spared. But consequently, longer usage of pesticide will have detrimental effects on human health. Apart from this, whitefly will become resistant to the dose and/or type of pesticide upon longer usage. Therefore, it is important that insecticides are managed in combination with other control methods.

14.5.1 NON-CONVENTIONAL METHOD

These methods are based on our recent understanding of system biology and plant-pathogen interactions at molecular level. Plant defence mechanism is regulated at molecular level to prevent the viral infection. The immune system of plant is suppressed in early stages of infection by viruses so they can command host cell for their replication, multiplication, and transport. In nonconventional approach, scientists try to overexpress or downregulate the prime molecule of cascade which is responsible for the occurrence of diseases condition, in which they try to focus on pathogen-derived resistance. Pathogen- or parasite-derived resistance is resistance conferred to plant by the gene/s derived from genetic arrangements of the pathogen itself. Coat protein and protein-mediated resistance are another two approaches for viral resistance. Many transgenic plants have been developed by using RNAi-mediated resistance. For example, Cassava plants were engineered to express small interfering RNAs (siRNA) derived from bidirectional transcription start site in the common region of *African cassava mosaic virus* (ACMV). Experimental plants showed significant attenuation to the viral symptoms upon viral infection with presence of 21-24 nt long siRNA.

RNAi-mediated technology is an important constraint because biosafety is an important concern to develop virus resistant transgenic plants. Recombination of different viral genome results in development of new symptom phenotypes that may express novel viral transgenes. Such novel transgenes may be transmitted by pollen to other nontransgenic plants (Snehi et al., 2015). CRISPR/Cas technology is a new emerging technology in this decade. It is very useful against plant as well as animal pathosystem. Some strategies are also growing to overcome begomovirus infection (Zahir et al., 2015). Kangquan et al. (2015) developed virus-based guide RNA system for CRISPR/Cas9-mediated plant genome editing to defend tomato leaf curl virus and cabbage leaf curl virus (CaLCuV). Moreover, the major advantage of genome editing is to get multiple pathogen resistance in the single plant can be achieved.

Integrated management can also be employed for management of virus diseases. Major concern of different methods is prevention of various methods which integrates prevention and setback of infection. These practices include managing alternative weed or change in the cropping cycle, insect vector management, and using resistant or tolerant varieties.

Although there are many effective approaches that exist for the management of begomovirus disease, it should be our prime focus to implement integrated strategy which should be (1) economical, (2) broad ranged and long lasting, (3) environmental friendly, and (4) can suppress onset of infection at initial stages and should not initiate dormant infections. An integrative control approach is best workable approach in many plant-pathogen models. It includes use of improved and/or genetically engineered plant material, use of insecticides, change in cropping pattern, removal of weed plants, and adequate irrigation during the dry season. Furthermore, reduction of vector load will significantly reduce virus inoculum and infection. Although it is very much clear that absolute control is difficult, yet compensatory measures will significantly reduce onset of infection.

14.5 CONCLUSION

Because of the favorable climate conditions of growing of cucurbits, they are more prone to attack with viral attack. During past few years, reports on begomovirus infection on cucurbit plants are increasing (Tiwari et al., 2008). This includes extending geographical boundaries, extending host range and infection by nonhost viruses. Since 1990, cucurbits were only infected with yellow vein mosaic disease of pumpkin. Later, leaf curl in musk melon, leaf distortion, and yellow mosaic disease of sponge gourd became a problem. At present, the begomovirus infections are commonly present in various cucurbitaceous hosts (Sohrab et al. 2006; Tiwari et al., 2012). Studies on molecular identification of begomoviruses infecting bottle gourd, bitter gourd, ivy gourd, pointed gourd, long melon, pumpkin, ridge gourd, water melon, and chayote have revealed the involvement of begomovirus species, *Tomato leaf curl New Delhi virus* (ToLCNDV), Squash leaf curl China virus (SLCCNV), Tomato leaf curl Palampur virus (ToLCPalV), Pepper leaf curl Bangladesh virus (PepLCBV), Ageratum enation virus (AgEV) (Sohrab et al., 2006; Singh et al., 2009; Raj et al., 2010; Chigurupati et al., 2012; Tiwari et al., 2012).

In conclusion, there were vast molecular diversity found in begomoviruses which infect cucurbit crops. Further, these viruses are not confined to particular geographical locations but are reported from distinct fields around the globe, which are geographically separated. They showed remarkably diverse molecular arrangements. Synergism, transcomplementation, and transreplication showed chances of mixed infection for production of recombinant strain. These findings will help in identifying and developing cucurbit cultivars with stable resistance against begomoviruses infection. Indeed, of their impact on agriculture, begomoviruses provide significant yield loss to vegetable crops, qualitatively and quantitatively. Furthermore, epidemiology of these viruses is important constraint to develop comprehensive integrated disease management system. Altogether, molecular characterization of viruses and their recombination profile is of utmost requirement to develop quick and reliable resistance strategies against virus outbreak. An in-depth knowledge of their host range, symptom diversity, and infection pattern will be useful to control their spread and minimize yield losses.

KEYWORDS

- **Begomovirus**
- **cucurbits**
- **geminiviruses**
- **mixed infection**
- **molecular diversity**
- **pathosystem**
- **weed**

REFERENCES

Adams, M. J. et al. Ratification Vote on taxonomic Taxonomic Proposals to the International Committee on Taxonomy of Viruses. *Arch. Virol.* **2016,** *161,* 2921–2949.

Adkins, S.; Polston, J. E.; Turechek, W. W. Cucurbit Leaf Crumple Virus Identified in Common Bean in Florida. *Plant Disease* **2009,** 93 (3), 320.

Al Rwahnih, M.; Alabi, O. J.; Westrick, N. M.; Golino, D.; Rowhani, A. Description of a Novel Monopartite Geminivirus and its Defective Subviral Genome in Grapevine. Phytopathology. **2017,** 107, 240–251.

Ali-Shtayeh, M. S.; Jamous, R. M.; Mallah, O. B.; Abu-Zeitoun S. Y. Molecular Characterization of Watermelon Chlorotic Stunt Virus (WmCSV) from Palestine. Viruses **2014,** 6(6), 2444-2462.

Anderson, P. K.; Cunningham, A. A.; Patel, N. G.; Morales, F. J.; Epstein, P. R.; Daszak, P. Emerging Infectious Diseases of Plants: Pathogen Pollution, Climate Change and Agrotechnology Drivers. **2004,** Trends Ecol. Evol. **2004,** 19(10), 535-544.

Bananej, K.; Dafalla, G. A.; Ahoonmanesh, A.; Kheyr-Pour, A. Host Range of an Iranian Isolate of Watermelon Chlorotic Stunt Virus as Determined by Whitefly-Mediated Inoculation and Agroinfection, and its Geographical Distribution. J. Phytopathol. **2002,** 150, 423-430.

Briddon, R. W. et al. Diversity of DNA β, a Satellite Molecule Associated with Some M Begomoviruses. Virology. **2003,** 312(1), 106-121.

Briddon, R. W.; Markham, P. G. Cotton Leaf Curl Virus Disease. Virus Res. **2001,** 71(1-2), 151-159.

Briddon, R.W.; Brown, J. K.; Moriones, E.; Stanely, J.; Zerbini, M.; Zhou, X.; Fauquet, C. M. Recommendation for the Classification and Nomenclature of the DNA β Satellites of Begomoviruses. Arch. Virol. **2008,** 153(4), 763-781.

Brown, J. K.; Idris, A. M.; Rogan, D. Melon Chlorotic Leaf Curl Virus, a New Begomovirus Associated with Bemisiatabaci Infestations in Guatemala. Plant Dis. **2001,** 85 (9), 1027.3.

Brown, J. K.; Idris, A. M.; Alteri, C.; Stenger, D. C. Emergence of a New Cucurbit-Infecting Begomovirus Species Capable of Forming Reassortants with Related Viruses in the Squash Leaf Curl Virus Cluster. Phytopathol. **2002,** 92(7), 734-742.

Brown, J. K. et al. Revision of Begomovirus Taxonomy Based on Pairwise Sequence Comparisons. Arch. Virol. **2015,** 160 (6), 1593-1619.

Brown, J. K.; Fauquet, C. M.; Briddon, R. W.; Zerbini, M.; Moriones, E.; Navas-Castillo, J. Geminiviridae. In Virus Taxonomy—Ninth Report of the International Committee on Taxonomy of Viruses. Eds. King, A.M.Q., Adams, M.J., Carstens, E.B. and Lefkowitz, E.J. Associated Press, Elsevier Inc.; London, 351–373.

Capoor, S. P.; Ahmad, R. U. Yellow Vein Mosaic Disease of Field Pumpkin and its Relationship with the Vector, Bemisiatabaci. Indian Phytopathol. **1975,** 28, 241–246.

Chigurupati, P.; Sambasiva R. K.; Jain, R. K.; Mandal B. Tomato Leaf Curl New Delhi Virus is Associated with Pumpkin Leaf Curl: A New Disease in Northern India. Ind. J. Virol. **2012,** 23 (1), 42–45.

Dorman, M.; Nelson, S. Root-Knot Nematodes on Cucurbits in Hawaiʻi. Plant Disease, College of Tropical Agriculture and Human Resource, **2012,** PD-84, 1–5.

Fauquet, C. M.; Briddon, R. W.; Brown, J. K.; Moriones, E.; Stanely, J. Geminivirus Strain Demarcation and nomenclature. Arch. Virol. **2008,** 153(4), 783–821.

García-Arenal, F.; Fraile, A.; Malpica, J. M Variability and Genetic Structure of Plant Virus Populations. Annual Rev. Phytopathol. **2001,** 39, 157–186.

Goodman, R. M. Single-stranded DNA Genome in a Whitefly Transmitted Plant Virus. Virol. **1977,** 83(1), 171–179.

Guzman, P.; Sudarshana, M. R.; Seo, Y. S.; Rojas, M. R. A New Bipartite Geminivirus (Begomovirus) Causing Leaf Curl and Crumpling in Cucurbits in the Imperial Valley of California. Plant Dis. **2000,** 84 (4), 488.3.

Howarth, A. J.; Vandemark, G. J. Phylogeny of Geminiviruses. J. Gen. Virol. **1989,** 70(10), 2717–2727.

Idris, A. M.; Mills-Lujan, K.; Martin, K.; Brown, J. K. Melon Chlorotic Leaf Curl Virus: Characterization and Differential Reassortment with Closest Relatives Reveal Adaptive Virulence in the Squash Leaf Curl Virus Clade and Host Shifting by the Host-Restricted bean calico mosaic virus. J. Virol. **2008,** 82(4), 1959–1967.

Jaiswal, N.; Saritha, R. K.; Datta, D.; Singh, M.; Dubey, R. S.; Rai, A. B.; ai, M. Mixed Infections of Begomoviruses in Pumpkins with Yellow Vein Mosaic Disease in North India. Arch. Phytopathol. Plant Protection.**2012**, 45(8), 938–941.

Jyothsna, P.; Rawat, R; Malathi, V. G. Molecular Characterization of a New Begomovirus Infecting a Leguminous Weed Rhynchosiaminima in India. Virus Genes **2011**, 42 (3), 407–14.

Kangquan, Y.; Ting, H.; Guang, L.; Tianyuan, C.; Ying, W.; Alice, Y.; Yu, L.; Yule, L. A Geminivirus-Based Guide RNA Delivery System for CRISPR/Cas9 Mediated Plant Genome Editing. Sci. Reporter 2015, 5, 14926, 1–10.

Khan, J. A.; Siddiqui, M. R.; Singh, B.P. Association of Begomovirus with Bitter Melon in India. Plant Dis. **2002**, 86, 328.

Khan, M.S.; Raj, S. K.; Singh, B. P. Some Weeds as New Hosts of Geminivirus as Evidenced by Molecular Probes. Ind. J. Plant Pathol.**2003**, 21,82–85.

Legg, J. P.; Fauquet, C. M. Cassava Mosaic Geminivirus in Africa. Plant Mol. Biol. **2004**, 56 (4), 585–99.

Leke, W. N.; Khatabi, B.; Fondong, V. N.; Brown, J. K. Complete Genome Sequence of a New Bipartite Begomovirus Infecting Fluted Pumpkin (Telfairiaoccidentalis) Plants in Cameroon. Arch. Virol. **2016**, 161(8), 2347–2350.

Mandal, B.; Mandal, S.; Sohrab, S. S.; Pun, K. B.; Varma, A. A New Yellow Mosaic Disease of Chayote in India. Plant Pathol. **2004**, 53, 797.

Mansoor, S.; Briddon, R. W.; Zafar, Y.; Stanely, J. Geminivirus Disease Complexes: An Emerging Threat. Trends in Plant Sc. **2003**, 8 (3), 128–134.

Maruthi, M. N.,; Rekha, A. R.; Muniyappa, V. Pumpkin Yellow Vein Mosaic Disease is Caused by Two Distinct Begomoviruses: Complete Viral Sequences and Comparative Transmission by an Indigenous Bemisiatabaci and the Introduced B-biotype. EPPO Bull. **2007**, 37, 412–419.

Mayer, R. T.; Inbar, M.; McKenzie, C. L.; Shatters, R.; Borowicz, V.; Albrecht, U.; Powell, C. A.; Doostdar, H. Multitrophic Interactions of the Silverleaf Whitefly, Host Plants, Competing Herbivores, and Phytopathogens. Arch. Insect Biochem. Physiol. **2002**, 51, 151–169.

Mnari-Hattab, M.; Zammouri, S; Hajlaoui, M. R. First Report of Hard Watermelon Syndrome in Tunisia Associated with Tomato Yellow Leaf Curl Virus Infection. New Dis. Reports **2014**, 30, 7.

Mnari-Hattab, M.; Zammouri, S.; Belkadhi, M. S.; Bellon Doña, D.; ben Nahia, E.; Hajlaoui, M. R. First Report of Tomato Leaf Curl New Delhi Virus Infecting Cucurbits in Tunisia. New Disease Reports **2015**, 31, 21.

Moffat, A. S. Geminiviruses Emerge as Serious Crop Threat. Sci.**1999**, 286 (5446), 1835.

Morales, F. J.; Anderson, P. K. The Emergence and Dissemination of Whitefly-Transmitted Geminiviruses in Latin America. Arch. Virol. **2001**, 146 (3), 415–441.

Mubin, M.; Shahid, M. S.; Tahir, M. N.; Briddon, R. W.; Mansoor, S. Characterization of Begomovirus Components from a Weed Suggests that Begomoviruses may Associate with Multiple Distinct DNA Satellites. Virus Genes. 40 (3), 452 – 57.

Muhire, B. M.; Varsani, A.; Martin, D. P. SDT: A Virus Classification Tool Based on Pairwise Sequence Alignment and Identity Calculation. PLoS ONE **2014**, 9(9), 1–8.

Muniyappa, V.; Maruthi, M. N.; Babitha, C. R.; Colvin, J.; Briddon, R.W.; Rangaswamy, K. T. Characterization of Pumpkin Yellow Vein Mosaic Virus from India. Ann. Appl. Biol. **2003**, 142, 323–331.

Nagendran, K.; Satya, V. K.; Mohankumar, S.; Karthikeyan, G. Molecular Characterization of a Distinct Bipartite Begomovirus Species Infecting Ivy Gourd (Cocciniagrandis L.) in Tamil Nadu, India. Virus Genes **2016**, 52 (1), 146–151.

Pelham, J.; Fletcher, J. T.; Hawkins, J. H. The Establishment of a New Strain of Tobacco Mosaic Virus Resulting from the Use of Resistant Varieties of Tomato. Annals Appl. Biol. **1970**, 75, 293–297.

Raj, S. K.; Khan, M. S.; Singh, R.; Kumari, N.; Prakash D. Occurrence of Yellow Mosaic Geminiviral Disease on Bitter Gourd (Momordicacharantia) and its Impact on Phytochemical Contents. Int. J. Food Sci. Nutr. **2005**, 56 (3), 1–7.

Raj, S. K.; Snehi, S. K.; Khan, M. S.; Tiwari A. K.; Rao G. P. Molecular Identification of Ageratum enation virus Associated with Mosaic Disease of Pointed Gourd (Trichosanthesdioica Roxb.) in India. Phytoparasitica **2011**, 39, 497–502.

Raj, S. K.; Snehi, S. K.; Khan, M. S.; Tiwari, A. K.; G. P. Rao. First report of Pepper Leaf Curl Bangladesh Virus Strain Associated with Bitter gourd (momordicacharantia l.) Yellow Mosaic Disease in India. Australasian Plant Dis.e Notes **2010**, 5, 14–16.

Raj, S. K.; Singh, B. P. Association of Geminiviral Infection with Yellow Mosaic Disease of Cucumissativus: Diagnosis by Nucleic Acid Probes. Ind. J. Experimental Biol. **1996**, 34, 603–605.

Rajeshwari, R.; Reddy, R. V. C.; Maruthi, M. N.; Colvin, J.; Seal, S. E.; Muniyappa, V. Host Range, Vector Relationships and Sequence Comparison of a Begomovirus Infecting Hibiscus in India. Annals Appl. Biol. **2005**, 147(1), 15–25.

Rajinimala, N.; Rabindran, R. First Report of Indian Cassava Mosaic Virus on Bitter Gourd (Momordicacharantia) in Tamil Nadu, India. Australasian Plant Dis. Notes. **2007**, 2, 81–82.

Raychaudhuri, S. P.; Verma, J. P. Therapy by Heat, Radiation and Meristem Culture, In Plant Diseases–An Advanced Treatise; Horsfall, J. G.; Cowling, E. B., Eds.; Acad. Press, New York, 1977.

Shahid, M. S.; Ikegami, M.; Briddon, R. W.; Natsuaki, K. T. Characterization of Tomato Yellow Leaf Curl Virus and Associated Alphasatellite Infecting Cucurbita maxima in Japan. J. Gen. Plant Pathol. **2015**, 81, 92.

Singh, A.; Mishra, K. K.; Chattopadhyay, B.; Chakraborty, S. Biological and Molecular Characterization of a Begomovirus Associated with Yellow Mosaic Vein Mosaic Disease of Pumpkin from Northern India. Virus Genes **2009**, 39 (3), 359–370.

Snehi, S. K.; Raj, S. K.; Prasad, V.; Singh, V. Recent Research Findings Related to Management Strategies of Begomoviruses. J. Plant Pathol. Microbiol. **2015**, 6, 273, 1–12.

Sohrab, S. S.; Mandal, B.; Ali, A.; Varma, A. Molecular Diagnosis of Emerging Begomovirus Diseases in Cucurbits Occurring in Northern India. Ind. J. Virol. **2006**, 17, 88–95.

Sohrab, S. S.; Mandal, B.; Ali, A.; Varma A. Chlorotic Curly Stunt: A Severe Begomovirus Disease of Bottle Gourd in Northern India. Ind. J. Virol. **2010**, 21, 56–63.

Subramanya S. K. Plant Virus and Viroid Diseases in the Tropics, 2014, Springer.

Suresh, L. M.; Malathi, V. G.; Shivanna, M. B. Molecular Detection of Begomoviruses Associated with a New Yellow Leaf Crumple Disease of Cucumber in Maharashtra, India. Ind. Phytopathol. **2013**, 66(3), 294–301.

Tahir, M.; Haider, M. S.; Briddon, R. W. Complete Nucleotide Sequences of a Distinct Bipartite Begomovirus, Bitter Gourd Yellow Vein Virus, Infecting Momordicacharantia. Arch. Virol. **2010**, 155, 1901–1905.

Thresh, J. M.; Cooter, R. J. Strategies for Controlling Cassava Mosaic Virus Disease in Africa. Plant Pathol. **2005**, 54(5), 587–614.

Tiwari, A. K.; Rao, G. P.; Khan, M. S.; Pandey, N.; Raj, S. K. Detection and Elimination of Begomovirus Infecting Trichosanthesdioica (pointed gourd) Plants in Uttar Pradesh, India. Arch. Phytopathol. Plant Protect. 45 (9), 1–6.

Tiwari, A. K.; Snehi, S. K.; Singh, R.; Raj, S. K.; Rao, G. P.; Sharma, P. K. Molecular Identification and Genetic Diversity Among Six Begomovirus Isolates Affecting Cultivation of Cucurbitaceous Crops in Uttar Pradesh, India. Arch. Phytopathol. Plant Prot. **2011,** 45 (1), 1–7.

Tiwari, A. K.; Snehi, S. K.; Rao, G. P; Raj, S. K. Begomoviruses: A Major Problem for Cucurbitaceous Crops in Eastern Uttar Pradesh. Ind. J. Virol. **2008,** 19 (1), 23–25.

Varma, A.; Malathi, V.G. Emerging Geminivirus Problems: A Serious threat to Crop Production. Annals of Appl. Biol. **2003,** 142 (2), 145–64.

Varma, P. M. Ability of the Whitefly to Carry More than One Virus Simultaneously. Current Sci. **1955,** 24, 317–318.

Varsani, A. et al. Capulavirus and Grablovirus: Two New Genera in the Family Geminiviridae. Arch. Virol. DOI: 10.1007/s00705-017-3268-6.

Vasudeva, R. S.; Lal, T. B. A Mosaic Disease of Bottle Gourd. Ind. J. Agri. Sci. **1943,** 3, 182–191.

Zahir, A.; Aala, A.; Ali, I.; Shakila, A.; Manal T.; Mahfouz, M. M. CRISPR/Cas9-Mediated Viral Interference in Plants. Genome Biol. **2015,** 16, 238, 1–11.

Zerbini, F. M. et al. ICTV Virus Taxonomy Profile: Geminiviridae. J. Gen. Virol. **2017,** 98(2), 131–133.

CHAPTER 15

Diversity of Potyviruses and Their Extent in Vegetable Pathosystem

BHAVIN S. BHATT[1], SANGEETA RATHORE[2], BRIJESH K. YADAV[2], FENISHA D. CHAHWALA[2], BIJENDRA SINGH[3], and ACHUIT K. SINGH[3*]

[1]*Department of Environmental Science, Shree Ramkrishna Institute of Computer Education and Applied Sciences, Surat 395001, Gujarat, India*

[2]*School of Life Sciences, Central University of Gujarat, Gandhinagar 382030, Gujarat, India*

[3]*Division of Crop Improvement, ICAR—Indian Institute of Vegetable Research, Varanasi 231304, Uttar Pradesh*

[*]*Corresponding author. E-mail: achuits@gmail.com*

ABSTRACT

In plant virus group family, *Potyviridae* represents the second largest family with single-strand positive sense RNA genome. They infect wide range of plants (monocotyledonous and dicotyledonous) such as fruits, vegetables, ornamental plants, and fiber crop those are now gradually being reported on new hosts of the tropical and subtropical regions of the world. They are characterized morphologically by filamentous virions (11×20 nm diameters) and contain either monopartite or bipartite linear single-strand positive sense, polygenic RNA (+RNA) genome of 9.3–10.0 kb in length. Among the family *Potyviridae*, viruses belonging to genus *Potyvirus* are most destructive. They are transmitted through aphid in nonpersistent and noncirculative manner. They cause diseases mainly in food, fiber, and ornamental plants. According to available database, potyvirus infection is more prevalent in crops belong to *Solanaceae*, *Fabaceae*, and *Cucurbitaceae* families. Infection of potyvirus produces different symptoms those were alike of other plant virus symptoms. Symptoms include mosaic, moteling, stipe, chlorosis, necrotic lesions, vein yellowing or vein clearing, stunting, and wilting, ultimately leading to quantitative yield loss. This review discusses the family *Potyviridae,* host range, and molecular biology of Potyvirus lifecycle. It also covers recent findings on the roles of potyviral proteins and its importance in successful disease incidence and progression.

15.1   INTRODUCTION

The ssRNA containing potyvirus distribution occurs worldwide, which is associated with severe disease complex in most significant vegetables crops as per economic point of view. Vegetables are the main source of dietary, micronutrients, and phytochemicals for living organisms but abiotic and biotic factors limit the consumption and production worldwide. Consumption of vegetable per capita is far below from the required intakes of recommended dietary micronutrient in Asia. Plants are affected by many pathogenic microorganisms such as fungi, bacteria, and viruses. The potyvirus are transmitted through aphid vector and cause significant crop loss in the field condition. There are many plant viruses that occur on the different crop grown in the fields. After severe infection, crop yield reduces up to 47% annually because of viral infection and in cases of severe infection, it goes up to 100%. The potyvirus infection goes up to 60–90 % in the fields. They can infect one plant to another plant and spread on whole field to whole region. This type of event occurs because of transmission of viruses.

The plant disease name often depends on development of symptoms in specific plant species. Plants such as leaf curl of papaya and leaf roll of potato are disease symptoms that cause particular kinds of distortion in leaf. Some of the plant viruses are not infected to only one particular plant (host) species, but may cause disease in several varieties of plants including tomato, pepper, cucumber, and tobacco infected by mosaic viruses. Some other viruses such as brome mosaic virus frequently cause disease in grains, bamboos, and grasses. Certain ornamental crops such as geraniums, tulips, dahlias, roses, Easter lilies are infected by viruses. Variety of symptoms are associated with virus infections such as reduced growth, stunting or dwarfing, mosaic pattern of yellow and green on the leaves, yellow streaking of leaves, yellow spotting on leaves, deformity of leaves or growing points, flower color breaking, cup-shaped leaves, yellowing of veins, leaf curling, and other deformities.

In this way, the most devastating plant viruses pathogens are in the field condition where they can cause huge economical loss. There are numerous types of plant viruses that exist in the world. These types include molecular structure, genomic nature, and mode of transmission, etc.; mainly, there are two types of viruses present, RNA virus and DNA virus. DNA viruses can be double-stranded or single-stranded DNA. Next one is RNA viruses, they are also double stranded or single stranded in nature. DNA viruses which affect the plants include Nanoviridae and Geminiviridae, where RNA viruses include TMV, potyvirus, etc. Potyviridae possesses the second biggest family among all RNA plant viruses. Here, in this chapter, we will discuss about the potyvirus infecting economically important vegetable crops.

15.2   POTYVIRUS

In Northern America, potyvirus was first revealed on the peas culture in Wisconsin and Washington in 1968. The outbreak of this disease was observed on USDA lines in 1973 in California (Rao and Huriki, 1985). In United Kingdom, Kenneth Smith was studying on potato viruses in the 1920s, and potato virus Y (PVY) was one of them. Afterwards, similar properties were shown in some other viruses. Later, in 1971, Harrison et al. suggested an acronym "potyvirus" for this group of viruses for an experimental period of 5 years. Potyviruses disease distribution occurs worldwide, which cause economically important crop damage. *Potyviridae* family comprises monopartite and bipartite plant viruses. Potyvirus is the largest genus with 160 species, which contains some economically important viruses such as PVY, plum pox virus, bean yellow mosaic virus, and Papaya ring spot virus. Morphology of these virions are nonenveloped, curved, filamentous capsid with pitch value about 3.4 nm and 11–20 nm diameter for helix. As per amino acid sequences of their coat proteins, the family *Potyiviridae* is categorized into eight genera, namely, potyvirus, poacevirus, macluravirus, ipomovirus, tritimovirus, rymovirus, brambyvirus, and bymovirus. Except bymovirus, all genera are monopartite in nature with particle lengths 650–950 nm (Table 15.1). Members of the genus bymovirus comprises bipartite with particles of two lengths of genome that are 250–300 and 500–600 nm. Among all genera of the *Potyviridae* family, only bymovirus contains a single type of genome which is 9.3–10.8 kb length positive sense, ssRNA (Fig. 15.1). In genome organization, a VPg (24 kDa) is covalently attached to the 5'-terminus nt. A polyadenylate tract contains 20–160 adenosines residues occur at the 3' terminus region. Genus bymoviruses contain two types of positive sense, ssRNA genomes (RNA-1 length 7.3–7.6 kb and RNA-2 length 3.5–3.7 kb). Both RNAs contain polyadenylate tracts at 3'-terminal and a VPg present at the 5' terminus. Virions comprise single type of coat protein of 28.5–47 kDa.

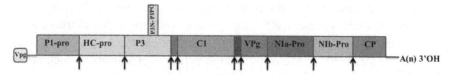

FIGURE 15.1   Genome organization of *Potyviridae*.

TABLE 15.1   Characteristics Features of Members of the Eight Genera in the Family *Potyviridae*.

Genus	Type species	Size of genome	Nature of genome	Name of vector	Host range	No. of total species
Potyvirus	*Potato virus Y*	9.4–11.0	Monopartite	Aphids	Wide	160
Rymovirus	*Ryegrass mosaic virus*	9.4–9.5	Monopartite	Eriophydid mites	Gramineae	3
Bymovirus	*Barley yellow mosaic virus*	RNA1: 7.2–7.6 RNA2 : 2.2–3.6	Bipartite	Chytrid fungus (*Polymyxa graminis*)	Gramineae	6
Macluravirus	*Maclura mosaic virus*	8.2	Monopartite	Aphids	Wide	8
Ipomovirus	*Sweet potato mild mottle virus*	9.0–10.8	Monopartite	Whiteflies	Wide	6
Tritimovirus	*Wheat streak mosaic virus*	9.2–9.6	Monopartite	Eriophydid mites	Gramineae	6
Poacevirus	*Triticum mosaic virus*	9.7–10.2	Monopartite	Wheat curl mite (*Aceria tosichella*)	Gramineae and Orchidaceae	3
Brambyvirus	*Blackberry virus Y*	10.8	Monopartite	Aerial vector	Rubus species	1

15.2.1 TAXONOMY OF POTYVIRIDAE

Recombination, pseudorecombination, mutation, transcomplementation synergism, etc. are molecular basis of profound diversity found in viruses. *Potyviridae* represent as the largest family among all plant viruses, its taxonomy and nomenclature is becoming complex because of rapid evolution of viruses and more numbers of sequence are being submitted in NCBI, GenBank. A committee was established as the International Committee on Taxonomy of Viruses (ICTV) for creating a universal taxonomic system for all the viruses infecting animals (vertebrates, invertebrates, and protozoa), plants (higher plants and algae), fungi, bacteria, and archaea. Over the previous two decades, the most appropriate and conventional way to classify the viruses is as per their name and arranged in viral entities order in a more relevant way and accepted universally. Viruses are categorized based on nature of genomic components, single or double stranded, positive or negative sense nucleic acid either RNA or DNA into approximately ~2284 species of viruses, viroid, divided into 349 genera, 87 families, 19 subfamilies, and 6 orders (ICTV 9th report). On the basis of biological criteria, molecular data and mainly vectors through transmission of viruses are found in plants (Adams et al., 2005, Gibbs and Ohshima, 2010). Demarcation criteria of genus are <46% nucleotide similarity for whole open reading frame but this given criteria does not distinct the genus rymoviruses from potyviruses, these are divided by means of their vectors. The threshold value for species demarcation criteria for nucleotide is set at <76% similarity, while for amino acid, the threshold for species distinction is set at <82% similarity. For species demarcation, the thresholds using nucleotide similarity values range from 58% for P1-coding region to 74–78% for the individual-coding regions.

15.2.2 MODE OF TRANSMISSION

Members of the *Potyviridae* are spread by different organisms. Genus potyvirus and macluravirus members are transmitted by aphid which transmits in a nonpersistent and noncirculative manner. Aphid transmissions are required a helper component and a specific coat protein amino acid triplet (i.e., DAG for potyviruses). Transmission occurs in the genera rymoviruses, poaceviruses, and tritimoviruses that are through eriophyid mites in a manner of semipersistent. Transmission of genus bymoviruses found by root-infecting vectors (order Plasmodiophorales), ipomoviruses genus are transmitted by whiteflies and vector of brambyvirus is unknown.

15.2.3   INFECTION CYCLE OF POTYVIRUS

Even plant viruses code a number of crucial proteins like movement proteins, coat proteins, replication enzymes, which is responsible for many functions; on the other hand, their capacity of coding is very limited that is why each step of infection cycle they must depend on their host factors. Infection cycle of potyvirus starts when viral particles enter into the plant host cells through wound or feeding stage of aphids. Then, these virus particles disassemble inside the host cell, resulting in the release of its RNA into the cytoplasm. Positive sense single-stranded RNA viruses (ssRNA+) of potyviruses require eIF4E host factor to complete their infection cycle (Wang and Krishnaswamy, 2012; Wang, 2015). These host factors are encoded by small multigene family and act as redundant manner in plants. Cellular mRNA contains cap structure while potyvirus has single virus-coded protein (VPg) attached covalently at 5' NTR. The plant initiation factors of translation like eIF4E and eIF(iso)4E interacted with virus-encoded protein (VPg) was reported (Wittmann et al., 1997; Leonard et al., 2000, 2004). The VPg role was not essential for able cap-independent translation mechanism of turnip mosaic virus (TuMV) disease (Basso et al., 1994; Niepel and Gallie, 1999). First time interaction was identified between the VPg protein and eIFiso4E of the TuMV in yeast two-hybrid screen (Wittmann et al., 1997); after that this is crucial for virus infection cycle (Leonard et al., 2000).

Recent studies showed that VPg protein is a crucial component for infection establishment. It acts as translation initiator protein. VPg protein of tobacco etch virus (TEV) and TuMV was found to interact with factor eIF(iso)4E of their susceptible host *Arabidopsis thaliana* (Wittmann et al., 1997; Lellis et al., 2002). Experimental studies also showed that *Arabidopsis* mutants with compromised expression of eIF(iso)4E factor lost their suscep-tibility to TuMV and TEV (Lellis et al., 2002). Hence, VPg plays role in the translation initiator protein which may be serving as recruitment site for translation initiator factors and consequently 40S ribosomal subunit (Lellis et al., 2002). Thus, Vpg has an analogous role to 5' cap structure of mature mRNA of eukaryotic cell.

15.2.4   HOST RANGE

Family potyviridae show a narrow range of host, few number of members infect upto only 30 families of plants, and some of the important crop are listed in Table 15.2. These potyviruses are spread through aphids in a nonpersistently manner and also transmissible by mechanical inoculation in the host plants. Some potyviruses are not transmitted by aphids vector because of mutation that occur within the helper component or CP cistrons. In some of the cases, distribution is also assisted by seed transmission mechanism.

TABLE 15.2 Potyviruses Isolated and Characterized from Diverse Host and Their Accession Nos.

Sr. no.	Virus name	Gen bank, NCBI accession no.	Host plant
1	Leek yellow stripe virus	KP168261	Leek
2	Soybean mosaic virus	KM979229	Soybean
3	Papaya ringspot virus W	EU475877	Papaya
4	Banana bract mosaic virus isolate TRY	HM131454	Banana
5	Papaya ringspot virus P isolate DEL	EF017707	Papaya
6	Bean yellow mosaic virus isolate CK-GL5	KF155420	Bean
7	Bean yellow mosaic virus isolate CK-GL4,	KF155419	Bean
8	Bean yellow mosaic virus isolate CK-GL3	KF155414	Bean
9	Bean yellow mosaic virus isolate CK-GL1	KF155409	Bean
10	Papaya ringspot virus isolate HYD	KP743981	Papaya
11	Cowpea aphid-borne mosaic virus isolate RR3	KM597165	Cowpea
12	Bean yellow mosaic virus isolate Vfaba2	JN692500	Bean
13	Onion yellow dwarf virus isolate RR1	KJ451436	Onion
14	Dasheen mosaic virus isolate CTCRI-II-14	KT026108	Dasheen
15	Zucchini yellow mosaic virus isolate AP Gherkin	KT778297	Zucchini
16	Dasheen mosaic virus isolate T10	KJ786965	Dasheen
17	Bean common mosaic virus strain NL-1n	KF114860	Bean
18	Narcissus yellow stripe virus	KU516386	Narcissus
19	Watermelon mosaic virus isolate RKG2	KM597071	Watermelon
20	Watermelon mosaic virus isolate RKG	KM597070	Watermelon
21	Vanilla distortion mosaic virus isolate VDMV-Cor	KF906523	Vanilla
22	Papaya ringspot virus isolate PRSVR3	KJ755852	Papaya
23	Chili veinal mottle virus isolate ChiVMV	GU170808	Chili
24	Chili veinal mottle virus isolate ChiVMV-Ch-Jal	GU170807	Chili
25	Bean common mosaic virus isolate NL7n	KY057338	Bean
26	Cyrtanthus elatus virus A isolate NBRI16	KX575832	Cyrtanthus
27	Sugarcane streak mosaic virus isolate TPT	GQ246187	Sugarcane
28	Cowpea aphid-borne mosaic	KM655833	Cowpea
29	Sugarcane streak mosaic virus isolate IND671	JN941985	Sugarcane
30	Lily mottle virus	GU440578	Lily

15.3   STATUS OF POTYVIRUS IN INDIA

As per NCBI, a total of approximately 55 potyvirus are recorded from India. These potyviruses infect a diverse host and therefore it emerged as an important pathogen for crop loss in India. Peanut stripe potyvirus occur predominantly in the groundnut growing state such as Andhra Pradesh, Gujarat, Karnataka, Maharashtra, and Tamil Nadu. During extensive survey in the year 1992–1994, observed peanut stripe potyvirus are more prominently present in other two states of Uttar Pradesh and Rajasthan, India (Jain et al., 2000). Chili (*Capsicum annuum* L.) crop infected by pepper veinal mottle virus results in significant yield loss (Nagaraju and Reddy, 1980). Potyvirus infection on chili was reported from Uttar Pradesh in 1997 and 1998. Distribution of disease was observed throughout eastern Uttar Pradesh and the incidence ranged from 5–75%. (Prakash et al., 2002). Chili veinal mottle virus was identified as causal agent of this disease. During survey, in 2013 and 2014, leaf samples of chili plants showing symptoms such as mottling, stunting, crinkling and malformed, and discolored fruits were collected from Punjab, India. Pepper mottle virus is responsible for disease and symptoms development to chili, which is an important and spice crop (Sukhjeet et al., 2014). Tuberose (*Polianthes tuberosa* L.) is an ornamental plant belonging to *Agavaceae* family affected by potyvirus. Occurrence of tuberose mild mosaic virus on tuberose plant in India was reported (Raj et al., 2009). Peace lily is commonly grown as foliage ornamental which is affected by Spathiphyllum chlorotic vein banding virus (SCVbV) from Andhra Pradesh, India (Padmavathi et al., 2011). From South India, a virus related with severe mosaic pattern disease of gherkin (*Cucumis anguria* L.) was identified (Srinivasulu et al., 2010). Gherkin is a significant cucurbitaceous vegetable crop utilized for slicing and pickling, which is grown in the southern states of India like Karnataka, Andhra Pradesh, and Tamil Nadu. Soybean (*Glycine max* L.) also infected by Tobacco streak virus exhibiting necrosis symptoms samples were collected from different regions of Maharashtra, India (Kumar et al., 2008). During survey in 2010, chlorotic spots on leaves, mosaic pattern, and floral deformations symptoms were observed in gladiolus plants. Potyvirus affect both quality and quantity of gladiolus flowers, significantly reducing the production (Kaur et al., 2011). Recently, the detection of jasmine potyvirus was reported in jasmine from Uttar Pradesh, India (Kaur et al., 2013). Jasmine (family *Oleaceae*) is a chief commercial floricultural crop which is infected by Jasmine yellow mosaic Virus-Andhra Pradesh

(JaYMV-AP), member of potyvirus (Sudheera et al., 2014). *Cyrtanthus elatus virus A* (CEVA,) a member of potyvirus, cause disease *Narcissus tazetta* L. cv. Paperwhite (family *Amaryllidaceae*), which is an ornamental plant (Kumar et al., 2015). The medicinally important herb *Amaranthus gracilis* was affected by potyvirus where disease incidence was observed 37.3% in India (Jahan et al., 2015). Dasheen mosaic virus (DsMV) is a member of potyvirus causing mosaic disease development in elephant foot yam (*Amorphophallus paeoniifolius*) in Kerala, India (Kamala et al., 2015).

15.3.1 DISEASE PREVALENCE OF POTYVIRUS

In China, potyvirus emerged as the significant plant pathogen causing reductions in the sweet potato yield. Because of the high rate of recontamination, in Vietnam, use of *Potato leaf roll virus* (PLRV)-free potato tuber seeds were not as much successful. Only in one or two season, the potyvirus infection reached up to 60–90% in the fields. In the case of cassava, viruses transmission from one plant to another plant occurred majorly by cutting method. The virus does not produce any apparent symptoms on the majority of potato cultivars. A few develop mosaic symptoms and virus particles are spread by transient vectors in a nonpersistent manner. Watermelon mosaic virus (formerly Watermelon mosaic virus 2, family *Potyviridae*) is one of the most damaging viral pathogen affecting production of watermelon in the United States and worldwide.

15.3.2 POTYVIRUSES INFECTION IN CROP

Potyviruses are the maximum devastating crop pathogen and found worldwide. Their transmission occurs by aphids and cause significant crop damage. Potyviruses infect species of utmost angiosperm taxa in all of the temperate and tropical climates. Infection of potyvirus is highly extant in *Solanaceae*, *Leguminosae*, and *Cucurbitaceae* families in India. Potyviruses infection were reported in crops *Solanum tuberosum* (potato), *Solanum lycopersicum* (tomato), tobacco (*Nicotiana tabaccum*) and pepper (*Capsicum annuum* L.), eggplant (*Solanum melongena*), cowpea (*Vigna unguiculata*), common bean (*Phaseolus vulgaris*), papaya (*Carica papaya*), soybean (*Glycine max*), til (*Sesamum indicum*), and ornamental plants (Gladiolus, Jasminum) (Shukla et al., 1994; Mali et al., 1980; Bhat et al., 1999; Sreenivasulu and Gopal, 2010; Babu et al., 2012; Sharma et al., 2013). Potyviruses can cause diversity of symptoms in infected host plants, comprising mosaic pattern on leaves (potato), ringspots (papaya), mottling (chili), necrotic or chlorotic lesions (cowpea, common bean), flower breaking (lily), necrosis, stripe stunting, wilting, vein clearing. Stunting and yield losses, which are most common with a combination of several symptoms, are usually occurring in both natural and experimental plants. Some strains of potyvirus can cause severe stunting, yellowing of leaf, systemic necrosis, petiole and stem necrosis, leaf curling and fruit distortion, terminal necrosis and defoliation which generally lead to plants death. Symptoms severity is determined by the various factors such as particular virus strain, specific host genotype, environmental conditions, and vectors.

Plum pox virus (PPV; genus potyvirus) is also of great economic impact as result to most destructive diseases of stone fruit occuring worldwide (Sochor et al., 2012). PPV can cause damage not only to plums (*Prunus domestica* L.) but also to apricots (*Prunus armeniaca* L.), peaches (*Prunus persica* L.), and nectarines (*Prunus persica* var. *nucipersica*) (Balan et al., 1995; Badenes et al., 1996). Potyviruses disease increase day-by-day because of recombination, mutation and are known to infect new host. Potyviruses have wide host range. For example, watermelon mosaic virus can infect 23 families of dicotyledonous and found naturally in cucurbit and legume crops. In tropical and subtropical regions including Indian subcontinent mainly, disease is caused by the papaya and watermelon biotypes of Papaya ring spot virus (PRSV-P & W) in papaya and cucurbits, respectively, resulting in noteworthy yield losses observed (Varma, 1988; Gonsalves, 1998). Continuously new hosts are reported regarding to potyvirus (Chen et al., 2003; Csorba et al., 2004; Arli-Sokmen et al., 2005; Parry and Persley, 2005; Morris et al., 2006). Important potyviruses and their accession number is listed in Table 15.2. Phylogenetic and molecular evolutionary relationships of potyviruses are shown in phylogram (Fig. 15.2).

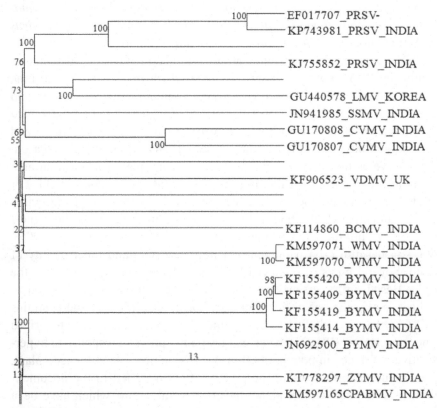

FIGURE 15.2 Phylogenetic tree of representative full length potyvirus genomes reconstructed using the complete nucleotide sequences. The sequences used for analysis were recovered from the GenBank, NCBI database and are identified by their accession no. and strain group affiliation. The scale bar represents a distance of 0.01 substitutions per site. The bootstrap values (1000 bootstrap resamplings) are pointed out on the branches.

15.4 ECONOMIC IMPORTANT CROPS AFFECTED BY POTYVIRUS

15.4.1 POTATO VIRUS Y (PVY)

Round the year, tropical and subtropical climates offer an appropriate climate for growth of vegetation and insects (vectors). Several of the crops serve as the reservoir for the plant viruses and insect vectors. The important viruses which infect potatoes include *Potato Virus X* (PVX; Genus *Potexvirus*), *Potato virus Y* (PVY; Genus *Potyvirus*), *Potato virus S* (PVS; Genus *Carlavirus*), *Potato leaf roll virus* (PLRV; Genus *Polerovirus*), *Potato virus*

V (PVV; Genus *Potyvirus*), *Andean potato latent virus* (APLV; Genus *Tymovirus*), *Andean potato mottle virus* (APMV; Genus *Comovirus*), *Potato virus T* (PVT; Genus *Trichovirus*), *Potato yellow mosaic virus.*

PVY is the most significant pathogen affecting potato crop damages worldwide. PVY shows a wide host range which infects more than nine families including 14 genera of *Solanaceae.* Tomato, pepper, tobacco, and eggplant are reported to be infected by PVY. PVY infection communicated by aphids species in a nonpersistently way which requires acquisition and inoculation times of less than 1 min (Bradley, 1954). In noncropping season, weeds serve as the reservoir and provide breeding platform, form an area for colonization earlier for the migration of aphids in the potato field. For aphids PVY type, member of the genus potyvirus together with potato virus A (PVA) and potato leaf roll virus (genus Polerovirus) emerged as the highest risk to potato production worldwide and reached crop yield loss up to 90%. PVY is a harmful crop pathogen which affects yield and tuber quality of potato. Several types of distinct PVY strains have been recognized in the world. Based on hypersensitive response, which strains were infected to potatoes is classified as ordinary or common strains (PVYO) and C strains (PVYC) (Mascia et al., 2010). Some other types of strains were categorized as necrotic strains (PVYN), whereas they do not provoke response to hypersensitive resistance in potato, but they cause vein necrosis in tobacco plant. The recombination events in the potatoes field have led to the documentation of a number of new pathotypes of several strains which have been differentiated like PVYN and PVYO, represented as PVYZ, PVYNTN, and PVYN Wilga causing different symptoms and level of symptoms expression on many commercial potato cultivars. PVY attack on other economically important solanaceous crops tomato (*Solanum lycopersicum* Mill.), tobacco (*Nicotiana species.*), and pepper (*Capsicum annuum* L.). PVY also infect to nonsolanaceous weeds which belongs to families *Amaranthaceae, Leguminosae, Chenopodiaceae,* and *Compositae.*

15.4.2 SOYBEAN MOSAIC VIRUS (SMV)

The molecular characteristic of SMV disease has been reported from several North American and East Asian countries like USA, Japan, Canada, Korea, China (Seo et al., 2009). As per NCBI (2016), 64 complete genome of SMV has been submitted. From India, only few reports have been done on SMV related to only molecular breeding but not at molecular level. The occurrence of SMV disease was reported on the basis of transmission mode, host range, and physical characteristics from New Delhi (Nariani and Pingaley, 1960), Uttar Pradesh (Singh et al., 1976), and Karnataka (Naik and Murthy, 1997). Soybean (Glycine max L.) crop is infected by tobacco streak virus in different locations of Maharashtra, India (Kumar et al., 2008). Seed borne nature of SMV was observed around 6–8%. After infection, soybean plants carried out typical mosaic symptoms such as dark and light green patches on upper leaves (Balgude et al., 2012). Some strains of SMV induce typical disease symptoms like systemic necrosis, yellowing leaf, terminal necrosis, severe stunting, and defoliation which lead to death of plants. In soybean, crop reduce yield upto 94% in severe condition. Sequencing data indicated that soybean sample showed maximum 99.0% nucleotide and amino acid identities for both domains with earlier reported isolate Ar13 of SMV from Iran. Recently, partial characterization of SMV based on CI and NIb-CP domains has been reported first time from Meghalaya, India. It has been possible that SMV disease introduced by seeds in northeast India (Banerjee et al., 2014). SMV disease also transmit mechanically. Among the vectors such as aphids and whiteflies, SMV disease transmitted by only aphids (*Myzus persicae* Sultz.) from soybean to soybean plants (80%).

15.4.3 SWEET POTATO VIRUS

Viruses emerged as main biotic factor on sweet potato (*Ipomoea batatas* (L.) Lam) production all over the world. During survey in 2005, disease occurrence was detected up to 10–60% on sweet potato growing fields and symptomatic plants exhibiting such as puckering, feathering, vein clearing and chlorotic spots, leaf curl with chlorotic and pink spots collected from Kerala, Orissa, and Andhra Pradesh. Vegetables of sweet potatoes are propagated from vines, root slips, or storage roots from their specific fields. Thus, if theses sweet potatoes viruses are existing in field for 1 year, in future they may be transmitted by propagation material resulting decrease in the yield. The sweet potato is very susceptible to infection by way of *sweet potato feathery mottle virus* (SPFMV). Numerous viruses infect sweet potatoes all over the world (Clark and Moyer, 1988; Cohen et al., 1997; Moyer and Salazar, 1989; Moyer and Larsen, 1991). There are many viruses which infect sweet potatoes including *sweet potato leaf curl virus* (SPLCV) (Green et al., 1992), *sweet potato chlorotic stunt virus* (Cohen, et al., 1992; Hoyer et al., 1996) and *Ipomoea crinkle leaf curl virus* (Cohen et al., 1997). Other five of these potyviruses have been reported in Japan, *Sweet potato feathery mottle* virus (SPFMV), *Sweet potato latent virus* (SPLV), *Sweet potato leaf curl virus* (SPLCV), Sweet *potato symptomless virus* (SPSLV), and *Sweet potato virus* G (SPVG) (Usugi et al., 1991, 1994; Sakai et. al., 1997).

15.5   POTYVIRUS INFECTION IN INDIA

Potyvirus infection is prevalent to range of economically important horticultural crops in India, including potato, eggplant, chili pepper, mungbean, cow pea, french bean, papaya, cucurbits, and groundnut. All such viruses have caused significant yield losses (Bhat et al., 1999; Sreenivasalu and Gopal, 2010). Potyviral infection symptoms of green mottling and distortion of leaves on chili plants was observed in Uttar Pradesh with highest disease incidence in Gorakhpur, upto 75% (Prakash et al. 2002). Nagaraju and Reddy, 1980 also reported pepper veinal mottle virus from south parts of India. Peanut stripe potyvirus infection on groundnut was prevalent in groundnut growing states of India, namely, Gujarat, Maharashtra, Tamil Nadu, and Karnataka (Jain et al., 2000). Kumar et al., 2008 reported incidence of tobacco streak virus infecting soybean (Glycine max L.) with typical disease symptoms of necrosis.

15.6   CONCLUSION

Vegetables such as sweet potato, chili, tomato, potato, cucurbits, etc. are shorty life cycle plants and are cultivated in overlapping seasons which increase the virus inoculum round the year. This led to a scenario of mixed infection under natural conditions. Mixed infections have biological and epidemiological effects as diverse viruses in a single host can exchange their genetic material and may lead to novel recombinant virus species. In the field conditions, polyphagous nature of aphid vector, genetic variation properties of viruses, association with other virus molecules probably increase the emergence of severe epidemics of disease.

Potyviruses multiplication inside host cell is complex process and divided under several cellular compartments. As other viruses, during early stages of an infection, potyvirus genome take a charge of host cellular machinery for its own multiplication. Polygenic +RNA serve as a template for both transcription (new RNA strand formation) and translation (viral protein formation). It is still an area of research what factors determine fate of newly synthesized RNA, whether it will be recruited for new RNA formation or serve as a template for protein formation or directed toward plasmodesmata for intercellular transport or encapsidated and release as new virus particles or targeted for degradation. It is also unknown how polygenic protein is cleaved into respective viral protein inside host cells and targeted to various organelles as per their functions.

Although we have many potyvirus sequence on hand, it is of utmost importance to identify molecular biology of potyvirus infection. The results of research finding will efficiently establish resistance strategies against potyvirus infection by targeting apex molecules that will efficiently block virus multiplication.

KEYWORDS

- **aphid**
- **filamentous virion**
- **pathosystem**
- **positive sense RNA**
- **potyvirus**
- **emerging viruses**

REFERENCES

Adams, M. J.; Antoniw, J. F.; Beaudoin, F. Overview and Analysis of the Polyprotein Cleavage Sites in the Family Potyviridae. *Mol. Plant Pathol.* **2005,** *6* (4), 471–487.

Arli-Sokmen, M.; Mennan, H.; Sevik, M. A.; Ecevit, O. Occurrence of Viruses in Field-Grown Pepper Crops and Some of Their Reservoir Weed Hosts in Samsun, Turkey. *Phytoparasitica.* **2005,** *33,* 347–358.

Babu, B.; Hegde, V.; Makeshkumar, T.; Jeeva, M. L. Rapid and Sensitive Detection of Potyvirus Infecting Tropical Tuber Crops Using Genus Specific Primers and Probes. *African J. Biotechnol.* **2012,** 11, 1023–1027.

Badenes, M. L.; Asins, M. J.; Carbonell, E. A.; Glacer, G. Genetic Diversity in Apricot, Prunusarmeniaca, Aimed at Improving Resistance to Plum Pox Virus. *Plant Breeding* **1996,** *115,* 133–139.

Balan, V.; Ivascu, A.; Toma, S. Susceptibility of Apricot, Nectarine and Peach Cultivars and Hybrids to Plum-Pox Virus. In: XVIth International Symposium on Fruit Tree Virus Diseases, Rome, 299–305.

Balgude, Y. S; Sawant, D. M; Gaikwad, A. P. Transmission Studies of Soybean Mosaic Virus. *J. Plant Dis. Sci.* **2012,** *7,* 52–54.

Banerjee A.; Chandra, S.; Swer, E. K. P.; Kumar S.; Sharma S. First Molecular Evidence of Soybean Mosaic Virus (SMV) Infection in Soybean from India. *Australasian Plant Dis. Notes* **2014,** *9,* 150.

Basso, J.; Dallaire, P.; Charest, P. J.; Devantier, S.; Laliberate, J. F. Evidence for an Internal Ribosome Entry Site Within the 5' Non-translated Region of Turnip Mosaic Potyvirus RNA. *J. Gen. Virol.* **1994**, *75*, 3157–3165.

Bhat, A. I.; Varma, A.; Pappu, H. R.; Rajamannar, M.; Jain, R. K.; Praveen, S. Characterization of a Potyvirus from Eggplant (Solanummelongena) as a Strain of Potato Virus Y by N-terminal Serology and Sequence Relationships. *Plant Pathol.* **1999**, *48*, 648–654.

Bradley, R. H. E. Studies of the Mechanism of Transmission of Potato Virus Y by the Green Peach Aphid, Myzuspersicae (Sulz.). *Canadian J. Zool.* **1954**, *32*, 64–73.

Chen, C. C.; Ko, W. F.; Lin, C. Y.; Jan, F. J.; Hsu, H. T. First Report of Carnation Mottle Virus in Calla Lily (Zantedeschia spp.). *Plant Dis.* **2003**, *87*, 1539.

Clark, C. A.; Moyer, J. W. Compendium of Sweet Potato Disease. *Am. Phytopathol. Soc.* **1988**, 74.

Cohen, J.; Franck A.; Vetten, H. J.; Lesemann, D. E; Loebenstein G. Closterovirus-like Particles Associated with a Whitefly-transmitted Disease of Sweet Potato. *Annals of Appl. Biol.* **1992**, *121*, 257–268.

Cohen, J.; Milgram, M.; Antignus, Y.; Pearlsman, M.; Lachman, O.; Loebenstein, G. Ipomoea Crinkle Leaf Curl Caused by a Whitefly-transmitted Gemini-like Virus. *Annals of Appl. Biol.* **1997**, *131*, 273–282.

Csorba R.; Kiss, E. F.; Molnar, I. Reactions of Some Cucurbitaceous Species to Zucchini Yellow Mosaic Virus (ZYMV). *Comm. Agri. Appl. Biol. Sci.* **2004**, *69*, 499–506.

Gonsalves, D. Control of Papaya Ringspot Virus in Papaya: A Case Study. *Annual Rev. Phytopathol.* **1998**, *36*, 415–437.

Green, S. K.; Luo, C. Y.; Wu, S. F. Elimination of Leaf Curl Virus of Sweet Potato by Meristem Tip Culture, Heat and Ribavirin. *Plant Protection Bull.* **1992**, Taiwan, 34, 1–7.

Harrison, B. D.; Finch, J. T.; Gibbs, A. J.; Hollings, M.; Shepherd, R. J. Sixteen Groups of Plant Viruses. *Virology* **1971**, *45*, 356–363.

Hoyer, U.; Maiss, E.; Jelkmann, W.; Lesemann, D. E.; Vetten, H. J. Identification of the Coat Protein Gene of a Sweet Potato Sunken Vein Closterovirus Isolate from Kenya and Evidence for a Serological Relationship Among Geographically Diverse Closterovirus Isolates from Sweet Potato. *Phytopathology* **1996**, *86*, 744–750.

Jahan, T.; Khan, A. A.; Naqvi, Q. A.; Mishra, V.; Khan, M. S. First Report of Distinct Potyvirus Associated with Amaranthusgracilis L. in India. *Agrica.* **2015**, *3*, 40–43.

Jain, R. K.; Lahiri, I.; Varma A. Peanut Stripe Potyvirus: Prevalence, Detection and Serological Relationships. *Ind. Phytopathol.* **2000**, *53*, 14–18.

Kamala, S.; Makeshkumar, T.; Sreekumar, J.; Chakrabarti, S. K. Whole Transcriptome Sequencing of Diseased Elephant Foot Yam Reveals Complete Genome Sequence of Dasheen Mosaic Virus. *Virology Reports* **2015**, *5*, 1–9.

Kaur, C.; Kumar, S.; Snehi, S. K.; Raj, S. K. Molecular Detection of Jasmine Potyvirus Associated with Yellow Mosaic Symptoms on Jasminumsambac L. in India. *Arch. Phytopathol. Plant Protection* **2013**, *46* (9), 1102–1107.

Kaur, C.; Raj, S. K.; Snehi, S. K.; Goel, A. K.; Roy, R. K. Natural Occurrence of Ornithogalum Mosaic Virus Newly Reported on Gladiolus in India. *New Dis. Reports* **2011**, *24*, 2.

Kumar S; Raj R.; Kaur C.; Raj, S. K. First Report of Cyrtanthus Elatus Virus A in Narcissustazetta in India. *Plant Dis.* **2015**, *99*, 1658.

Kumar, N. A.; Narasu, M. L.; Zehr, U. B.; Ravi, K. S. Molecular Characterization of Tobacco Streak Virus Causing Soybean Necrosis in India. *Ind. J. Biotechnol.* **2008**, *7*, 214–217.

Lellis, A. D.; Kasschau, K. D.; Whitham, S. A.; Carrington, J. C. Loss-of-Susceptibility Mutants of Arabidopsisthaliana Reveal an Essential role for eIF(iso)4E during Potyvirus Infection. *Curr. Biol.* **2002**, *12*, 1046–1051.

Leonard, S.; Plante, D.; Wittmann, S.; Daigneault, N.; Fortin, M. G.; Laliberté, J. F. Complex Formation Between Potyvirus VPg and Translation Eukaryotic Initiation Factor 4E Correlates with Virus Infectivity. *J. Virol.* **2000**, 74, 7730–7737.

Leonard, S.; Viel, C.; Beauchemin, C.; Daigneault, N.; Fortin, M. G.; Laliberte, J. F. Interaction of VPg-Pro of Turnip Mosaic Virus with the Translation Initiation Factor 4E and the Poly(A)-binding Protein in Planta. *J. Gen. Virol.* **2004**, *85*, 1055–1063.

Mali, V. R.; Kulthe, K. S. A Seedborne Potyvirus Causing Mosaic of Cowpea in India. *Plant Dis.* **1980**, *64*, 925–928.

Mascia, T.; Finetti-Sialer, M. M.; Cillo, F.; Gallitelli, D. Biological and Molecular Characterization of a Recombinant Isolate of Potato Virus Y Associated with a Tomato Necrotic Disease Occurring in Italy. *J. Plant Pathol.* **2010**, *92* (1), 131–138.

Morris, J.; Steel, E.; Smith, P.; Boonham, N.; Spiegel, H.; Barker, I. Host Range Studies for Tomato Chlorosis Virus, and Cucumber Vein Yellowing Virus Transmitted by Bemisiatabaci (Gennadius). *Eur. J. Plant Pathol.* **2006**, *114*, 265–273.

Moyer, J. W.; Larsen, R. C. Management of Insect Vectors of Viruses Infecting Sweet potato. In: *Sweet Potato Pest Management, A Global Perspective*; Jansson, R. K.; Raman, K. V., Eds.;, Westview Press, Boulder, CO, 1991; pp 341–358.

Moyer, J. W.; Salazar, L. F. Viruses and Virus Like Diseases of Sweet Potato. *Plant Dis.* **1989**, *73*, 451–455.

Nagaraju, R.; Reddy, H. R. Occurrence and Distribution of Bell Pepper Viruses Around Bangalore. *Curr. Res.* **1980**, *10*, 155–156.

Naik, R. G.; Murthy, K. K. V. Transmission of Soybean Mosaic Virus Through Sap, Seed and Aphids. *Karnataka J. Agric. Sci.* **1997**, *10*, 565–568.

Nariani, T. K.; Pingaley, K. V. A Mosaic Disease of Soybean (Glycinemax L.). *Ind. Phytopathol.* **1960**, *13*, 130–136.

Niepel, M.; Gallie, D. R. Identification and Characterization of the Functional Elements Within the Tobacco Etch Virus 5' Leader Required for Cap-independent Translation. *J. Viroly.* **1999**, *73*, 9080–9088.

Padmavathi, M.; Srinivas, K.; Subba R. P.; Ramesh, B.; Navodayam, K.; Krishnaprasadji, J.; Babu, R.; Sreenivasulu, P. Identification of a New Potyvirus Associated with Chlorotic Vein Banding Disease of Spathiphyllum spp., in Andhra Pradesh, India. *Plant Pathol. J.* **2011**, *27* (1), 33–36.

Parry, J. N.' Persley, D. M. Carrot as a Natural Host of Watermelon Mosaic Virus. *Australasian Plant Pathol.* **2005**, *34*, 283–284.

Prakash, S.; Singh, S. J.; Singh, R. K.; Upadhyaya, P. P. Distribution, Incidence and Detection of a Potyvirus on Chili from Eastern Uttar Pradesh. *Ind. Phytopathol.* **2005**, *55*, 294–298.

Raj, S. K.; Snehi, S. K.; Kumar A; Rambabu, T.; Goel, A. K. First Report of Tuberose Mild Mosaic Potyvirus from Tuberose (Polianthestuberosa L.) in India. *Australasian Plant Dis. Notes* **2009**, *4*, 93–95.

Rao, A. L. N.; Huriki C. Clover Primary Leaf Necrosis Virus, a Strain of Red Clover necrotic mottle Virus. *Plant Dis.* **1985**, *69*, 959–961.

Sakai, J.; Yamasaki, S; Ogawa, T; Ohnuki, M; Hanada, K. Detection of Sweet Potato Virus G by RT-PCR from Sweet potato in Kyushu. *Annals Phytopathol. Soc. Japan.* **1997**, *63*, 484.

Seo, J. K.; Ohshima, K.; Lee, H. G.; Son, M.; Choi, H. S.; Lee, S. H.; Sohn, S. H.; Kim, K. H. Molecular Variability and Genetic Structure of the Population of Soybean Mosaic Virus Based on the Analysis of Complete Genome Sequences. *Virology*. **2009**, *393*, 91–103.

Sharma, P.; Sahu, A. K.; Verma, R. K.; Mishra, R.; Choudhary, D. K.; Gaur, R. K. Current Status of Potyvirus in India. *Arch. Phytopathol. Plant Protection.* **2013**, *47* (8), 906–918.

Shukla, D. D.; Ward, C. W.; Brunt, A. A. The Potyviridae. CAB International, Wallingford, UK, 1994.

Singh, B. R.; Singh, D. R.; Saksena, H. K. A Mosaic Disease of Soybean at Kanpur, India. *Scientific Culture*. **1976**, *42* (1), 53–54.

Sochor, J.; Babula, P.; Adam, V.; Krska, B; Kizek, R. Sharka: The Past, the Present and the Future. *Viruses*. **2012**, *4*, 2853–2901.

Sreenivasulu, M.; Gopal, D. Development of Recombinant Coat Protein Antibody Based IC-RTPCR and Comparison of its Sensitivity with Other Immunoassays for the Detection of Papaya Ringspot Virus Isolates from India. *J. Plant Pathol*. **2010**, *26*, 25–31.

Srinivasulu, M.; Sarovar, B.; Anthony , J. A. M.; Sai Gopal, D. V. R. Association of a Potyvirus with Mosaic Disease of Gherkin (Cucumisanguria L.) in India. *Ind. J. Microbiol.* **2010**, *50*, 221–224.

Sudheera, Y.; Vishnu, G. P.; Vardhan, M.; Hema, K.; Krishna R. M.; Sreenivasulu, P. Characterization of a Potyvirus Associated with Yellow Mosaic Disease of Jasmine (Jasminumsambac L.) in Andhra Pradesh, India. *Virus Dis*. **2014**, *25*, 394–397.

Sukhjeet Kaur, S.S. Kang, Abhishek Sharma and Shikha Sharma (2014). First report of Pepper mottle virus infecting Chili pepper in India. New Disease Report, 30, 14.

Usugi, T.; Nakano, M; Shinkai, A; Hayashi, T. Three Filamentous Viruses Isolated from Sweet Potato in Japan. *Annals Phytopatho. Soc. Japan*. **1991**, *57*, 512–521.

Varma, A. The Economic Impact of Filamentous Plant Viruses—The Indian Sub-continent. In: *The Plant Viruses*; Milne, R.G. Ed.. Plenum Press; 1988, 371–376.

Wang, A. Dissecting the Molecular Network of Virus–Plant Interactions: The Complex Roles of Host Factors. *Annual Rev. Phytopathol*. **2015**, *53*, 45–66.

Wang, A.; Krishnaswamy, S. Eukaryotic Translation Initiation Factor 4e-mediated Recessive Resistance to Plant Viruses and its Utility in Crop Improvement: eif4e-mediated Resistance to Plant Viruses. *Mol. Plant Pathol*. **2012**, *13*, 795–803.

Wittmann, S.; Chatel, H.; Fortin, M. G.; Laliberte, J. F. Interaction of the Viral Protein Genome Linked of Turnip Mosaic potyvirus with the Translational Eukaryotic Initiation Factor (iso)4E of Arabidopsisthaliana Using the Yeast Two-Hybrid System. *Virology*. **1997**, *234*, 84–92.

CHAPTER 16

Soilborne Microbes: A Culprit of Juvenile Plants in Nurseries

PUJA PANDEY[1*] and BINDESH PRAJAPATI[2]

[1]*Department of Plant Pathology, Anand Agricultural University, Anand 388110, Gujarat, India*

[2]*Assistant Research Scientist, Directorate of Research, S. D. Agricultural University, Gujarat, India*

Corresponding author. E-mail: pujapandey41124@gmail.com

ABSTRACT

Soilborne diseases are found to be very severe in intensive systems because of frequent cropping. Hence, destruction of the inoculum surviving between crops is beneficial. In the management of soilborne diseases, cultural practices play a very important role. Soilborne diseases play a very critical role in reducing the productivity of new cultivars in some agricultural crops. These diseases are difficult to manage because of their highly heterogeneous incidence and lack of information regarding the epidemiological aspects of these pathogens. Only through study on survival, dissemination of soilborne pathogens, its effective management is possible. Effect of climatic conditions with role of cultural practices and host resistance–susceptibility will play a major role in management of disease.

16.1 INTRODUCTION

Soil inhabiting microbes play a chief role in damaging juvenile plants under nursery. Pre-emergence or post-emergence nature of damage may lead to complete damage. However, many serious diseases are associated with soilborne plant pathogens, which results in damping off, root rot, crown rot, and wilts in various field and horticultural crops, especially in nursery stage. Soilborne diseases are a common problem in nursery production, which leads to sustainable economic loss by attacking the juvenile plants at early stage of life. Practical application of this knowledge will lead to healthier plants and improved plant growth with greater uniformity. It also helps to minimize biosecurity risks. Several serious soilborne plant pathogens are being spread around the world with nursery plants. Many tropical soils harbor numerous soilborne pathogens for several years which often limit crop production. These pathogens includes various fungal, bacterial, and some nematodes. Some soilborne fungi and bacteria such as *Sclerotium rolfsii* Sacc. and *Pseudomonas solanacearum* E. F. Smith have broad host ranges and are difficult to manage because of their ability to reside and persist in soil. Root rots are especially detrimental to nursery plants. Free water (from rain, dew, and irrigation), high levels of humidity, favorable temperatures, and suscep-tible plant tissue from rich fertilization (young, succulent, rapidly growing, unsuberized roots) is always available for infection. Most nursery crops are monocultures with limited genetic diversity and thus they are extremely vulnerable to disease epidemics such as those caused by *Phytophthora* and *Cylindrocladium*. The management of soilborne pathogens may have to rely more heavily on sustainable practices that have been used for centuries in traditional farming systems (Thurston, 1992).

For effective management of soilborne plant pathogens, we should have good knowledge of crop husbandry, physical, chemical, and biological properties of soil, ecology, and epidemiology of root diseases (Naik, 2003). Therefore, this chapter gives an overview about soilborne plant pathogens, their host range, epidemiology, types of symptoms they produce, their distri-bution in relation to climate, and its management.

16.2   SOILBORNE PATHOGENS AND THEIR ECONOMIC IMPORTANCE

Soilborne diseases are major limiting factor in the production. There are two major players in the soilborne diseases: fungi and nematodes (Agrios, 2005). Few groups of bacteria are soilborne because bacteria which are not forming spores cannot survive in soil for a long time. Bacteria require penetration point such as wound or natural opening to penetrate the plant and cause infection,that is, *Ralstonia solanacearum*, cause of bacterial wilt of tomato (Genin and Boucher, 2004). Most nematodes in soil are free-living, consuming bacteria, fungi, and other nematodes, but some can infect plants. Some are migratory ectoparasitic, migratory endoparasitic, and sedentary endoparasites. Fungi are eukaryotic, filamentous, multicellular, heterotrophic organisms that produce a network of hyphae called the mycelium and absorb nutrients from the surrounding substrate (Alexopoulos et al., 1996). Oomycetes produce zoospores and contain cellulose as component of cell wall. The type of parasitism of almost all soilborne fungi is necrotrophic.

Pathogens survive in the soil in modified form, such as sclerotia, chlamydospores, thick walled conidia, and in-crop residues (Bruehl, 1987). When climatic conditions are favorable, the fungus germinates by root or seed exudates and chemotactically grows toward the plant. The germ tube attach to the surface of the root penetrate and infect the epidermal cells of the root tips, secondary roots, and root hairs, or attack the emerging shoots and radicles of seedlings. Fungi use cell wall degrading enzymes and mechanical turgor pressure, and colonize the root cortex. After the roots have been killed and the fungus ramifies through the cortex, it reproduces and forms spores within the root tissue. Mycelium spread up the root, internally or externally. Few pathogens that cause wilt diseases, for example, *Fusarium oxysporum*, *Verticillium dahlia*, can penetrate through the endodermis into the vascular tissue and move up the xylem to above-ground parts of the plant (Beckman, 1987).

16.3   CRITERIA OF DEFINING PLANT PATHOGENS

Soilborne plant pathogens can be defined as those which undertake the major part of their life span in the soil. Because of their microscopic size and the nonspecific symptoms of an infection, these organisms are out of mind of the growers. They infect roots or stem bases in nursery of vegetable crops and results in an economic loss.

16.3.1   CRITERIA DEFINING FUNGI

Garrett (1956, 1970) has suggested two ways of grouping the soilborne fungi (Table 16.1). One is based on their mode of parasitism and divides them into specialized and nonspecialized parasites. The other is based on ecological criteria and groups them into soil inhabitors (specialized parasites) and soil invaders (true soil inhabitors).

TABLE 16.1   Difference Between Soil Invading and Soil Inhabiting Pathogens.

Soil invading or root inhabiting (specialized parasites)	Soil inhabiting (nonspecialized parasites)
Invade only living root tissues and having limited host range.	Colonize dead plant tissue as competitive saprophytes and unable to invade living, undamaged root tissues with wide host range.
Survive in the soil for relatively shorter periods.	Survive in soil for relatively longer periods.
Fungi	**Fungi**
Verticillium dahaliae	*Rhizoctonia solani*
Fusarium oxysporum	*Pythium* spp.
Protozoa	*Phytopthora* spp.
Plasmodiphora brassicae	*Fusarium* spp.
Spongospora subterranae	*Sclerotium rolfsii*
Synchytrium endobioticum	*Macrophomina phaseolina*
Bacteria	**Bacteria**
Erwinia Spp.	*Pseudomonas* spp.
	Xanthomonas spp.
	Burkholderia solanacearum
	Agrobacterium spp.
	All plant parasitic nematodes

16.3.2   CRITERIA DEFINING BACTERIA

Fewer soilborne diseases are caused by bacteria than by fungi and very few of them are capable of saprophytic survival in soil. Budenhagen (1965) has grouped the soilborne by dividing them into three groups of the soilborne phase in their life cycle.

1.   A pathogen whose population is produced almost exclusively in the host but the bacterium may overwinter in the host residues in the soil.
2.   A pathogen that builds up its population in the host and is returned to the crop residues where the population declined only gradually.
3.   Bacteria whose populations are produced in the soil and have only a transitory relation to plant disease.

Pathogen causing angular leaf spot of cotton (*Xanthomonas campestris* pv. *malvacearum*) has limited survival in crop residues and can be readily controlled by crop rotation, although *Agrobacterium tumefaciens* (now *Rhizobium radiobacter*) have the capacity for saprophytic survival.

16.3.3   CRITERIA DEFINING NEMATODES

Phytonematodes, the hidden enemies of crops, are real threat to successful and profitable cultivation of field and horticultural crops. Endoparasitic nematodes such as *Meloidogyne* spp. and *Heterodera* spp. are dreadful causing havoc to modern agriculture. *Meloidogyne* spp. is noticed to attack over 2500 hosts. The most preferred hosts are dicot vegetables, pulses, and fruit crops (Naik and Rani, 2008). The feeding of nematodes on crop roots increases nutrient leakage and provides wounds through which soilborne plant pathogens can gain entry to the roots.

16.4   CULPRITS OF SEEDLINGS DISEASES

Various plant pathogens attack nursery seedlings and cause damping off, root rots, root knot, and seedling decay which includes some fungi, bacteria, and nematodes.

16.4.1   FUNGAL PATHOGENS

Pythium: Species of *Pythium* cause seed rot and damping off of seedlings of all cultivated crops. In India, serious problem of *Pythium* is only in rainy season in nursery conditions. Root rot caused by *Pythium* spp. is a common crop disease. When this pathogen kills newly emerged seedlings, it is known as damping off. This disease makes complex with other pathogens such as *Phytophthora* and *Rhizoctonia*. Pythium wilt is caused by zoospore infection in older plants, leading to biotrophic infections that become necrotrophic in response to colonization pressures or environmental stress (Jarvis, 1992).

Rhizoctonia: This is an important species infecting 500 plant species. It causes different types of diseases viz., stem blight, stalk rot, fruit rot, seedling decay, and leaf blight in crop plants (Dhingra and Sinclair, 1975). The fungus survives in many weed hosts. This fungus is prevalent in arid, subtropical, and tropical climates. The fungus survives in the soil as hyphae and sclerotia. This fungus does not produce any spores except in its sexual stage, which occurs very rarely. *Rhizoctonia solani* can survive in the soil for several years as sclerotia. It has thick outer layers for survival and acts as the overwintering structure of the pathogen. The fungus is attracted toward plant by chemical stimuli released by a growing plant and/or decomposing plant residues (Ref: https://en.wikipedia.org/wiki/*Rhizoctonia_solani*).

Sclerotium: This fungus survives and attacks plants near the soil line. Before it penetrates host tissue, it produces mass of mycelium on the plant surface. Pathogen produces an enzyme which deteriorates the host's outer cell layer and it penetrates within it which leads to decaying of tissue, colonization of mycelium and the formation of *sclerotia. Sclerotium* is able to survive for a wide range of environmental conditions. The optimum pH range for mycelial growth is 3.0–5.0, and for *sclerotial* germination is 2.0–5.0. Germination is inhibited at pH above 7.0. Maxiumum mycelial growth occurs between 25°C and 35°C but it retards at 10°C or 40°C. Mycelium is generally killed at 0°C, but *sclerotia* can survive up to −10°C temperatures. High moisture is required for optimal growth of this fungus. Sclerotia germinate best at relative humidity of 25–35% (http://www.extento.hawaii.edu/Kbase/crop/type/s_rolfs.htm).

Phytopthora: It is one of the important genera of soilborne fungi belonging to oomycetes group. These are primary invaders of plant tissue with limited saprophytic ability. They are usually associated with root rots of established plants but are also involved in damping off. These species enter the root tips and result into water soaked brown to black rot similar to *Pythium*. The disease often occurs after wet weather and in low, poorly drained areas, particularly on clay, compacted soils.

Macrophomina: The genus *Macrophomina* is represented by single species, *phaseolina. Macrophomina phaseolina* (Tassi) Goidanich causes root rot, seedling blight, and charcoal rot of many crop species (Smith and Carvil, 1977). It is an important pathogen where high temperatures and water stress occurs during the growing season.

Fusarium: The greatest number of fungal wilt diseases in higher plants is caused by *Fusarium. Fusarium oxysporum* generally produces symptoms such as wilting, chlorosis, necrosis, premature leaf drop, browning of the vascular system, stunting, and damping off (Agrios, 1988). *F. oxysporum* f. sp. *melonis* attacks muskmelon. It causes damping off in seedlings and causes chlorosis, stuting, and wilting in older plants. Necrotic streaks appear on the stem portions (Booth, 1971). The optimum temperature for growth under laboratory condition is between 25°C and 30°C, and the optimum soil temperature for infection in root is 30°C.

Verticillium: The genus *Verticillium* was established in 1816 by Nees Von Esenbeck on the basis of the characteristic of verticillate conidiophores. Verticillium wilt is a disease of over 350 species of eudicot host plants caused by total six species of Verticillium genus (*V. dahliae, V. albo-atrum, V. longisporum, V. nubilum, V. theobromae,* and *V. tricorpus*). *Verticillium* can survive cold weather. The modified resting structures of *Verticillium* are able to survive freezing, thawing, heat shock, dehydration, and are quite vigorous and difficult to get rid of. During extended periods of anerobic conditions, it becomes difficult for resting structures to survive. *Verticillium* will grow best between 20°C and 28°C. Water is essential requirement for germination of resting structures (Pegg, 2002).

16.4.2 BACTERIA PATHOGENS

Pseudomonas: *Pseudomonas syringae* pv. *glycinea causes bacterial blight of soybean.* Bacterial blight infects on all above-ground plant parts, infected seedlings becomes stunted and sometimes killed in severe cases. Plants can be infected at any time during the growing season. Disease development is favored by cool, wet weather, and rain storms. Disease progress stops during dry, hot conditions. Bacterial blight is spread through wind and rain (Anon, 2016). Another important disease of rice is panicle blight caused by *Burkholderia glumae* and *Burkholderia gladioli* (Severin) Yabuchi et al. It is a major disease of rice that infects during hot and dry weather. Losses include reduced yield and milling and losses reported in commercial rice fields range from a trace to as high as 70% (Anon, 2015). *Pseudomonas glumae* cause seedling rot of rice (Kurita et Tabei) Tominaga. *Pseudomonas glumae*, which is known as the pathogen of bacterial grain rot of rice, sometimes causes browning on rice seedlings (Uematsu et al., 1976).

Xanthomonas: Black rot of cabbage is usually of minor importance, when conditions are suitable it may become serious and many growers sustain severe economic loss. In some cases, the crop may be a total loss. Black rot of cabbage is caused by *Xanthomonas campestris*. The bacterium enters the plant principally through natural openings such as hydathodes, stomata, and injuries on the leaves. In time, the bacterium spreads in the vascular system of the leaf and stem. The bacteria spread and cause most damage in wet, warm weather. It does not usually spread in dry weather and is inactive at temperatures below 10°C (Anon, 2000).

16.4.3   NEMATODE ACT AS ROOT-ASSOCIATED PARASITE

Root-knot nematode (*Meloidogyne* spp.): Root-knot nematodes are plant-parasitic nematodes from the genus *Meloidogyne*. Worldwide, many host plants are susceptible to infection by root-knot nematodes and they cause approximately 5% of global crop loss (Sasser and Carter, 1985). It leads to the development of root-knots that drain the plant's photosynthate and nutrients. This genus includes more than 90 species (Moens et al., 2009) with some species having several races. Four *Meloidogyne* species (*M. javanica, M. arenaria, M. incognita, and M. hapla*) are major pests worldwide.

Stunt **(*Tylenchorhynchus* spp.):** *Tylenchorhynchus* are soil dwelling stunt nematodes. They can cause stressing or disease in plants. About 8% of the studied species of *Tylenchorhynchus* are parasitic.

Reniform (*Rotaylenchulus* spp.): The reniform nematode (*Rotylenchulus reniformis*) is a species of parasitic nematode of plants with a worldwide distribution in the tropical and subtropical regions. This nematode has a wide host range, infecting many species of host plants around the world.

16.5   HOST RANGE OF SOILBORNE PLANT PATHOGENS

16.5.1   FUNGI

Pythium: *Pythium* has a wide host range and can have an economic impact on the cultivation of peppers, chrysanthemum, soybeans, cucurbits, beets, cotton, and turf-grasses. It is major cause of root rot in papaya production in subtropical areas (Anon, 2016b).

Rhizoctonia: It is one of the fungi responsible for Brown patch (a turfgrass disease), damping off (e.g., in soybean seedlings as well as black scurf of potatoes, bare patch of cereals, root rot of sugar beet, belly rot of cucumber, sheath blight of rice) (Anon, 2016c).

Sclerotium: *Sclerotium* has an extensive host range, almost 500 species in 100 families are susceptible. The most common hosts are the legumes, cucurbits, and crucifers. This fungi is prevalent in arid, subtropical, and tropical climates and other warm temperate regions (Anon, 2016d).

Phytopthora: These are relatively host-specific parasites. *Phytophthora cinnamomi*, although, infects thousand of species ranging from ferns, cycads, conifers, grasses, lilies to members of many dicotyledonous families. Many species of *Phytophthora* are plant pathogens of considerable economic importance such as late blight of potato caused by *Phytopthora infestasns* which caused the Great Irish Famine (1845–1849), and still remains the most destructive pathogen of solanaceous crops, including tomato and potato, and the soybean root and stem rot agent, *Phytophthora sojae* (Anon., 2016e).

Macrophomina: It has a very wide host distribution covering most of the tropics and subtropics, extending well into temperate zones (Songa, 1995).

Fusarium: The fungal pathogen *Fusarium oxysporum* cause disease to a wide variety of hosts of any age. Tomato, tobacco, legumes, cucurbits, sweet potatoes, and banana are a few of the most susceptible plants to this pathogen.

Verticillium: Many economically important plants are susceptible including cotton, tomato, potato, brinjal, pepper, and ornamentals.

16.5.2 BACTERIA

Pseudomonas: *Pseudomonas syringae attacks various crop plants such as peas, beet, beans, and soybean (Anon, 2016f).*

16.5.3 NEMATODE

Root-knot nematode (*Meloidogyne* spp.): Root-knot nematodes are distributed all over the world, and are obligate parasites of the roots of many plant species, including monocotyledonous and dicotyledonous, woody and herbaceous plants. It is a serious problem in nurseries of mostly all vegetable crops during warm climate, resulting into severe losses in nursery of crops.

Tylenchorhynchus spp.: *Tylenchorhynchus* spp. affect many host plants such as soybean, tobacco, tea, oat, alfalfa, sweet potato, sorghum, rose, lettuce, grape, elms, and citrus (Anderson and Potter, 1991). In which, *Tylenchorhynchus claytoni* is a major problems in tobacco growing areas during nursery stage.

Rotaylenchulus spp.: This nematode attacks during nursery stage of vegetable seedlings and has a wide host range, infecting many species of plants around the world including fruit trees, lentil, cotton, pigeonpea, tea, tobacco, soybean, pineapple, banana, tomato, okra, coconut, cabbage, sweetpotato, radish, alfalfa, corn, asparagus, palm, cucumber, eggplant, squash, cassava, guava, melon, and ginger (Robinson et al., 1997).

16.6   EPIDEMIOLOGY OF SOILBORNE DISEASES

Many epidemics of root diseases involving soil fungi depend on the interplay between fungal growth and the spatial and temporal heterogeneity of the soil environment. Colonization of a root occurs at fine scales with growth and movement of fungal mycelia through soil. However, epidemics are observed at coarser scales and depend on a cascading spread through populations of roots.

16.6.1   FACTORS INFLUENCING THE SOILBORNE PLANT PATHOGENS

Moisture: Moisture is of utmost importance for pathogenic fungi and bacteria. Rain splash plays an important role in the dispersal of some fungi and nearly all bacteria, and a period of leaf wetness is necessary for the germination of spores. Propagules germinate in the presence of water and are dispersed. Because the process of germination and infection takes time, the duration of leaf wetness also affects the occurrence of the infection. The duration necessary for infection varies with temperature. Usually, a longer period of leaf wetness is needed to establish an infection in cooler temperatures, as germination and infection are generally accelerated in warmer conditions.

Temperature: Temperature affects the incubation or latent period (the time between infection and the appearance of disease symptoms), the generation time (the time between infection and sporulation), and the infectious period (the time during which the pathogen keeps producing propagules). The period of leaf wetness and temperature information is being used to predict outbreaks of diseases (infection periods) and can be helpful in prophylactic spray.

The edaphic (soil) factor: Soil factors play important role in the development of plant disease. Soilborne diseases are largely affected by amount of **moisture** available to pathogens for germination, survival, and dispersal. Germination and infection process also depend on the temperature of soil. Plant defenses are weakened by nutrient deficiency. The organic matter content of soil also affects the disease development.

Soil pH: Different factors are very important in the occurrence and severity of plant diseases. Plants and pathogens have an optimum pH level for their growth and reproduction. When these optimum pH levels are known for plants and pathogens, diseases could be reduced by changing pH levels to be an optimum to plants not to pathogens.

16.7 DISTRIBUTION OF SOILBORNE DISEASES IN RELATION TO CLIMATE

Climate change is the now the biggest threat of the present century. Climate change is affecting plants in natural and agricultural ecosystems throughout the world (Stern, 2007). The impact of climate change has been observed in many dimensions such as effects on biodiversity, food grain production, insects, and plant diseases. Out of this, impact on plant diseases is one of the important dimension that has to be seen in broader prospective. Plant pathogens are ubiquitous in nature. Therefore, they constitute a fundamental group of biological indicators that needs to be evaluated regarding climate change impacts besides being responsible for production losses and potentially threat to sustainability of an agroecosystem. Changes in environmental conditions are strongly associated with differences in the level of losses caused by a disease because the environment directly or indirectly influences plants, pathogens, and their antagonists. The changes associated with global warming (i.e., increased temperatures, changes in the quantity and pattern of precipitation, increased CO_2 and ozone levels, drought, etc.) thus may affect the incidence and severity of plant disease and influence the further coevolution of plants and their pathogens (Eastburn et al., 2011). Moisture is particularly important to pathogenic bacteria and fungi. The rain duration necessary for infection varies with temperature. Usually, a longer period of leaf wetness needed to establish an infection in cooler temperatures as germination and infection are generally accelerated in warmer conditions.

16.7.1 EFFECTS OF CLIMATE CHANGE ON BACTERIAL PATHOGENS

Debris-borne bacteria survives on and in host tissues. On annual hosts, bacteria such as *Pseudomonas syringae* pv. *phaseolicola* may survive in host debris in soil or on soil surface. On perennial hosts, bacteria such as *Erwinia amylovora* overwinter on infected host tissue, and inoculum get spread from host to host in the next season.. Vector-borne bacterial pathogens, such as *Erwinia stewartii* survive in its insect vectors and these vectors act as the source of primary inoculum in the next Bacterial pathogens, such as *Pseudomonas syringae* pv. tomato and *Xanthomonas campestris* pv. *vesicatoria*, arise from infected seed and possibly also survive in debris and weeds. Bacteria generally spread to their host plants mainly through water, usually in the form of rain splash and insects. Therefore, in summers it is expected with limited bacterial diseases. However, bacteria often enter hosts through wounds and the increase in frequency and intensity of summer storms with high winds, rain, and hail increase wounding of plants and provide moisture for the spread of bacteria (Anon., 2014).

16.7.2 EFFECTS OF CLIMATE CHANGE ON NEMATODES

Majority of plant pathogenic nematodes spends part of their lives in soil and therefore soil is the source of primary inoculums. Lifecycle of a nematode can be completed within 24 weeks under optimum environmental conditions. Temperature is the most important factor and development is slower with cooler soil temperatures. Warmer soil temperatures are expected to accelerate nematode development, perhaps resulting in additional generations per season and drier temperatures are expected to increase symptoms of water stress in plants infected with nematodes such as the soybean cyst nematode. (Anon., 2014).

16.8   SYMPTOMS

The major soilborne pathogens causing root disease in nursery plants are *Phytophthora, Pythium, Rhizoctonia, Chalara, Fusarium* and *Aphanomyces, Sclerotinia,* and *Sclerotium* species. Some root rot pathogens are also well adapted to attack the aerial parts of plants. For example, *Rhizoctonia* also cause leaf spots, shoot and seedling blight and stem cankers as well as root rot and damping off. The symptoms caused by these soilborne root pathogens are remarkably similar. Initially, plants may appear stunted and slow growing. This is followed by leaves yellowing, wilting, and leaf necrosis. Affected root tissue is soft and water soaked, becoming discolored to dark brown or black. Symptoms will vary depending on the pathogen, host plant, and the extent of root rot. It is difficult to determine which pathogen is causing such nonspecific symptoms. These symptoms looks similar as it caused by vascular wilt pathogens (*Verticillium, Fusarium, Ralstonia*). However, vascular wilt pathogens also cause a browning of the vascular bundles in the stem. Plants infected by the bacterium *Ralstonia* can be identified by the presence of bacterial ooze in addition to vascular browning. This ooze may be difficult to detect in some hosts during early stages of infection. Plants can have root infection but not show symptoms until they are stressed. Stresses include transport from the nursery, planting in the field, repotting, or the application of extra water or fertilizer. It is not unusual for plants with root rot to regenerate new roots to replace those killed by the pathogen, but the newly developed root system may only support the plant when adequate moisture is available. Such plants may not show above ground symptoms well enough to be culled in the nursery prior to sale. It is difficult to determine which pathogen is causing such nonspecific symptoms. These symptoms can also be easily confused with those caused by vascular wilt pathogens (*Verticillium, Fusarium, Ralstonia*) and anoxia. However, vascular wilt pathogens also cause a browning of the vascular tissue in the stem (vascular browning). Plants affected by the bacterium *Ralstonia* can be characterized by the presence of bacterial ooze in addition to vascular browning. This ooze may be difficult to detect in some hosts especially in the early stages of infection. Plants can have root infection but not show symptoms until they are stressed. Stresses include transport from the nursery, planting in the field, repotting, or the application of extra water or fertilizer. It is not unusual for plants with root rot to regenerate new roots to replace those killed by the pathogen, but the newly developed root system may only support the plant when adequate moisture is available. Such plants may not show above ground symptoms well enough to be culled in the nursery prior to sale. As several pathogens can produce the same symptoms, and two or more pathogens may be affecting the nursery crop, nurseries need access to a professional diagnostic laboratory. Accurate identification of pathogens is essential to allow the correct decision to be made regarding disease management (Anon., 2016g).

16.8.1   *PHYTOPHTHORA* spp.

Phytophthora spp. generally attacks roots, often entering at injury points (Gray and Hine, 1976). It infects living plant tissue, grows into and between plant cells, and leaves a necrotic lesion behind. Infection by *Phytophthora* usually starts at the root tip and rapidly involves all below ground parts of the plant and it is able to destroy the entire root system of mature plants and also cause trunk cankers near the soil surface.

16.8.2   *PYTHIUM* spp.

Pythium can cause seed rots and death of young seedlings (damping off) but in older plants they generally attack undifferentiated feeder roots and disrupt the uptake of water and nutrients. They may cause a minor disruption in growth but rarely cause mortality in older plants. Most *Pythium* species have a wide host range. Infection of young seedlings by *Pythium* spp. occurs at two points on seedling plants—at the point of emergence from the seed as saprophyte and then move to the healthy seedling tissue of both root and shoot (Pratt and Mitchell, 1976). Pre-emergence damping off is caused by total infection of the seedling prior to emergence. Post-emergence damping off is characterized by a later or, perhaps, slower infection which eventually engulfs the stem of the seedling above ground level and produces collapse of the seedling. *Pythium* infection normally produces a soft wet rot of the roots and/or stem, which turn tan to mid-brown in color. If drier conditions prevail following a late infection of a seedling, the lesion on the stem may appear drier, and collapse of the cell walls give the lower stem a slightly constricted appearance. In a seed box or seed bed, *Pythium* infection is usually seen in patches of plants which may develop symptoms, often in a wetter area of the box. Alternatively, individual plants with slight infections may remain stunted.

The following factors make *Phytophthora and Pythium* spp. an efficient nursery pathogen:

(a) Growing conditions in nurseries where plants are young, richly fertilized, and often overwatered are ideal for water moulds.

(b) Water moulds have the capacity to increase propagule density rapidly. If an infection is successful, another generation of sporangia and zoospores (secondary inoculum) are produced on the host surface within 3–5 days; this rapid cycling of infection followed by another generation of sporangia and zoospores will result in an explosive epidemic if environmental conditions are favorable.

(c) They have the ability to survive adverse conditions without a host plant for long periods.

(d) They may be present in nursery plants long before symptoms appear depending upon host plant susceptibility.

16.8.3   *APHANOMYCES* spp.

The genus contains several destructive root pathogens, the two main ones are *Aphanomyces enteiches* which attacks legume seedlings, and *Aphanomyces cochlioides* which infects beetroot and spinach seedlings. Affected roots are initially grey and watersoaked, eventually becoming soft and brown to black. Disease is favored by warm, wet conditions and slow draining potting mixes. When cultured in a laboratory, it can be distinguished from *Phytophthora* and *Pythium,* which also have nonseptate hyaline hyphae by its delicate, sparse, wandering growth on media.

16.8.4   *Rhizoctonia solani*

R. solani is an important disease causing organism capable of infecting plants at most stages of growth from early germination and seedling stage (where pre- and post-emergence damping off can result) through to mature plants where root and crown rots may develop through *Rhizoctonia* infection. Dingley (1969) recorded the pathogen on grasses, cereals, fodder crops, vegetables (e.g., cabbage, rhubarb, potato), bedding plants (e.g., carnation and anemone), and trees and shrubs (e.g., *Pinus radiate* and *Citrus sinensis).*

The following make *Rhizoctonia* an efficient nursery pathogen:

a) The fungus has a wide host range, is disseminated by contaminated soil, water-splashed sclerotia, contact with diseased plant parts, with dust and by insects, for example, fungus gnats.

b) Rhizoctonia survives as a competitive saprophyte (aggressively colonises organic debris) and produces sclerotia and hyphae in soil and host plant residues. Both propagules can directly infect host tissue.

c) Sclerotes can survive on nursery equipment (pots, labels, and stakes).

d) In addition to damping off and cutting decay, Rhizoctonia can attack seedlings to produce reddish brown lesions just below the soil line. When these are transplanted into the field, lesions enlarge and plants fall over at the soil line.

e) Hyphae from the surface of the potting media can grow up the stem to infect leaves and produce web blight, a serious disease which can spread widely within 1 week. It is important to set plants with adequate spacing to avoid crowding and formation of high.

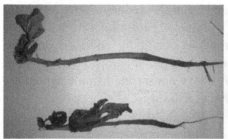

FIGURE 16.1   (See color insert.) Root rot caused by *Rhizoctonia solani* in okra. *Courtesy:* Aravind T., Anand Agricultural University, Anand.

16.8.5   *FUSARIUM* Spp.

A large number of *Fusarium* spp. are present in soil, often occupying a saprophytic role from where susceptible host tissue may be attacked. *Fusarium avenaceum, F. culmorum, F. oxysporum,* and *F. solani* are the more important species involved in diseases of seedlings and cuttings. These species are generally not host specific at the seedling stage and have a wide host range. *Fusarium* attack can result in seedling death. *Fusarium* species associated with plants can be pathogens, endophytes, or saprophytes. They are widespread and can be a significant threat to production nurseries. They can cause damping off, root and bulb rots, crown rots, stem and cutting rots, leaf spots and vascular wilts. Arguably. the two most important species which impact nursery production are *Fusarium oxysporum* and *Fusarium solani.*

Fusarium spp. can cause damping off of seedlings, root, and crown rots as well as vascular wilt diseases. The fungus contains a complex group of strains some of which are pathogenic and others are not. Most pathogenic strains have excellent saprophytic capabilities and can live in alternative hosts as endophytes. Some strains cause a root and crown rot, whereas others colonize the vascular tissue causing discolouration and a characteristic wilt symptom. The latter are placed in forma speciales based on pathogenicity, and strains often have a very narrow host range.

Factors that make *Fusarium* an efficient nursery pathogen:

(a) *Fusarium* species can produce up to five types of propagules that can serve as inoculum sources—macroconidia, microconidia, chlamydospores, ascospores, and mycelium. Conidia and spores are normally dispersed by water splash although some are dispersed by air currents.

(b) Species generally possess excellent saprophytic capabilities, and can survive for long periods as chlamydospores in host tissues, and as nonpathogenic parasites on alternative hosts.

(c) *Fusarium* can be introduced into nurseries in a number of ways but most commonly with infected seed, corms or bulbs, cuttings, and transplants. Seed coats can be contaminated with spores, or pieces of infected plant tissue, or there may be internal seed infection. Infected cuttings can be taken from apparently healthy plants.

16.8.6   *PSEUDOMONAS* Spp.

Wet and soft rot that affects any part of vegetable crops including heads, curds, edible roots, stems, and leaves.

16.9   SURVIVAL OF SOILBORNE PLANT PATHOGENS

a)   Fungi survive as saprobes in last year's infected host plant or other organic debris.

b)   Free-living organisms directly in the soil.

c)   Produce on the host crop that are released during tillage and decomposition-resistant structures:

(i) **Sclerotia:** Withstand temperature extremes, dry conditions, periods in absence of suitable hosts (Figs. 16.2a and b).

FIGURE 16.2   (a) Sclerotia on okra. (b) Sclerotia (microscopic image).
Courtesy: Aravind T., Anand Agricultural University, Anand.

(ii) **Oospores (Oomycetes)** – *Phytophthora* spp., *Pythium* spp., *Aphano-myces* spp. (Fig. 16.3)

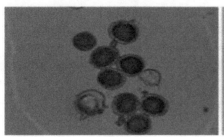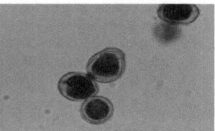

FIGURE 16.3   Oospores.
Courtesy: Puja Pandey., G.B.P.U.A. and T., Pantnagar

(iii) **Chlamydospores** – *Fusarium* spp., oomycetes

16.10   MANAGEMENT OF SOILBORNE PATHOGENS IN NURSERY PRODUCTION

A fundamental principle for disease management in nurseries is that it is better to avoid disease than to have to apply controls after a disease outbreak. Excluding soilborne pathogens from production nurseries can be difficult at times, but is the most cost-effective disease management strategy. Operations based on strict hygiene guidelines can be used to maintain a nursery free of soilborne pathogens. Although nursery diseases are best excluded, the management of a disease outbreak in a production nursery will depend on early recognition and a detailed knowledge of the pathogen and how disease develops.

It is important to ask questions. Where did it come from? How does it spread? What environment and host factors favor development of disease? How does it respond to fungicides and biocontrol agents? As discussed extensively above, soilborne pathogens are most likely to enter the nursery through soil, water, or plant material. The importance of following factors cannot be emphasized strongly enough:

(1) Good hygiene is a fundamental component of effective pest and disease control.

(2) Poorly-drained and aerated potting mixes will aggravate soilborne diseases.

(3) The introduction of plants or cuttings into a production nursery comes with inherent risks.
(4) Take actions to only bring clean plants into the growing area.
(5) Fungicides should not be used to compensate for poor nursery practices and poor plant growth.

There is no "quick fix" to a disease problem in a nursery. Disease control generally requires a holistic approach which combines host resistance, cultural control measures, sensible application of appropriate fungicides to protect healthy plants and biological control measures.

16.10.1   CULTURAL CONTROL

The term "culture" refers to "cultivation," and when we modify or manipulate our cultivation practices, it normally leads to change in the agroecosystem within which the pest develops. This change affects the lifecycle of the pest and thus pest problem reduces. A control method that reduces pest problems by manipulating the cultivation practice is therefore known as cultural control. Subsoiling prior to planting was found increasing the green pea yields of root rot susceptible and tolerant cultivars planted in the soil infested with *F. solani f. sp. pisi* and *Pythium ultimum* (Singh, 2001). It is important that production nurseries produce healthy plants free of soilborne pathogens. This may be difficult and challenging to some nurseries because of the complex disease cycles of the many soil pathogens, with their diverse spore forms, but it is achievable through following practices:

(a) Limit nursery access to essential staff. Those entering a nursery should walk through footbaths which are regularly cleaned and replenished. Nursery staff should then wash their hands with soap and water or an approved hand-washing biocide.
(b) Nursery floors should preferably be sealed with concrete or bitumen. Coarse gravel (75 mm deep) can be used where the ground is well drained. Floors should be washed and drenched with disinfectant solutions on a regular basis.
(c) As some soilborne organisms (e.g., *Rhizoctonia, Phytophthora nicotianae*) can survive in dust, dust from adjacent roads must not enter production or propagating areas.

(d) Ideally, plants should be on raised benches (slatted or with mesh) to provide drainage and aeration, otherwise on raised beds of coarse gravel making sure water does not accumulate in ponds.

(e) Use irrigation water that is free of soilborne pathogens. Surface supplies (dams, rivers) are nearly always contaminated with soil-borne pathogens especially following heavy rainfall and associated surface run-off.

16.10.1.1   FERTIGATION

Fertigation is a method of applying fertilizers, soil amendments, and other water-soluble products required by the plant during its growth stages through drip or sprinkler irrigation system. In this system, fertilizer solution is distributed evenly in irrigation. The availability of nutrients is very high therefore the efficiency is more. In this method, liquid fertilizer as well as water-soluble fertilizers are used (Indu et al., 2015).

16.10.1.1.1   Benefits of Fertigation

(a) **Higher nutrient use efficiency:** Nutrient use efficiency by crops is greater under fertigation as compared with conventional application of fertilizers to the soil.

(b) **Less water pollution:** Intensification of agriculture led by use of irrigation water and indiscriminate use of fertilizers has led to the pollution of surface and ground waters by chemicals. Fertigation helps lessen pollution of water bodies through the leaching of nutrients such as nitrogen (N) and potassium (K) out of agricultural fields.

(c) **Higher resource conservation:** Fertigation helps in saving of water, nutrients, energy, labor, and time.

(d) **More flexibility in farm operations:** Fertigation provides flex-ibility in field operations, for example, nutrients can be applied to the soil otherwise it prohibits entry into the field with conventional equipment.

(e) **Efficient delivery of micronutrients:** Fertigation provides oppor-tunity for efficient use of compound and ready-mix nutrient solu-tions containing small concentrations of micronutrients, which are otherwise very difficult to apply accurately to the soil when applied alone.

(f) **Healthy crop growth:** When fertigation is applied through the drip irrigation system, crop foliage can be kept dry thus avoiding leaf burn and delaying the development of plant pathogens.

(g) **Helps in effective weed management:** Fertigation helps to reduce weed menace particularly between the crop rows. Use of plastic mulch along with fertigation through drip system allows effective weed control in widely spaced crops.

16.10.1.1.2   Chemical and Biological Aspects in Fertigation

Effective fertigation requires an understanding of rate of plant growth including nutrient requirements and rooting patterns, soil chemistry such as solubility and mobility of the nutrients, fertilizers chemistry (mixing compatibility, precipitation, clogging, and corrosion), and the quality of water used especially pH, electrical conductivity, salt and sodium hazards, and toxic ions (Imas, 1999).

16.10.1.2   TILLAGE PRACTICES

Tillage incorporates various types of organic matter including crop residues, manure, green manure, volunteer crop plants, and weeds into the soil. Tillage practices tend to have indirect effects on the spread of plant pathogens, although some forms of inoculums can be widely dispersed by implements. Tillage reduces populations of weeds and volunteer crop plants that harbor pathogens between crops. It also buries plant pathogens from the top soil into deeper layers of the soil where they cause less or no disease. Practices involved in the preparation of seed beds can greatly physical properties of soils such as moisture characteristics, bulk density, aeration and temperature profiles which in turn influence the incidence of disease. Forming the soil into hills, ridges, or raised beds provides better drainage and irrigation. Tillage may also influence nutrient release mechanisms and the total effect is often expressed as increased crop vigor. Healthy plants may be more resistant to some pathogens but the more humid microclimate within the crop can be conducive to the spread of other pathogens. Successive tillage operations can reduce inoculums levels of some pathogens through exposure of the inoculums to desiccation by the sun. Modern agriculture has moved away from regular cultivation of soil between crops toward a system of minimum tillage, or even no-tillage. Minimum tillage is a method of planting crops that involves no seed bed preparation other than opening the soil to place the seed at the intended depth. Usually the soil is not cultivated during crop production and chemicals are used for weed control. Minimum tillage reduces injury to the roots of crop plants caused by mechanical tillage or hand weeding, reducing the opportunities for opportunistic pathogens to infect. It also reduces the spread of pathogens by tillage practices. However, it may favor the development of some diseases such as those that can be controlled by burying inoculums too deep for infection of plant roots to occur. In no-tillage systems, crop residues are left on the surface of the soil. The residues may become a food source for pathogens or alter the physical environment occupied by both the host and its pathogens. The effect of tillage practices on the soil micro flora is often overlooked. Many of these organisms are competitors or antagonists of soilborne pathogens. Minimum tillage practices are thought to promote greater microbial antagonism in the vicinity of cereal roots than normal cultivation practices. However, little is known about the influence of cultivation on these activities although incorporation of organic matter (e.g., green manure crops) is known to reduce the incidence of some diseases. Plant residue increases the microbial activity of the soil which may result in the formation of fungitoxic or even phytotoxic compounds.

16.10.1.3   SOWING AND HARVESTING PRACTICES

Many crop plants tend to be more susceptible to attacks by various parasites at certain stages of their development. Changing the usual sowing time of a crop can exploit weather conditions which are unfavorable for the spread of pathogens or their vectors and reduce crop losses. These practices often necessitate introductions of new cultivars which are adapted to the selected growing period. However, the introduction of a "new" cultivar with different growth characteristics may also create other disease problems. In temperate climates, the crop growing season is largely influenced by the prevailing temperature regime, providing that rainfall is adequate or irrigation is possible. Other strategies may be applied in tropical climates so that a susceptible growth phase occurs in relatively dry periods. Greatest variation in sowing times is possible in greenhouses where considerable manipulation of crop microclimates is practicable. Methods of sowing vary considerably between crops and even with the same crop in different regions. It is therefore difficult to generalize about the effect of sowing methods on the spread of diseases. Farmers aim to optimize germination and early growth at densities which give maximum yields in the particular environment. Depth of sowing is usually determined by seed size and the moisture status of the soil. Deeper sowing may promote germination but it also lengthens the susceptible pre-emergence seedling phase. Smuts and seedling diseases caused by *Fusarium* spp. and *Rhizoctonia* spp. are more serious if seeds are planted deeply. Similarly, potato seed pieces are more readily attacked by *Rhizoctonia* if planted too deeply. Crop density can influence disease development. Generally, as the density of a crop increases, the incidence of disease also increases. For example, in the 1950s, because of an increased demand for seedlings, nurseries expanded in size and increased the density of seedlings in their nursery beds. This practice leads to crowding of the seedlings and serious epidemics of damping off caused by fungi such as *Pythium, Fusarium, Rhtzoctonia*. There are a number of reasons why high plant densities increase disease incidence. Most are associated with the ease of transferring inoculum from one plant to another when the plants are close together. For example, the distance pathogens or their vectors have to travel from one host to the next is reduced, splash dispersal of inoculum is easier and there is also increased contact between the leaves or roots of neighboring plants. In addition, with denser plantings, more plants may be wounded during cultivation, creating more opportunities for weak pathogens to infect. Finally, the microenvironment within the crop is altered. Temperatures are more uniform, the relative humidity increases and leaves stay wet longer after rain or dew. All these conditions favor the development of disease. Sometimes, dense stands may reduce disease incidence and increase yields.

16.10.1.4   IRRIGATION

Irrigation can have a major influence on the spread of some pathogens and on disease development. Irrigation applied during dry seasons means the propagules of pathogens are not exposed to desiccation during periods of drought. Consequently, the level of inoculums increases. The seriousness of this situation is compounded in areas where, because of irrigation, it is possible to grow two susceptible crops in the same field in 1 year. In addition, irrigation water may contain propagules of pathogens which it carries from one place to another unless carefully treated before use. Overhead watering may promote disease by increasing the period of time a layer of free moisture remains on leaf surfaces (leaf wetness period). The longer leaf wetness periods increase the likelihood that sufficient time will be available for fungal spores to germinate, form infection structures, and penetrate the plant surface to the relatively constant and favorable environment within the leaf. Irrigated crops may become a green island in an otherwise dry environment and attract insect vectors of virus diseases. In such cases, it is often better to delay sowing an irrigated crop for some time after other vegetation has dried up and the vector population has been reduced by desiccation. On the other hand, irrigation can be used as a tool to reduce the level of inoculum and to retard disease development. Alternately, drying and rewetting soil encourages the activity of micro-organisms that destroy sclerotia. Overhead irrigation can reduce or inactivate airborne inoculum by washing it out of the atmosphere. Short daily watering encourage the germination of powdery mildew spores but the plants do not stay wet enough for long enough for the fungus to penetrate. Overhead sprinkling of dormant fruit trees reduces the incidence of apple scab because the short-lived ascospores are released in response to temperature changes while the tree is dormant and cannot survive until leaves are present. Flood, furrow, and overhead (spray, sprinkler) irrigation can facilitate the spread of pathogens. Flood irrigation can spread soilborne inoculum all over an area whereas furrow irrigation disperses inoculum along rows. The action of overhead irrigation systems washes inoculum out of the air and facilitates the spread of pathogens that rely on water splash for dispersal. Many important foliage and fruit pathogens such as *Phytophthora infestans* and *Alternaria solani* form their spores at night and release them during the day. Overhead irrigation in the early part of the day washes these spores from the air and splashes them about. If overhead irrigation is delayed until the evening or early night, spores of *Phytophthora infestans* dry out on the plant and cannot infect. However, spores of *A. solani* are resistant to drying out and can survive until dew forms at night so the timing of overhead watering has little influence on disease development. Overhead irrigation plays a role in the distribution of inoculum within a crop as it washes inoculum from higher to lower parts of the plant. Trickle or drip irrigation, developed in response to the need to conserve water, supplies water directly to the root zones of individual plants and the rate of application is insufficient to disperse pathogens. Moreover, drip irrigation produces a mosaic of soil moisture conditions, rather than uniformly moist conditions, which probably inhibits the spread of root pathogens.

16.10.2 BIOLOGICAL CONTROL

Biological control does offer an environmentally friendly approach to the management of soilborne pathogens. However, the introduction of specific organisms that work against pathogens in potting mixes will fail unless it is combined with cultural and physical controls, a degree of host resistance, and limited chemical control. They do not totally suppress disease but can provide a partial level of control. They do not work as quickly as chemical methods. In field studies with eggplant, fruit numbers went from an average of 3.5 per plant to an average of 5.8 per plant when inoculated with Gigaspora margarita mycorrhizal fungi. Average fruit weight per plant went from 258 g to 437 g. A lower incidence of Verticillium wilt was also realized in the mycorrhizal plants (Matsubara et al., 1995).

Mechanism: Understanding the mode of action of a biocontrol agent may improve the consistency of control either by improving the mechanism or by using the biocontrol agents under conditions where it is predicted to be more successful (Zhang et al., 2002). Biocontrol agents involve a bewildering array of mechanisms in achieving disease control. However, the conclusive evidences for the involvement of a particular factor in biological control is determined by the strict correlation between the appearance of factor and the biological control (Handelsman et al., 1989).

According to Punja and Raymond (2003), among the four biocontrol products evaluated against the damping off and root rot of tomato, the biocontrol products were effective in reducing the mortality of tomatoes, with presstop WP and presstop mix being most effective in both the seasons. This was followed by Mycostop. The study revealed that the plant mortality was reduced significantly by pressstop mix andpresstop WP. The reason for the improved activity of both these formulations over the control can be attributed to the mode of action. The mode of action of these formulations is believed to involve mycoparasitism (McQuilken et al., 2001). Also in vitro studies of the fungus was shown to produce cell wall degrading enzyme (McQuilken et al., 2001). A list of antibiotics produced by the bioagents in suppressing the activity if the plant pathogen is given in Table 16.2.

TABLE 16.2 List of Antibiotics.

Sr. no.	Antibiotic	Source	Target pathogen	Disease	Reference
1	2,4 - Diacetyl-pholoroglucinol	*Pseudomonas fluorescence* F113	*Pythium*	Damping off	(Shanahan et al., 1992)
2	Bacillomycin D	*Bacillus amylo liquefaciens* strain FZB42	*Fussorium oxysporium*	Wilt	(Koumoutsi et al., 2004)

TABLE 16.2   *(Continued)*

Sr. no.	Antibiotic	Source	Target pathogen	Disease	Reference
3	Xanthobacin A	*Lyco*bacter sp. Strain K88	*Aphanomyces cochlioides*	Damping off	(Islam et al., 2005)
4	Gliotoxin	*Trichoderma virens*	*Rhizoctonia solani*	Root rot	(Wilhite et al., 2001)
5	Mycostubilin	*Bascillus* BBG100	*Pythium aphanidermatum*	Damping off	(Leclere et al., 2005)
6	Iturin	*Bascillus subtillus* QST713	*Botrytis, Rhizoctonia solani*	Damping off	(Paulitz and Blanger, 2001)

16.10.3   CHEMICAL CONTROL

Fungicides to control soilborne pathogens in the nursery should be used only as a last resort to help prevent the spread of an existing infestation. They rarely eradicate a pathogen or eliminate disease, and should not be used to compensate for poor nursery hygiene and adverse growing conditions. Inappropriate use of chemicals can lead to serious problems further down the supply chain, as chemicals will generally only mask symptoms and the pathogen will still be present. Temporary suppression of symptoms is not control. Chemicals to control soilborne organisms are best used as protectants to healthy plants at vulnerable times, for example, when roots are damaged during repotting.

Since water molds such as *Phytophthora* and *Pythium* do not belong to the true fungi, fungicides used to control soilborne fungi are sometimes ineffective (Drenth and Guest, 2004). The chemicals metalaxyl and phosphonate have been very effective in controlling water molds. Metalaxyl is very water soluble, moves readily in potting mixes, and is absorbed by plant roots. It is xylem mobile and rapidly translocated to new tissues. Metalaxyl will kill some but not all water mould inoculum in the growing media. Resistance to metalaxyl has become a major problem. Phosphonates are xylem and phloem mobile and can be used as both a drench and foliar spray, where registered. Phosphonates persist well in plant tissues but in perennial crops sequential applications are needed to maintain effective concentrations. Phosphonates do not have any effect on the population of water molds in the potting mix. They act as a fungistat and activate the defence responses of the host plant. Resistance to phosphonates is unlikely but reduced sensitivity to the chemical has been found. This has not affected disease control. They inhibit spore production and chlamydospore germination. The chemical etradiazole also reduces root rot caused by water molds in nursery plants. Common fungicides used to control the soilborne fungi in nurseries are thiophanate–methyl and PCNB (quintozene). These active ingredients are less likely to be effective against water molds than etridiazole, metalaxyl, and phosphonates. Always refer to the label or minor use permit to ensure that it is suitable for use. Application of lime (2500 kg/ha) was found to effectively control the club root of cabbage by increasing soil pH to 8.5 (Utkhede and Guptha, 1996). Similarly application of sulphur (900 kg/ha) to soil reduced the soil pH to 5.2 and hence the incidence of common scab of potato cause by streptomyces scabies was also reduced (Davis et al., 1974).

16.10.4   RESISTANT CULTIVARS

A safe and low input way to manage plant diseases is to grow resistant cultivars of a crop. If a particular disease is prevalent in a geographic area, determine if appropriate resistant cultivars are available or not.

16.10.5   BIOSECURITY

The spread of many soilborne pathogens has been linked to the international nursery trade. For example, *Phytophthora ramorum* has spread from being a minor foliar blight pathogen of ornamentals in nurseries to cause serious and widespread damage in natural woodland ecosystems in Europe and North America. In contrast to *Phytophthora cinnamomi,* which is a root pathogen transmitted by soil and water, *Phytophthora ramorum* is primarily an aerial pathogen with a soilborne survival phase. If introduced, it would become a major threat to Australian forests and woodlands (Erwin and Ribeiro, 1996). Nursery plants have also played a major role in the evolution and spread of other pathogens (e.g., *Phytophthora alni*). Nurseries need to adopt best management practices to minimize biosecurity threats, and to improve pest and disease management. This will help prevent infested and infected asymptomatic plants from entering the supply chain, and perhaps the international trade in plant products.

KEYWORDS

- diseases
- soilborne
- management
- nursery
- pathogens

REFERENCES

Agrios, G. N. *Plant Pathology*. Academic Press: San Diego, 2005.

Agrios, G. N. *Plant Pathology*. Academic Press: San Diego, 1988.

Alexopoulos C .J., Mims C. W., Blackwell M. *Introductory Mycology*, 4th ed; Wiley, New York, 1996.

Anderson, R. V.; Potter,J. Stunt Nematodes: *Tylenchorhynchus, Merlinius*, and Related genera. In *Manual of Agricultural Nematology*; Nicle, W. R., Ed.; CRC Press: New York, 1991; pp 359–360.

Anonymous, 2000. https://www.ces.ncsu.edu/depts/pp/notes/oldnotes/vg16.html

Anonymous, 2015. http://www.lsuagcenter.com/NR/rdonlyres/676D0BE2-0CD8-4EE0-B8FA-331C11D92577/74885/pub3106BacterialPanicleBlightHIGHRES.pdf

Anonymous, 2014. http://www.biotecharticles.com/Agriculture-Article/Impact-of-Climate-Change-on-Plant-Diseases-3275.html

Anonymous, 2016b. https://en.wikipedia.org/wiki/Pythium_aphanidermatum

Anonymous, 2016c. https://en.wikipedia.org/wiki/Rhizoctonia_solani

Anonymous, 2016d. http://www.extento.hawaii.edu/Kbase/crop/type/s_rolfs.html

Anonymous, 2016e. https://en.wikipedia.org/wiki/Phytophthora

Anonymous, 2016f. https://en.wikipedia.org/wiki/Pseudomonas_syringae

Anonymous, 2016g.https://www.ngia.com.au

Anonymous, 2016a. http://www.extension.umn.edu/agriculture/cropdiseases/soybean/bacterialblight.html

Arora, I., et al. Am. Int. J. Res. *Formal, App. Nat. Sci.* **2015,** *10* (1), 14–17.

Beckman C. H. *The Nature of Wilt Diseases of Plants*. APS: St Paul, MN, 1987.

Booth, C. *The Genus Fusarium*. Commonwealth Agricultural Bureaux,1971, 146.

Bruehl G. W. Soilborne Plant Pathogens. Macmillan, NY, 1987.

Darin, E.1996. Illinois Fruit and Vegetable News. 2 (1). February 21, 1996 (Darin Eastburn) [Darin Eastburn, Illinois Fruit and Vegetable News (Part 2) Vol. 2, No. 4, April 10, 1996].

Davis J. R., Garner J. G., Callihan R. H. Effects of Gypsum, Sulfur, Terraclor and Terraclorsuper-x for Potato Scab Control. *Am. Potato J.* **1974,** *51* (2), 35–43.

Dingley, J. M. Records of Plant Diseases in New Zealand. D.S.I.R. *Bulletin* **1969,** *192*, pp 298.

Drenth, A.; Guest D. I. *Diversity and Management of Phytophthora in Southeast Asia*. ACIAR: Canberra, 2004..

Eastburn, D. M.; Degennaro, M. M.; Delucial, E. H.; Demody, O.; McElrone, A. J. Elevated Atmospheric Carbon Dioxide and Ozone alter Soybean Diseases at Global Change Biology. **2011**, *16*, 320–330.

Erwin, D. C.; Ribeiro, O. K. *Phytophthora Diseases Worldwide* APS Press: Minnesota, 1996.

Genin S.; Boucher, C. Lessons Learned from the Genome Analysis of *Ralstonia solanacearum*. *Ann. Rev. Phytopathol.* **2004**, *42*, 107–134.

Gray, F.A.; Hine, R.B. Development of Phytophthora Root Rot of Alfalfa in the Field and the Association of Rhizobium Nodules with Early Root Infections. *Phytopathology* **1976**, *66*, 1413–1417.

Handelsman J.; Parke J. L. Mechanism in Biocontrol of Soil Borne Plant Pathogens. In: *Plant Microbe Interactions, Molecular and Genetic Perspective*; Kosuge, T, Nester, E. W., Eds.; McGrawHill: New York, 1989; Vol. 3, pp. 27–61.

Imas, P. Recent Techniques in Fertigation of Horticultural Crops in Israel. In *Recent Trends in Nutrition Management in Horticultural Crops*, IPI-PRII-KKV Workshop, Dapoli, Maharashtra, India, February 11–12, 1999. http://www/ipipotash.org/presentn/qaknhc.html

Islam T. M.; Hashidoko, Y.,; Deora, A.; Ito, T.; Tahara S. Suppression of Damping-off Disease in Host Plants by the Rhizoplane Bacterium *lysobacter* sp. Strain sb-k88 is Linked to Plant Colonization and Antibiosis Against Soilborne Peronosporomycetes. *Appl. Environ. Microbiol.* **2005**, *71*, 3786–3796.

Jan, M. J.; Nisar, A. D.; Tariq. A. B.; Arif. H. B.; Mudasir, A. B. Commercial Biocontrol Agents and their Mechanism of Action in the Management of Plant Pathogens. *Int. J. Modern Plant Animal Sci.* **2013**, *1* (2): 39–57.

Jarvis, W. R. Managing Diseases in Greenhouse Crops. St Paul, Minnesota: APS Press. ISBN 978-0-89054-122-7

Koumoutsi, A.; Chen, X. H.; Henne, A.; Liesegang, H.; Gabriele, H.; Franke, P.; Vater, J.; Borris, R. Structural and Functional Characterization of Gene Clusters Directing Nonribosomal Synthesis of Bioactive Lipopeptides in *Bacillus amyloliquefaciens* Strain FZB42. *J. Bacteriol.* **2004**, *86*, 1084–1096.

Matsubara, Y.-I.; Haruto T.; Takashi H. Growth Enhancement and Verticillium Wilt Control by Vesicular-Arbuscular Mycorrhizal Fungus Inoculation in Eggplant. *J. Japanese Horticultural Soc.* **1995**, *64* (3), 555–561.

McQuilken, M. P.; Gemmell, J.; Lahdenpera, M. L. *Gliocladium catenunulatum* as a Potential Biocontrol Agent of Damping off in Bedding Plants. *J. Phytopathol.* **2001**, *49*, 171–178.

Moens, M.; Roland, N. P.; James, L. S. 2009. "Meloidogyne Species: A Diverse Group of Novel and Important Plant Parasites." In *Root-knot Nematodes*; Roland, N. P., Moens, M., James, L. S., Eds. CABI Publishing: Wallingford, UK, 2009, pp 1–17.

Naik, M. K. Challenges and Opportunities for Research in Soil-Borne Plant Pathogens with Special Reference to *Fusarium* Species. *J. Mycol. Pl. Pathol.* **2003**, *33* (1), 1–14.

Naik, M. K.; Rani, D. *Advances in Soil Borne Plant Diseases*. New India Publishing Agency: Pitam Pura, New Delhi, 262.

Nees Von Esenbeck, C. G. *"Dan system der pilze and Schwamme"*. Stahelschen Buchhandlung: Wurzburg, 1816.

Paulitz, T. C. Effect of *Pseudomonas putida* on the Stimulation of *Pythium ultimum* by Seed Volatiles of Pea and Soybean. *Phytopathology* **1991**, *81*, 1282–1287.

Pegg, G. F.; Brady, B. L. *Verticillium Wilts*. CABI Publishing: New York, NY, 2002.

Pratt, R. G.; Mitchell, J. E. Interrelationships bf Seedling Age, Inoculum, Soil Moisture Level, Temperature and Host and Pathogen Genotype in Phytophthora Root Rot of Alfalfa. *Phytopathology* 1976, *66,* 81–85.

Punja, Z. K.; Yip, R. Biological Control of Damping-off and Root Rot caused by *Phythium aphanidermatum* on Greenhouse Cucumbers. *Canadian J. Plant Pathol.* **2003,** *25*, 411–417.

Ram A. J.; Suhas, P. W.; Kanwar, L. S.; Piara, S.; Dhaka, B. L. Fertigation In Vegetables Crops: Fertigation in Vegetable Crops for Higher Productivity and Resource Use Efficiency. *Ind. J. Fertilizer* **2011,** *7* (3), 22–37.

Sasser, J. N.; Carter, C. C. Overview of the International *Meloidogyne* Project 1975–1984. In: *An Advanced Treatise on Meloidogyne*; Sasser, J. N., Carter, C. C., Eds.; Raleigh: North Carolina State University Graphics, 1985, 19–24.

Shanahan, P.; O'Sullivan, D. J.; Simpson, P.; Glennon, J. D.; O'Gara, F. Isolation of 2, 4-Diacetylphloroglucinol from a Fluorescent Pseudomonad and Investigation of Physiological Parameters Influencing its Production. *Appl. Environ. Microbiol.* **1992,** *58*, 353–358.

Singh, R. S. *Disease Management—The Practices. Introduction to Principles of Plant Pathology.* Oxford & IBH Publishing Co. Pvt. Ltd: New Delhi, pp 310.

Smith, G. S.; Carvil, O. N. Field Screening of Commercial and Experimental Soybean Cultivars for their Reaction to *Macrophomina phaseolina. Plant Dis.* **1977,** *81*, 363–368.

Songa, A. Variation and Survival of Macrophomina Phaseolina in Relation to Screening Common Bean (*Phaseolus vulgaris* L.) for Resistance. Ph.D. Thesis, 1995.

Stern, N. The Economics of Climate Change: The Stern Review. Cambridge University Press: Cambridge, UK. University of Reading 1995, 2007.

Thurston, H.D. Sustainable Practices for Plant Disease Management in Traditional Farming Systems. Westview Press, Boulder, CO, 1992.

Tsutomu, U.; Daizaburo, Y.; Koushi, N.; Tadao, I.; Hiroshi, F. Pathogenic Bacterium Causing Seedling Rot of Rice. *Japanese J. Phytopathol.* **1976,** *42* (4), 464–471.

Utkhede, R. S., Guptha, V. K. Management of Soil Borne Diseases. Kalyani Publishers. New Delhi, 1996.

Wilhite, S. E.; Lunsden, R. D.; Strancy, D. C. Peptide Synthetase Gene *in Trichoderma virens*. *Appl. Environ. Microbiol.* **2001,** *67*, 5055–5062.

Zhang, S.; Moyne, A. L.; Reddy, M. S.; Kloepper, J. W. The Role of Salicylic Acid in the Induced Systemic Resistance Elicited by Plant Growth Promoting Rhizobacteria Against Blue of Tobacco. *Biological Control* **2002,** *25*, 288–296.

CHAPTER 17

Biological Control of Postharvest Diseases in Vegetables

PARDEEP KUMAR[1*] and R. K. SINGH[2]

[1]*Krishi Vigyan Kendra, KVK, Siddharthnagar, Narendra Dev University of Agriculture and Technology, Faizabad 224229, Uttar Pradesh, India*

[2]*Department of Plant Pathology, R. V. S. Krishi Vishwa Vidyalaya, College of Agriculture, Indore 452001, Madhya Pradesh, India*

Corresponding author. E-mail: drpardeepviro@gmail.com

ABSTRACT

Postharvest diseases of vegetables may be incited by pre- or postinfection by number of pathogens and cause greater losses. Postharvest diseases are mainly incited by bacteria and fungi, which developed on vegetables between harvesting and consumption. During the last two decades, huge information and advances concerning the selection of antagonist, mode of action, different approaches to enhance biocontrol activity, formulation, and production have been achieved. A few biofungicides are available in the market that are commercially utilized by growers. Other promising biological control measures that include the resistance are available, which can be used in combination with biotic antagonists to enhance a greater stability and effectiveness than the approach of utilizing single microbial biocontrol agents. Microbial antagonists are applied either pre- or postharvest, but postharvest applications are more effective than pre harvest applications. Mixed cultures of the microbial antagonists appear to provide better control of postharvest diseases over individual cultures or strains. Similarly, the effectiveness of the microbial antagonist(s) can be enhanced if they are used with low doses of fungicides, salt additives, radioactive, and physical treatments like hot water dips, irradiation with ultraviolet light, and so on. At the international level, different microbial antagonists like *Bacillus subtilis* (Ehrenberg) Cohn and *Trichoderma harzianum* Rifai, *Saccharomyces cerevisiae* are being used. Biocontrol products like Aspire, BioSave, Shemer, and so on have also been developed and registered.

17.1   INTRODUCTION

Losses in crop yield due to postharvest disease are affected by a great number of factors including: commodity type, commodity susceptibility to postharvest diseases, the postharvest environment (temperature, relative humidity, atmosphere composition, etc.), produce maturity and ripeness stage treatments used for disease control, produce handling methods, and postharvest hygiene. These factors will be discussed in detail later in this chapter. Virtually all postharvest diseases of vegetables are caused by fungi and bacteria. In some root crops and brassicas, viral infections present before harvest can sometimes develop more rapidly after harvest. In general, however, viruses are not predominantly causing much more postharvest diseases. Postharvest diseases are generally classified on the basis of the infection initiated. Postharvest infection may be "quiescent" or "latent" infections, which are those where the pathogen initiates infection of the host at some point in time (usually before harvest), but then enters a period of inactivity or dormancy until the physiological status of the host tissue changes in such a way that infection can proceed. The dramatic physiological changes that occur during fruit ripening are often the trigger for reactivation of quiescent infections. All types of infections occur through surface wounds created by mechanical or insect injury and lack of proper handling at the time of transportation. Intensity of wounds in different vegetables is not uniform in all types of vegetables but varied according to the nature of pathogens and availability of the inoculum load of the specific pathogens. Postharvest diseases are interaction of availability of virulent and optimum inoculum of pathogen, susceptibility conditions of the vegetable, and also predisposing factors of storage conditions. Localized-based right time picking of vegetables, handling at the time of picking, pre-existing sign or symptoms, conditions of the storage, way of transportation, and shelf life of the vegetables really play an important role for diseases incited during postharvest. Common postharvest diseases resulting from wound infections include blue and green mould. Major postinflectional pathogens that results in diseases belongs to saprophytic of facultative in nature and moreover are dependent on congenial conditions of storage and availability of entry point on the surface of vegetables. The availability of optimum conditions (proper environmental and wound) for saprophytes may enhance the comparative colonization in very short span of time during storage due to faster rate of multiplication.

India is the second largest producer of fruits and vegetables, that is, 90 million metric tonnes and 169 million metric tonnes, respectively, after China (137 million metric tonnes of fruits and 573 million metric tonnes of vegetables). Vegetables are susceptible to pathogens by varieties of micro-organisms, which leads to substantial decay and losses during postharvest handling. After the vegetables are harvested, the value of vegetables is added in successive stages and remains in living phase. Losses due to postharvest disease may occur at any time during postharvest handling. Major objective of postharvest management is to maximize the keeping quality and storage value. The postharvest vegetables are living and respiring materials; and from field to consumer, several factors influence quality and there are many microorganisms (fungal and bacterial pathogens) that affect them. The quali-tative value of these delicate products peaks at harvest when they are green, fresh, and crisp, but quality attributed deterioration may be realized during handling and storage, resulting in vegetables of inferior and deteriorated quality when they reach to the consumers. Furthermore, in India, most of the vegetables are generally grown by farmers in remote villages and are transported to consumers who reside in town or cities. Transportation from producer to consumer is done by mainly through railway wagons, trucks, and even through traditional bullock or horse carts. Unhygienic and poor handling may lead to breakage or injury of the products that creates suitable conditions for fungal and bacterial infection. Important genera of anamor-phic postharvest pathogens include *Penicillium, Aspergillus, Geotrichum, Botrytis, Fusarium, Alternaria, Colletotrichum, Dothiorella, Rhizopus, Lasiodiplodia,* and *Phomopsis.* Some of these fungi also form fruiting structures. For management of postharvest losses, integrated approaches of management should be implemented to reduce the qualitative and quanti-tative loss of vegetables when fresh and during storage. Chemical sprays, as the last option, have also been utilized to reduce postharvest losses in different diseases of postharvest.

17.2 CAUSES OF LOSS AFTER HARVEST

There are many a slips between the crops and the consumer. Postharvest loss escalates the noncultivation cost of a crop, reduces the nutritional securities of the food that society inspires for, leads to serious national malnutrition problem in the country, increases retail price drastically, cuts the quantum of money value of exports and influences postharvest technology and management efforts. An estimate indicates that the postharvest loss of perishable crops is ranged as 25–50% for fruits, 30–32% for vegetables, and 12–32% for root and stem tubers. In India, 20–30% loss occurs within the marketing chain. Fungal and bacterial spectra differ among various types of vegetables species and also among the same types of fruits. Some studies indicate succession in favor of pathogens toward the harvesting time. Some pathogen entities are present throughout the year but are not carried on the produce and cannot infect unwounded or unripe tissue. On contrary, surface organisms can cause the spoilage. In addition, a few pathogenic fungi produce combined effect. For instance, *R. solani* and *F. oxysporum* are nonpathogenic on pumpkin, but together they produce spongy rot. Postharvest vegetables are living entities that contain 65–95% water. Naturally, when food and water reserves are exhausted, the produce dies and decays. Anything that increases the rate at which a product's food and water reserves are used up increases the likelihood of losses. Normal physiological changes could be increased due to high temperature, low atmospheric humidity, and physical injury. These types of injury often results from careless handling causing internal bruising, splitting, and skin breaks, which leads to increased water loss.

Breathing in growing plants or harvested produce is a continuous process and cannot be stopped without damage or external interference. It uses stored starch or sugar and stops when its reserves are exhausted, thereby leading to ageing. Respiration directly or indirectly depends on good air availability, when the air supply is restricted, fermentation starts instead of respiration. Poor ventilation in storage also leads to the accumulation of high concentration of carbon dioxide. High concentration of carbon dioxide leads to quick damage of produce.

Generally, fresh postharvested produce continues to lose water, which causes shrinkage and loss of weight. The rate of loss of water varies according to the product. Leafy vegetables loose water quickly because they have a thin skin with enormous stomata; whereas, potatoes have a thick skin with few pores. The rate of water loss may be minimal or maximum and is totally dependent on types of product or storage conditions. The most significant factor is the dimension of the surface area of the vegetable to its volume; produces having greater dimension have more rapid loss of water. The rate of loss is related to the osmotic pressure between the water vapor pressure inside the produce and in the air. To extend the shelf life of the produce, it must, therefore, be kept in a moist environment.

Fungi and bacteria are the major inciting agent for postharvest diseases, but viruses don't cause major diseases in postharvest; however, viruses are biotrophic and common only in growing crops. Deep expansion of decay makes infected produce totally unusable; such decay often results from infection of the produce in the field before harvest (preinfection). Quality loss occurs when the disease affects only the surface; however, skin blemishes may lower the selling price but do not render a fruit or vegetable inedible. Spread of fungal and bacterial diseases are generally caused by microscopic sexual or asexual spores, which are distributed in the air and soil and via decaying plant material. Infection of product by pathogens after harvest can occur at any time. It is usually the result of harvesting or handling injuries.

Ripeness of young vegetables is followed by senescence and breakdown of the vegetables. Nonclimacteric fruit only ripen while still attached to the parent plant. Their eating quality suffers if they are harvested before getting fully ripe as their sugar and acid content does not increase further. Early harvesting is often carried out for export shipments to minimize loss during transport, but a consequence of this is that the flavor suffers. Climacteric vegetables are those that can be harvested when mature but before ripening, like tomato. In commercial vegetables, the rate of ripening is controlled artificially, thus enabling careful planning of transport and distribution. Ethylene gas is produced in most plant tissues and is important in starting off the ripening process. It can be used commercially for the ripening of climacteric vegetables. However, natural ethylene produced by vegetables can lead to *in-storage* losses. For example, ethylene destroys the green color of plants. Leafy vegetables will be damaged if stored with ripening fruit. Ethylene production is increased when fruits are injured or decaying and this can cause early ripening of climacteric vegetable during transport.

TABLE 17.1 Maturity Standards of Different Vegetables.

Vegetable crops	Maturity indices
Tomato	Well-mature green, pink, breaker, and red ripe; pulp surrounding the seeds is jelly-like, seeds slip away from the knife, for long-distance shipment, is harvested at mature green stage, ripe stage indicates that most of the surface is pink or red, and they are firm.
Okra	Pods are still young, tender, exhibiting maximum growth, mature pods are fibrous, and tough *Asparagus*. Spears grow above the ground.
Cauliflower	Head size and condition and before they become discolored, loose, racy, and blemished. Over mature flowers became too long, flower stocks elongate, resulting in being fuzzy and racy.

TABLE 17.1   *(Continued)*

Vegetable crops	Maturity indices
Carrots	Size is the primary consideration and at least three-fourth diameters, proper color development without zoning.
Peas	Sugar content: 5–6% in maturity sugar decline, Increase in starch/protein. Tenderness and appearance of pods: should be well filled with young tender peas. Change in color from dark to light green with firmness of 5 kg/cm².

17.3   POSTHARVEST PHYSIOLOGY

The development of postharvest disease is intimately associated with the physiological status of the host tissue, therefore, creating the right environment may minimize postharvest losses due to disease. It is important to understand the physiological changes that occur after produce is harvested.

All plant organs undergo the physiological processes of growth, development, and senescence. Growth and development only occur while the organ is attached to the plant (with the exceptions of seed germination and sprouting of storage organs), but senescence will occur regardless of whether the organ is attached or not. When an organ, such as a fruit, is harvested from a plant, it continues to respire and transpire depleting both food reserves and water; such changes ultimately lead to senescence. Treatments that slow respiration and water loss, such as cold storage, help to delay senescence. Several commodity types vary greatly in the processes that precede senescence, making it difficult to draw general conclusions. For this reason, the following discussion will be limited to vegetables. "Maturity" is a term of completion used in reference to fruits and vegetables and is frequently confused with the term "ripeness." In simple terms, physiological maturity is attained when the process of natural growth and development is complete. In practice, fruits are considered to be mature when they have reached the stage where after harvesting and postharvest handling, they ripen to an acceptable quality. In contrast, a ripe fruit or vegetable is one which is ready to eat. The process of ripening basically signifies the end of development and the beginning of senescence. It may involve a number of changes in the fruit, such as conversion of starch to sugars, increase in pH, increase in softness, development of aromas, reduction in chlorophyll content, and corresponding increase in carotenoid levels (yellow and orange pigments). Fruits and vegetables are often classified into two groups on the basis of how they ripen.

Climacteric fruits exhibit pronounced increase in respiration and ethylene production coincidentally with ripening. Climacteric fruits or vegetables can be harvested in an unripe state and, providing they are sufficiently mature, will ripen to an acceptable quality. Nonclimacteric fruit cannot exhibit a rapid increase in respiration during the ripening process. The eating quality of nonclimacteric fruit does not improve after harvest, although they may undergo some changes in color development and softening. For this reason, they should not be harvested until they are ready to eat.

17.4   INFECTION PROCESS

Infection of fruits and vegetables by postharvest pathogens can occur before, during, or after harvest. Infections that occur before harvest and remain quiescent until some point during ripening are particularly common among tropical crops. Anthracnose, which is the most serious postharvest disease of a wide range of tropical and sub-tropical crops such as mango, banana, papaya, pawpaw, and avocado, is an example of a disease arising from quiescent infection established pear to harvest. Anthracnose is also an important postharvest disease of a number of vegetables. Various species of *Colleotrichum* can cause anthracnose; some species (e.g., *C. musae*) are hosts to specific fruits or vegetables, whereas others can attack a wide range of fruits and vegetables, for example, *C. gloeosporioides*. The infection process is occurring when conidia germinate on the surface of host tissue to produce a germ tube followed by an appressorium. Although it is known that there is a quiescent phase in the life cycle of the fungus, it is not entirely clear whether the un-germinated or the germinated appressorium represents the quiescent stage, as different studies have reported conflicting results. It may be that the fungus behaves differently on different hosts, or perhaps some researchers have been unable to detect appressorial germination due to the limitations of the techniques used.

In avocado, for example, early study reported that un-germinated appressoria were the quiescent phase of *C. gloeosporioides*. Studies conducted two decades later, however, showed that appressoria germinated to produce infections hyphae prior to the onset of quiescence. In any case, the fungus ceases growth soon after appressorium formation and remains in a quiescent state until fruit ripening occurs. During ripening, the fungus resumes activity and colonizes the tissue, leading to the development of typical anthracnose symptoms. Natural antifungal compounds present in fruits and vegetable tissue may be involved in regulating the quiescence of *Colletotrichum* infections. In avocado fruits, antifungal defence compounds are present in peel under unripe condition which is inhibitory to *Colletotrichum gloesporides*. After ripening of these defence compounds declines to subfungitoxic level coincidently which leads to development of symptoms. Treatments that stimulate the production of defence compounds also delay symptom development, which suggests that these compounds have a role in regulating quiescence. Grey mould of strawberry caused by *Botrytis cinerea* is another important postharvest disease sometimes arising from quiescent infections established before harvest. Conidia of *B. cinerea* on the surface of necrotic flower parts germinate in the presence of moisture. The fungus colonizes the necrotic tissue and then remains quiescent in the base of the floral receptacle. Several months later when fruits are harvested, infections develop as a stem end rots in a ripe fruit. Many postharvest diseases develop from the stem end of fruit. The mode of infection involved in this group of disease can, however, vary considerably in the example of *B. cinerea*; lesions occurring at the stem end of fruit arise from quiescent floral infections. Stem end rotting of citrus fruits are caused by *Lasiodiplodia theobromae* and *Phomopsis citri* that results from quiescent infections in the stem button of fruit. These infections can be initiated at any state of fruit development, and remain quiescent until the button beings to separate from the fruit during abscission. In other stem end rot diseases, infections occur during and after harvest through the wound created by severing the fruit from the plant (e.g., banana crown rot). Endophytic infection, whereby the fungus symptomless and systemically colonizes the stem, inflorescence and fruit pedicel tissue is important in a number of stem end rot diseases of tropical vegetable and fruit. Mango stem end rot caused by *Dothiorella dominicana* is one example of postharvest disease arising from entophytic colonization of fruit pedicel tissue. In this case, the fungus colonizes the pedicel and stem end tissue of unripe fruit, where it remains commences. A considerable number of postharvest diseases develop from infections that occur shortly before harvest. Such infections may not be visible at the time of harvest, yet at the same time are not necessarily inactive. Symptoms may develop more rapidly after harvest, particularly if storage conditions favor pathogen development. Examples of postharvest diseases, which can arise from late season infections include brown rot of peach (*Monilinia fructicola*), grey mould of grape (*Botrytis cinerea*), yeasty of tomato (caused by *Geotrichum candidum*), and sclerotium rot of various vegetables (caused by *Sclerotium rolfsii*). Many of the postharvest pathogens are not able to penetrate directly to the host cuticle, these pathogens, therefore, infect through surface wounds or natural openings such as stomata and lenticels. Injuries can vary in size from microscopic to clearly visible and may arise in a number of ways. Mechanical injuries such as cuts, abrasions, pressure damage, and impact damage commonly occur at the time of harvesting and handling. Insect injuries may occur before harvest and yet remain undetected at the time of grading, providing ideal infections sites for postharvest pathogens. Infections of lychee fruit by yeasts commonly occur through insect injuries, which are difficult to notice during harvest. Some chemical treatments used after harvest, such as fumigants used in insect disinfestations and disinfectants such as chlorine, may also injure produce if applied incorrectly. Various types of physiological injuries such as chilling and heat injury can predispose produce to infection by postharvest pathogens. Tropical fruits and vegetables in particular are very sensitive to low temperatures and many develop symptoms of chilling injury below 13°C (depending on storage duration). Symptoms of major chilling injuries include abnormal ripening, peel and flesh discoloration, water soaking, and pitting. Injury caused by chilling exposure can increase the susceptibility of produce to postharvest disease considerably. For instance, the incidence of *Alternaria rot* in pawpaw, apple, and various fruit crops is increased by exposure to excessive cold. Similarly, storage high temperatures exposure could also increase susceptibility of harvested produce to disease. For instance, hot water dipping of mangoes for excessive times and temperatures can result in increased levels of stem end rot (caused by *Dothiorella* spp.). The natural resistance of fruit and vegetables to disease generally declines with storage duration and ripeness. Weak pathogens (saprophytes), which normally require a wound or scratching on external surface in order to infect, can become a problem in produce that has been stored for long period of time. Treatments that help to maintain the natural "vitality" of fruits and vegetables aid in delaying the onset of disease in stored produce. The infection of pathogen that happen shortly before the harvest may cause considerable number of postharvest diseases and such infections may not express any visible symptoms at the time of harvest, but only while in storage conditions. Such type of preinfected pathogen expresses their unique symptoms more rapidly. Examples of postharvest diseases that can arise from late season infections include *Monilinia fructicola* incited brown rot of peach, grey mould of grape (*Botrytis cinerea*), yeasty rot of tomato (*Geotrichum candidum*), and *Sclerotium rot* of various vegetables caused by *Sclerotium rolfsii*. Some very common weak parasite postharvest pathogens are unable to penetrate directly to the host cuticle; these type of pathogen are always dependent on availability of surface injuries or natural openings such as stomata and lenticels. Mechanical injuries such as cuts, abrasions, pressure damage, and impact damage commonly occur during harvesting and handling and the intensity of injuries vary from microscopic to naked eye visible stretches. Often the injuries caused by insect may occur before harvest and yet remain undetected at the time of grading, thereby providing ideal infection sites for many postharvest pathogens, for instance, yeasts infection in litchi fruits commonly occurs through insect injuries which are difficult to see at harvest. Some chemical treatments used after harvest, such as fumigants used in insect disinfestations and disinfectants by chlorine may also injure vegetables, if applied incorrectly. Many of the physiological injury due to chilling and heat can predispose produce to infection of postharvest pathogens.

17.5   POSTHARVEST DISEASES

Correct diagnosis of the pathogen causing postharvest disease is a central dogma to the selection of an appropriate disease control strategy. Table 17.2 enlists some common postharvest diseases and pathogens of fruit and vegetables. Majority of the fungi that cause postharvest disease belong to the phylum Ascomycota and the associated Fungi Anamorphic (*Fungi Imperfecti*). Often in Ascomycota, the asexual stage of fungus (the anamorph) is usually encountered more frequently in postharvest diseases than the sexual stage of the fungus (the *teleomorph*). Leading genera of anamorphic postharvest pathogens include *Penicillium, Aspergillus, Geotrichum, Botrytis, Fusarium, Alternaria, Colletotrichum, Dothiorella, Lasiodiplodia,* and *Phomopsis.* Some of these fungi included ascomycete's sexual stages. In the phylum Oomycota, the genera *Phytophthora* and *Pythtium* are important postharvest pathogens, causing a number of diseases such as brown rot in citrus (*Phytophthora citrophthora* and *P. parasitica*) and cottony leak of cucurbits (*Pythium spp.*). *Rhizopus* and *Mucor* are important genera of postharvest pathogens in the phylum Zygomycota. *R. stolonifer* is a common wound pathogen of a very wide range of fruit and vegetables, causing a rapidly spreading watery soft rot. Genera within the phylum Basidiomycota are generally not important causal agents of postharvest disease, although fungi such as *Sclerotium rolfsii* and *Rhizoctonia solani*, which have basidiomycete sexual stages, can cause significant postharvest losses of vegetable crops such as tomato and potato. So far as the pre-inflectional pathogens at field and postharvest infection is considered, total 33 postharvest diseases of fruits and vegetables caused, while 535 diseases caused by these pathogens are primarily field diseases. The development of symptoms often accelerates after harvest. The major causal agents of bacterial soft rots are various species of *Erwinia, Pseudomonas, Bacillus, Lactobacillus,* and *Xanthomonas*. Bacterial soft rots are very important postharvest diseases of many vegetables, although they are generally of less importance in fruits. This is because most fruits have a low pH, which is inhibitory to the majority of bacterial plant pathogens.

TABLE 17.2   Examples of Common Postharvest Diseases and Pathogens of Vegetables.

Disease	Pathogens		Remarks
	Anamorph	Teleomorph	
Cucurbits			
Charcoal rot	*Macropphomina phaseolina*		Melon
Cottony leak	*Pythium* spp.		Cucumber
Rhizopus rot	*Rhizopus* spp.		Bitter gourd
Anthracnose	*Colletotrichum lagenarium*		Ash gourd, bitter gourd, bottle gourd
Fruit rot	*Pythium* and *Phytophthora*		Ash gourd
Soft rot	*Pythium aphanidermatum*		Bitter gourd
Cottony white rot	*Fusarium* spp.		Bitter gourd, bottle gourd
Diplodia rot	*Diplodia natalensis*		Bottle gourd
Tomato, Eggplant, and Capsicum			
Grey mould	*Botrytis cinerea*	*Botryotinia fuckeliana*	Tomato
Fusarium rot	*Fusarium* spp.,		Tomato
Alternaria rot	*Alternaria alternata*		Brinjal, chilli
Cladosporium rot	*Cladosporium* spp.		Brinjal
Rhizopus rot	*Rhizopus* spp.		Brinjal
Watery soft rot	*Sclerotinia* spp.		Brinjal
Cottony leak	*Pythium* spp.		Brinjal
Sclerotium rot	*Sclerotium rolfsii (sclerotial state)*	*Athelia rolfsii*	Brinjal
Phytophthora rot	*Phytophthora parasitica*		Brinjal, chilli
Phomopsis rot	*P. vexans*		Brinjal
Black mould rot	*Aspergillus niger*		Brinjal
Anthracnose	*C. capsici*		Chilli

TABLE 17.2   *(Continued)*

Disease	Pathogens		Remarks
	Anamorph	**Teleomorph**	
Ripe rot	*Gloeosporium piperatum*		Chilli
Black rot	*Rhixoctonia bataticola*		Chilli
Phoma rot	*P.capsici*		Chilli
Soft rot	*F. diversisporum*		Chilli
Diplodia rot	*Diplodia sp*		Chilli
Pink rot	*Trichothecium rosium*		Chilli
Cow pea, pea			
Anthracnose	*Colletotrichum landemuthianum*		Cow pea, pea
Phoma rot	*P. vignae*		Cow pea
Cottony pink rot	*F. moniliforme*		Cow pea
Snuffy mould	*Aspergillus esculeatus*		Cow pea
Cottony leak	*P. parasitica*		Pea
Pod spot	*Alternaria* spp.		Pea
French bean/ common bean, lima bean			
Anthracnose	*Colletotrichum lindemuthianum*		French bean, lima bean
Cottony leak	*Phytophthora parasitica*		French bean
Diplodia dry rot	*D. natalensis*		Lima bean
Okra			
Brown rot	*F. oxysporum*		Okra
Phytophthora rot	*Phytophthora palmivora*		Okra
Pythium rot	*Pythium aphanidermatum*		Okra
Cottony leak	*Pythium butleri*		Okra
Gray blight rot	*Sclerotium rolfsii*		Okra
Onions			
Bacterial soft rot	Various *Erwinia* spp. *Lactobacillus* spp. and *Pseudomonas* spp.		Onion
Black mould rot	*Aspergillus niger*		Onion
Fusarium basal rot	*Fusarium oxysporum f. sp.cepae*		Onion

TABLE 17.2 *(Continued)*

Disease	Pathogens		Remarks
	Anamorph	**Teleomorph**	
Smudge	*Colletotrichum circinans*		Onion
Tuber crops			
Bacterial soft rot	Various *Erwinia* spp. and *Pseudomonas* spp.		Carrot, ginger
Rhizopus rot		*Rhizopus* spp.	Carrot
Grey mould	*Botrytis cinerea*	*Botryotinia fuckeliana*	Carrot
Watery soft rot		*Sclerotinia* spp.	Carrot
Sclerotium rot	*Sclerotium rolfsii* (sclerotial state)	*Athelia rolfsii*	Carrot, elephant foot yam
Chalara and	*Chalara thielavioides*		Carrot
Thielaviopsis rots	*Thielaviopsis basicola*		Carrot
Black rot	*Aletrnaria radicina*		Carrot
Cottony white rot	*Fusarium* spp.		Carrot, ginger
Botryodiplodia rot	*Botryodiplodia theobromae*		Cassava
Anthracnose	*C. gloeosporoides*		cassava
Mucor rot	*Mucor hiemalis*		Cassava
Soft rot	*Rhizopus spp*		Sweet potato
Charcoal rot	*Macrophomina phaseolina*		Sweet potato
Brown rot	*Sclerotium rolfsii*		Sweet potato
Java black rot	*Botryodiplodia theobromae*		Sweet potato
Corm rot	*Cylindrocarpon lichenicola*		Elephant ear
Dry rot	*Fusarium coeruleum*		Elephant ear
Dirty grey rot	*S. rolfsii*		Elephant foot yam
Potato, tomato			
Bacterial soft rot	*Erwinia* spp.		Potato
Dry rot	*Fusarium* spp.	*Gibberella* spp.	Potato
Phytophthora rot	*P. infestans*		Potato
Brown rot	*Alternaria solani*		Potato
Powedery scab	*Spongospora subterranea*		Potato
Gangrene	*Phoma exigua* var *exigua* and var *foveata*		Potato

TABLE 17.2   *(Continued)*

Disease	Pathogens		Remarks
	Anamorph	**Teleomorph**	
Black scurf	*Rhizoctonia solani.* (sclerotial state)	*Thanatephorus cucumeris*	Potato
Silver scurf	*Helminthosporium solani*		Potato
Skin spot	*Polyscytalum pustulans*		Potato
Brown rot	*Phytophthora parasitica*		Tomato
Alternaria rot	*alternata*		Tomato
Waxy rot/sour rot	*Geotrichum candidum*		Tomato
Soft rot	*Rhizopus* spp.		Tomato
Sunken black spot	*Phoma destructiva*		Tomato
Black mould rot	*Aspergillus niger*		Tomato
Yellow mould	*Aspergillus flavus*		Tomato
Woody black rot	*Myrothecium roridum*		Tomato

17.6   MAINTENANCE OF HOST RESISTANCE TO INFECTION THROUGH MANIPULATION OF THE POSTHARVEST

17.6.1   ENVIRONMENT

Colonization of postharvest pathogens is not only determined by the availability of sufficient inoculum load on infection site, but also successful infection and intensity of pathogen on produce is governed by the micro environmental conditions of storage. The ability to control the postharvest environment provides a tremendous opportunity to delay senescence. Temperature and relative humidity is perhaps the most important factor influencing disease development after harvest. Temperature directly influences not only the rate of pathogen growth, but also the rate of fruit ripening. The development of many postharvest diseases is closely associated with fruit ripeness, so treatments that delay ripening tends to also delay disease development. Low temperature storage of fruits and vegetables is used extensively to delay ripening and the development of disease, although the temperatures commonly used for storage are not lethal to the pathogen. For this reason, cool stored produce, which is transferred to ambient temperatures for ripening and/or sale, may rapidly breakdown with postharvest disease. For example, many temperate fruits and vegetables (e.g., apples, peaches, and broccoli) can be stored at 0°C, whereas many tropical fruits cannot be stored below 10°C without developing symptoms of chilling injury. Modifying the storage atmosphere is sometimes used to delay produce senescence. The rate of fruit respiration can be reduced by increasing CO_2 and decreasing O_2 levels in the storage environment. Storage atmosphere can also have a direct effect on pathogen growth, although levels of CO_2 or O_2 required to achieve this are often damaging to the produce if applied for extended periods. A notable exception to this is strawberry, which for extended storage periods can tolerate the high levels of CO_2 (20–30 ppm) required to inhibit the development of grey mould caused by *Botrytis cinerea*. Short-term exposure to very high levels of CO_2 has shown some potential for delaying the onset of anthracnose caused by *Colletotrichum gloeosporioides* symptoms in avocado, although the treatment is not currently in commercial use. The relative humidity of the storage environment can have a major influence on the development of postharvest disease. High humidity is often used to minimize water loss of produce. This, however, can increase disease levels, particularly if free moisture accumulates in storage containers. The humidity chosen for storing produce is frequently a trade-off between minimizing water loss and minimizing disease. Maintenance of hygiene at all stages during production and postharvest handling is critical in minimizing sources of inoculum for postharvest diseases. To effectively reduce inoculum, a good knowledge of the life cycle of the pathogen is essential. Sources of inoculum for postharvest diseases depend largely on the pathogen and when infection occurs. In the case of postharvest diseases which arise from pre-harvest infections, practices which make the crop environment less favorable to pathogens will help reduce the amount of infection that occurs during the growing season. For instance, in tree crops, pruning and skirting can increase ventilation within the tree canopy, making conditions less favorable for fungi and bacteria. Removal of dead branches and leaves entangled in the tree canopy is also an important way to minimize inoculum build-up. In many diseases, overhead irrigation can encourage pathogen spread and infection; trickle or micro-sprinkler irrigation systems may be more appropriate. As many pathogens are soil-borne, minimizing contact of leaves and fruit with the soil is desirable. Inoculum for infections occurring after harvest commonly originates from the packing shed and storage environments. Water used for washing or cooling produce can become contaminated with pathogen propagules if not changed on a regular basis and if a disinfectant such as chlorine is not incorporated. Water temperature also has an important influence in the transfer of inoculum in some situations. For example, tomatoes harvested during hot weather may have a higher temperature than the water used to wash them. In this scenario, inoculum present in the washing water can be taken in by the fruit tissue, causing higher levels of diseases such as bacterial soft rot. Rejected produce that has not been discarded from the packing shed or storage environment provides an ideal substrate for postharvest pathogens. The implements and utensils generally utilized for different purposes in storage and transportation of produce, the hygienic of utilized utensils must be maintained to reduce the infection of pathogens. Similarly packing and grading equipment, particularly brushes and rollers, which is not cleaned and disinfected on a regular basis, can also be a major source of inoculums. Containers used for storing and transporting fruit can harbor pathogen propagules, particularly if recycled a number of times without proper cleaning. Wide ranges of preharvest factors congenial for disease development influence the dispersal or colonization of postharvest disease, these include the locally available weather factors (rainfall, temperature, etc.), required for production, choice of cultivar, cultural practices (pesticide application, fertilization, irrigation, planting density, pruning, mulching, fruit bagging, human activity, etc.) and planting material. These factors may have a direct influence on the development of disease by reducing inoculums sources or by discouraging infection. Alternatively, they may affect the physiology of the produce in a way that impacts on disease development after harvest. For example, the application of certain nutrients may improve the "'strength" of the fruit skin so that it is less susceptible to injury after harvest and, therefore, less prone to invasion by wound pathogens. Every produce have their localized strength, need to explore those strength for long-term transportation and storage, and make sure the supply of fresh and healthy postharvested produce to consumer without utilizing much more effort and measures, by which cost of the produce must be economical and price friendly.

17.6.2   EMERGING TECHNOLOGIES FOR POSTHARVEST DISEASE CONTROL

Fruits and vegetables are directly or indirectly used by the consumer for eating, and therefore, produces should not have excessive use of any pesticides; it may cause health hazards or other diseases in human being. Increasing consumer concerns over the presence of pesticide residues in food have prompted the search for nonchemical disease control measures. The fruits and vegetables are generally grown by the farmers in remote areas and without proper guidance, and some time the farmers uses many of the unknown and nonprescribed pesticides to manage preharvest diseases. The time of application of pesticides is also very important for inhibition of pathogens. Fungicides used after harvests are of particular concern because they are applied close to the time of consumption. A number of new approaches to control postharvest diseases are currently under investigation, including biological control, constitutive or induced host resistance, and natural fungicides.

17.7 BIOLOGICAL CONTROL

Biological management of postharvest diseases through biological measures is ecofriendly and don't leave any residual impact on edible product. The use of such intervention is very promising and sustainable, and favors the health of consumer. In recent years, there has been considerable interest and innovations in the use of antagonistic microorganisms for the control of postharvest diseases. Such organisms can be isolated from a variety of sources including fermented food products and the surfaces of leaves, fruits, and vegetables. Once isolated, organisms such as bacteria, yeasts, or filamentous fungi can be screened in various ways for inhibition of selected pathogens in vitro or in vivo. In most reported cases, pathogen inhibition is greater when the antagonist is applied before the infection taking place. For this reason, control of quiescent field infections (e.g., *Colletotrichum* spp.) using postharvest applications of antagonists is often more difficult to achieve than control of infections occurring after harvest (e.g., *Penicillium* spp.). Unless an antagonist has suppressive activity or has some effect on host defense responses, field applications are often necessary to achieve control of quiescent infections. In South Africa, both field and postharvest applications of *Bacillus subtilis* and *B. Licheniformis* suppress anthracnose (caused by *Colletotrichum gloeosporioides*) and stem end rot (caused by *Dothiorella* spp. and other fungi) development in avocado. A number of modes of action are involved in these pathogen–antagonist interactions, including site exclusion, nutrient and space competition, and antibiotic production. In Australia, however, only field applications of a nonantibiotic producing strain of *Bacillus* spp. have shown potential for the control of anthracnose in avocado. There are numerous reports in the literature concerning the biological control of wound pathogens in various fruit and vegetables. An antagonist successfully colonize on wound site, by which pathogen may not get space for colonization and further infection on wound site and create exclusion. Antagonists, which act against postharvest pathogens by competitive inhibition at wound sites include the yeasts *Pichia guitllier-mondii, Cryptococcus laurentii*, and *Candida* spp. While the potential for biological control of postharvest diseases clearly exists, future success relies on the ability to achieve consistent results in the field and after harvest. It will be necessary to enhance the efficacy of biological control agents against postharvest disease and commercialize the technology involved. Ideally, an antagonist should be effective against a broad spectrum of pathogens on a wide range of fruit and vegetables; it should be unique and be able to be produced on inexpensive growth media. Formulations incorporating such antagonists should have a long shelf life and be able to be manufactured at low cost. Many are quite specific in their activity against pathogens and for this reason may not be particularly appealing to prospective investors. Antagonists are quite dependent on environmental conditions and are often influenced by the external conditions of the microclimatic conditions prevailed in the surrounding of infection site; therefore, success rate of bio-agent application is somewhat dependent on these complications, so user must have the proper training and expertise on their usage.

17.8   TYPES OF BIOLOGICAL CONTROL

Many types of biological control have been researched and utilized for the successful management of insect pest. Most of them are clearly related to the control of pests, mainly insect and mites, and have been extrapolated to disease control. Because plant disease may be suppressed by the activities of one or more plant associated microbes, researchers have attempted to characterize the organisms involved in biological control. Primarily, this has been done through isolation, characterization, application of individual organisms, and commercialization.

17.8.1   CONSERVATION

Because plant diseases may be suppressed by the activities of one or more plant-associated microbes, researchers have attempted to characterize the organisms involved in biological control. Conservation biological control involves manipulation of the environment to enhance the survival, fecundity, longevity, and behavior of natural enemies to increase their effectiveness (Landis et al., 2000). Conservation of beneficial microbial to suppress the diseases could be obtained through cultural practices, suppress soils general suppression, and specific suppressions. Specific suppressions result from the activities of one or just a few microbial antagonists. This type of suppression is thought to be occurring when inoculation of a biocontrol agent results in substantial level of disease suppressive mess. Its occurrence in natural system may also occur from time to time. For instance, the introduction of *pseudomonas fluorescens* that produce the antibiotic 2,4-diaectylphloroglucinol can result in the suppression of various soil-borne pathogens. However, specific agents must compete with other soil- and root–associated microbes to survive, propagate, and express their antagonistic potential during those times when the targeted pathogens pose an active threat to plant health. In contrast, general suppression is more frequently invoked to explain the reduced coincidence or severity of plant disease because the activities of multiple organisms can contribute to a reduction in disease pressure. High soil organic matter supports a large and diverse mass of microbes resulting in the availability of fewer ecological niches for which a pathogens competes. The extent of general suppression will vary substantially depending on the quantity and quality of organic matter present in a soil. Functional redundancy within different microbial communities allows for rapid depletion of the available nutrient pool under a large variety of conditions before they can utilize them to proliferate and cause diseases; this is only due to vigorous saprophytic ability of antagonists. For example, diverse seed–colonizing bacteria can consume nutrients that are released into the soil during germination thereby suppressing pathogen germination and growth.

17.8.2   AUGMENTATION/AUGMENTATIVE BIOLOGICAL CONTROL

Biocontrol control agent (BCA) involves periodic release and reestablishment of a BCA that dies out each year, but which can rapidly expand its population when the conditions are suitable; while in inundative biocontrol, mass release of the BCA that can not reproduce and thus can not attain adequate population size without human interventions. Natural enemies (usually parasites or pathogens, but occasionally predators) are mass-reared and released in huge numbers in the field, at an appropriate stage of the pest's life cycle. They swamp the pests and bring them under control, then die back themselves because of lack of food. Therefore, unlike with classical biocontrol, this exercise has to be repeated every year. Costs are, therefore, comparable to chemical for control, though environmental and health damage is minimal, for example, release of *Trichogramma* species for insect pest and release of *Agrobacterium radiobacter* against the pathogen *A. tumefaciens* causing crown gall disease, fire blight, and postharvest disease.

It is a defined intentional introduction of an exotic, usually co-evolved, biological control agent for permanent establishment and long-term pest control. It's usually targeted at an imported alien pest species. It employs natural enemy introduced from the region where the pest species is native. This strategy is most suited to organisms introduced from overseas (exotics). One or more natural enemies (usually parasites or pathogens) are introduced from the center of origin, after rigorous host-specificity testing under quarantine. This approach can be the highly successful and very cost effective, but in some cases, the control is only partial, such as when the control agents adapt less well to the new country than the pest first did. Once the BCA is introduced and established, BCA usually is self-sustaining. The great advantage of this method is that when it is successful, there is no further cost after initial establishment—the pest and enemy species remain in balance; with lessgovernment support to institutions at national or regional level, BCA isgenerally self-sustaining following a single release of a "national enemy."

17.8.3   CONSTITUTIVE AND INDUCED HOST RESISTANCE

Plants possess various biochemical and structural pre- or postinfectional defense mechanisms that protect them against infection. Some of these mechanisms are in place before arrival of the pathogen (i.e., constitutive resistance), while others are only activated in response to infection (i.e., induced resistance). Compared to intact plants, relatively little is known about host defense responses in harvested commodities, although there is growing interest in this area. In avocado, for example, research in Israel and Sri Lanka is uncovering the biochemical mechanisms associated with host resistance to anthracnose. Preformed antifungal compounds present in fungitoxic concentrations in unripe fruit decline to sub-lethal concentrations in ripe fruit allowing the pathogen to develop once fruit ripening commences. Levels of the diene compounds can be increased by applying various treatments such as challenge inoculation with either pathogenic or nonpathogenic strains of *Colletotrichum* or by treatment with certain antioxidants or high concentrations of CO_2. In contrast to preformed antimicrobial compounds, *phytoalexins* are only produced in response to pathogen invasion, although in some cases they can be elicited by certain chemical and physical treatments. For example, nonionizing ultraviolet-C radiation is known to induce production of phytoalexins in various crops. UV treatment of carrot slices induces production of 6-methoxymellen, which is inhibitory to *Botrytis cinerea* and *Sclerotinia sclerotiorum*. Chitosan, which is a natural compound present in the cell wall of many fungi, is another elicitor of host defense responses. Chitosan can stimulate a number of processes including production of chitinase, accumulation of *phytoalexins*, and increase in lignification. A wide range of other compounds (e.g., salicylic acid, methyl jasmonate, and phosphonates) and treatments (e.g., heat) can induce host defenses in harvested commodities.

17.8.4 NATURAL FUNGICIDES

Many compounds produced naturally by microorganisms and plants have antifungal properties. Chitosan, for example, is not only an elicitor of host defense responses but also has direct fungicidal action against a range of postharvest pathogens. Antibiotics produced by various species of Trichoderrna have potent antifungal activity against *Botrytis cineria, Sclerotinia sclerotiorum, Corticium rolfsii*, and many other important plant pathogens. There are many other natural compounds that have been isolated and shown to possess considerable antifungal activity. Although these compounds may be more desirable than synthetic chemicals from a consumer viewpoint, their potential toxicity to humans needs to be evaluated before useable products are developed. On a positive note, the high specificity of many of these compounds also means that they biodegrade readily.

17.8.5 NATURAL PLANT PRODUCTS

The study of green plants of their antimicrobial activity has received little attention though such activity was known since ancient times. Neem leaves were kept in woolly cloths and grain in store houses for preventing deterioration by moulds and pests and still today it is a common practice. Baskets of fruits and vegetables are lined with neem leaves for their protection against microbial attacks. It is believed that toxic substance emitted by the leaves keep the air remarkably free of pathogenic microorganisms. Like the same neem oil is too good as the neem leaf extract is for the management of the pathogens. Many researchers have shown the benefits of different garlic concentrations for controlling *Aspergillus niger, Gliocladium roseum*, and *Sclerotium rolfsii* rots of apple, *bottle gourd mosaic virus*, and *cucumber mosaic virus* (Kumar and Awasthi, 2003; Singh and Tripathi, 2013).

17.8.6   STRATEGIES NEEDED BY SCIENTISTS

- Utilization of conventional and biotechnological available processes for development of improved varieties having high production potential with high keeping quality attributes and also that showes resistance to biotic and abiotic stresses;
- Modernization of processing unit through mechanization and intro-duction of need-based interventions are demand of present era to reduce the pre- and postharvested mishandling;
- Diversification and introduction of economic methods for utilizing fruits and vegetables processing waste;
- Upgradation of research on traditional Indian foods with commercial value;
- Explore the possibility of biodegradable and zero oxygen perme-ability packaging materials;
- Introduction of economical methods formonitoring temperature and relative humidity in CA/MA/MAP/Low pressure storage;
- Technological improvement of the minimal processing of fresh produce;
- Complete and practicable integrated package of practice must be validated and prescribed to reduce the economic loss.

17.8.7   STRATEGIES NEEDED BY GROWERS

- Adaptation of the technique of high-density planting in order to increase productivity and quality for crops such as mango, pineapple, banana, tomatoes, onion, potatoes, etc.;
- Efficient land and input use programs—arid cultivation, fertigation, and water harvesting;
- Mechanization of efficient harvesting techniques particularly for large orchards and plantations;
- Knowledge on postharvest biology of the produce;
- Adaptation of integrated package of practice as suggested locally by the scientist;
- Area must be identified for particular crop cultivation to reduce more infection of the pest and diseases.

17.8.8 STRATEGIES NEEDED BY INDUSTRIES

• Technologies, industrial plants, and machinery must be designed to suit the processing requirements of available raw materials at specific locations;

• Facilitate localized industrialization at nearby area of production to reduce the harmful effect of transportation;

• Industry should make provision for guaranteeing stable prices to horticulturalists and reliable supplies must be provided at a reasonable price to the consumer;

• Improvement of low cost appropriate packaging material;

• Focus on the utilization of the wastage from the processing industry as by products;

• Social and innovative programme must be sponsored by industrialist to update the knowledge of local growers.

17.9 FUTURE OUTLOOK

Greater emphasis must be placed on problem-oriented research that would employ integrated approaches to solving postharvest issues. Apart from missing links in our understanding and implementation of postharvest technology, proper linkages must be established with the processing sector. While mechanical harvesting of horticultural crops increases efficiency, it results in considerable wastage. Attention must, therefore, be given by food scientists and engineers, to the development of techniques that would minimize wastage. Hygienic handling of raw materials and proper sanitation of equipment necessitates continuous assessment. Innovations in postharvest technology and particularly in the development of infrastructure could help achieve this goal. The processing of fruits and vegetables for export and for domestic markets requires utmost attention in developing countries. Efforts have been made for a quantum jump in the utilization of fruits and vegetables by the processing industry. Standardization of maturity indices for the harvesting of fruits and vegetables in order to control raw material quality prior to processing is needed (Table 17.1). Waste generated by processing factories must be put to profitable usage, either through conventional technologies or through the adoption of biological processes (Verma and Joshi, 2000). The quality and safety of fruits and vegetable products (freedom from microbial toxin and the pesticide residues) must continue to receive greater attention in view of its significant implications for human health. Attention must be paid to maintain the microbiological quality of processed products and to the development of realistic standards including rapid microbiological methods to ensure food safety. Diversification of the processed product base, including the production of low alcoholic fermented beverages to make use of surplus quantities of fruit, could be one of the several approaches to reduce the postharvest losses in developing countries. Beverages of this type are gaining popularity in view of their role in coronary heart disease. Apart from emerging technologies such as high electric field pulses, oscillating magnetic fields, intense high pulses, high pressure treatments, ohmic heating, irradiation processing, modified atmospheric packaging, edible coating, low calorie substitutes for fat replacement, solid state fermentation, and biotechnological approaches must be assessed and assimilated by the food industry for success in the future (Arya, 1998).

17.10   CONCLUSION

India has a good resource base for adequate research and development infrastructure and excellence in several areas of horticultural interest. Production and postharvest processing activities within the country have, therefore, been changing at a rapid rate. Of course, India is leading in production of fresh fruits and vegetables but ground level processing based industries are not vigorously grown in nearby area of production. The demand for horticultural produce in India is on the rise, owing to increasing populations; changing food habits; the nutritional value of horticultural crops; and a greater emphasis on postharvest management, processing and value addition. Advantages of growth can be harnessed with well-planned strategies so as to ensure a positive future outlook. Food safety is of growing importance in food production, processing, and marketing. Investment in food safety, while assuring risk-free ventures related to horticulture, will increase the cost of fresh produce and the processing of fruits and vegetables. Efforts geared toward improving production technologies for fruits and vegetables will become meaningful only if the effective postharvest technology is developed and wastage is reduced. Postharvest management and technology adoption is a continuous uninterrupted active process undertaken by a chain of researchers, extension workers, growers, and end users. A wide variety of fungal and bacterial pathogens cause postharvest disease in fruits and vegetables. Some of these infect produce before harvest and then remain quiescent until conditions are more favorable for disease development after harvest. Other pathogens infect produce during and after harvest through surface injuries. In the development of strategies for postharvest disease control, it is imperative to take a step back and consider the production and postharvest handling systems in their entirety. Many preharvest factors directly and indirectly influence the development of postharvest disease, even in the case of infections initiated after harvest. Traditionally, fungicides have played a central role in postharvest disease control. However, trends toward reduced chemical usage in horticulture are forcing the development of new strategies. This provides an exciting challenge for the 21st century. Biological-based ecofriendly, sustainable, and suited to human health interventions must be explored against the losses caused by several postharvest stresses. To avoid preharvest infection, proper and recommended practices must be ensured to eliminate the chance of pathogen survival in storage.

KEYWORDS

- **postharvest diseases**
- **pathogen**
- **antagonists**
- **biological control**

REFERENCES

Arya, S. S. Emerging Trends in Food Processing—Indian Scenario. *Process. Food Industries* **1998,** *1* (11), 19.

Australian United Fresh Fruit and Vegetable Association Ltd. *Fresh Produce Manual: Handling and Storage Practices for Fresh Produce;* Australian United Fresh Fruit and Vegetable Association Ltd: Footscray, Victoria, 1989.

Beattie, B. B.; Mc Glasson, W. B.; Wade, N. L. Postharvest Diseases of Horticultural Produce. In *Temperate Fruit;* CSIRO: Melbourne, 1989; Vol. 7.

Champ, R.; Highley, E.; Johnson, G. L. In *Postharvest Handling of Tropical Fruit,* Proceedings of an International Conference, Chiang Mai, Thailand, July 19–23, 1993, Australian Centre for International Agricultural Research, Canberra, ACT, 1984.

Coates, L.; Cooke, A.; Persley, D.; Beattie, B.; Wade, N.; Ridgeway, R. Postharvest Diseases of Horticultural Produce. In *Tropical Fruit;* DPI: Queensland, 1995; Vol. 2.

Kader, A. *Postharvest Technology of Horticultural Crops,* Publication 3311, University of California, Davis, California, 1982.

Kumar, P.; Awasthi, L. P. Management of Infection and Spread of Bottle Gourd Mosaic Virus Disease in Bottle Gourd Through Botanicals. *Indian Phytopath.* **2003,** *56,* 361.

Singh, Rinki; Tripathi, Pramila. A Review on Studies of Post-harvest Fungicides Diseases of Some Fruits. *Inter. J. Agric. Food. Sci. Technol.* **2013,** *4* (7), 721–724.

Snowden, A. L. A Colour Atlas of Post-Harvest Diseases and Disorders of Fruits and Vegetables. In *General Introduction of Fruits, Volume 2Vegetables;* Wolfe Scientific: London, (1990, 1991); Vol. 7.

Verma, L. R.; Joshi, V. K. Postharvest Technology of Fruits and Vegetables. An Overview. In *Postharvest Technology of Fruits and Vegetables,* 2000; Vol. 1, 68.

Wills, R. B. H.; McGlasson, W. E.; Graham, D.; Lee, T. H.; Hall, E. G. *Postharvest: An Introduction to the Storage and Handling of Fruit and Vegetables,* 3rd ed; University of New South Wales Press: Sydney, 1989.

CHAPTER 18

Postharvest Handling, Diseases and Disorders in Bulb Vegetables

SANGEETA SHREE* and AMRITA KUMARI

Department of Horticulture (Vegetable and Floriculture), Bihar Agricultural University, Sabour, Bhagalpur 813210, Bihar, India

Corresponding author. E-mail: sangeetashreee@gmail.com

ABSTRACT

Bulb vegetables have underground economic part modified into thickened and fleshy leaves that are used as food. Bulb vegetables belong to the genus *Allium,* which includes onion, garlic, leek, shallots, chives, and scallions. However, garlic and onion are the most common among them, and in general consumed throughout the world. It is estimated that the magnitude of losses because of inadequate postharvest handling, transportation, and storage in fresh vegetables is relatively higher, 20–50% in developing countries, in comparison to 5–25% in developed countries. The quality of both the crops can be improved after harvest by proper handling and storage. The only postharvest treatment required for the long storage of bulb crop is thorough adequate curing. The growing conditions also have an effect on quality of the onions in storage. Soil containing fingal spores or bacterium get washed or blown into the neck or base of the crop during the growing phase of the onion and cause disease. During suitable storage condition, wide range of fungal and bacterial pathogen survives on bulbs rendering them unfit for consumption and severe postharvest loss.

18.1 INTRODUCTION

Vegetables form the most important component of a balanced diet owing to their precious nutritional components such as vitamins, minerals, antioxidants, and a series of micronutrients indispensable for human health (Bozzini, 2002). Vegetables are potent protective supplementary food since they are the reservoirs for several vitamins, minerals, and amino acids. They are exclusive source of fibers required to keep the body fit and healthy. The vast range of photochemical present in fruits and vegetables provide them antioxidant, antibacterial, antifungal, antiviral, and anticarcinogenic properties.

Nutrients which are present in vegetables vary from crop to crop. Some contain useful amounts of protein although generally they contain little fat (Frodin, 2004). Peas and beans are enriched with proteins. Root crops such as tapioca, sweet potato, and potato are well known for carbohydrates. Some important minerals such as calcium, potassium, iron, magnesium are the important minerals which are present in abundant quantities in the vegetables such as peas, beans, spinach, and okra but are lacking in cereal cops. Onion and garlic are found to possess antibacterial, antifungal, and antiviral properties. They are also highly recognized for their medicinal value to human health. Vegetables may be considered to be the staple food next to cereal. Besides being cheap source, vegetables are capable of providing principle food nutrients to human beings. These are important and primary source of income of small and marginal farmers as the benefit cost ratio of vegetable cultivation is very high. With only very little investments, farmers can expect huge benefit. Vegetable cultivation generates employment round the year as several crops can be taken in 1 year. It is an intensive farming operation. More income per unit area is generated than in the cereal crops.

They have high aesthetic value and aids to overall increase in country's economy, growth, and development. However, the perishable nature of vegetables demands comprehensive planning for movement, storage, processing, and distribution of vegetable products, particularly for postharvest losses caused by various diseases. The huge postharvest losses (30–40%) occur every year because of the improper postharvest management practices. Due to high moisture content and tender nature, vegetable tends to pose postharvest problems. They are injured easily and are metabolically active than the durables, these features significantly limit the postharvest life and are easily prone to diseases. In fresh vegetables transpiration, respiration and sprouting of tuber and bulb crops leads to direct food loss. Vegetables are prone to microbial spoilage caused by fungi, bacteria yeast, and molds. Many serious postharvest diseases of fresh vegetables occur rapidly and causes extensive breakdown of commodity. Inadequate harvesting, transportation, storage, marketing facilities, and legislation lead to condition for postharvest losses in vegetables. They are spoilable and highly prone to losses because they are composed of living tissues. These tissues must be kept alive and healthy throughout the process of marketing. These are framed of thousands of living cells which require care and maintenance. Therefore, the reduction of postharvest losses of vegetables is a complementary means for increasing production. It is always wise to prevent loss after harvest as it costs lesser than that to produce the amount of the crop lost. The food availability can be increased to a considerable extent by minimizing the postharvest food loss.

18.2   BULB VEGETABLES

Bulb vegetables are produced underground. They have vertical shoots that have modified leaves (or thickened leaf bases) which are used as food storage organs by the dormant plants. Important bulb vegetables include chives, garlic, leeks, onion, and shallots. These vegetables belong to the genus *Allium*. Among vegetables, *Alliums* are perhaps the most important genus and are consumed throughout the world. The genus *Allium* comprises more than 500 known species distributed over Europe, North America, North Africa, and Asia (Frodin, 2004). The number of species belonging to the genus has grown steadily over the time as botanist has discovered new species throughout the world. The general characteristics of the *Allium* species are that the plants are almost exclusively herbaceous, perennials, and usually form bulbs. Some species, however, often forms thickened rhizomes. In genus *Allium*, garlic (*A. sativum*) and onion (*A. cepa*) are mostly consumed throughout the globe. Apart from being used as a prime vegetable, garlic also has numerous medicinal properties. In India, references about cultivation of onion and garlic are found from ancient times onwards as is evidenced by Charaka-Samhita, a famous early medical treatise of India. The varied medicinal uses and pharmacological actions like the anti-inflammatory, antihelmintic, and heart friendly nature of garlic were known many centuries ago (Kirtikar and Basu, 1975).

Garlic (*Allium sativum*) is an annual, vegetatively grown crop, which can be sown in mild and cold climates. There are different types or subspecies of garlic, and most notably are hard neck garlic and soft neck garlic (Volk, 2004). Right kind of garlic has to be recommended for a given latitude, since it can be day-length sensitive. In general, hard neck garlic is generally grown in cool climates, whereas the soft neck garlic is grown close to the equator. Garlic is a monocot, having flat leaves, slender scape, long-beaked spathes, and heads bearing bulbils. The small bulbs are always mixed with the flowers of the inflorescence and the flowers usually abort at the bud stage. Commonly, garlic is propagated through cloves. Cloves are the sole organs of storage and these are the modified axillary buds of the foliage leaves. At maturity, the main stem of the bulb, roots, and leaves attached to it all die, only the cloves remain to carry the plant on to the next generation (Brewster, 2008). Garlic is completely sterile,that is, garlic has no complete flower for performing pollination and unable to set seed, therefore propagated asexually only from cloves (Shemesh, 2008; Kamenetsky, 2007).

Onion (*Allium cepa*) is an important vegetable and is the most widely cultivated species of the *Allium* (Black, 2010). The onion is normally a biennial or a perennial plant, but for commercial purposes it is taken usually as an annual crop. The onion plant has bluish-green leaves which are tubular and hollow and the underground bulb is used for vegetable and salad purposes. Bulb formation starts when a certain day length is reached. Onion with multiple bulbs is also known as shallots and potato onions. At maturity, the foliage dies down and the outer layers of the bulb become dry and brittle. The crop is harvested and cured and the onions are ready for use or storage. The crop is attacked by several pests and diseases such as thrips, mites, nematodes, and various fungi cause rotting. Several postharvest diseases also spoil the postharvest quality of these vegetables and also reduce the storage life. Both abiotic and biotic causes are responsible for postharvest losses in vegetable.

Physiological changes occur after harvesting. Physiological changes can be caused by high temperature, low atmospheric humidity, and physical injury which lead to produce losses. Vegetables are highly susceptible to mechanical injury, which is generally received by the use of poor harvesting practices such as the use of rusted equipments for harvesting, unsuitable containers used at harvest time, or during the marketing process, for example, containers that can be easily squashed or have splintered wood, sharp edges or poor nailing, overpacking or underpacking of containers, and careless handling of containers, etc. This results in damage of fruits, internal bruising, and crushing of skin of produce. Vegetables contain a wide range of organic substrates and high water activity, and thus are good substrates for microbial spoilage. The most common pathogens causing decays in fruits and vegetables are species of the fungi *Alternaria, Botrytis, Botryosphaeria, Collectotrichum, Diplodia, Monilinia, Penicillium, Phomopsis, Rhizophus,* and *Sclerotinia* and of the bacteria *Erwinia* and *Pseudomonas* (Wills et al., 2007).

18.3 POSTHARVEST BIOLOGY AND TECHNOLOGY OF VEGETABLES

Harvested vegetables are living products. They are highly spoilable in nature and may be unacceptable for consumption if not handled properly. Furthermore, fresh horticultural products are important wares of international commerce after the globalization of trade and free trade agreements. Longer shipments and transport periods may finally increase the potential of heavy losses; hence, the importance of proper cares and techniques for handling of fresh produce after harvest has been recognized. Postharvest, the connecting link between the grower and the consumer is concerned with the biology of harvested plant materials and use of this knowledge to develop efficient and feasible handling technologies that retard the rate of senescence. The main purposes of applying postharvest technology to harvested vegetable crops are to minimize losses between harvest and utilization, to maintain possible quality with respect to appearance, texture, flavor, and nutrition.

All fresh vegetables are living tissues. Due to high moisture content, succulent nature, active metabolisms, tender nature, and rich in nutrients, they are vulnerable to dehydration, biotic and abiotic stresses, and mechanical injury and are usually highly perishable. These are the limiting factors for long term storage life of vegetables and cause significant deterioration following harvest. Postharvest losses can occur at any point in the production and marketing chain. It is estimated that the magnitude of these losses due to inadequate postharvest handling, transportation, and storage in fresh vegetables is relatively higher, 20–50% in developing countries when compared with 5–25% in developed countries. Kader (2005) reported that complete elimination of postharvest losses may be impossible and uneconomical, but to minimize them by 50% is possible and desirable.

A very effective way to increase food availability without further boosting crop production is by minimizing postharvest losses of produce because a lot of inputs have been invested in producing them. To reduce these losses, understanding the causes of deterioration in vegetables is the fundamental step, and followed by utilizing appropriate and affordable technological procedures to delay senescence and maintain quality of produce.

18.4   POSTHARVEST OF BULB VEGETABLES: FACTORS AFFECTING THE POSTHARVEST PERFORMANCE OF ONIONS AND GARLIC

There are several causes of postharvest losses in fresh vegetables. Since horticultural products are diverse in morphological structure, composition, developmental stages, and general physiology, the losses in fruits and vegetables are commodity specific (Wills et al., 2007). However, the main causes of produce deterioration are growth and active metabolism, water loss, physiological disorders, mechanical damage, and pathological breakdown (Kader, 2002; Wills et al., 2007).

18.4.1   GROWTH AND METABOLISM

Since vegetables are living entity, they are subject to metabolic and developmental changes even after harvesting. It is well known that the quality of the harvested commodities cannot be improved naturally after harvest but it can be maintained as such without further deterioration till their consumption if the rate of metabolic activities is slowed down by adopting the proper postharvest handling procedure. Respiration rate is a significant parameter determining the metabolic activity of a horticultural product, which is usually associated with the decline in quality of produce. Respiration which involves enzymatic oxidation of organic substrates with energy production results in O_2 consumption, and CO_2 and water production, which represents sum of all the metabolic activities of the tissue (Kader and Saltveit, 2003; Kays and Paull, 2004). Respiratory rate of produce after harvest have negative correlation with its storage life, that is, the higher the rate of respiration the shorter is the storability. It is because of the produce which is detached from its source of photosynthates and is entirely dependent on its own food reserves. Respiration rate of a horticultural commodity is affected by different biotic and abiotic factors. The abiotic factors affecting respiratory rate of fresh vegetables after harvest, temperature and gas composition, such as O_2, CO_2, or ethylene, surrounding the postharvest produce, is considered the most important in amending this physiological parameter. The postharvest technologists are mainly concerned with slowing down the rate of respiration for maintaining quality and maximizing storage life. Exposure to ethylene can be detrimental to quality of most fresh horticultural commodities; therefore, ethylene is of major concern to all produce handlers during postharvest period (Lin, 2009; Saltveit, 1999).

Elongation and curvature of stem vegetables and sprouting or rooting of tuberous and bulbous crops are undesirable and lead to a great reduction in market quality and accelerates deterioration. Most vegetables contain 80–90% water in fresh weight basis. As little as 5% losses in water have adverse effects on appearance, salable weight, and texture quality of many perishable commodities. Shriveling and physiological loss in weight leads to quality loss, especially in bulb crops, being stored for longer duration. Stomatal transpiration accounts for major of moisture loss in fresh fruits and vegetables commodities and it is more common in leafy vegetables.

18.4.2 PHYSIOLOGICAL DISORDERS

Physiological disorders in vegetables are because of exposure to unsuitable postharvest and preharvest environmental factors or nutritional imbalance arising during growth (Kader, 2002; Kays et al., 2004). The improper temperatures may lead to the interference in the normal metabolism of the harvested products. Chilling injury (CI) by low ($< 10–13°C$) and nonfreezing temperature is observed common with tropical and some subtropical vegetables. Common symptoms of CI are development of surface lesions, external and internal discoloration, water soaking of tissues, uneven ripening, and enhanced decay. The symptom becomes conspicuous after the commodity is transferred to room temperature. However, freezing injury results from holding the commodities below their freezing temperatures. The damage from ice crystals formed in tissues usually results in immediate collapse of the tissues and total loss of the commodity (Wang, 1993). Because of freezing injury, soft tissues lose their integrity and become translucent and watery in appearance and texture in bulb crops. When freezing is followed by thawing, the freeze damaged scales become a grayish yellow color (Conn et al., 2012).

Nutritional disorders arising from preharvest mineral imbalance are sometimes appearing only in postharvest products. Calcium, sulphur, boron are linked with postharvest-related deficiency disorders more than any other mineral. Sprouting in onion and garlic on storage has been noticed because of higher nitrogen application after bulb formation has initiated in the field. Respiratory disorders are associated with very low O_2 ($< 1\%$) and/or high CO_2 ($> 20\%$) concentrations in and/or around harvested produce in storage or packaging condition (Kader, 2003). Waxy breakdown in garlic is a storage disorder associated with storage at low oxygen concentration or poor ventilation condition.

18.4.3   MECHANICAL DAMAGE

Mechanical damage of fruits and vegetables caused by improper and unsuitable harvesting and postharvest handling methods is one of the most common and widespread reason for loss of quality and yield of horticultural products. Tuberous and bulbous vegetable crops, which grow underground, need extra care and attentiveness while being harvested. Mechanical damage not only directly affects skin and flesh lesions and browning but also creates sites for pathogen infection and water loss. Furthermore, physical injury stimulates ethylene production and respiration in plant tissues, which can lead to acceleration of senescence (Kader, 2002; Kays and Paull, 2004; Wills et al., 2007). Many times mechanical injury received by the vegetable during the transportation leads to cracks and breakage of inner tissue and cells. Various processing operations like grading, sorting, excessive polishing, peeling or trimming, packaging add to the loss of the commodity.

18.4.4   PATHOLOGICAL DECAY

Vegetables typically contain a wide range of organic substrates and high moisture content, which are good substrates for microbial spoilage. Major fraction of losses of vegetables after harvest is attributed to diseases caused by pathogens which flourish because of high water content of fruits and vegetables. Many serious postharvest diseases of fresh vegetables and fruits occur and spread very rapidly and cause their spoilage and loss. About 36% of vegetables perish because of infections caused by soft rot bacteria. Fungal and bacterial diseases are disseminated in the air and soil and via decaying plant material. Infection can happen any time once the crop is harvested, and also more frequently the result of improper harvesting or handling injuries (Wikipedia). The most common pathogens causing decays in vegetables are species of the fungi *Alternaria, Botrytis, Botryosphaeria, Collectotrichum, Diplodia, Monilinia, Penicillium, Phomopsis, Rhizophus,* and *Sclerotinia,* and bacteria like *Erwinia* and *Pseudomonas* (Wills et al., 2007). Both bacteria and fungi invade vegetables having pH above 4.5.The phenomenon of senescence, ripening, various abiotic stresses or cracking or splitting of fruits because of mechanical damage and CI may render the fruits susceptible to infections by all kinds of pathogens. In spite of the fact that such situations favor the attack of most of the pathogens, few such as *Colletotrichum* can actively penetrate the skin of healthy product (Kader, 2002; Wills et al., 2007).Bacterial soft rot occurs widely and causes serious diseases of vegetable crops in the field, in transit and especially in the storage. It causes a greater total loss of produce than any other bacterial disease (Agrios, 2006). Onion suffers from a great number of diseases during its lifecycle, right from sowing until harvest. As per international survey, it is evident that damages caused because of different diseases amounts to about 35–40% loss in onion crop (Gupta and Verma, 2002). Microbial infection can occur before and/or after harvest. Latent infection, or quiescent infection, is the state in which a product is infected prior to harvest with no obvious symptom developing until the pathogens are reactivated by onset of favorable conditions, such as fruit ripening or favorable temperatures.

18.5   TECHNOLOGIES TO IMPROVE POSTHARVEST QUALITY

The main objective of postharvest technology is to restrict deterioration of produce and to ensure that maximum market value. The technologies involved in postharvest handling of vegetables are enormously complicated because the products are divergent in their structural origin, developmental stage, physiological status, and nature of perishability. However, in order to protect the harvested products proper packaging, minimizing their respiration lowering storage temperature, or manipulating their physiology, eliminating or suppressing microbial activities, which are all the basis of postharvest techniques. Some commonly used and fundamental postharvest technologies are summarized below.

18.5.1   TEMPERATURE MANAGEMENT

In order to maintain quality and extend the shelf-life of horticultural crops after harvest, management of temperature is very crucial and effective, as temperature is responsible for the rate of most biochemical, physiological, physical, and microbiological reactions contributing to postharvest deterioration. Most of the fruits and vegetables have an optimum shelf-life at temperatures of around 0°C and with every 10°C increase in temperature the rate of deterioration of perishable products increases three times (Kader and Rolle, 2004).

Thus, storage at low temperature has been the main strategy to keep the harvested horticultural products. The major effect of the low temperature application between harvest and end use is a reduction of the produce metabolism and consequently a delay of the evolution of the parameters related to quality loss and senescence. The harvested product must be chilled and cooled as quickly as possible in order to maintain the quality of the products. Precooling is the first step for good temperature management. It is the process that removes field heat from freshly harvested products by a cooling treatment before shipment, storage, or processing. Because of slowing down the respiration rate and water loss, as well as the growth of decay microorganisms around, prompt cooling after harvest is desirable for retarding the deterioration of fruits and vegetable. There are different methods of precooling comprising room cooling, forced air cooling, hydrocooling, package icing, and vacuum cooling, each with specific advantage for particular produce. Bulb crops have to be necessarily cooled to room temperature with cool air forced through storage piles or bins before being delivered to market. Cooling may also be done with cold ambient air or with air cooled by mechanical refrigeration. Temperature management begins with the time of harvest. It is often good practice to harvest during the coolest part of the day to reduce product warming. Protect harvested produce from exposure to direct sunlight when accumulating fruits or vegetables in the field, then rapidly deliver them to packing house for precooling.

18.5.2   CONTROL OF WATER LOSS

Structure of vegetable and the physiological age at harvest largely influence the rate of water loss. Loss of water from fruits and vegetables depend upon the relative humidity, temperature, air movement, and atmospheric pressure of the environment. In general, larger the surface area of vegetables more rapid is loss of moisture. Water loss is directly proportional to the vapor pressure difference (VPD) between the commodity and its environment.

The basic principle of minimizing water loss from horticultural crops during postharvest period is to minimize the capacity of surrounding air to hold additional water (Wills et al., 2007). It can be achieved by commodity with the help of different treatments, such as curing, surface waxing or coating and plastic film wrapping, or by environment manipulations, such as reduction of vapor pressure deficit between the product and air via lowering temperature or by raising relative humidity (RH), or control of air movement (Kader, 2002; Kays and Paull, 2004; Wills et al., 2007), despite the fact that maintenance of a high relative humidity atmosphere is necessary to arrest water loss.

18.5.3 ATMOSPHERE MODIFICATION

The modified atmospheres best maintain the quality and shelf-life of produce. Alternation in the concentrations of the gases around horticultural products can importantly increase their storage life, resulting from reduction in rate of respiration of produce, retardation of senescence, and growth inhibition of many spoilage microorganisms. The terms controlled atmospheres (CA) or modified atmospheres (MA) create an atmospheric constitution around the produce which is different from normal air by addition or removal of gases. The levels of O_2, CO_2, N_2, and ethylene in the atmosphere may be manipulated. In CA and MA usually involve reducing O_2 levels below 5% and elevating CO_2 levels above 3%. Gas control is more precise in CA than in MA. The tolerance of produce to the damage because of declined O_2 and increased CO_2 concentration is a significant factor for successful development of CA and MA technology. Modified atmosphere packaging (MAP) is usually designed to maintain 2–5% O_2 levels and 8–12% CO_2 levels, which is widely used in extending the shelf-life of the produce. Controlled and modified atmospheric storage of horticultural crops extend their shelf-life by checking respiration and senescence, slowing down of various metabolic processes like ethylene biosynthesis and sensitivity of produce, reduction in incidence and severity of decay and control of fungi, bacteria and pests infestations in some commodities. Mismanagement of CA and MA technology could aggravate the physiological disorder, irregular ripening of fruits, development of off-flavor, and increase susceptibility to decay (Kader, 2002; Kader and Saltveit, 2003; Wills et al., 2007).

18.5.4 HEAT TREATMENT

Heat treatment is considered as environment friendly method to control deterioration. The most common methods of heat treatment are hot water immersion, forced-hot air treatment, and vapor heat treatment, etc. Hot water immersion has been used for fungal control and vapor heat treatment, specifically for insect control while forced-hot air treatment is used for both fungal and insect management according to Lurei (1998). The tolerance of produce to heat treatments must be carefully evaluated, inappropriate heat treatments may lead to heat injury. The benefit and damage caused because of heat treatment have to be studied very specifically and one has to be very precise and observant in its application as it a matter of difference of only a few degrees between its beneficial and injurious effect on the commodity on which it is applied. During the entire procedure of heat treatment, the threshold temperature and uniformity in space are the parameters of significant considerations.

16.6   POSTHARVEST HANDLING OF BULB VEGETABLES

Quality cannot be improved but can only be maintained after harvest. Therefore, postharvest handling of bulb crops require adequate attention .Vegetables should be harvested at the proper stages of maturity and should have attained appropriate size and be at the peak quality. Harvesting index is an important parameter and has prominent role in deciding the postharvest quality and shelf-life of horticulture commodity. Immature or overmature produce may not last as long in storage. Harvest should be done during the coolest part of the day and products be handled gently. For storage purposes, crops should be free from bruises, blemishes, rotten spots, decay, and other defects. These defects degrade the physical appearances of the commodity besides providing gateway for decay organisms as well.

Bulb crop, onion, and garlic are everyday household need but they are harvested once or twice in a year only. Therefore to make them available in the kitchen round the year, or to meet the demand for domestic as well as export market, they need to be handled with great care and concern immediately after their harvest. In India, *Kharif* or the rainy season onion is harvested in rainy and late rainy onion season and is consumed within 1 or 2 months as there is great demand during those months and therefore does not require storage. However, those harvested in winter is high in quantity and have to be stored so that it becomes available till next October–November. As for garlic, it is harvested only once in a year and it is harvested during March–April and the same product is utilized in the kitchen as spices or vegetable adjunct for almost a year. Therefore, adequate postharvest treatment must be given so that it can be stored for a longer period. Curing is a common postharvest practice followed in the bulb crops in which is a drying process to dry off the necks and outer scale leaves of the bulbs to check the moisture loss and microbial rot during storage. During the course of drying formation of new epidermal tissue called wound periderm takes place. This suberized epiderm also called periderm is responsible for safe storage of the bulb crops by preventing the infestation of disease pathogens and insect pest. Onions and garlic are often cured in the field. The bulbs are allowed to dry in field from 2 to 5 days or longer for curing. Field curing allows the inhibitors presence in the leaves to slowly move down in the bulbs. These inhibiters play vital role in dormancy of bulbs and also in inhibiting the sprouting. For curing the onion, it should be dried along with leaves for 3–4 days after harvesting. Thereafter, leaves should be topped off leaving 2–3 cm long neck along with the bulbs. These bulbs should be spread under shade for 2–3 weeks for proper drying and further curing. Where prevalent weather conditions are adverse, curing may be done artificially with warm forced air with temperatures around 25–32°C (77–90°F) for the development of attractive scale color.

The essentials factors for curing are heat and good ventilation, preferably with low humidity. Wright and Grant (1997) in his work on onion have observed that the outermost layer of onion, which were in direct contact with soil generally peel off during the curing and dry under layer is exposed to the surface, which bears an attractive appearance. Sorting and grading should be done at field level to minimize postharvest losses at subsequent stages. According to consumer preference, use of packing material for graded bulbs, avoiding drop of bulbs from more than 30 cm height, avoiding sunscald by eliminating overdrying of outer scales directly in sun, etc., improve shelf-life of onions.

Perforated hessian bags and plastic woven bags are used for onion packing to allow proper ventilation. Tier system of transportation on roads, restriction of loading height in trucks and wagons, providing good ventilation in railway wagons, and quick movement of onion wagons or truck loads are other factors which can help in minimizing the postharvest losses of onions.

Exposure of onion bulbs after harvesting, with 60–90 Gy irradiation, inhibit their sprouting regardless of crop season, environmental condition, and type of storage and also reduce the microbial and other losses.

18.7 POSTHARVEST PATHOLOGY

Losses caused by postharvest diseases are greater than generally realized because they have to go through several channels while passing from the field to the consumer. Postharvest losses (10–30%) are estimated per year despite the use of modern storage facilities and techniques (Harvey, 1978). Postharvest diseases affect a wide variety of crops, particularly in developing countries that lack postharvest storage facilities (Jeffries and Jeger, 1990). Infection by fungi and bacteria during the growing season at harvest time, during handling, storage, transport, and marketing, or even after purchase by the consumer (Dennis, 1983) causes losses produce in larger amount. Reduction of losses in food crops resulting from postharvest diseases has become a major objective of international organizations (Kelman, 1989).

18.7.1   FACTORS THAT INFLUENCE POSTHARVEST PATHOLOGY

Crops are grown in open conditions and are thus subjective to the influence of several biotic and abiotic factors. These conditions make them vulnerable to several diseases. Crop maturity and harvesting index, harvesting methods, and postharvest handling of the products have large impact on the storage and further processing of the crops, particularly the horticultural produce.

Parameters described below are the deciding factors for disease development in specific crop.

18.7.1.1   WEATHER

Weather is an important parameter that affects plant diseases. The amount of inoculums causing disease as well as the quantity of pesticide residue that remains on the crop at harvest depends upon the prevailing weather conditions (Conway, 1984). Plentiful inoculums and conducive conditions for contamination during the season leads to serious infection till the crop reaches maturity or are fit to be harvested. Postharvest decay accounts for further development of preharvest infections along with new infections that come up from germination of spores on the bulbs.

18.7.1.2   PHYSIOLOGICAL CONDITION

Physiological condition of produce at harvest determines how long the crop can be safely stored. For example, bulb vegetables which have been picked up at overmature stage and which have got cut and bruised during harvest perform poorly during storage. The stages of ripening and senescence are the most vulnerable phase as far as the contaminations by pathogens is concerned (Kader, 1985). However, management of crop nutrition can help in reducing the decay and rot in horticultural crops. Sulphur application on the standing crop is very helpful in developing disease resistance in onion and garlic.

18.7.1.3   FUNGICIDES

Postharvest application of fungicides to control decay are used on several major crops which are either stored or undergo long periods of transport to distant markets. A number of pathogens that cause significant postharvest losses in produce are of preharvest in origin. After harvest, the crops which need to be stored for later usage are treated with fungicides. There are several methods of fungicides usage like drenching, dipping, smoke or fumigant, hot water treatment, etc., which varies with the nature of the fungicide type and the type of the crop. Fungicides along with some growth retardants too are helpful in reducing the susceptibility of plants to various diseases. It is a established fact that the preharvest spray of maleic hydrazide@ 2000 ppm + carbendazim@ 1000 ppm at 30 days before harvest of onion bulb reduces the rotting and physiological losses and also enhances the shelf-life of onion bulbs (up to 6 months) and improved the quality parameters like TSS content, total sugar, reducing sugar, and sulphur content (Anbukkarasi, 2010).

18.7.1.4   PACKING AND SANITATION

It is necessary to maintain sanitary conditions in all areas where packing of the final produce is done. Organic matter like culls, extraneous plant parts, soil, etc., can act as substrate for decay-causing pathogens. Chlorine is helpful in killing microorganisms suspended in dump tanks and flumes if the amount of available chlorine is adequate. In recent times, chlorine dioxide was found to work against common postharvest decay fungi (Roberts and Reymond, 1994). Ozone treatment is nowadays being used for maintaining sanitation in food processing units. Ozone is helpful in killing foodborne pathogens and microbes responsible for spoilage and decay of food of food materials. The use of gaseous acetic acid for possible use as a sanitizing agent on several crop is still being experimented (Sholberg, 1998).

18.7.1.5   POSTHARVEST TREATMENTS

Postharvest treatments to control postharvest decay of fruits and vegetables may be categorized as biological or chemical treatments, depending upon several consideration like type of pathogen involved in the decay, location of the pathogen in the produce, best time for application of the treatment, maturity of the host, environment during storage, transportation, and marketing of produce (Ogawa and Manji, 1984)

18.7.1.6   CHEMICAL CONTROL

A number of chemicals are applied to horticultural products in order to obtain a desirable postharvest life. Most of these are applied after harvest but few are applied in the field even before harvest in order to obtain a specific result.

Several fungicides are presently used as postharvest managements for control of a varied range of decay-causing microorganisms. Bulb crops are generally treated with fungicide before they are stored for long term use. Growth retardant chemical like maleic hydrazideare sprayed about 15 days before harvest inhibits sprouting in onion. Ethylene dibromide and methyl bromide are generally used as fumigants to control insects or molds. Many products, mostly the products containing benomyl which were formerly used as postharvest treatment to enhance shelf-life are no longer permitted because of their alarming residual toxicity effect. Host of other products have lost their credibility as the target pathogen has developed resistance against them by their sustained long term usage.

18.7.1.7   BIOLOGICAL CONTROL

Postharvest biological control is a relatively a recent introduction and has several benefits as compared with the existing conventional biological control (Wilson and Pusey, 1985; Pusey, 1996) where the precise environmental conditions can be created and conserved, the biocontrol agent can be targeted much more efficiently, and the whole control systems and techniques are cost-effective. Biocontrols are undoubtedly effective but they do not always give steady results. The possible reason for this may be that the biocontrol efficacy is largely dependent on the amount of pathogen inoculum present (Roberts, 1994). Compatibility with chemicals used during handling is also important. Biological control agents must be combined with other strategies for effective disease control. Holistic approach using integrated management system is quite useful.

18.7.1.8   IRRADIATION FOR POSTHARVEST DECAY CONTROL

Ultraviolet light has no role in reducing decay of fruits and vegetables on storage but certainly has a lethal effect on bacteria and fungi that are directly exposed to the ultraviolet rays (Hardenburg et al., 1986). Gamma radiation effectively controls decay and enhance the storage life and shelf-life of fresh fruits and vegetables. In the future when methyl bromide is no longer available to control insect infestation in stored products, gamma irradiation may be used instead. In case of onion, application of irradiation is useful at very low dose levels (60–90 Gy) and inhibits sprouting when properly cured bulbs are irradiated within 2–3 weeks of harvesting.

18.7.2   POSTHARVEST DISEASES AND PESTS OF ONION AND GARLIC

A wide range of fungal and bacterial pathogens feed on and damage bulbs of onion, garlic, and other member of *Allium* species (Table 18.1 and Fig. 18.1). Usually these are organisms which are commonly found in the field or may be seed transmitted, as was found with *Botrytis allii* and certain bacterial pathogens.

Each pathogen has its own temperature requirements for its growth and development, whereas all are encouraged to grow by high humidity (>80% RH) during storage. They include low temperature pathogens such as neck rot, and those needing high temperatures like most of the bacteria, black mold or *Aspergillus niger*, and some other fungi. The cultivation and usage of bulb crop is gradually becoming popular among the growers as well as the consumers. The growers prefer to store the crop until the next planting season and the consumers also like the crop having long shelf-life. Therefore, it becomes imperative to have proper understanding and information about the pests and pathogens causing damage of the crop on storage.

Some of the more serious postharvest diseases and disorders which cause greater loss of produce are following:

(i) Fusarium basal rot of onion and garlic

Fusarium basal rot disease affects onion, garlic, leek, shallots, and also some other Allium members and are caused by *Fusarium oxysporumcepae*. The fungus persists in the soil debris and infects the bulbs which are injured either by insects or by faulty handling. During later half of the crop season, the symptoms appear on the standing crop as yellowing or tanning of leaves, wilting and dieback. In onion, rotting and discoloration of the basal disc appears in the beginning which gradually advances up into the spongy scale of bulb. The affected bulbs when cut appear brown and watery. Bulbs appear soft spongy or sunken and the rot progresses on storage. In garlic, white or light pink or reddish fungal growth may be noticed on clove and even shattering of the clove of affected bulbs has been seen. Such cloves are small, reduced in size and somewhat dry.

For control of this disease, long crop rotation of not less than 4 years duration should be followed with non-host crop such as cowpea, barley, or

TABLE 18.1   Important Postharvest Diseases and Pests of Onion and Garlic.

Disease	Causal agent	Class
Bulbs bacterial soft rot	*Erwinia caratovora*	Bacterium
Bacterial brown rot	*Pseudomnas aeruginosa*	Bacterium
Black rot	*Aspergillus niger*	Hyphomycete
Blue mold rot	*Penicillium* spp.	Hyphomycete
Fusarium basal rot	*Fusarium oxysporum*	Hyphomycete
Neck rot	*Botrytis* spp.	Hyphomycete
Purple blotch	*Alternaria porri*	Hyphomycete
Sclerotium rot	*Sclerotium rolfsii*	Agonomycete
Smudge	*Colletotrichum circinans*	Coelomycete
Mites	*Aceria tulipae*	Arachinids

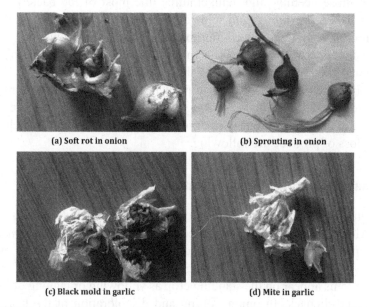

(a) Soft rot in onion   (b) Sprouting in onion

(c) Black mold in garlic   (d) Mite in garlic

FIGURE 18.1   (See color insert.) Important Postharvest Diseases and Pests of Onion and Garlic.

wheat. Resistant onion cultivars (Arka Pitamber and Arka Lalima) should be used. Plants should be protected from all types of damages because of insects, improper cultural practices, and various types of mechanical injuries. Bulbs should be cured properly and should be dried enough before topping. Proper handling and curing is essentially required for safe storage of the bulbs. Storage of damaged and injured bulb must be avoided. Bulb should only be stored in dry conditions with proper ventilations.

(ii) Botrytis neck rot of onion and garlic

Botrytis neck rot is a serious storage disease of onion and garlic of fungal origin, mainly Botrytis *allii* which grows optimally at 21°C and is therefore a problem in temperate climates, such as Northern Europe and Canada. The pathogen is soilborne as well as seed borne. The infection survives in dormant stage on plant debris, in diseased bulb, and also as sclerotia in soil. The Botrytis neck rot on storage is caused by different species than the Botrytis leaf spot, seen on the standing crop. Infections of Botrytis neck rot disease spread through neck tissue or wounds in bulbs. Bulbs which do not have very clear and visible symptoms of this disease become infected and symptomatic during topping of the necks, when the fungus directly enters the neck via airborne spores. The neck rot disease in onion is more common and clear and conspicuous after harvest which aggravates and progresses rapidly on storage. Symptoms occur after 8–10 weeks of storage, with a softening and rotting of neck tissues. Numerous small and small black sclerotia develop beneath the outer dry skins (Hayden and Maude, 1997). As the infection advances, the rotting and discoloration moves in to the inner scales and finally bulb ends up into a soft mass.

In garlic, the disease generally becomes visible at initial stage on necks near the soil surface. The fungus travels swiftly into the neck region of succulent garlic bulb, producing water soaked appearance. A grey mould develops on the surface of the bulb or between garlic scales. Later, black bodies called sclerotia develop around the neck. The infected plant may not survive at all or infected bulb may be reduced into a soft mass and saprophytes may destroy it further.

Neck rot is more severe if infection occurs early in the growing season and after this artificial curing may not be effective (De Visser et al., 1994). Moist, cool, and humid conditions favor disease development. There is evidence to show that *B. allii* conidia produced at low temperatures cause more rapid and destructive rots than the conidia produced at higher temperatures (Bertrolini and Tian, 1997).

Onion and garlic bulbs tops should be allowed to mature well before lifting them. In dry weather, bulbs should be cured on the ground for about 6–12 days. Only mature, dry bulbs with tight feel and appearance should be harvested. Bacterial diseases and Botrytis neck rot fungus do not move in dry tissue. Hence, proper curing and drying prevents infection of these diseases. Care should be taken to ensure minimum bruising and least mechanical injury in topping and storing. They should be stored in well ventilated houses at 32°F or slightly higher. They can be stored at higher temperatures if humidity cannot be held below 75%. The other means of discouraging pathogen development include postharvest treatment by heated air drying at 30–32°C during the early stage of storage. Crop rotation should be adopted. Frequent and excessive irrigation which creates moist condition should be avoided.

(iii)   Purple blotch of onion and garlic

Purple blotch of onion and garlic is caused by a fungus *Alternaria porri,* The pathogens of this fungus survive on infected bulbs and debris in the field and also in seeds. This disease appears on the leaves as large bleached lesions with purple center that rapidly gets enlarged, eventually leading to rot of infected bulbs. The most favorable temperature is 28–30°C with 80–90% RH. Another fungus, *Stemphylium vesicarium,* attacks both onion and garlic. This is a major fungal disease on leaves and seed crop in Northern India. It can also cause purple blotch in onion. Both these diseases are the most common as leaf diseases, but can also affect bulbs in storage. Secondary infection often follows as injury caused either by other fungi, bacteria, viruses, insects, or by sand or dust particles on windy days. Mature leaves and plants are more susceptible than young and juvenile plants. Spores require moist and humid conditions and continual dew to cause infection. Optimum temperature to induce disease epidemics is ranges between 77°F and 81°F (Cavanagh and Hazzard, 2013.) Infection does not occur below 55°F. Symptoms appear as small yellow to orange spots or streaks in the middle of leaves and on flower stalks on one side. In moist weather condition, these spots become covered with a brownish black, powdery fungus growth. Leaves with large spots turn yellow and are blown over by the wind. Bulbs may decompose or grow moldy during or after harvest. At first, watery rot around the neck appears which further progresses into yellowish to wine red discoloration in the neck region and eventually the whole bulb dies. Cultural control like crop rotations, lesser hours of leaf wetness, wider plant spacing, and good cultural control should be practiced to avoid and evade such fungal infections. Old onion culls should be destroyed and debris must be buried or burnt. Wounded, bruised, and damaged bulbs should be removed. Proper curing of bulbs in the field is recommended. Optimum drying of bulbs before lifting and topping is advised. Tolerant or resistant varieties must be planted. Eexcessive irrigation is often the culprit. Fungicides and some insecticides in case of secondary infections can be used to fight this disease.

(iv) White rot of onion and garlic

White rot of onion and garlic is caused by a fungus, *Sclerotium cepivorum* which is soil borne. All members of *Allium* family such as onion, leek, garlic, and shallot can be infected by this fungus but onion and garlic are the most susceptible. White rot disease spreads through sclerotia which live in the soil for a long period and even one sclerotia can infect a large number of adjoining plants. Sclerotia can infect plants from 25 to 30 cm below the soil surface and symptoms appear as watery rot of bulbs and roots with some fuzzy white material, which is nothing but the fungal mycelium. Above the ground, symptoms first appear on the outer leaves which proceed inward as the disease advances usually anytime from mid-season to harvest. Even 100% plants can die because of this disease. Infected bulbs become unmarketable or fetch very poor price. Once the disease comes in a field, it is very not an easy task to grow onion and garlic productively. Disease spreads with infected seeds, sets or transplants, water, equipment, shoes, grazing cattle (movement of infested soil), and in the wind. Fungal activity is favored by cool weather and is restricted above 75°F. High humidity also favors the disease. Decay of infected bulbs because of white rot in storage can persist if humidity continues to be high. White rot thrives under the same climatic condition that are conducive for onions and garlic (cool weather and moist soil). Hence, it is difficult to evade the pathogens by altering planting time. Only disease-free seeds or sets should be planted in disease-free soil. Rouging for infected or the diseased plants from field of the healthy plants must be practiced. Soil solarization practices in infected areas can reduce the number of sclerotia. All equipment, boots, etc., must be washed with water so that all soil is washed off. Safe disposal of the material should be practiced. Diallyl disulfide (DADS) can be applied artificially in the field in the absence of Alliums, so that sclerotia germinate and in the absence of *Allium* host, they die, rather than lying dormant (Bob Ehn et al., 2012). Three fungicides, tebuconazole, fludioxonil, and boscalid, which are currently registered for white rot control can be useful.

(v) Blue mold of garlic

Blue mold of garlic is caused by *Penicillium* spp. fungi. It is also known as Penicillium mold. Common species of *Penicillium* that cause blue mold in garlic are *P. hirsutum* (syn. *P. corymbiferum*), *P. aurantiogriseum* (syn. *P. cyclopium*), *P. citrinum, P. digitatum, P. expansum,* and *P. funiculosum.* Infection first occurs on wounds or bruises that occur when cloves are separated from the parent bulb. The fungus is transferred from soil to the bulbs and the cloves. The symptoms become prominent on storage of bulbs. At start, yellowish lesions appear which are soon covered with the characteristic blue–green spores. In extreme cases, the entire clove may be reduced to a mass of spores. The primary source of inoculums is the infected planting material. Airborne spores contaminate the cloves when the bulbs are broken for planting. Wounded and bruised cloves are very much susceptible to this disease. Cloves are often invaded by saprophytes which further aggravate the disease. Spore germination and disease development is favored during warm weather conditions. Planting garlic early when soil temperatures are still high (above 25°C) favors disease incidence and augments disease severity. Bulb should be harvested carefully. Care should be taken to maintain minimum bruising and wounding while harvesting the bulbs. The bulbs must be cured and dried promptly. Hot water seed should be done before planting of the cloves else the germination would be adversely affected. Bulb should be stored at 4–5°C with low relative humidity for safe and longer storage.

(vi) Black mold

Black mold, caused by *Aspergillus* spp. fungi, is a postharvest disease under hot climate which cause significant losses. The disease is common in onion and garlic stored in hot climates where the temperature ranges between 30°C and 40°C. Symptomatically, bulbs show black tint at the neck and streaks of black mycelium and conidia beneath the outer dry scales. The black discoloration is because of black dust of spore clusters of fungus. The bulb has cheap market value because of the black appearance on the outer as well as the inner scales of the bulb. When the garlic scales are thin, these spores mass is usually visible through the scales. In severe stage of disease development, all the cloves get infected and the bulbs wither. For management of the disease, the bulb should be stored only after proper drying. Onion should be stored at 1–15°C. Bruising should be avoided when bulbs are harvested, stored, or transported. The crop should be sprayed with fungicides 10–15 days before harvesting. Before planting, garlic cloves should be treated with suitable fungicides.

(vii) Bulb canker/skin blotch

Bulb canker/skin blotch is a one of the major problems of onion garlic on storage caused by *Embellisia alli* (Campan). The symptoms include the manifestation of greyish spots on the outer scales of the bulb in the initial stage and slowly the entire bulb is coated with dark blackish color. To evade this problem, bulb should be stored essentially only after proper curing and drying with forced heated air at 27–35°C. Application of suitable fungicide during the period of growth reduces the severity in the symptoms of stored bulbs of onion and garlic. Irrigation should be strictly discontinued and tops should be allowed to dry down as the time of harvest comes nearer.

(viii) Bacterial brown rot

Bacterial brown rot is caused by bacterium *Pseudomnas aeruginosa*. It is a serious storage disease and infection is spread through the oozing of the neck. Dark brown discoloration in bulb scale is the typical feature of this disease. Browning and rotting of inner scale is an important symptom of this disease. The inner scales are first affected and show rotting symptoms which is soon passed on to the outer scales. Affected bulb should not be stored along with the healthy bulbs. Rain at maturity aggravates the problem. Streptocyclin 200 ppm is recommended at weekly interval, if rain occurs at maturity stage, to minimize diseases incidence percentage during storage. Neck cutting about 2.5–3.0 cm long above the bulb must be practiced to reduce the bacterial infection incidence. Light irrigation should be provided during entire cropping period.

(ix) Mites

The most common species of mites which infects wild and cultivated members of the genus *Allium,* including onion, garlic, and leeks is *Aceria tulipae* or *Eriophyes tulipae*. This mite is different from wheat curl mite, which infects cereal crops. Mites attack the bulb crops in the field as well as in storage, where it is more troublesome. They suck the plants and give them pale and sickly appearance. Infected plants may lack vigor and bear stunted, deformed, and weak leaves. In storage, mites feed on the bulbs of garlic and as a result sunken brown spots are formed on the cloves. The injured cloves later dry and desiccate and cause huge loss to the garlic bulbs in storage. The wounds caused by feeding of mites also serve as doorway for soft rot bacterial and fungal pathogens and may lead to rotting. The eggs, nymphs, and adults of mite overwinter in soil and also in infected garlic during storage. Hot water treatment of bulbs at 140°F for 10–15 min prior to planting is effective in reducing mite populations, but can also affect germination adversely. Good control was reported with soaking affected cloves for 24 h in 2% soap and 2% mineral oil before planting. When the infestations are not very high, it can be controlled with the normal curing and drying process before storage. Rotation of crops out of *Allium* for at least 4 years help in reducing incidence of mite attack.

(x) Waxy breakdown

Waxy breakdown is a physiological disorder that affects garlic during later stages of growth, particularly near harvest or afterwards in storage and is usually associated with high temperature during these periods. Poor ventilation and lack of oxygen supply during storage results in waxy breakdown in garlic. Early symptoms are small, light-yellow sunken areas in the clove flesh that darken to amber color. Later, the clove becomes translucent, soft, rubbery, gummy, and waxy. The outer scale covering of the bulb is not affected and it often obscures the inner symptoms until very advance stage is reached and cloves begin to shrink and shrivel and the yellow waxy cloves becomes noticeable through the outer flimsy scales. Waxy breakdown in garlic normally occurs in storage and during long distance and hardly ever in the field. Proper aeration and sufficient oxygen levels during storage may reduce the cause of waxy breakdown in garlic.

KEYWORDS

- **bulb**
- **onion**
- **postharvest**
- **storage**
- **vegetables**

REFRENCES

Abdalla, A. A.; Mann, L. K. Bulb Development in the Onion (*Allium Cepal.*) and the Effect of Storage Temperature on Bulb Rest. *Hilgardia* **1963,** *35,* 85–112.

Agrios, G. N.*Bacterial Soft Rots.* 5th ed; Academic Press: San Diego, 2006.

Anbukkasari, V. Studies on Pre and Post Harvest Treatments for Extending Shelf Life of Onion (*Allium Cepal. Varaggregatumdon*).Cv Co On 5. Ph.D. Thesis, Department of Vegetable Crops, Tamil Nadu Agriculture University, Coimbatore, 2010.

Ara, M. A, M.; Khatun, M. L.; Ashrafuzzaman, M. Fungi Causing Rots in Onions at Storage and Market. *J. Bangladesh Agril. Univ.* **2008,** *6* (2), 245–251.

Bahnasawy, A. H.; Dabee, S. A. Technological Studies on Garlic Storage, 2006. www.bu.edu. eg/portal/uploads/.

Bertrolini, P.; Tian, S. P. Effect of Temperature of Production of B. *Allii* Conidia on Their Pathogenicity to Harvested White Onion Bulbs. *Plant Pathol.* **1997,** *46,* 432–438.

Ehn, B.; Ferry, A.; Turini, T; Crow, F. efile:///C:/Users/welcome/ Downloads/Fall%20 2012%20Newsletter-insert.pd.

Bozzini, A. Vegetable Research and Poverty Alleviation. In *Vegetables for Sustainable Food and Nutrition Security in the New Millennium.* Souvenir International Conference on Vegetables: Bangalore, India, 2002, pp 32–37.

Brewster, J. L. *Onions and Other Alliums.* CABI Publishing. ISBN 978-1-84593-399-9.

Brice, J.; Currah, L.; Malins, A.; Bancroft, R. *Onion Storage in the Tropics.* NRI Publications: University of Greenwich, UK, 1997.

Brickell, C. *The Royal Horticultural Society Encyclopedia of Gardening.* Dorling Kindersley, 1992, pp 345. ISBN 978-0-86318-979-1.

Büchner, F. L.; Bueno-de-Mesquita, H. B.; Ros, M. M.; Overvad, K.; Dahm, C. C.; Hansen, L.; Tjønneland, A.; Clavel-Chapelon, F.; Boutron-Ruault, M. -C. "Variety in Fruit and Vegetable Consumption and the Risk of Lung Cancer in the European Prospective Investigation into Cancer and Nutrition". Cancer Epidemiology, Biomarkers & Prevention: A Publication of the American Association for Cancer Research, Cosponsored by the American Society of Preventive Oncology, 2010, Vol. *19* (9), *pp 2278–2286.* doi:10.1158/1055-9965.EPI-10-0489. ISSN 1538-7755. PMID 20807832.

Bufler, G. Quality Comparison for Seed and Plant Shallot Cultivars. *Gemüse (München)* **1998,** *34,* 460–462 (In German).

Cavanagh A.; Hazzard, R. Information & Images from Oregon State Extension, 2013, http:// ipmnet.org/plantdisease/.

Ceponis, M. J.; Butterfield, J. E. Dry Onion Losses at the Retail and Consumer Levels in Metropolitan New York. *Hort. Sci.* **1981,** *16,* 531–533.

Conn, K. E.; Lutton, J. S.; Rosenberger, S. A. Onion Disease Guide (A Practical Guide for Seedsmen, Growers and Agricultural Advisors). Seminis Vegetable Seeds, USA, 2012.

Couey, H. M. Heat Treatment for Control of Postharvest Diseases and Insect Pests of Fruits. *Hort. Sci.* **1989,** *24,* 198–202.

Currah, L.; Proctor, F. J. *Onions in Tropical Regions.* Natural Resources Institute: Chatham, UK,1990, pp 232.

De Visser, C. L. M.; Hoekstra, L.; Hoek, D. *Research into Effective Chemical Control of Leaf Spot and Neck Rot and into Methods to Predict Neck Rot in Onions.* Verslag Proefstation voor de Akkerbouwende Groenteteelt in de Vollegrond No. 178, Proefstation voor de

Akkerbouwende Groenteteelt in de Vollegrond, Lelystad: The Netherlands, 1994, pp 85 (in Dutch).

Dennis, C. *Postharvest Pathology of Fruits and Vegetables*. Academic Press: London, UK, 1983.

Eric, B. "*Garlic and Other Alliums: The Lore and the Science.*" Royal Society of Chemistry: Cambridge, 2010

Fritsch, R. M.; Friesen, N."Chapter 1: Evolution, Domestication, and Taxonomy". In: *Allium Crop Science: Recent Advances*; Rabinowitch, H. D.; Currah. L., Eds.; CABI Publishing: Wallingford, UK, 2002, pp 9–10. ISBN 0-85199-510-1.

Frodin, D. G. "History and Concepts Of Big Plant Genera". *Taxonomy* **2004,** *53* (3), 753–776. doi:1 00.2307/4135449.

Germplasm Resources Information Network (GRIN). "*Allium cepa* information from NPGS/ GRIN". USDA, ARS, National Genetic Resources Program. Retrieved 22 April 2011.

Gregory, E. W. *Vegetable Production and Practices.*

Gupta, R. P.; Verma, L. R. Problem of Diseases during Storage in Onion and Garlic and Their Strategic Management. In: *Implication of Plant Diseases on Produce Quality*, Singh, D. P., Ed.; Kalyani Publishers: Ludhiana, pp 55–62.

Hardenburg, R. E.; Watada, A.E.; Wang, C. Y. *The Commercial Storage of Fruits, Vegetables, and Florist and Nursery Stocks. Agriculture Handbook 66, U.S. Department of Agriculture, Agricultural Research Service*: Washington, DC, 1986.

Harvey, J. M. Reduction of Losses in Fresh Fruits and Vegetables. *Ann. Rev. Phytopath.* **1978,** *16*, 321–341.

Hayden, N. J.; Maude, R. B. The Use of Integrated Pre and Post Harvest Strategies for the Control of Fungal Pathogens of Stored Temperate Onions. *Acta Hort.* **1997,** *433*, 475–479.

Hayden, N. J.; Maude, R. B.; Proctor, F. J. Studies on the Biology of Black Mould (*Aspergillus Niger*) on Temperate and Tropical Onions. 1. A Comparison of Sources of the Disease in Temperate and Tropical Field Crops. *Plant Pathol.* **1994,** *43*, 562–569.

http://www.cdc.gov./nutrition/everyone/fruitsvegetables/index.html "Fruits and vegetables". *Nutrition for everyone.Centers for Disease Control and Prevention.* Retrieved 2015-03-30. https://en.wikipedia.org/wiki/Vegetable

Jeffries, P.; Jeger, M. J. The Biological Control of Postharvest Diseases of Fruit. *Biocontrol News Info.* **1990,** 11, 333–336.

Jones, H. A.; Mann, L. K. *Onions and Their Allies*. Leonard Hill: London, UK, 1963.

Kader, A. A.; Saltveit, M. E. Atmosphere Modification. In *Postharvest Physiology and Pathology of Vegetables*; Bartz, J. A.; Brecht, J. K., Eds., 2nd ed; Marcel Dekker, 2003, pp 229–246.

Kader, A. A. Postharvest Biology and Technology: An Overview. In: *Postharvest Technology of Horticultural Crops*; Kader, A. A., Ed., 3rd ed; Agriculture & Natural Resources: University of California, 2002pp 39–47.

Kader, A. A. Increasing Food Availability by Reducing Postharvest Losses of Fresh Produce. *Acta Hort.* **2005,** *682*, 2169–2175.

Kader, A. A. 1985. Biochemical and Physiological Basis for Effects of Controlled and Modified Atmospheres on Fruits and Vegetables. *Food Technol.* **1985,** *40*, 99–104.

Kader, A. A.; Saltveit, M. E. Respiration and Gas Exchange. In: *Postharvest Physiology and Pathology of Vegetables*, 2nd ed; Bartz, J. A.; Brecht, J. K., Eds.; Marcel Dekker, Inc.: New York, USA, 2003, pp 7–23.

Kader, A. A.; Rolle, R. S. *The Role of Post-Harvest Management in Assuring the Quality and Safety of Horticultural Produce*. FAO Agricultural Services Bulletin: FAO Rome, 2004.

Kamenetsky, R. Garlic: Botany and Horticulture. Hort. Rev. *Am. Soc. Hort. Sci.* **2007**, *33*, 123–171.

Karmarkar, D. V.; Joshi, B. M. Respiration Studies of the Alphonse Mango. *Ind. J. Agric. Sci.* **1941**, *11*, 993–1005.

Kashmire, F. R.; Cantwell, M. Postharvest Handling Systems: Underground Vegetables (Rotos, Tubers and Bulbs). In: *Postharvest Technology of Horticultural Crops*, 2nd ed; Kader, A., Ed.; University of California, Davis CA, USA, 1992, pp 271.

Kays, S. J.; Paull, R. E. *Postharvest Biology*. Exon Press: USA, pp 568.

Kelman, A. Introduction: The Importance of Research on the Control of Postharvest Diseases of Perishable Food Crops. *Phytopathology* **1989**, *79*, 1374.

Kirtikar, K. R.; Basu, B. D. *Indian Medicinal Plants,* Vol. 4, Bishen Singh Mahendra Pal Singh Periodical Experts: New Delhi, 1975.

Li, T. S. C. Vegetables and Fruits: Nutritional and Therapeutic Values. CRC Press, 2008, pp 1–2. ISBN 978-1-4200-6873-3.

Lin, Z.; Zhong, S.; Grierson, D. Recent Advances in Ethylene Research. *J. Exp. Botany* **2009**, *60*, 3311–3336.

Lurei, S. Postharvest Heat Treatments. *Postharvest Biol. Technol.* **1998**, *14*, 257–269.

Magruder, R.; Webster, R. E.; Jones, H. A.; Randall, T. E.; Snyder, G. B.; Brown, H. D.; Hawthorn, L. R; Wilson, A. L. *Descriptions of Types of Principal American Varieties of Onions*.USDA: Washington, DC, USA, 1941a.

Magruder, R.; Webster, R. E.; Jones, H. A.; Randall, T. E.; Snyder, G. B.; Brown, H. D.; Hawthorn, L. R. *Storage Quality of the Principal American Varieties of Onions*. USDA: Washington DC, USA, 1941b, pp 8.

Mahmud, M.S.; Monjil, M. S. Storage Diseases of Onion under Variable Conditions. *Progressive Agri.* **2015**, *26*, 45–50.

Mari, M. T.; Cembali, E. B.; Casalini, L. Peracetic Acid and Chlorine Dioxide for Postharvest Control of *Monilinialaxa* in Stone Fruits. *Plant Dis.* **1999**, *83* 773–776.

Maude, R. B.; Shipway, M. R.; Presly, A. H.; O'Connor, D. The Effects of Direct Harvesting and Drying Systems on the Incidence and Control of Neck Rot (*Botrytis Allii*) in Onions. *Plant Pathol.* **1984**, *33*, 263–268.

Noureddine, B. Effect of Maleic Hydrazide on Respiratory Parameters of Stored Onion Bulbs (*Allium cepa* L.) *Braz. J. Plant Physiol.* **2004**, *16* (1).

Ogawa, J. M.; Manji, B. T. Control of Postharvest Diseases by Chemical and Physical Means. In: *Postharvest Pathology of Fruits and Vegetables: Postharvest Losses in Perishable Crops*; Moline, H. E., Ed University of California: California Agricultural Experiment Station, Davis, CA, 1984, pp 55–66.

Pusey, P. L. Micro-organisms As Agents in Plant Disease Control. In: *Crop Protection Agents from Nature: Natural Products and Analogues*; Copping, L. G., Ed; Royal Society of Chemistry: Cambridge, United Kingdom, 1996, pp 426–443.

Rees, D.; Farrell, G.; John. Orchard Crop Post-Harvest: Science and Technology: Perishables. Blackwell Publishing Ltd, 2012.

Roberts, R. Integrating Biological Control into Postharvest Disease Management Strategies. *Hort. Sci.* **1994**, *29*, 758–762.

Saltveit, M. E. Effect of Ethylene on Quality of Fresh Fruits and Vegetables. *Postharvest Biology and Technology* **1999**, *15*, 279–292.

Schery, R. W. *Plants for Man* (Adapted from Vavilov). Prentice Hall: Englewood Cliffs, New Jersey, 1954.

Shemesh, E.; Scholten, O.; Rabinowitch, H. D.; Kamenetsky, R. Unlocking Variability: Inherent Variation and Developmental Traits of Garlic Plants Originated from Sexual Reproduction. *Planta* **2008,** *227,* 1013–1024.

Sholberg, P. L. 1998. Fumigation of Fruit with Short-Chain Organic Acids to Reduce the Potential of Postharvest Decay. *Plant Dis.* **1998,** *82,* 689–693.

Spotts, R. A.; Peters, B. B. Chlorine and Chlorine Dioxide for Control Of 'd'Anjou' Pear Decay. *Plant Dis.* **1980,** *64,* 1095–1097.

Stow, J. Resistance of 20 Cultivars of Onion (*Allium cepa*) to Rotting and Sprouting During High Temperature Storage. *Expt. Agric.* **1975,** *11,* 201–207.

Terry, L.. *Health-Promoting Properties of Fruits and Vegetables. CABI, 2011, pp 2–4.* ISBN 978-1-84593-529-0.

Thompson, A. K. *The Storage and Handling of Onions.* Report of the Tropical Products Institute: London, 1982, pp G160.14.

Thompson, A. K.; Booth, R. H.; Proctor, F. J. Onion Storage in the Tropics. *Trop. Sci.* **1972,** *14,* 19–43.

Utkhede, R. S.; Sholberg, P. L. Biological Control of Diseases of Temperate Fruit Trees. *Curr. Top. Bot. Res.* **1993,** *1,* 313–334.

Volk, G. M.; Henk, A. D.; Richards, C. M. Genetic Diversity Among Us Garlic Clones as Detected Using Aflp Methods. *J. Amer. Soc. Hort. Sci.* **2004,** *129,* 559–569.

Wang, C. Y. Approaches to reduce Chilling Injury of Fruits and Vegetables. *Hort. Rev.* **1993,** *15,* 63–95.

Wills, R. B. H.; McGlasson, W. B.; Graham, D.; Joyce, D. C. *Postharvest—An Introduction to the Physiology and Handling of Fruits, Vegetables and Ornamentals*, 5th ed, CAB International: Oxfordshire, UK, pp 227.

Wilson, C. L.; Pusey, P. L. Potential for Biological Control of Postharvest Plant Diseases. *Plant Dis.* **1985,** *69,* 375–378.

Wisniewski, M. E.; Wilson, C. L. Biological Control of Postharvest Diseases of Fruits and Vegetables: Recent Advances. *Hort. Sci.* **1992,** *27,* 94–98.

Wright, P. J.; Grant, D. G. Effects of Cultural Practices at Harvest on Onion Bulb Quality and Incidence of Rots in Storage. *New Zealand J. Crop Hort. Sci.* **1997,** *25,* 353–358.

Wright, P. J. ; Grant, D. G.; Triggs, C. M. Effects of Onion (Allium Cepa) Plant Maturity at Harvest and Method of Topping on Bulb Quality and Incidence of Rots in Storage. *New Zealand J. Crop Hort. Sci.* **2001,** *29,* 85.

Wright, R. C.; Lauritzen, J. I.; Whiteman, T. M. *Influence of Storage Temperatures on Keeping Qualities of Onion & Onion Sets.* Technical Bulletin of the US Department of Agriculture No. 475. US Department of Agriculture: Washington, DC, 1935.

www.dogr.res.in. Directorate Of Onion And Garlic Research (DOGR):ICAR.

Zohary, D.; Hopf, M. *Domestication of Plants in the Old World: The Origin and Spread of Cultivated Plants in West Asia, Europe, and the Nile Valley*, 3rd ed. Clarendon, Oxford, UK, 1988.

CHAPTER 19

Management of Soilborne Diseases of Vegetable Crops Through Spent Mushroom Substrate

DURGA PRASAD* AND SRINIVASARAGHAVAN A.

Department of Plant Pathology, Bihar Agricultural University, Sabour 813210, Bihar, India

Corresponding author. E-mail: dp.shubh@gmail.com

ABSTRACT

Edible mushrooms are cultivated mainly on substrates, natural plant wastes are due to their capability in utilizing lignin, cellulose, hemicelluloses for the production of fruit bodies with excellent dietary and therapeutic attributes. Mushroom cultivation is an age-old tradition and is considered to be an ecofriendly activity as it employs the byproducts of agriculture and allied activities. The substrate left out after the harvest of mushroom is referred as "spent mushroom." The substrate is exhausted by the fungi for its growth and production of fruiting body. The exhausted substrate is recomposted by several methods and it is useful to several crop plants with respect to plant growth as well as suppression of plant disease. The mushroom waste is known to be an excellent source of carbon, nitrogen, and other essential elements. After a few cycles of fungal growth, the natural plant waste used as substrate becomes a simpler form of protein-rich component, which can be used as very good soil conditioners for the cultivation of crops and also for the production of microbial bioagents. The use of substrate as soil amendment is known to play a vital role in reducing the severity of soil-borne diseases in vegetable crops. Soil amendment promotes a population of antagonistic microorganisms, which reduces the population of pathogenic fungi. Aged compost, upon recolonization with heterotrophic fungi, actinomycetes, and mesophilic bacteria mitigates plant diseases as well. It also stimulates a natural plant defense system and the organic substances produced from substrate in soil are known to suppress the pathogens. Soil application singly or in combination with bioagents is useful to increase the population of antagonists and minimize different soilborne fungal pathogen. The application promotes to reduce the soilborne bacterial pathogens, various nematodes, and a few soilborne viruses.

19.1   INTRODUCTION

The burgeoning population poses a great challenge in terms of achieving food and nutritional security. In order to achieve the set goals of food production, alternate food sources need to be encouraged. At the same time, the demand for protein-rich food and the incapability of conventional methods in fulfilling this goal compel the need to explore alternatives for producing protein-rich food at low production cost (Mukherjee and Nandi, 2004). In this context, mushroom finds itself highly suitable because of its simplicity of cultivation methodology. The cultivation of mushroom requires minimal resources in terms of land, revenue, and labor. India has great potential for mushroom cultivation because of climatic suitability for growing all commercial edible and medicinal mushrooms. Currently, mushroom industry has an output greater than 25 m tons across the globe. The current largest producer of mushrooms in the world, China, produces more than 20 m tons of mushroom, which accounts for over 80% of the world's production (Li, 2012). The substrate left after the harvest of one full crop of mushroom beyond which extension of crop becomes unremunerative is called as spent mushroom substrate (SMS) or spent mushroom compost (SMC). The compost for mushroom cultivation is generally made from crop residue, such as hay, straw, straw horse bedding, poultry litter, cottonseed meal, cocoa shells, and gypsum. Each kilogram of mushroom production leaves behind 5 kg of SMS. The treatment and disposal of SMS is a major environmental issue in the mushroom cultivation. An obvious solution to the disposal problem is to increase the demand for SMS through the exploration of new application as in organic cultivation.

It would be highly economical and favorable if SMS is recycled and reused efficiently. Often SMS is considered as an agricultural waste product with little inherent value, yet there is much that can be utilized in the substrate as it is rich in nutrients and organic content and can provide benefits to other agricultural or nonagricultural sectors. SMS released after mushroom crop harvesting may cause various environmental problems because the large dumped piles of this substrate become anerobic and give off offensive odor and also contaminate ground water, leading to nuisance (Beyer, 1996). Mushroom industry needs to dispose-off over 50 m tons of used mushroom compost each year (Fox and Chorover, 1999). SMS is known for its quality as an organic matter because of slow mineralization rate and it is also has rich nutrient status (Dann, 1996). Hence, SMS possesses the quality of good organic manure for raising healthy crops such as food grains, fruits, vegetables, and ornamentals. SMS, in addition to its capacity of soil reclamation, is utilized for nutritionally poor soil, neutralizing acidic soil, and in other cases for improving polluted sites (Pannier, 1993; Ahlawat and Rai, 2002; Ahlawat et al., 2005). The management of diseases using chemicals is not safer to environment because of residual problem. Increasing concern regarding food safety and environmental pollution has generated an interest in ecofriendly practices like soil amendment and application of biocontrol agents to manage the plant diseases. Soil amendment with SMS is useful to increase the population of antagonists and it also reduces the different soilborne pathogens.

19.2 PHYSIOCHEMICAL PROPERTIES OF SMS

The recomposted SMS should be used singly as basal application or in combination with inorganic fertilizer (Maher, 1994). SMS being rich in nutrients adds nutrition to the soil, helps in neutralizing acidic soils, facilitates plant growth in barren areas and in some cases, it improves water quality along with bioremediation of contaminated industrial sites. In order to improve the physicochemical characteristics of SMS for its use as manure, recomposting method can be employed (Buswell, 1994). Fidanza et al. (2010) found that pH of SMS fall in range of 6.7–7.8 and EC was very high ranging from 10.5 to 14.9 mmhos cm^{-1}. SMS has got an initial pH of about 7.28, which increases during weathering (Wuest and Fahy, 1991). Guo et al. (2001) found that the leachate from 24-month-old naturally weathered piles of 3' and 5' SMS depth contains 0.8–11.0% of dissolved organic carbon. SMS normally contains 1.9:0.4:2.4%, N-P-K before decomposition and 1.9:0.6:1.0, N-P-K after decomposition for 8–16 months. Nitrogen and phosphorus do not leach out during weathering but potassium being more leachable is lost in significant amount during weathering (Gupta et al., 2004). They reported that 6-months-old SMS contain 2.73% nitrogen and 0.31% phosphorus, whereas the potassium was 0.32%. SMS is nutrient rich and has about 80% of the total nitrogen in bound form with fractions of high molecular weight lignin and humic substances. The absorption capacity of stable "humus" available in SMS helps in retaining of nitrogen in the top soil (Grabbe, 1978). The organic mineral fertilizers are made with three diverse formulae having different levels of nitrogen (2, 7, and 10%), and each supplemented with 2% phosphate (P_2O_5) and 2% of potassium (K_2O) has the potential of a balanced organic fertilizer (Sochtig and Grabbe, 1995). Similarly, Lohr et al. (1984) observed that the Fe, Mn, and Zn content of fresh SMS was 67, 7.5, and 1.7 mmol kg^{-1}, whereas that of aged SMS was 71, 7.1, and 2.1 mmol kg^{-1}, respectively.

Gonani et al. (2011) reported that SMC is rich in organic matter and constitutes an important source of micronutrients and macronutrients for plants and microbes, thereby increasing the soil microflora, biological activity, and soil enzyme activity (Debosz et al., 2002; Crecchio et al., 2001). It contains calcium carbonate ($CaCO_3$), which provides short-term buffering of the acidic waters and elevates soil pH (Rupert, 1994). The SMS by virtue of its ability to bind mineral particles together improves aeration and moisture retention, thereby providing good soil structure (Piccolo and Mbagwu, 1994). Use of SMC in crop production reduces the amount of biodegradable waste disposed in landfill sites and helps in transforming them into cost-effective useful agricultural products (Szmidt, 1994). Jonathan et al. (2011) reported that SMC contains nutrients useful for photosynthetic plants, which can be used in crop production. The waste generated in mushroom cultivation is generally considered as nontoxic to crop plants; hence, it can be effectively used as soil amendment for different crops. Aslam et al. (2013) reported that spent mushroom is an important source of humus, even after its nitrogen content being depleted by the composting and growing mushrooms. It also remains a fair source of common nutrients (0.7% N, 0.3% P, 0.3% K, and trace elements) and a soil conditioner. Wisniewska and Pankiewirz (1989) found that SMS treatment in soil elevated the P, K, Ca, and Mg contents in soil. Maher (1991) found that the phosphorous and potassium requirements of crop plants can be fulfilled by just incorporating 5% of SMC v/v, whereas required nitrogen by 25% of v/v. Stewart et al. (1998) pointed out that SMS treatment has affected crop yield and plant nutrition positively by reducing the volume weight, soil clod, slide layer formation and increasing infiltration rate, and water holding capacity of soil. Medina et al. (2009) showed that the addition of SMC to the growing media increased pH values, salt contents, and macronutrients and micronutrients concentrations.

According to Sendi et al. (2013), SMS is known as an important source for humus, which provides nutrition to plants; water holding capacity of soil enhances soil aeration and promotes favorable soil structure. In addition to primary nutrients, SMS also supplies secondary, micronutrients and improves physical property. Increased activity of bacteria, actinomycetes, was increased due to humic acid in rhizosphere, where soil was amended with paddy straw (Zadrazil, 1976; Shukry et al., 1999). According to Rinker et al. (2004), SMS contains higher percentage of three primary nutrients, namely, nitrogen, phosphorus, and potassium as a fertilizer. Castro et al. (2008) stated that Pleurotuswaste was adequate to sustain the growth of *Salvia officinalis* by improving air porosity and mineral content of the soil. Phan and Sabaratnam (2012) concluded that SMS was no longer regarded as a waste but as a renewable resource from the mushroom industry. SMS can be employed in a number of green technology endeavors. The three enzymes characterized from SMS are employed in various purposes such as bioremediation of pollutants and biotechnology industrial purposes, for example, lignocelluloses, namely, laccase, xylanase, lignin peroxidase, cellulase, and hemicellulase. Polat et al. (2009) concluded that application of spent mushroom as organic amendment in greenhouse soil gave positive effect in quality cucumber production and helped for efficient recycling of the SMC. Bindhu (2010) reported that organic carbon high in SMS than soil. Medina et al. (2012) concluded that the application of SMS to soil improved the soil fertility, since soil organic C and N, available P were increased significantly. Fidanza et al. (2012) reported that pH of fresh mushroom compost was around 6.6, with a C:N ratio of 13:1. According to Sendi et al. (2013), SMS contains nutrients which could be effectively used for the crop production. SMS, generally nontoxic to crop plants, could be used as soil amendment for various crops.

TABLE 19.1   Major Mineral and Nutritional Composition of Spent Mushroom Substrate.

Content	Fresh SMS8	8–16 months old weathered SMS	Fresh SMS	16 months old weathered SMS
Sodium	0.72	0.22	0.21–0.33	0.06
Potassium	2.35	1.03	1.93–2.58	0.43
Magnesium	0.71	0.91	0.45–0.82	0.88
Calcium	4.93	6.16	3.63–5.15	6.27
Aluminum	0.40	0.80	0.17–0.28	0.58
Iron	0.44	0.92	0.18–0.34	0.58
Phosphorus	0.36	0.55	0.45–0.69	0.84
Ammonia-Nitrogen	0.11	0.03	0.06–0.24	0.00
Organic-Nitrogen	1.83	1.89	1.25–2.15	2.72
Total-Nitrogen	1.93	1.92	1.42–2.05	2.72

The unit of content is represented as % dry weight.
Source: Ahlawat and Sagar, 2007.

TABLE 19.2   Additional Advantages of Application of SMS for Reclamation of Affected Soils.

Type of SMS	Type of soil/waste	Effect
Shiitake spent Substrate	Pentachlorophenol contaminated soils	44.4–60% breakdown of PCP
Oyster mushroom spent substrate	Three- and 4-ring compounds contaminated soils	50–87% reduction in 3-ring and 34–43% reduction in 4-ring compounds
Button mushroom SMS	Commercial and industrial sites	Stabilization of sites
Button mushroom SMS	Fungicides, pesticides, and heavy metal contaminated soils	Increased degradation of pesticides and heavy metals
Button mushroom SMS	Land sites used for hazardous waste disposal	Reduction in chlorophenols, polycyclic aromatic hydrocarbons and aromatic monomers

Source: Khan, R., 2016

19.3   ANTAGONISTS ASSOCIATED WITH SMS

The SMS released after button mushroom cultivation contains all the essential nutrients needed for harboring fungal biomass and large population of heterotrophic microbes (Pill et al., 1993). Weathering causes a slow reduction in organic matter contents (volatile solids) and leads to different characteristics of weathered SMS because of ongoing microbial activity (Beyer, 1999). The inhabitant fungi, bacteria andactinomycetes in the compost, not only play role in its decomposition but also exercise antagonism against pathogens surviving and multiplying in vicinity. The biological analysis of SMS extract by Yohalem et al. (1994) showed that it contains *Pseudomonas, Trichoderma,* and *Bacillus.* SMS harbors different mycoflora and shows differences in its effect on inhibition of conidial germination and disease suppression. Spent substrate from *Pleurotusflorida* (oyster mushroom) harbors 5–23-fold higher fungal population than other spent substrates. Among the fungi, *Trichoderma* spp. followed by *Aspergillus* spp. and *Mucor* spp. dominate in different spent substrates. *Trichoderma* dominates in all spent substrates, whereas *Mucor* in paddy straw mushroom and *Aspergillus* in both paddy straw mushroom and oyster mushroom spent substrates. Huang et al. (1995, 1996) has also reported that SMS stimulates the microbial population of

Aspergillus spp., *Penicillium* spp., *Trichoderma* spp., and maintains it for 28 days. Butt et al. (2001) and Lucas (1998) observed that organic manure like SMS are known to enhance the early growth and establishment of the bioagents including *Trichoderma* spp. Romaine and Holcomb (2001) found that organic manure including SMC encouraged a population of antagonists like *Trichoderma* spp., *Bacillus* spp. that interfere with the activity of plant pathogens.

19.4 MANAGEMENT OF SOILBORNE PLANT PATHOGENS THROUGH SMS

Soilborne plant pathogens mainly affect the root system of crop plants or the basal region of the stem, in some cases symptoms can be seen developing on aerial plant parts. Such pathogens usually cause widespread damage by means of root necrosis, vascular disease, root or tuber rot, gall, proliferation, etc. (McDonald, 1994). Soilborne plant pathogens cause significant reduction in quantity and quality of crops. Among them, root rots are considered to be most familiar diseases affecting below ground tissues and the vascular wilts are initiated through root infections (Table 19.3). Among the soilborne plant pathogens, soil inhabitants are known to survive for a relatively long time compared with soil transients in soil. Survival of these phytopathogens in soil mainly depends on the survival structure (resting spores, sclerotia, oospores, cysts) and competitive saprophytic ability. The soilborne plant pathogenic fungi belong to different classes, namely, plasmodiophoramycetes, oomycetes, zygomycetes, ascomycetes, basidiomycetes, and deuteromycetes. Only few soilborne bacterial pathogens are considered as serious. For example, species of *Pseudomonas, Erwinia, Ralstonia,* and *Xanthomonas* usually persist in the soil for only a short time. Various nematodes and a few soilborne viruses affect crops. The SMS contains a diverse range of soil microorganisms. This fact is proven by its disease suppressing properties and its effectiveness in bioremediation (Koike et al., 2003).

Remya (2012) found that among the various SMS used, paddy straw SMS of P. sajor-caju and isolated antagonists from SMS were found to be most effective for the management of soilborne diseases of ginger. Roshna (2013) reported that the isolates of *T. hamatum* and *B.subtilis* from pleurotus SMS provided luxuriant growth and was found effective against *Phytophthora* diseases of pepper cuttings. Grebus et al. (1994) reported that a positive effect of SMS, as observed for other composts, is disease suppression because of a stimulus of the soil microbial antagonism to rapid development of newly colonizing microbial species like pathogens.

Donald et al. (2005) reported that mushroom compost containing initial populations of antagonistic fungi at fairly high level shows lower incidence of many fungal diseases in field condition. Several researchers had reported lower incidence of disease on using SMS as manure in vegetable crops (Ahlawat et al., 2007). Romaine and Holcomb (2001) reported a general trend for increasing seedling survival as the proportion of compost increased up to 100% and that compost at a level of 50% or greater provided highly effective disease control.

TABLE 19.3   Important Soilborne Diseases of Vegetable Crop.

Disease	Pathogen
Bacterial soft rots	*Erwiniacaratovora, E. chrysanthemi*
Cottony rot/pink rot	*Sclerotiniasclerotiorum*
Crown rot/root rots	*Rhizoctoniasolani*
Phytophthora root rot	*Phytophthora sp.*
Root dieback/wire stem	*Pythium sp.*
Root knot nematode	*Meloidogyne sp.*
Southern blight/collar rot	*Sclerotium rolfsii*
Fusarium yellows/wilts	*Fusarium oxysporum*
Pink rot	*Sclerotinia sclerotiorum*
Anthracnose	*Microdochiump anattonianum*
Verticillium wilt	*Verticillium dahlia*
Clubroot	*Plasmodiophora brassicae*
Cyst nematode	*Heterodera species*
White rust	*Albugo candida*
Wirestem	*Rhizoctoniasolani*
Damping-off	*Fusarium oxysporum, Pythium sp., Rhizoctonia solani*
Charcoal rot	*Macrophomina phaseolina*
Powdery scab	*Spongospora subterranean*
Root knot nematode	*Meloidogyne species*
Silver scurf	*Helminthosporium solani*
White mold	*Sclerotinia sclerotiorum*
Fusarium root rot	*Fusarium solani*

Source: Koike et al., 2003.

TABLE 19.4   Response of Different Plant Species Toward Different Doses and Age of SMS.

Crop	Age of SMS	Quantity (q ha⁻¹)	Impact		
			Yield	Quality	Disease/insect pests
Capsicum	12 months	250	Plant growth promotion	Superior quality	2–20% reduction in insect pests and diseases
Cauliflower	12 months	250	Increased curd size	Superior quality	40–60 % reduction in insect pests and diseases
Pea	12 months	250	Plant growth promotion	Superior quality	Reduction in Fusarium wilt and Powdery mildew
Brinjal	12–24 months	250	Increased fruit size	Superior quality	NR

Source: Khan, R. (2016)

19.5   EFFECT OF SMS ON SOILBORNE FUNGAL PATHOGENS

The fungi associated with SMS are *Aspergillus flavus*, *Aspergillus nidulans*, *Aspergillus terreus*, *Aspergillus versicolor group*, *Chrysosporium luteum*, *Mucor* spp., *Nigrospora* spp., *Oidiodendron* spp., *Paecilomyces* spp., *Penicillium chrysogenum*, *Penicillumexpansum*, *Trichoderma viride*, and *Trichurus* spp. (Kleyn and Wetzler, 1981). SMS exhibited suppressive characteristics against various fungi, as well as against plant diseases caused by fungi (Davis et al., 2005). Ntougias et al. (2008) reported that the disease damping off caused *P. aphanidermatum* to suppress by SMS. Parada et al. (2011) found reduced severity of cucumber anthracnose because of application of sterilized or nonsterilized SMS water extract. Shitole et al. (2013) found that application of combination of bioagents (*Trichoderma viride* and *Pseudomonas fluorescens*) + SMS was very effective to reduce the incidence of pre- and post-emergence damping off in tomato.

Paddy straw SMS of *Pleurotussajor-caju* is possessing antagonistic property against soil inhabiting plant pathogens of ginger, namely, *Pythiumaphanidermatum*, *Fusariumoxysporum*, *Rhizoctoniasolani*, *Sclerotiumrolfsii, and Ralstoniasolanacearum* under in vitro conditions (Remya and Beena, 2015). Hoitink et al. (1993) reported that the soilborne phytopathogens comprising *Rhizoctonia* are inhibited and restricted by a number of microorganisms, for example, *Trichoderma* spp. and *Gliocladium spp.* present in lingocellulosic matter. Chiu and Huang (1997) reported that SMC inhibited the incidence of different root diseases of watermelon, cabbage, tomato, and pepper. SMS has property to suppress the soilborne fungal pathogens (Davis et al., 2005) and plant diseases (Segarra, 2007), and to increase the microbial densities in soils (Perez-Piqueres et al., 2006). Many researchers reported that the organic substrates such as SMC can suppress a variety of plant pathogenic fungi including soilborne pathogens like *Rhizoctonia* spp. *and Pythium spp.* (Craft and Nelson, 1996; Grebus et al., 1994; Hoitink et al., 1997, Hoitink and Fahy, 1986; Nelson and Craft, 1992; Phae et al., 1990; Viji et al., 2003; Zhang et al., 1998; Philippoussis et al., 2004). Harender et al. (1997) found that mushroom was most effective in reducing the mean inoculum load and increasing total fungal and bacterial population in the soil, indicating enhanced microbial activity. Verma et al. (2017) reported that SMS contains bioagents (*T. harzianum* and *T. viride*) and it is also rich nutritionally. They found that SMS was effective against R. solani infecting roots in tomato.

19.6   EFFECT OF SMS ON SOILBORNE BACTERIAL PATHOGENS

Kwak et al. (2015) reported the antibacterial property of culture filtrates of different mushrooms like *Hericiumerinaceus, Lentinula edodes, Grifolafrondosa,* and *Hypsizygusmarmoreus* against *Ralstonia solanacearum,* causing wilt of tomato. Ethyl acetate, *n*-butanol, and water extracts of SMS of *H. erinaceus* possesses antibacterial property against many plant pathogenic bacteria, that is, *R. solanacearum, Xanthomonasoryzae*pv. *oryzae, X. campestris* pv. *campestris, Pectobacteriumcarotovorum* subsp. *carotovorum, Agrobacterium tumefaciens, X. axonopodis* pv. *vesicatoria, X. axonopodis* pv. *Glycine, and X. axonopodis* pv. *citiri.* Quantitative real-time PCR resulted that expressions of plant defense genes encoding β-,3-glucanase (*GluA*) and pathogenesis-related protein-1a (*PR-1a*), associated with systemic acquired resistance were induced because of water extracts of SMS (WESMS) of *H. erinaceus*. WESMS also exhibited reduced wilt and increased yield attributes in tomato. Kleyn and Wetzler (1981) reported that the most common bacterial pathogens associated with SMS are *Bacillus licheniformis, Thermoactinomyces vulgaris, Streptomyces diastaticus, Streptomyces albus, Streptomyces griseus, Thermoactinomyces thalpophilis, Thermomonospora fusca,* and *Thermomonospora chromogena.* Ntougias et al. (2004) reported that SMC is affected by the thermal pasteurization treatment and it contains G^+ bacteria belonging to genera

Arthrobacter, Microbacterium Bacillus, Paenibacillus, Exiguobacterium, Staphylococcus, Desemzia, Carnobacterium, and Brevibacterium. A study was done by Zeeshan et al. in 2016 and he found that spent composts have property to reduce the *Ralstonia solanacearum* causing tomato wilt under in vitro and in vivo conditions.

19.7 EFFECT OF SMS ON NEMATODES

Aslam et al. (2013) reported that SMC contains antimicrobial phenolic compounds against *Meloidogynespp,* causing root knot in tomato. He also noticed the antimicrobial property of water extracts of mushrooms like *Agaricusbisporus, Lentinusedodes, Boletusedulis,* and *Pleurotusostreatus.* Kaul and Chhabra (1993) observed reduced incidence of *Meloidogyne* root knot infections in tomato because of soil application of SMS. It exhibited higher efficiency than the application of carbofuran (2 kg a.i. ha⁻¹). Mushroom compost water extract at 25% concentration paralyzed approximately 40% of nematodes. Repeated drenching might be needed for long-term nematode control (Ching and Wang, 2014). Aslam and Saifullah (2013) found that soil application of SMS of oyster and button mushroom were effectively inhibiting the galling and production of egg masses of *M. incognita* in the roots. SMS of oyster mushroom exhibited that its increased concentration can reduce nematode population significantly up to 70% (Abbasi et al., 2014). The ability of SMS of Pleurotus spp. to inhibit or kill the sugar beet cyst nematode (*Heteroderaschachtii*) is well known. *P. ostreatus and P. eryngii* SMS extracts can paralyze about 90% and 50% nematodes, respectively (Palizi et al., 2009).

19.8 CONCLUSION

The substrate left after the harvest of mushroom is termed as SMS. SMS have many potential uses as it is rich with many potential antagonists and contains a diverse range of soil microorganisms. Among them, the disease controlling property is quite interesting. The application of SMS plays vital role in disease management in crops. Soil amendment with SMS promotes a population of antagonistic microorganisms, which interfere with the activities of soilborne pathogens. SMS recomposted by different methods is beneficial to several crop plants in respective of growth promotion as well as disease suppression. It stimulates a natural disease defense system in plants and the organic substances produced from SMS in soil are also suppressing the pathogens. Hence, application of SMS alone or along with bioagents in soil is an ecofriendly and sustainable approach to increase the extent of antagonists and reduce the soilborne pathogens without damaging the environment.

KEYWORDS

- **bioagents**
- **compost**
- **soil amendment**
- **soilborne diseases**
- **spent mushroom substrate**
- **WESMS**

REFERENCES

Abbasi, N., Torkashvan, A. M.; Rahanandeh, H. Evaluation of Mushroom Compost fror the Bio Control Root-Knot Nematode. *Int. J. Biosci.* **2014,** *5*, 147–153

Ahlawat, O. P.; Sagar, M. P. *Management of Spent Mushroom Substrate.* National Research Centre for Mushroom, ICAR: Himachal Pradesh, India, 2007, pp 48.

Ahlawat, O. P.; Indurani, C.; Sagar, M. P. *Spent Mushroom Substrate Properties and Recycling for Beneficial Purposes.* National Research Centre for Mushroom: Himachal Pradesh, India, 2005, pp 314–334.

Ahlawat, O. P.; Gupta, P.; Kumar, S. Spent Mushroom Substrate—A Tool for Bioremediation. In: *Mushroom Biology and Biotechnology*; Rai, R. D., Singh, S. K., Yadav, M. C., Tewari, R. P., Eds.; Proceedings of International Conference, Solan, National Research Centre for Mushroom: Solan, India, 2007, pp 341–366.

Ahlawat, O. P.; Rai, R. D. Recycling of Spent Mushroom Substrate. In: *Recent Advances in the Cultivation Technology of Edible Mushrooms*; Verma, R. N., Vijay, B., Eds.; National Research Centre for Mushroom: Solan, India, 2002, pp 261–282.

Aslam, S; Saifullah Organic Management of Root Knot Nematodes in Tomato with Spent Mushroom Compost. *Sarhad J. Agric.* **2013,** *29* (1), 63–69.

Beyer, D. M. Spent Mushroom Substrate Fact Sheet. http:/mushroom spawn.cas.psu.edu/ spent.htm.

Bindhu, C. J. Calcium Dynamics in Substrate Wormcast-Mushroom-Plant Continuum. M.Sc. (Ag) Thesis, Kerala Agriculture University, Thrissur 2010, pp 105.

Buswell, J. A. Potential of Spent Mushroom Substrate for Bioremediation Purposes. *Mushroom News* **1994,** *43* (5), 28–34.

Butt, T. M.; Jackson, C. W.; Magan, N. *Fungi as Biocontrol Agent: Progress, Problems and Potential.* CABI, Press: Oxon, UK, 2001, pp 390.

Castro, R. I. L.; Delmastro, S.; Curvetto, N. R. Spent Oyster Mushroom Substrate in aMix with Organic Soil for Plant Pot Cultivation. *Mycologia Appl. Int.* **2008,** *20* (1), 17–26.

Ching, S.; Wang, K. H. Mushroom Compost to Battle Against Nematode Pests on Vegetable Crops. Hānai'Ai/The Food Provider, 2014, pp 1–7.

Chiu, A. L.; Huang, J. W. Effect of Composted Agricultural and Industrial Wastes on the Growth of Vegetable Seedlings and Suppression of Their Root Diseases. *Plant Pathol. Bull.* **1997,** *6,* 67–75.

Craft, C. M; Nelson, E. B. Microbial Properties of Composts that Suppress Damping-Off and Root Rot of Creeping Bentgrass Caused by *Pythium graminicola. Appl. Environ. Microbiol.* **1996,** *62* (5),1550–1557.

Crecchio C.; Curci. M.; Mininni, R.; Ricciuti, P.; Ruggiero, P. Short Term Effects of Municipal Solid Waste Compost Amendments on Soil Carbon and Nitrogen Content, Some Enzyme Activities and Genetic Diversity. *Biol. Fertility of Soils* **2001,** (34), 311–318.

Dann, M. S. The Many Uses of Spent Mushroom Substrate. *Mushroom News* **1996,** *44* (8), 24–27.

Davis, D. D.; Kuhns, L. J.; Harpster, T. L. Use of Mushroom Compost to Suppress Artillery Fungi. *J. Environ. Hort.* **2005,** *23* (4), 212–215.

Debosz, K.; Petersen S. O.; Kure, L. K.; Ambus, P. Evaluating Effects of Sewage Sludge and Household Compost on Soil Physical, Chemical and Microbiological Properties. *Appl. Soil Ecol.* **2002,** (19), 237–248.

Donald, D. D.; Larry, J. K.; Tracey, L. H. Use of Mushroom Compost to Suppress Artillery Fungi. *J. Environ. Hort.* **2005,** *23* (4), 212–215.

Fidanza, M. A.; Sanford, D. L.; Beyer, D. M.; Aurentz, D. J. Analysis of Fresh Mushroom Compost. *Hort. Technol.* **2012,** *20* (2), 133–138.

Fidanza, M. A.; Sanford, D. L.; Beyer, D. M.; Aurentz, D. J. Analysis of Fresh Mushroom Compost. *Hort. Technol.* **2010,** *20* (2), 449–453.

Fox, R.; Chorover, J. Seeking a Cash Crop in Spent Substrate. http:/ aginfo.psu.edu/psa/sgg/mushrooms.html, 1999.

Gonani Z.; Riahi, H.; Sharifi, K. Impact of Using Leached Spent Mushroom Compost as a Partial Growing Media for Horticultural Plants. *J. Plant Nutr.* **2011,** *34* (3), 337–344.

Grabbe, K. Verfahren zur Herstellungeines Dung emittels. BRD Patent NR 2831583, 1978.

Grebus, M. E.; Watson, M. E.; Hoitink, H. A. J. Biological, Chemical, and Physical Properties. *Compost. Sci. Util.* **1994,** (3), 80–83.

Guo, M.; Chorover, J.; Fox, R. H. Effects of Spent Mushroom Substrate Weathering on the Chemistry of Underlying Soils. *J. Environ. Quality* **2001,** *30,* 2127–2134.

Gupta, P.; Indurani, C.; Ahlawat, O. P.; Vijay, B.; Mediratta, V. Physico-chemical Properties of Spent Mushroom Substrates of *Agaricusbisporus. Mushroom Res.* **2004,** *13* (2), 84–94.

Harender, R. J.; Kapoor, Raj, H. Possible Management of *Fusarium* Wilts of Capsicum by Soil Amendments with Composts. *Ind. Phytopathol.* **1997,** *50,* 387–394.

Hoitink, H.; Fahy P. C. Basis for the Control of Soil Borne Plant Pathogens with Composts. *Ann. Rev. Phytopathol.* **1986,** *24,* 93–114.

Hoitink, H. A. J.; Boehm, M. J.; Hadar Y. Mechanisms of Suppression of Soil-Borne Plant Pathogens in Compost-Amended Substrates. In: *Science and Engineering of composting: Design, Environmental, microbiological and Utilization Aspects.* Renaissance Publications: Worthington, Ohio, 1993, pp 601–621.

Hoitink, H. A. J.; Stone, A. G.; Han, D. Y. Suppression of Plant Diseases by Composts. *Hort. Sci.* **1997**, *32*, 184–187.

Huang, J. W.; Hu, C. K.; Shih, S. D. The Role of Soil Microorganisms in Alleviation of Root Injury o Garden Pea Seedlings by Alachlor with Spent Golden Mushroom Compost. *Plant Pathol. Bull.* **1996**, *5* (3), 137–145.

Huang, J. W.; Hu, C. K.; Tzeng, D. D. S.; Ng, K. H. Effect of Soil Amended with Spent Golden Mushroom Compost on Alleviating Phytotoxicity of Alachlor to Seedling of Garden Pea. *Plant Pathol. Bull.* **1995**, *4* (2), 76–82.

Jonathan S. G.; Lawal, M. M.; Oyetunji, O. J. Effect of Spent Mushroom Compost of *Pleurotuspulmonarius* on Growth Performance of Four Nigerian Vegetables. *Mycobiol.* **2011**, *39* (3), 164–169.

Kaul, V. K.; Chhabra, H. K. Control of *Meloidogyne incognita* by Incorporation of Organic Wastes. *Plant Dis. Res.* **1993**, *8*, 35–41. http:/aginto.psu.edu/psa/sgg/mushroom5.html.

Khan, R. Studies on Comparative Efficiency of Spent Mushroom for Managing Wilt and Wilt Like Diseases of Chickpea. M.Sc. (Ag.) Thesis, Dept of Plant Pathology, JNKVV, Jabalpur, 2016.

Kleyn, J. G.; Wetzler, T. F. The Microbiology of Spent Mushroom Compost and Its Dust. *Can. J. Microbiol.* **1981**, *27* (8), 748–753.

Koike, S. T.; Subbarao, K. V.; Davis, R. M.; Turini, T. A. *Vegetable Diseases Caused by Soil Borne Pathogens*. ANR Publication, University of California: California, 2003, pp 13.

Kwak, A. M.; Min, K. J.; Lee, S. Y.; Kang, H. W. Water Extract from Spent Mushroom Substrate of *Hericiumerinaceus* Suppresses Bacterial Wilt Disease of Tomato. *Mycobiol.* **2015**, *43* (3), 311–318.

Li, P. P.; Mao, H. P.; Wang, D. H. Effect of Medium Residue From Mushroom Culture as a Soilless Culture Medium for Vegetables Crops. *Chine Veg.* **1998**, *5*, 12–15.

Li, Y. Present Development Situation and Tendency of Edible Mushroom Industry in China. In: Physical and Chemical Characteristics of Fresh and Aged Spent Mushroom Compost; Lohr, V. I., Wang, S. H., Wolt, J. D., Eds. *Hort. Sci.*. **2012**, *19* (5), 681–683.

Lucas, J. A. *Plant Pathology and Plant Pathogens*, 3rd ed, 1998, pp 273.

Maher, M. J. Spent Mushroom Compost (SMC) as a Nutrient Source in Peat Based Potting Substrate. *Mushroom Sci.* **1991**, *13* (2), 645–650.

Maher, M. J. The Use of SMS as an Organic Manure and Plant Substrate Component. *Compost Sci. Util.* **1994**, *2* (3), 37–44.

McDonald, B. A.; Miles, J.; Nelson, L. R.; Pettaway, R. E. Genetic Variability in Nuclear DNA in Field Population of Stagonosporanodorum. *Phtopathol.* **1994**, *84*, 250–255.

Medina, E., C.; Paredes, M. D.; Perez-Murcia, M. A.; Bustamante, Moral, R. Spent Mushroom Substrates as Component of Growing Media for Germination and Growth of Horticulture Plants. *Biores. Technol.* **2009**, *100* (18), 4227–4232.

Medina, E.; Paredes, C.; Bustamante, M. A.; Moral, R.; Moreno-Caselles, J. Relationships Between Soil Physico-chemical, Chemical and Biological Properties in a Soil Amended with Spent Mushroom Substrate. *Geoderma* **2012**, 152–161.

Mukherjee, R.; Nandi, B. Improvement of *in vitro* Digestibility Through Biological Treatment of Water Hyacinth Biomass by Two *Pleurotus*spp. *Int. Biodeterioration Biodegradation.* **2004**, *53*, 7–12.

Nelson, E. B.; Craft, C. M. Suppression of Dollar Spot on Creeping Bentgrass and Annual Bluegrass Turf with Compost-Amended Topdressings. *Plant Dis.* **1992**, *76*, 954–958.

Ntougias, S.; Papadopoulou, K. K.; Zervakis, G. I.; Kavroulakis, N.; Ehaliotis, C. Suppression of Soil-borne Pathogens of Tomato by Composts Derived from Agro-Industrial Wastes Abundant in Mediterranean Regions. *Biol. Fertil. Soils* **2008**, *44*, 1081–1090.

Ntougias, S.; Zervakis, G. I.; Kavroulakis, N.; Ehaliotis, C.; Papadopoulou, K. K. Bacterial Diversity in Spent Mushroom Compost Assessed by Amplified Rdna Restriction Analysis and Sequencing of Cultivated Isolates. *Syst. Appl. Microbiol.* **2004**, *27*, 746–754.

Palizi, P.; Goltapeh, E. M.; Pourjam, E.; Naser, S. Potential of Oyster Mushroomfor the Biocontrol of Sugar Beet Nematode (*Heteroderaschachtii*). *J. Plant Prot. Res.* **2009**, *49* (1), 10045-009-0004-6.

Pannier, W. Spent mushroom compost: A Natural Resource that Provides Solution to Environmental Problems. *Mushroom News* **1993**, *41*, 10–11.

Parada, R. Y.; Murakami, S.; Shimomura, N.; Egusa, M.; Otani, H. Autoclaved Spent Substrate of Hatakeshimeji Mushroom (*Lyophyllumdecastes* Sing.) and its Water Extract Protect Cucumber from Anthracnose. *Crop Prot.* **2011**, *30*, 443–450.

Perez-Piqueres, A.; Edel-Hermann, V.; Alabouvette, C.; Steinberg, C. Response of Soil Microbial Communities to Compost Amendments. *Soil Biol. Biochem.* **2006**, *38*, 460–470.

Phae, C. G.; Sasaki, M.; Shoda, M.; Kubota, H. Characteristics of *Bacillus subtilis* Isolated from Composts Suppressing Phytopathogenic Microorganisms. *Soil Sci. Plant Nutr.* **1990**, *36*, 575–586.

Phan, C.; Sabaratnam, V. Potential Uses of Spent Mushroom Substrate and its Associated Lignocellulosic Enzymes. *Appl. Microbiol. Biotechnol.* **2012**, *96*, 863–873.

Philippoussis, A.; Zervakis, G. I.; Diamantpoulou, P.; Papadopoulou, K.; Ehaliotis, C. Use of Spent Mushroom Compost as a Substrate for Plant Growth and Against Plant Infections Caused by *Phytophthora*. *Mushroom Sci.* **2004**, *16*, 579–584.

Piccolo, A; Mbagwu, J. S. C. Humic Substances and Surfactants Effects on the Stability of Two Tropical Soils. *Soil Sci. Soc. Am. J.* **1994**, (58), 950–955.

Pill, W. G.; Evans, T. A.; Garrison, S. A. Forcing White Button Asparagus in Various Substrates Under Cool and Warm Regimes. *Hort. Sci.* **1993**, *28* (10), 996–998.

Polat, E., H.; Uzun, H. I.; Topcuo-lu, B.; Onal, A. K.; Onus, N.; Karaca, M. Effects of Spent Mushroom Compost on Quality and Productivity of Cucumber Grown in Greenhouses. *Afr. J. Biotech.* **2009**, *8* (2), 176–218.

Remya, J. S. Exploitation of Spent Mushroom Substrate as Mulch for the Management of Rhizome Rot Complex Disease of Ginger. M.Sc. (Ag.) Thesis, 2012.

Remya, J. S.; Beena, S. Antagonistic Potential of Fungal Microflora from Spent Mushroom Substrate Against Soil Borne Pathogens Of Ginger. *Int. J. Appl. Pure Sc. Agri.* **2015**, *1* (12), 1–10.

Rinker, D. L.; Zeri; Kang, S. W. Recycling of Oyster Mushroom Substrate. *Mushroom Growers' Handbook* **2004**, *9*, 187–191.

Romaine, C. P.; Holcomb, E. J. Spent Mushroom Substrate: A Novel Multifunctional Constituent of Potting Medium for Plants. *Mushroom News* **2001**, *49*, 4–15.

Roshna, S. Potential of SMS for the Management of Nursery Diseases of Black Pepper. M.Sc. (Ag) Thesis, Kerala Agriculture University, Thrissur, pp 165.

Rupert, D. R. Use of SMS in Stabilizing Disturbed and Commercial Sites, 1994.

Segarra, G.; Casanova, E.; Borrero, C.; Aviles, M.; Trillas, I. The Suppressive Effects of Composts Used as Growth Media Against *Botrytis Cinerea* in Cucumber Plants. *Eur. J. Plant Pathol.* **2007**, *117*, 393–402.

Sendi, H.; Mohamed, M. T. M.; Anwar, M. P.; Saud, H. M. *Spent mushroom waste as a Media Replacement for Peat Mossin Kai-Lan (Brassica oleracea var. Alboglabra) Production.* Hindawi Publishing Corporation, The Scientific World Journal, 2013, pp 8.

Shitole, A. V.; Gade, R. M.; Zalte, A.; Bandgar, M. S. Utilization of Spent Mushroom Substrate gor Management of Tomato Damping Off. *J. Pl. Dis. Sci.* **2013,** *8* (2), 196–199.

Shukry, W. M.; El-Fallal, A. A.; El-Bassiouny, H. M. S. Effect of Spent Wheat Straw Growth, Growth Hormones, Metabolism and Rhizosphere of Cucumis Sativa. *Egyptian J. Physiol. Sci.* **1999,** *23*, 39–69.

Sochtig, H.; Grabbe, K. The Production and Utilization of Organic-Mineral Fertilizer from Spent Mushroom Compost. *Mushroom Sci.* **1995,** *14* (II), 907–915.

Stewart, D. P. C.; Cameron, K. C.; Cornforth I. S.; Sedcole, J. R. Effects of Spent Mushroom Substrate on Soil Physical Conditions and Plant Growth in an Intensive Horticultural System. *Austr. J. Soil Res.* **1998,** *36* (6), 899–912.

Szmidt R. A. K. Recycling of Spent Mushroom Substrates by Aerobic Composting to Produce Novel Horticultural Substrates. *Compost Sci. Util.* **1994,** (2), 63–72.

Verma, S.; Kumar, A.; Singh, J.; Singh, S.; Rana, R. S. Antagonistic Effect of Trichoderma Spp. Present in Spent Mushroom Substrate Against Rhizoctoniasolani Causing Damping-Off and Root Rot in Tomato. *Int. J. Agric. Sci. Res.* **2017,** *7*, 149–158.

Viji, G.; Uddin, W.; Romaine, C. P. Suppression of Gray Leaf Spot (Blast) of Perennial Ryegrass Turf by *Pseudomonas aeruginosa* from Spent Mushroom Substrate. *Biol. Control* **2003,** *26*, 233–243.

Wisniewska, G. H.; Pankiewiez, T. Evaluation of the Suitability of Spent Mushroom Substrate for Tulip Cultivation. *Prace Instytutu Sadownictura Kwiaciarstwa Skerniewicack* **1989,** *14*, 7–13.

Wuest, P. J.; Fahy, H. K. Spent Mushroom Compost Traits and Uses. *Mushroom News* **1991,** *39* (12), 9–15.

Yohalem, D. S.; Harris, R. F.; Andrews, J. H. Aqueous Extracts of Spent Mushroom Substrate for Foliar Disease Control. *Compost Sci. Util.* **1994,** *2* (4), 67–74.

Zeeshan, A. M.; Khan, I.; Shah, B.; Naeem, A.; Khan, N.; Ullah, W.; Adnan, M.; Ali, S. R.; Shah, J. S. K.; Iqbal, M. Study on the Management of *Ralstonia solanacearum* (Smith) with Spent Mushroom Compost. *J. Entomol. Zool. Studies* **2016,** *4* (3), 114–121.

Zhang, W.; Han, D. Y.; Dick, W. A.; Davis, K. R.; Hoitink, H. A. J. Compost and Compost Water Extract-induced Systemic Acquired Resistance in Cucumber and *Arabidopsis.* *Phytopathology* **1998,** *88*, 450–455.

Index

A

Abelmoschus esculentus
source of resistance, 239
Alphasatellite, 351
Alternaria leaf blight
causal organisms, 215
disease cycle, 215–216
disease management
chemical controls, 216–217
cultural practices, 216
disease symptoms, 214–215
distribution, 214
Anthracnose, 25
causal organism, 218–219
disease cycle, 219
disease management
chemical control, 220
cultural practices, 220
resistance, 220
disease symptoms, 217–218
dispersal/transmission, 25
distribution, 217
ecology of pathogen, 25
epidemiology, 25
pathogen biology, 219
survival, 25
Argentina
legumoviruses, 356–357
Auxins, 279–280

B

Bacteria
root-knot nematodes, 131
Bacterial pathogens
nature of infection, 281–285
role of bacterial secretion system, 286
Bacterial wilt, 85–86
disease cycle and epidemiology, 90–91
economic importance, 89–90
geographical distributions, 89
host range, 88–89
sign and symptoms, 87–88

Bangladesh
begomovirus
diversity of legume-infection in
different countries, 355
Begomovirus, 327–328
associated with DNA-satellites, 350–351
diversity of legume-infection in different
countries, 353
Bangladesh, 355
India, 353–354
Pakistan, 354–355
Vietnam, 355
factors affecting cross-continent genetic
diversity, 363
diversification, 363–364
mutation, 364–365
pseudorecombination, 367–369
recombination, 365–367
list of infecting legumes and, 359
role of virus–vector–host interactions
host plant, 371–372
whiteflies, 369–370
Begomovirus infection, management, 396
color coding representation, 399
conventional method
infected plants, removal, 401
resistant/tolerant varieties,
development, 401
vector, 400
weeds, 400–401
whitefly control strategies, 401–402
non-conventional method
CRISPR/Cas technology, 402
curl virus and cabbage leaf curl virus
(CaLCuV), 402
integrated management, 403
plant defence mechanism, 402
RNAi-mediated technology, 402
phylogenetic tree, 397–398
strategies, 399
Betasatellites, 351–352
Biocontrol control agent (BCA), 475

Brazil
    legumoviruses, 356–357
Bulb vegetables, 483
    Alliums, 485
    bacterial brown rot, 505
    blue mold of garlic, 504
    botrytis neck rot, 501–502
    canker/skin blotch, 505
    Charaka-Samhita, 485
    cloves, 486
    fusarium basal rot disease, 499–501
    garlic, 485
        growth and metabolism, 488–489
        mechanical damage, 490
        pathological decay, 490–491
        physiological disorders, 489
        purple blotch, 502–503
        white rot of, 503
    losses, causes, 495
        biological control, 498
        chemicals, applied, 497–498
        fungicides, 496–497
        irradiation for, 498–499
        onion and garlic, and pests, 499–506
        packing and sanitation, 497
        physiological condition, 496
        postharvest diseases, 499–506
        postharvest treatments, 497
        weather, 496
    mites, 505–506
    mold, 504–505
    nutrients, 484
    onions, 486
        growth and metabolism, 488–489
        mechanical damage, 490
        pathological decay, 490–491
        physiological disorders, 489
        purple blotch, 502–503
        white rot of, 503
    peas and beans, 484
    physiological changes, 486
    postharvest technology
        atmosphere modification, 493
        curing, factors, 494–495
        heat treatment, 493–494
        living tissues, 487
        loss of water, 492
        losses, 487

        onions and garlic, 494
        perforated hessian bags, 495
        periderm, 494
        plastic woven bags, 495
        quality, 494
        sorting and grading, 495
        temperature management, 491–492
        vapor pressure difference (VPD), 492
        wound periderm, 494
    tissues, 485
    waxy breakdown, 506
Burkholdeira solanacearum, 86

C

Candidatus Phytoplasma, 50
Cauliflower mosaic virus, 317
Cell wall degrading enzymes (CWDEs),
    101, 273–274
Cellulases, 277
Cercospora canescens
    dispersal/transmission, 26
    ecology of pathogen, 26
    epidemiology, 26
    survival, 26
Cercospora leaf spot, 26
    causal organisms, 222
    disease cycle, 222–223
    disease management
        chemical control, 223
        cultural practices, 223
    dispersal/transmission, 26
    distribution, 221
    ecology of pathogen, 26
    epidemiology, 26
    survival, 26
    symptoms, 221–222
Chili/capsicum
    management, 153
    root-knot nematode, 151–152
Chilli plants
    symptoms of leaf, 65
China
    legumoviruses, 356
Colletotrichum lindemuthianum, 25
    dispersal/transmission, 25
    ecology of pathogen, 25
    epidemiology, 25
    survival, 25

Colletotrichum orbiculare, 218–219
Colombia
  legumoviruses, 356–357
Coronatine, 186
Cowpea, 23–24, 34
  bacterial blight, 32
  fungal diseases, 25–31
  IDM, 32–33
  important diseases, 24
  viral diseases, 31–32
Crop pathogens threat, 46–47
  ecological and population genetics
    information depicting model, 47
Crop placement, 375
Crucifers, 68
Cuba
  legumoviruses, 359–360
Cucumber mosaic virus (CMV), 317–318,
  334–335
  infecting vegetable crops, 318–319
Cucurbit crops, pathosystem
  bacterial diseases, 387
  begomovirus infection
    symptoms, 389
  bitter melon leaf curl disease (BMLCD),
    396
  bittergourd yellow vein virus (BGYVV),
    394
  chayote yellow mosaic virus (ChaYMV),
    392
  chlorotic curly stunt disease (CCSD), 394
  coccinia mosaic virus, 393
  crumple virus, 389–390
  cucurbit leaf curl virus (CuLCV), 391
  disease symptoms, 395
  Indian cassava mosaic virus (ICMV), 396
  melon chlorotic leaf curl virus
    (MCLCuV), 390–391
  non-host viruses infecting cucurbits, 394
  Parwal, 396
  pepper golden mosaic virus (PepGMV),
    395
  pepper leaf curl Bangladesh virus
    (PepLCBV), 395
  pumpkin yellow vein mosaic virus
    (PYVMV), 392
  telfairia mosaic virus, 393
  tomato leaf curl disease (TLCD), 394

tomato leaf curl New Delhi virus
  (ToLCNDV), 394–395
tomato leaf curl Palampur virus
  (ToLCPMV), 394–395
tomato yellow leaf curl virus (TYLCV),
  394
watermelon chlorotic stunt virus
  (WmCSV), 394
Cucurbita, 203–204
  downy mildew
    causal organism, 206–207
    disease cycle, 208
    disease management, 208–209
    distribution, 205
    pathogen biology, 207
    symptoms, 206
  ecological impediments, 204–205
  foliar fungal diseases, 205
  powdery mildew
    causal organism, 211
    cultural practices, 212–213
    disease cycle, 212
    disease management, 212
    distribution, 209–210
    fungicide applications, 213–214
    pathogen biology, 211–212
    resistant varieties, 213
Cucurbits, 66–68
  disease management, 208–209
    chemical control, 209
    cultural practices, 208
    resistant varieties, 208–209
  distribution, 205
  downy mildew
    causal organism, 206–207
    disease cycle, 208
  pathogen biology, 207
  symptoms, 206
Cutinases, 275–276
Cytokinins, 280

D

Deltasatellites, 352–353
Disease dynamics of vegetable crops,
  255–257
  occurrence in, 256
Disease resistance (NBS-LRR) gene
  homologs, 377

E

Eggplant, phyllody
  natural symptoms, 63
  reduction of leaf size, 63
Ethylene, 281
Euphorbia cyparissias, 174
Extracellular polysaccharides, 100–101
Extra-chromosomal DNA, 59

F

Foliar diseases
  management, 33
Forecasting system, 13–14
Fungi
  root-knot nematodes, 131–132
Fusarium oxysporum f. sp. *tracheiphilum*
  dispersal/transmission, 30
  ecology of pathogen, 30
  epidemiology, 30
  survival, 30

G

Geminiviridae, 348
  alphasatellite, 351
  begomovirus, 348–349
    associated with DNA-satellites,
      350–351
  betasatellites, 351–352
  deltasatellites, 352–353
  genome organization, 355
Geminiviruses, 343–344
Giant cells, 130
  mechanism, 130–131
  reniform nematode, 130–131
Gibberellins, 280
Ginger Chlorotic Fleck Virus, 320
Glucose-6-phosphate dehydrogenase, 172
Glycolysis, 171
  phosphofructokinase (PFK), 171
Grain amaranth, 69–70

H

Hidden enemy of farmers, 125
HOST-specific toxin, 278
  ACR-toxin, 183–184
  AM-toxin, 182–183
  HC-toxin, 180
  HS-toxin, 181–182

PM-toxin, 181
  T-toxin, 180–181
  Victoria, 181
hrpB and *hrpG* genes, 98–99

I

India
  begomovirus
    diversity of legume-infection in
      different countries, 353–354
  second largest producer of vegetables and
    fruits, 39–40
Indoleacetic acid (IAA), 273
Infection, 272
  CWDEs, 273–274
  fungal pathogens, 274–275
  toxic molecules and lytic enzymes,
    273–274
  toxins, 273
Insect vectors
  phytoplasmas, 52
Integrated disease management (IDM),
  32–33
Integrated pest management, 373
  whitefly
    cultural practices, 374
    insecticide, 373–374
Intercropping, 375

J

Japan
  legumoviruses, 356

L

Legumes, 69, 344
  major producing countries, 346
  worldwide production, 349
Legumoviruses, 347
  Argentina, 356–357
  Brazil, 356–357
  China, 356
  Colombia, 356–357
  Cuba, 359–360
  geographical diversity and distribution,
    358
  host plant resistance, 375–377
  management strategies to limit, 372
    integrated pest management, 373–374

Mexico, 357–358
neighbor-joining phylogenetic analysis, 362
Nigeria, 360–362
Puerto Rico, 358
Sri Lanka, 356
Loofa yellow mosaic virus, 386

M

M. incognita and *Rhizoctonia bataticola*
root-knot nematodes, 131–132
Macrophomna phaseolina, 28
dispersal/transmission, 28–29
ecology of pathogen, 28–29
epidemiology, 29
survival, 28–29
Meloidogyne spp., 126
associated with tomato crop, 127–141
economic losses, 127
Mexico
legumoviruses, 357–358
Microbial pathogenicity, 282
essential virulence factors, 284
pathogens, 282
production of phytohormones auxins, 284–285
virulence factors, 282
Microbial toxins
involved in plant disease development, 278–279
Mitigation strategies, 14
Molecular biology approaches, 377–378
Mungbean, 345
Mushrooms
compost, 512
cultivation, 512
management of diseases, 513
treatment and disposal, 512
waste, 511
Mycoplasma-like organisms (MLOs), 50

N

Nematode associated with brinjal crop
bacterial interaction, 141
fungal interaction, 141
management
biological methods, 143–144
cultural methods, 142–143

integrated methods, 144–145
Nematode associated with potato crop
lesion nematode, 147–148
interaction with bacteria, 148
interaction with fungi, 148
mechanism of lesion formation, 148
management
botanicals, 151
cultural methods, 149–150
integrated methods, 151
physical methods, 149
regulatory methods, 148
potato cyst nematode, 145–146
root-knot nematodes, 146–147
Nepal
legumoviruses, 356
Nigeria
legumoviruses, 360–362
Non-host specific toxin, 278
coronatine, 186
phaseolotoxin, 185–186
syringomycin, 187–188
tabtoxin, 185
tagetitoxin, 188
tentoxin, 187

O

Okra, 230
wild relative and their occurrence in India, 230
Onion yellows (OY)
phytoplasma genome, 59
mild strain, 61
phytoplasmaassociated Amp, 60
Oxidative pentose phosphate (OPP)
pathway, 172

P

Pakistan
begomovirus
diversity of legume-infection in different countries, 354–355
Pathogen infection, 176
Pathogenesis
auxins, 279–280
cytokinins, 280
ethylene, 281
fungal toxin, role, 277–279

host-specific, 278
non-host specific, 278
gibberellins, 280
pathogen produced growth regulators,
    role, 279–281
    abnormalities in growth, 281
pathogen secreted enzymes, role,
    275–276
    cellulases, 277
    cutinases, 275–276
    pectinases, 276–277
Pathosystem, 314
    significance for integrated management,
        334–336
Pathosystem A 1 and 2, 316–317
    cauliflower mosaic virus, 317
    CMV, 317–318
        bean, 319
        cucurbits, 318
        lettuce, 318
        pepper, 318
        spinach, 318
        tomato, 319
Pathosystem-B1
    Ginger Chlorotic Fleck Virus, 320
    Potato virus X, 320
Pathosystem-C1, 320–321
    southern bean mosaic virus, 321–322
    TMV, 321
    tomato bushy stunt virus, 322
    viroids, 322–323
Pathosystem-C2, 323–324
Pathosystem-D1, 324–325
    PVMV, 326
    ZYMV, 325
Pathosystem-D3, 327–329
Pathosystem-D4, 329–331
Pathosystem-E1, 331–333
Pathosystem-F1, 333–334
Pea seed borne mosaic virus, 334–335
Pectinases, 276–277
Pepper veinal mottle virus (PVMV), 326
Phanarogemic plants
    nature of infection, 291–293
Phaseolotoxin, 185–186
Phenols, 238
Phytonematodes, 125
Phytophthora cactorum, 30–31

dispersal/transmission, 31
ecology of pathogen, 31
epidemiology, 31
survival, 31
Phytophthora infestans (Mont.) de Bary,
    1, 6
    effect of temperature on the efficacy, 16
Phytoplasmas, 49–50
    association with brinjal crop, 52
    biology and genomics, 59–60
    detection and identification, 55–59
    groups infecting vegetable crops, 53–55
    host–phytoplasma interaction, 60
    insect vectors, 52
    management, 70–71
    proliferation of axillary shoots, 51
    strategies, 71
    symptoms, 51
    virulence and pathogenicity, 60–62
Plant diseases, 271–272
Plant extracts, 263
Plant growth-promoting rhizobacteria
    (PGPR), 261–262
Plant parasitic nematodes, 125–126
    nature of infection, 293–298
Plant viruses, 314
    description of pathosystem, 316
    epidemics and pathosystem, 315–316
Plasmids, 59
Pointed gourd, 41
    infected, 45
    pathoecosystems, 44–45
Polycyclic diseases, 4–5
    primary inoculums, 5
Polyethylene soil mulches, 374–375
Postharvest loss, 459
    biological management, 473–474
        augmentation/augmentative, 475–476
        BCA, 475
        conservation, 474–475
        constitutive, 476–477
        and induced host resistance, 476–477
        natural fungicides, 477
        natural plant products, 477
        strategies, 478–479
    breathing, 460
    breathing leafy vegetables, 460
    climacteric fruits, 463

climacteric vegetables, 461
diseases and pathogens of vegetables,
    467–470
diseases, correct diagnosis
    bacterial soft rots, 467
    sexual stages, 466
    symptoms, 467
fungi and bacteria, 461
host resistance, maintenance
    emerging technologies for, 472–473
    environment, 470–472
infection
    anthracnose, 463
    C. gloeosporioides, 463
    grey mould of strawberry, 464
    mango stem end rot, 464
    mechanical injuries, 465, 466
    natural resistance, 465
    symptoms, 465
maturity standards vegetables, 461–462
plant organs, 462
ripeness, 461
Potato, 1, 2, 41
    impact of climate change, 2–3
    late blight disease, 4–5
        distribution of appearance, 5
        elevated CO_2, 10
        future impact, 14
        impact of climate change on
            management, 11–14
        influence of climate change, 6–11
        ozone, 10–11
        rainfall impact on management, 13–14
        rainfall/high humidity, 9
        symptomatic variations, 15
        temperature effect, 6–8
        temperature impact on management,
            11–12
    losses in production, 4
    pathoecosystems, 41–43
    polycyclic diseases, 4–5
    rot nematode and yam nematode, 151
    warming, 2
Potato virus Y (PVY), 320
    host range, 414–415
    India, status, 416
        disease prevalence of, 417
    infection cycle of, 414

infection in crop, 417–418
mode of transmission, 413
pathogen, 420
polyadenylate tract, 411–412
potyviridae taxonomy of, 413
soybean mosaic virus (SMV), 420–421
sweet potato virus, 421–422
tropical and subtropical climates, 419
Potyviridae, viruses, 409
Powdery mildew
    causal organism, 211
    cultural practices, 212–213
    disease cycle, 212
    disease management, 212
    disease symptoms, 210–211
    fungicide applications, 213–214
    pathogen biology, 211–212
    resistant varieties, 213
Pseudocercospora cruenta
    dispersal/transmission, 26
    ecology of pathogen, 26
    epidemiology, 26
    survival, 26
Pseudomonas solanacearum, 86
Puerto Rico
    golden mosaic disease (GMD), 358
    legumoviruses, 358

R

Race 3 biovar 2, 85
Ralstonia solanacearum, 85–86
    activation of metabolic pathways, 99
    aerotaxis, 104
    antimicrobial peptides, 112–113
    candidate genes, 95
    chemical control, 111
    chemotaxis, 104
    classification and geographic origins, 94
    cultural practices, 113
    detection, 91–94
    essential oils, 112
    gene sequences, 95
    host–pathogen interaction, 95–99
    hrpB and *hrpG* genes, 98–99
    identification, 91–94
    management options, 105–114
    motility, 102–103
    pathogenicity determinants, 99–104

prophages and genome evolution, 104
root border cells as a defense system, 105
sanitary and phytosanitary control
    measures, 114
sign and symptoms, 87–88
silicon (Si) and chitosan induced
    resistance, 112
taxis, 103–104
use of bacteriophages, 110–111
Reniform nematode, 130–131
life cycle, 131
Respiration in plants, 171–173
accumulation of lipids and carotenoids,
    173
Rhizoctonia solani
dispersal/transmission, 27
ecology of pathogen, 27
epidemiology, 27–28
survival, 27
Root border cells, 105
biological control, 108
breeding for wilt disease resistance, 107
executor-mediated disease resistance
    deployment, 106–107
grafting, 108
host resistance, 106
immune system receptors deployment,
    106
management options, 105–114
nonpathogenic strain as potential
    bioagent, 108–109
QTL mapping, 107–108
use of bacteriophages, 110
Root rot of cowpea, 28
dispersal/transmission, 28–29
ecology of pathogen, 28–29
epidemiology, 29
survival, 28–29
Root-knot nematodes, 126
associated with
    brinjal crop, 141–145
    potato crop, 145–151
    tomato crop, 127–141
chili/capsicum, 151–153
giant cells, 130
heavy galling on tomato roots, 128
interaction with other pathogens,
    131–132

life cycle, 128–129
management, 133–142
    biological methods of management,
        137–138
    cultural methods, 133–136
    host resistance, 138–139
    integrated methods, 139–141
    physical methods, 133
    in protected cultivation, 154–156
    rubbing, 133
    soil solarization, 133
tomato
    biocontrol agents, effect, 154–155
    treatments, effects, 155
white mature female, 128
Rust of cowpea
dispersal/transmission, 29
ecology of pathogen, 29
epidemiology, 29
survival, 29

S

Sec system (*SecA, SecY,* and *SecE* genes),
    61
Serological and nucleic-acid-based assay, 53
Single-stranded RNA viruses (ssRNA)
plant disease, 410
Soilborne bacterium, 86
Soilborne diseases, 428
antibiotics, 451–452
bacteria, criteria defining, 430–431
biological control, 450
biosecurity, 453
chemical control, 452–453
distribution
    climate change, effects, 437–438
    nematodes, 438
economic importance, 428–429
factors
    moisture, 436
    soil factors, 436
    soil pH, 436–437
    temperature, 436
fungi, criteria defining, 429
host range
    bacteria, 435
    fungi, 434–435
    root-knot nematode, 435

Rotaylenchulus spp., 435
Tylenchorhynchus spp., 435
nematodes, criteria defining, 431
nursery production, management, 444
  cultural control, 445
  fertigation, 446–447
  irrigation, 449–450
  sowing and harvesting practices, 448–449
  tillage practices, 447–448
resistant cultivars, 453
seedlings diseases, culprits
  bacteria pathogens, 433–434
  fungal pathogens, 431–432
  reniform, 434
  root-knot nematode, 434
soil inhabiting pathogens, 430
soil invading, 430
survival
  chlamydospores, 444
  oospores, 444
  sclerotia, 443
symptoms, 438
  Aphanomyces spp., 440–441
  Fusarium spp., 442–443
  Phytophthora spp., 439
  Pseudomonas spp., 443
  Pythium spp., 440
  Rhizoctonia solani, 441
Solanaceous vegetables, 62
Solanum tuberosum, 1, 2
  impact of climate change, 2–3
  late blight disease, 4–5
  losses in production, 4
  polycyclic diseases, 4–5
  warming, 2
Southern bean mosaic virus, 321–322
Soyabeans, 345
  major producing countries, 346
  worldwide production, 349
Spent mushroom substrate (SMS), 512
  absorption capacity, 513
  antagonists associated
    Aspergillus spp., 516
    biological analysis, 516
    Mucor spp., 516
    organic manure, 517
    Pleurotusflorida, 516

    Trichoderma spp., 516
    weathering, 516
  bacterial pathogens, effect
    antibacterial property, 520
    quantitative real-time, 520
    Ralstonia solanacearum, 521
    water extracts of SMS (WESMS), 520
  calcium carbonate ($CaCO_3$), 514
  crop production, 514
  doses and age, 518
  fungal pathogens, effect
    and plant diseases, 520
    Pleurotussajor-caju, 519
  mineral and nutritional composition, 515
  nematodes, effect
    Meloidogyne spp., 521
  nitrogen and phosphorus, 513
  organic mineral fertilizers, 514
  plant species, 518
  Pleurotuswaste, 515
  reclamation of affected, 516
  soilborne plant pathogens, management
    initial populations, 518
    management, 517
    root system, 517
    survival, 517
    vegetable crop, 518
  treatment, 514
Squash plant
  stunting and virescence, 67
Sri Lanka
  legumoviruses, 356
16SrII-D phytoplasma subgroup
  stunted tomato plants, 64
Stem rot, 30–31
  dispersal/transmission, 31
  ecology of pathogen, 31
  epidemiology, 31
  survival, 31
Stolbur phytoplasma
  infection causing big bud on tomato, 64
Syringomycin, 187–188

T

Tabtoxin, 185
Tagetitoxin, 188
Taxis
  aerotaxis, 103

chemotaxis, 104
Tengu-su inducer (TENGU), 61
Thailand
    legumoviruses, 356
Tobacco mosaic virus (TMV), 321
Tomato, 41
    HRPB and HRPG genes, 98–99
    infection and colonization, 97–98
    pathoecosystems, 41–43
Tomato bushy stunt virus, 322
Tomato spotted wilt virus (TSWV), 329
Tombusvirus, 322
Tospovirus, 329–330
    geographical distribution, 331
    infecting vegetable crops, 332
Toxins, 176
    ACR-toxin, 183–184
    AM-toxin, 182–183
    HC-toxin, 180
    HOST-specific, 178–184
        ACR-toxin, 183–184
        AM-toxin, 182–183
        HC-toxin, 180
        HS-toxin, 181–182
        PM-toxin, 181
        T-toxin, 180–181
        Victoria, 181
    HS-toxin, 181–182
    non–HOST-specific, 185–188
        coronatine, 186
        phaseolotoxin, 185–186
        syringomycin, 187–188
        tabtoxin, 185
        tagetitoxin, 188
        tentoxin, 187
    plant pathogenic
        fungi, 176
        microbes, 176
    PM-toxin, 181
    tentoxin, 187
    T-toxin, 180–181
Tri carboxylic acid (TCA)
    cycle, 171
Type III secretion system, 101–102

U

Uromyces phaseoli vignae
    dispersal/transmission, 29

ecology of pathogen, 29
epidemiology, 29
survival, 29

V

Vegetable pathosystem
    age-related resistance, 263–264
    associated virus groups
        pathosystem A1 and 2, 316–319
        pathosystem B1, 319–320
        pathosystem C1, 320–323
        pathosystem C2, 323–324
        pathosystem D1, 324–326
        pathosystem D3, 327–329
        pathosystem D4, 329–331
        pathosystem E1, 331–333
        pathosystem F1, 333–334
    begomovirus
        Geminivirideae, 388
        Subgroup III, 388
    biocontrol method
        PGPR, 261–262
        plant extracts, 263
        seed treatment, 261
    chemical method, 258–261
        foliar application, 259, 261
        seed treatment, 258–260
        soil application, 258
    disease development, 265
    management
        age-related resistance, 263–264
        biocontrol method, 261–263
        chemical method, 258–261
    physical method, 257–258
Vegetables, 163–164, 271–272, 459
    association between host and pathogen,
        165
    changes in concentration and distribution
        photosynthesis and carbohydrate,
            174–176
        toxin, 176–188
    crops, 336
        biological detection, 55–56
        microscopic techniques, 56
        molecular detection, 57–59
        phytoplasma diseases, 62–72
        serological detection, 57
    fungal infection

changes in concentration and
distribution, 171–188
early events at cell membrane, 165
calcium and ion channels, 166
protein kinases, 166–167
reactive oxygen species, 168
issues in production, 164
nitrogen and amino acid metabolism
increase, 169–171
primary pathways, 170
parasitic relationship, 164–165
respiration, 171–173
Victoria, 181
Vietnam
begomovirus
diversity of legume-infection in
different countries, 355
Vigna unguiculata (L.) Walp.), 23–24
bacterial blight, 32
fungal diseases, 25–31
IDM, 32–33
important diseases, 24
viral diseases, 31–32
Viral diseases, 315
management, 33
Viral pathogens
nature of infection, 286–289
Viriod pathogens
nature of infection, 289–291
Viroids, 322–323
Viruses, 313
characterization, 313

ecology of pathogen, 27
epidemiology, 27–28
survival, 27
Whitefly, 347
management through CI, 378
Wilt diseases, 29–30
ecology of pathogen, 30
epidemiology, 30
management, 33

Y

Yellow vein mosaic (YVMD) okra,
229–231
genetics, 236–239
influence of climatic condition
vector population and viral incidence,
236–239
pathogen/viral biology, 231–233
resistant varieties, 247
RNAI strategy against, 247
role of vector in disease spreading, 233
alternative host, 234
source of resistance, 239–241
breeding, 244–246
crop management practices, 242–244
cross-ability and fruit setting, 246
diverse begomoviruses, 241–242
mutation breeding, 246–247
wild, 241
symptoms and economic importance,
234–235
YidC system, 61

W

Web blight disease, 27
dispersal/transmission, 27

Z

Zucchini yellow mosaic virus (ZYMV), 325